ORAL BIOAVAILABILITY ASSESSMENT

WILEY SERIES ON PHARMACEUTICAL SCIENCE AND BIOTECHNOLOGY: PRACTICES, APPLICATIONS, AND METHODS

Series Editor:

Mike S. Lee
Milestone Development Services

Mike S. Lee • *Integrated Strategies for Drug Discovery Using Mass Spectrometry*

Birendra Pramanik, Mike S. Lee, and Guodong Chen • *Characterization of Impurities and Degradants Using Mass Spectrometry*

Mike S. Lee and Mingshe Zhu • *Mass Spectrometry in Drug Metabolism and Disposition: Basic Principles and Applications*

Mike S. Lee (editor) • *Mass Spectrometry Handbook*

Wenkui Li and Mike S. Lee • *Dried Blood Spots—Applications and Techniques*

Mike S. Lee and Qin C. Ji • *Protein Analysis using Mass Spectrometry: Accelerating Protein Biotherapeutics from Lab to Patient*

Ayman F. El-Kattan • *Oral Bioavailability Assessment: Basics and Strategies for Drug Discovery and Development*

ORAL BIOAVAILABILITY ASSESSMENT

Basics and Strategies for Drug Discovery and Development

AYMAN F. EL-KATTAN

WILEY

This edition first published 2017

Edition History

Registered Office
John Wiley & Sons, Inc., 111 River Street, Hoboken, NJ 07030, USA

Editorial Office
111 River Street, Hoboken, NJ 07030, USA

For details of our global editorial offices, customer services, and more information about Wiley products visit us at www.wiley.com.

Wiley also publishes its books in a variety of electronic formats and by print-on-demand. Some content that appears in standard print versions of this book may not be available in other formats.

Library of Congress Cataloging-in-Publication Data

Names: El-Kattan, Ayman F., author.
Title: Oral bioavailability assessment : basics and strategies for drug
discovery and development / Ayman F. El-Kattan.
Description: Hoboken, NJ : John Wiley & Sons Inc., 2017. | Includes bibliographical references and index.
Identifiers: LCCN 2016055388 (print) | LCCN 2016056137 (ebook) | ISBN 9781118916698 (cloth)
| ISBN 9781118916940 (Adobe PDF) | ISBN 9781118916933 (ePub)
Subjects: | MESH: Biological Availability | Drug Evaluation, Preclinical | Toxicokinetics | Administration, Oral
Classification: LCC RS403 (print) | LCC RS403 (ebook) | NLM QV 38 |
DDC 615.1/9–dc23
LC record available at https://lccn.loc.gov/2016055388

Cover image:
Background: © Pobytov/Gettyimages
Figure: Courtesy of Ayman F. El-Kattan
Cover design by Wiley

Set in 9/11pt TimesLTStd by SPi Global, Chennai, India

Printer in the United States of America

10 9 8 7 6 5 4 3 2 1

For my parents
My kids, Yasmine, Fawzi, Salma, and Zachary
My best friend, companion and wife Sirine

I love you

CONTENTS

1

DRUG PHARMACOKINETICS AND TOXICOKINETICS

1.1 INTRODUCTION

Pharmacokinetics (PK) is the science that describes the time-course of drug concentration in the body resulting from administration of a certain drug dose. Similarly, toxicokinetics (TK) is the science that investigates how the body handles toxicants as illustrated by its plasma profile at various time points. In comparison, pharmacodynamics (PD) is the science that describes the relationship of the time-course of drug concentration and its effects in the body [1, 2].

PK is considered a biomarker of drug exposure as well as marker of efficacy and safety. Key determinants of the pharmacokinetics of a drug include absorption, distribution, metabolism, and elimination (ADME) [3]. Discovering novel therapeutic agents is an increasingly time-consuming and costly process. Most estimates indicate that it takes approximately 10–15 years and more than $1.2 billion to discover and develop a successful drug product [4]. It is well established that poor drug PK is one of the leading causes of compounds failure in preclinical and clinical drug development [5]. For example, attrition due to poor pharmacokinetics contributed to 10% of the attrition reported for compounds developed by the pharmaceutical industry in 2001 (Figure 1.1) [6].

Compounds with poor PK profile tend to have low oral systemic plasma exposure and high interindividual variability, which limits their therapeutic utility (Figure 1.2) [7]. Therefore, a better understanding of the PK profile early on enables the discovery of compounds with drug-like properties [8]. In drug discovery settings, the main outcomes of PK/TK assessments are to

- select compounds with the maximum potential of reaching the target;
- determine the appropriate route of administration to deliver the drug (typically oral);

Oral Bioavailability Assessment: Basics and Strategies for Drug Discovery and Development, First Edition.
Ayman F. El-Kattan.

- understand how the drug blood levels relate to efficacy or toxicity in order to choose efficacious and safe doses;
- facilitate appropriate dose sections for rodent and/or nonrodent species in toxicology testing and drug safety evaluation;
- decide on the frequency and duration of dosing in order to maintain adequate drug concentration at target for disease modification; and
- accurately predict the PK in humans profile prior to clinical studies.

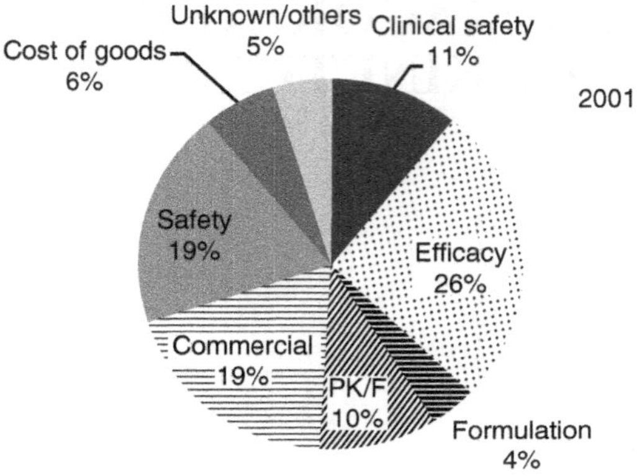

FIGURE 1.1 The contribution of various factors to the overall attrition of NCEs in year 2001. Kola and Landis 2004 [6]. Reproduced with permission of Nature Publishing Group.

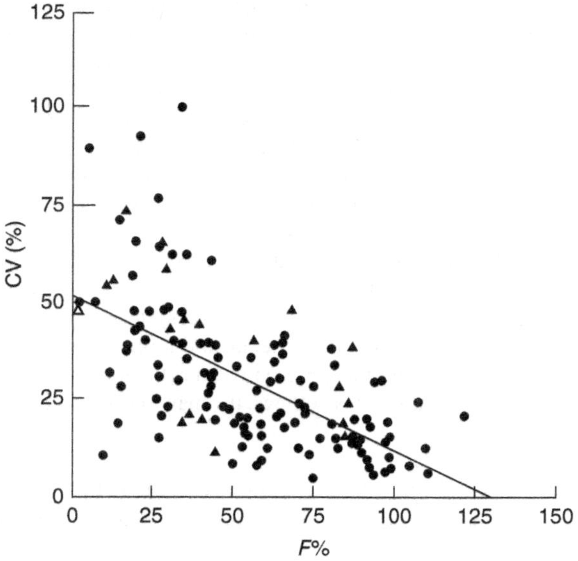

FIGURE 1.2 The relationship between drug oral bioavailability and interindividual variability reported as coefficient of variation (%). Hellriegel et al. 1996 [7]. Reproduced with permission of John Wiley & Sons.

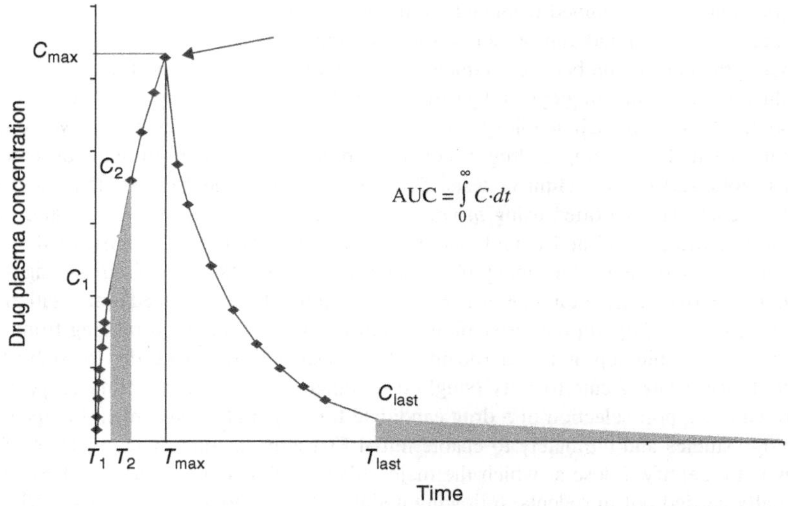

FIGURE 1.3 Estimation of the area under the plasma concentration–time curve (AUC).

A PK/TK study involves dosing animals or humans with NCE and collect blood samples at predefined time points. After sample preparation and quantification, a concentration–time profile is generated (Figure 1.3). In drug discovery, preliminary PK studies are usually conducted in rodents to evaluate the extent of drug exposure *in vivo*. These rodent studies are commonly followed by studies in nonrodents such as dogs or monkeys to better characterize the PK profile of the compound and to support safety risk assessment studies. Pharmacokinetic scaling, also known as allometry, is a discipline that was extensively used in the past to predict human PK profile using preclinical data and in predicting the drug human half-life, dose, and extent of absorption. This approach is based on empirical observations that various physiological parameters are a function of body size. The allometric methods assume that the same metabolic and disposition processes in the species evaluated are correlated with those observed in humans. However, the cytochrome P450 enzymes in the rat are not the same as those in humans, and thus, may exhibit altered disposition of the compound or even produce different metabolite patterns (see Chapter 2) [9, 10]. Similarly, uptake and efflux transporters in the animal species may differ in substrate specificity or rate, as compared to humans, and thus may confound predictions of human PK [11]. Accurate prediction of human pharmacokinetic profile is imperative to minimize drug failure in development due to pharmacokinetic liability. More detailed description of methods in predicting human PK is beyond the scope of this chapter, but can be found in many excellent reviews [12–15]. An in-depth discussion of various PK concepts and their applications can be found in various references [16, 17].

1.2 TOXICITY ASSESSMENT IN DRUG DISCOVERY AND DEVELOPMENT

Several toxicology studies are conducted during early drug discovery and all the way to the late stages of drug development before a new drug application (NDA) filing is made. In spite of comprehensive toxicity assessment in early- and late-stage discovery, attrition of NCEs in clinical studies is not uncommon owing to disconnect in predictions of risk in humans based upon preclinical data obtained from cell culture and animal models. Nevertheless, extensive preclinical assessment and appropriate scaling and modeling tools will improve predictions. In general, the correlation between human and animal toxicities is good for conditions such as cardiovascular, hematological, and gastrointestinal diseases and the poorest correlation for adverse drug reactions such as idiosyncratic reactions, skin rash, hypersensitivity, and hepatotoxicity. Toxicology testing in drug discovery is initiated by the high-throughput screening, which is followed up by definitive tests. Screening refers to the methods that yield rapid and comprehensive data often using *in vitro* tools. The origin of any toxicological or safety outcome is multifactorial and complex and thus demands for use of sophisticated systems for definitive assessment. Thus, many pharmaceutical companies are also introducing *in vivo* (i.e., animals) toxicology studies as early as possible, quite often in the lead optimization (LO) stage. Extensive and appropriate toxicology studies of varying duration ranging from acute, single dose to chronic, repeat dose in rodent and nonrodent species are needed to establish safe human clinical trials. Acute toxicity (single dose-ranging) studies in preclinical species are performed to support selection of a drug candidate for potential advancement to repeat-dose toxicology studies and ultimately to enable initial FIH clinical trials. The objective of such studies is to identify a dose at which the major adverse effects are observed. These studies are usually carried out in rodents, following a single dose up to a limit of 2000 mg/kg. The information obtained may be translated to select the dose levels for the first in-human studies and also to give an indication of potential effects of acute overdose in humans.

Early drug development starts with candidate compound selection. Repeat-dose toxicity studies (7–14 days in duration) in both rodent and nonrodent species are used to better refine safety margins, PK/PD modeling, and set appropriate dosages for the subsequent good laboratory practice (GLP) 1-month general toxicology and safety pharmacology (i.e., cardiovascular testing in a nonrodent; CNS and respiratory function tests in a rodent) studies that proceeds the investigational new drug (IND) application before starting FIH clinical trials. Toxicokinetic assessment is based on the multiple samples obtained throughout the duration of the study along with the PK data. Such data are critical to define a margin of safety between the no observed adverse effect level (NOAEL) and the projected plasma concentrations achieved in human. It is generally considered that a 100-fold safety factor (rodent-to-human exposure ratio) from the most sensitive species NOAEL provides good safety margin in clinical studies. However, our enhanced capability of understanding interspecies sensitivity and detecting more and more subtle effects may warrant a more flexible approach. The toxicology assessment profile includes, for example, the maximum tolerated dose (MTD), safety margins and therapeutic index, target organ toxicities, most sensitive preclinical species, and reversibility of an effect/toxicity. Biomarkers characterization and preclinical to clinical translation can also be investigated in these GLP toxicology studies.

Later drug development includes Phases I–IV. Phase I (FIH) starts with a single dose escalation, then multiple dosing in normal healthy subjects. These studies are used to establish human safety profile and MTD. Phase II defines the efficacy/safety of candidate profile in target patient population (e.g., rheumatoid arthritis), drug–drug interactions, and proof of concept (POC) before proceeding into Phase III. Several repeat-dose toxicology studies (general toxicology, embryo-fetal and developmental, fertility, juvenile, carcinogenicity) of longer duration (3 months and up to 2 years) in both rodent and nonrodent species are conducted to support clinical trials of longer duration in patients.

The purpose of this chapter is to introduce the fundamentals of PK and TK, and their applications to drug discovery and development. It also presents the fundamentals of computational analysis of the data derived from the estimated concentrations in the biological matrices such as plasma. Finally, the implications of species differences, genomics, and exposure of the metabolites in determining the safe dose in the first in human (FIH) clinical trials and further identification of clinical dosage regimen are discussed.

1.3 PARAMETERS THAT DEFINE PHARMACOKINETIC PROFILE

1.3.1 Area Under the Curve (AUC)

The first step in a pharmacokinetic experiment is to dose animals or humans with NCE and collect blood samples at predefined time points. Animals are generally dosed intravenously (IV) and/or orally (po). After sample preparation and quantification usually using LC/MS/MS, a plasma concentration–time profile is generated (Figure 1.3) [18].

Mathematically, area under the plasma (or blood) concentration–time curve (AUC) can be calculated from the obtained concentration–time profile by

$$\mathrm{AUC} = \int_0^{\infty} C \cdot dt \tag{1.1}$$

AUC is a primary measure of the extent of drug availability to the systemic circulation (i.e., reflects the total amount of unchanged drug that reaches the systemic circulation following intravenous or extravascular administration). The unit for AUC is concentration per unit time (e.g., ng*h/mL). AUC is determined using simple integration method as shown in Equation 1.1 or a linear trapezoidal method, which is the most widely used approach (Figure 1.3).

The area of each trapezoid is calculated using the following equation:

$$\mathrm{AUC}_{t_1 \to t_2} = \frac{(C_2 + C_1)}{2} \times (t_2 - t_1) \tag{1.2}$$

The extrapolated area from t_{last} to ∞ is estimated as

$$\mathrm{AUC}_{t_{\mathrm{last}} \to \infty} = C_{\mathrm{last}} / K_{\mathrm{e}} \tag{1.3}$$

where C_{last} is the last observed concentration at t_{last} and K_{e} the slope obtained from the terminal portion of the curve, representing the terminal elimination rate constant. The total AUC ($\mathrm{AUC}_{0 \to \infty}$) is determined as

$$\mathrm{AUC}_{0 \to \infty} = \mathrm{AUC}_{0 \to t_{\mathrm{last}}} + \mathrm{AUC}_{t_{\mathrm{last}} \to \infty} \tag{1.4}$$

AUC is used in the calculation of clearance, apparent volume of distribution, and bioavailability (see Sections 1.3.2, 1.3.3, and 1.3.5) and reflects the general extent of exposure over time.

1.3.2 Mean Residence Time (MRT)

Mean residence time (MRT) is the average time for all drug molecules to exist in the body. MRT is another measure of drug elimination and its unit is time (e.g., hour). Following intravenous dosing, $\mathrm{MRT}_{\mathrm{iv}}$ is calculated as

$$\mathrm{MRT} = \frac{\mathrm{AUMC}}{\mathrm{AUC}} = \frac{\int_0^{\infty} C \cdot t \cdot dt}{\int_0^{\infty} C \cdot dt} \tag{1.5}$$

where AUMC is the area under the first moment versus time curve from time $t = 0$ to ∞ and calculated using trapezoidal rule similar to AUC.

In some cases, MRT can be a better parameter to assess drug elimination compared to half-life ($t_{1/2}$) This assessment can be attributed to the greater analytical sensitivity shown with various analytical systems such as LC/MS/MS, the lower drug concentrations measured following drug administration appeared to yield longer terminal half-life, which are not related to the drug's pharmacologically relevant half-life. In a case like this, it would be recommended to measure MRT rather than half-life to assess drug elimination.

1.3.3 Clearance (CL)

Clearance (CL) is a primary pharmacokinetic parameter that describes the process of irreversible elimination of a drug from the systemic circulation. CL is defined as the volume of blood or plasma that is totally cleared of its content of drug per unit time. Thus, CL measures the removal of drug from blood or plasma. However, CL does not indicate the amount of drug that is being removed, but instead represents the rate of drug elimination from blood. Therefore, CL unit is given as mL/min or mL/min/kg (normalized to body weight).

The most widely used approach to evaluate plasma (total) CL involves intravenous administration of a single dose and measuring its plasma concentration at different time points to calculate its AUC (Figure 1.3). In this manner, the calculated CL (Equation 1.6) will not be confounded by complex absorption and distribution phenomena, which is commonly observed during oral dosing [7].

$$CL_{tot} = Dose_{IV}/AUC_{IV} \tag{1.6}$$

In general, a drug is either eliminated unchanged through excretion in the urine and/or bile, or by metabolic conversion into more polar metabolite(s) that can be readily excreted in urine and/or bile. Therefore, total body clearance is an additive parameter and the sum of all clearances by various mechanisms. Mathematically, it is also expressed as shown in Equation 1.7 (Figure 1.4),

$$CL_{tot} = CL_{hep} + CL_{ren} + CL_{bil} \tag{1.7}$$

where CL_{tot} is the total body clearance from all different organs and mechanisms, CL_{hep} the hepatic blood clearance, CL_{ren} the renal clearance, and CL_{bil} the biliary clearance.

It is interesting to note that around three quarters of the top 200 prescribed drugs in the United States are primarily cleared by hepatic metabolism [19]. The hepatic extraction ratio (E_h) is a pharmacokinetic parameter that is widely used to assess the liver's ability to extract

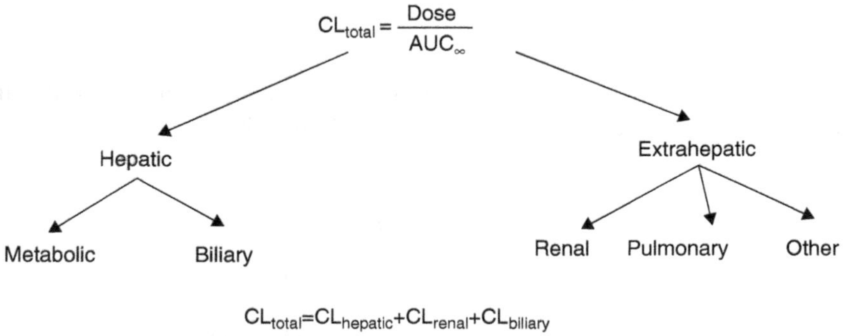

FIGURE 1.4 Various routes/mechanisms of eliminations that contribute to drug CL_{total}.

TABLE 1.1 Typical Body Weight and Hepatic Blood Flow for Various Preclinical Species and Human

Species	Body Weight (kg)	Liver Blood Flow (mL/min/kg)
Mouse	0.02–0.025	90
Rat	0.25	55
Rabbit	2.5–3	71
Cynomolgus monkey	4–5	44
Dog	10–12	31
Human	70	21

drug from the systemic circulation [17]. E_h is defined as the fraction of a drug in the blood that is cleared (extracted) on each passage through the liver and is a function of CL_{hep} and the hepatic blood flow (Q) [17]:

$$E_h = \frac{CL_{hep}}{Q} \tag{1.8}$$

Typical values for the hepatic blood flow in various preclinical species and human are summarized in Table 1.1.

If the predominant clearance mechanism for a compound is via hepatic metabolism, then it is reasonable to assume that the CL_{tot} is equal to CL_{hep}. Thus,

$$E_h = \frac{CL_{hep}}{Q} = \frac{CL_{total}}{Q} \tag{1.9}$$

Compounds that undergo hepatic metabolism can be classified according to their E_h. Compounds with $E_h > 0.7$ are considered high extraction drugs, whereas, compounds with $E_h < 0.3$ are considered low extraction drugs. E_h has a major impact on oral drug bioavailability.

1.3.4 Plasma versus Blood Clearance

Calculation of E_h from drug clearance in blood requires the determination of drug concentration in whole blood. Since determination of drug concentration is usually performed in plasma or serum, knowledge of the blood/plasma concentration ratio is necessary to estimate the blood clearance. Blood clearance is calculated using this equation:

TIP:

Various factors can lead to a total clearance of an investigated compound that is higher than hepatic blood flow (Table 1.1). For example, extrahepatic elimination pathways can play a key role in the elimination of xenobiotics, although hepatic clearance is commonly the main route of elimination [20]. Compounds with high blood to plasma ratio are preferentially distributed in red blood cells. Therefore, their plasma clearance would overestimate blood clearance. Furthermore, compounds with poor stability in blood/plasma tend to have high clearance. Overall, these factors should be considered and investigated when this trend is observed.

$$\frac{\text{Plasma clearance}}{\text{Blood clearance}} = \frac{\text{Blood concentration } (C_b)}{\text{Plasma concentration } (C_p)} \tag{1.10}$$

1.3.5 Apparent Volume of Distribution (V_d)

Volume of distribution is a proportionality factor that relates the amount of a drug in the body to its blood or plasma concentrations at a particular time,

$$\text{Amount of drug in the body at time } t = V_d \times C \text{ plasma at time } t \tag{1.11}$$

Following intravenous dosing and at $t = 0$ h, the amount of drug in the body is equal to the administered intravenous dose. V_d at $t = 0$ is termed volume of the central compartment (V_c).

TIP:

Need to always remember that volume of distribution has *no* physiological relevance. There are compounds that have a V_d that is significantly lower (e.g., acetyl salicylic acid $V_d = 0.15$ L/kg) than total body water (0.6 L/kg) and ones that are significantly higher (e.g., loratidine $V_d = 120$ L/kg). This question usually arises when V_d is smaller than total body water. The answer is simple: V_d is *not* physiologically relevant.

Similar to CL, V_d is a primary independent pharmacokinetic parameter and its unit is volume (e.g., L/kg). V_d is a mathematical constant that has no physiological relevance. V_d is used to assess the extent of drug distribution within or outside the total body water. In the literature, V_d ranges from 3 to more than 40,000 L per 70 kg human body weight. For example, if the drug has a V_d that is smaller than the total body water (human total body water = 42 L per 70 kg human body weight, which is equivalent to 0.6 L/kg), then the drug would be expected to have limited tissue distribution (e.g., acetyl salicylic acid has a $V_d = 10.5$ L per 70 kg human body weight, which is equivalent to 0.15 L/kg) [21]. On the other hand, if a drug has a V_d larger than the total body water, then the drug is likely able to distribute to body tissues (e.g., loratidine has a $V_d = 8400$ L per 70 kg human body weight, which is equivalent to 120 L/kg) (Figure 1.5) [22]. Therefore, the term apparent volume of distribution is usually used.

It should be emphasized that binding to both blood and tissue components such as lipids and proteins has a significant impact on the drug volume of distribution as outlined in the following equation:

$$V_d = V_{\text{blood}} + V_{\text{tissue}} \cdot \frac{f_{\text{u,blood}}}{f_{\text{u,tissue}}} \tag{1.12}$$

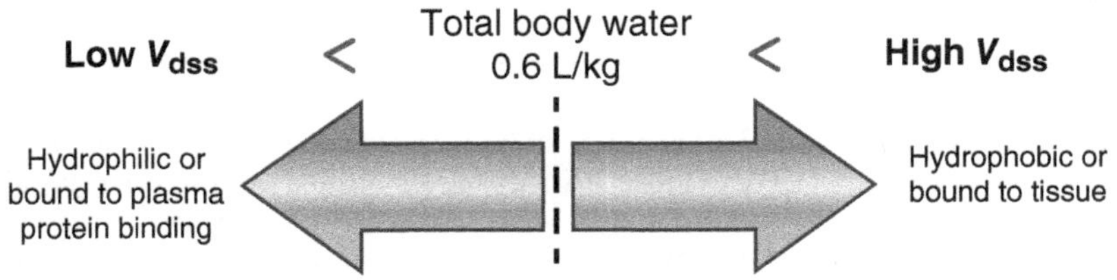

Compound	V_{dss} (L/kg)	Compound	V_{dss} (L/kg)
Acetylsalicylic acid	0.15	Amiodarone	66
Vancomycin	0.39	Loratidine	120
Warfarin	0.14	Vinorelbine	76

FIGURE 1.5 Volume of distribution and its relation with the extent of drug distribution in blood and tissues.

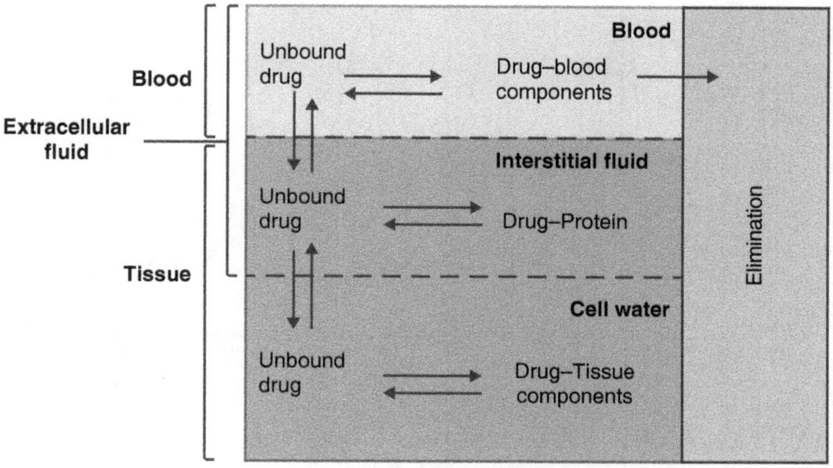

FIGURE 1.6 Tissue and blood binding and their impact on drug volume of distribution.

where $f_{u,blood}$ is the free fraction of the drug in blood, $f_{u,tissue}$ the free fraction of the drug in tissue, V_{blood} the volume of drug in blood, and V_{tissue} the volume of drug in tissue. As depicted in Figure 1.6, an increase in $f_{u,blood}$ is associated with an increase in drug V_d, whereas an increase in $f_{u,tissue}$ is associated with a decrease in drug V_d. Furthermore, increasing drug lipophilicity is associated with a decrease in $f_{u,tissue}$, which usually leads to an increase in the drug V_d.

1.3.5.1 Apparent Volume of Distribution at Steady State (V_{dss}) V_{dss} is the volume of distribution that is determined when plasma concentrations are measured at steady state and in equilibrium with the drug concentration in the tissue compartment.

$$V_{dss} = \frac{\text{Amount of drug in the body at equilibrium conditions}}{\text{Steady state plasma concentrations } (C_{ss})} \tag{1.13}$$

Although V_{dss} is a steady-state parameter, it can be calculated using non-steady-state data as

$$V_{dss} = \text{CL} \times \text{MRT} \tag{1.14}$$

Furthermore, V_{dss} is used in the calculation of a loading dose as

$$\text{Loading dose} = \frac{V_{dss} \times C_{ss}}{F} \tag{1.15}$$

Use of loading dose is important especially for those drugs in which it is desirable to immediately or rapidly reach the steady-state plasma concentration (C_{ss}) (e.g., anticoagulant, antiepileptic, antiarrhythmic, and antimicrobial therapy).

1.3.6 Half-life ($t_{1/2}$)

$t_{1/2}$ is the time that is required for the amount (or plasma concentration) of a drug to decrease by one half. It is calculated by the following equation:

$$t_{1/2} = \frac{0.693 \times V_d}{\text{CL}} \tag{1.16}$$

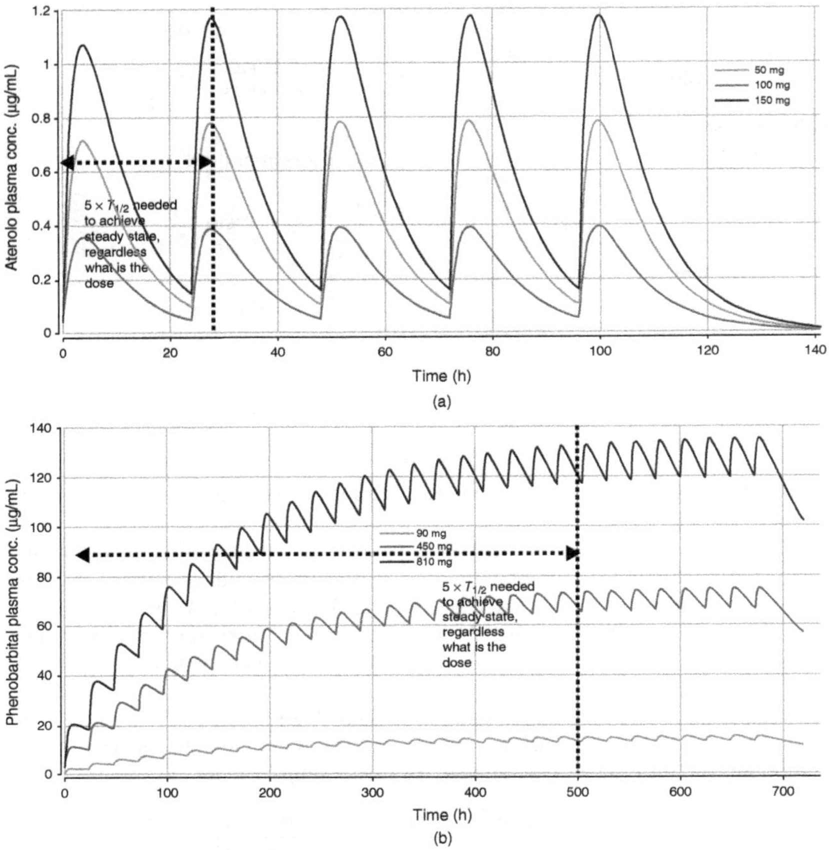

FIGURE 1.7 Plasma concentration–time profiles for drugs with half-lives of 6, 36, or 99 h administered once daily. Simulations were performed using Berkeley Madonna Software®. (a) Half-life is 6 h (e.g., atenolol); (b) half-life is 99 h (e.g., phenobarbital).

$t_{1/2}$ is a dependent pharmacokinetic parameter that is determined by both CL and V_d, which are independent primary pharmacokinetic parameters. Therefore, $t_{1/2}$ is increased by a decrease in CL or increase in V_d and vice versa. $t_{1/2}$ is the most widely reported pharmacokinetic parameter since it may constitute a major determinant of the duration of action after single and multiple dosing. The unit for $t_{1/2}$ is time (e.g., h). In addition, $t_{1/2}$ plays a key role in determining the time that is required to reach steady state following multiple dosing and the frequency with which doses can be given. In general, for a drug that follows one compartment kinetics, it takes five half lives for it to reach steady-state concentrations after multiple dosing and under linear conditions. For example, for a drug with a half-life of 6 h (e.g., atenolol), steady-state concentrations are reached in about 30 h regardless of its dose or dosage regimen. Similarly, a drug such as phenobarbital with a $t_{1/2}$ of 99 h, would take 495 h to reach its steady-state concentrations (Figure 1.7).

If a drug follows one compartment model following intravenous dosing, then its $t_{1/2}$ is calculated as follows:

$$t_{1/2} = 0.693 \cdot \text{MRT}_{\text{iv}} \tag{1.17}$$

where MRT_{iv} is the mean residence time following intravenous dosing. This calculation assumes that $t_{1/2}$ is proportional to MRT_{iv}.

The elimination rate constant k_e is a first-order rate constant that is used to describe drug elimination from the body. The k_e can be calculated directly from the slope of the straight line or from biologic $t_{1/2}$ using Equation 1.18.

$$k_e = \frac{0.693}{t_{1/2}} \tag{1.18}$$

It is interesting to note that in light of the major advancements realized in the field of drug analysis and as greater analytical sensitivity has been achieved, lower concentrations are being detected, therefore, using the $t_{1/2}$ calculated from the terminal elimination phase resulted in significantly longer terminal $t_{1/2}$. For example, a $t_{1/2}$ of 120 h was calculated with indomethacin, whereas 2.4 h pharmacologically relevant $t_{1/2}$ is reported. Therefore, scientists are recommended to determine the most biologically relevant $t_{1/2}$ by using Equation 1.16 where $t_{1/2}$ is defined by the drug clearance and volume of distribution.

TIP:

Develop a habit of double-checking the $t_{1/2}$ calculated from the terminal elimination phase following intravenous dosing by comparing it with that calculated using Equation 1.16. If the two numbers are similar, then this is the pharmacologically relevant $t_{1/2}$. Otherwise, report the value determined using Equation 1.16.

1.3.7 Maximum Plasma Concentration (C_{max}) and Time of Maximum Concentration (t_{max})

C_{max} is defined as the maximum observed drug concentration in the plasma concentration–time profile following intravenous or oral dosing. Most commonly, C_{max} is obtained by direct observation of the plasma concentration–time profile (Figure 1.3). For some drugs, the biological effect is dependent on the C_{max}. For example, aminoglycosides, which are widely used antibiotics, need to achieve a C_{max} that is at least 8- to 10-fold higher than the minimum inhibitory concentration (MIC) to obtain a clinical response $\geq 90\%$ [23, 24]. The unit of C_{max} is concentration unit (e.g., ng/mL).

t_{max} is the time required to reach C_{max}. As with C_{max}, t_{max} is usually determined from direct observation of the plasma concentration–time profile and its unit is time (e.g., h) (Figure 1.3). As depicted in Equation 1.18, t_{max} is independent of drug dose, bioavailability, or volume of distribution and is only determined by the rate constants of absorption (k_a) and elimination (k_e).

1.3.8 Absorption Rate Constant (k_a)

The k_a for a drug administered by a route other than intravenous is the rate of absorption of a drug absorbed from its site of administration. The rate of absorption usually follows first-order kinetics. Many approaches are used to calculate this parameter. For example, rate of absorption can be calculated from the following equation:

$$t_{max} = \frac{2.303}{k_a - k_e} \log \frac{k_a}{k_e} \tag{1.19}$$

The k_a can also be calculated using the method of residuals also known as feathering. The calculation is made with the assumption that the pharmacokinetics of the compound tested follows one compartment model with first-order input and output and is described using Bateman equation (Equation 1.19). The shape of the compound plasma profile is described by k_a and k_e. In general, k_a is larger than k_e and suggests that the compound absorption is faster than its overall elimination rate:

$$C_p = \frac{F \cdot \text{Dose} \cdot k_a}{V \cdot (k_a - k_e)} \cdot [e^{-k_e \cdot t} - e^{-k_a \cdot t}] \tag{1.20}$$

The following steps can be used to calculate k_a:

1. Graph measured plasma concentration in semilog scale plot.
2. If $k_a > k_e$, then $e^{-k_a \cdot t}$ achieves zero faster than $e^{-k_e \cdot t}$. As a result, the plasma concentration (C') is described by

$$C'_p = \frac{F \cdot \text{Dose} \cdot k_a}{V \cdot (k_a - k_e)} \cdot e^{-k_e \cdot t} \tag{1.21}$$

3. Determine the intercept (Equation 1.22) and k_e (slope) of the terminal linear portion of the graph using either linear regression or graphically (Figure 1.8).

$$\text{Intercept} = \frac{F \cdot \text{Dose} \cdot k_a}{V \cdot (k_a - k_e)} \tag{1.22}$$

4. Calculate the difference between C' that depicts the terminal phase of the oral plasma profile and C (Bateman equation).
5. Plot ($C' - C$) values in the same semilog scale plot.
6. Calculate the k_a from the slope using either linear regression or graphically (Figure 1.9).

Finally, k_a can also be calculated using the moment method:

$$k_a = \frac{1}{\text{MRT}_{po} - \text{MRT}_{iv}} \tag{1.23}$$

where MRT_{po} is the mean residence time after oral dosing and MRT_{iv} the mean residence time after intravenous dosing.

1.3.9 Flip–Flop Kinetics

TIP:

To determine if a drug undergoes flip–flop kinetics following oral administration, both intravenous and oral plasma profiles for the drug should be characterized. If observed, the cause, usually associated with poor solubility, dissolution, and/or permeability of the tested article, may need to be investigated [25].

Flip–flop kinetics is a phenomenon where the terminal phase of the plasma profile of a drug following its oral administration is determined by the drug absorption. Here, the drug k_a is much slower than its k_e. This condition is usually associated with sustained absorption characterized by a decrease in C_{max} and increase in t_{max} (Figure 1.10). It should be emphasized that the obtained AUC stays the same with similar oral dose.

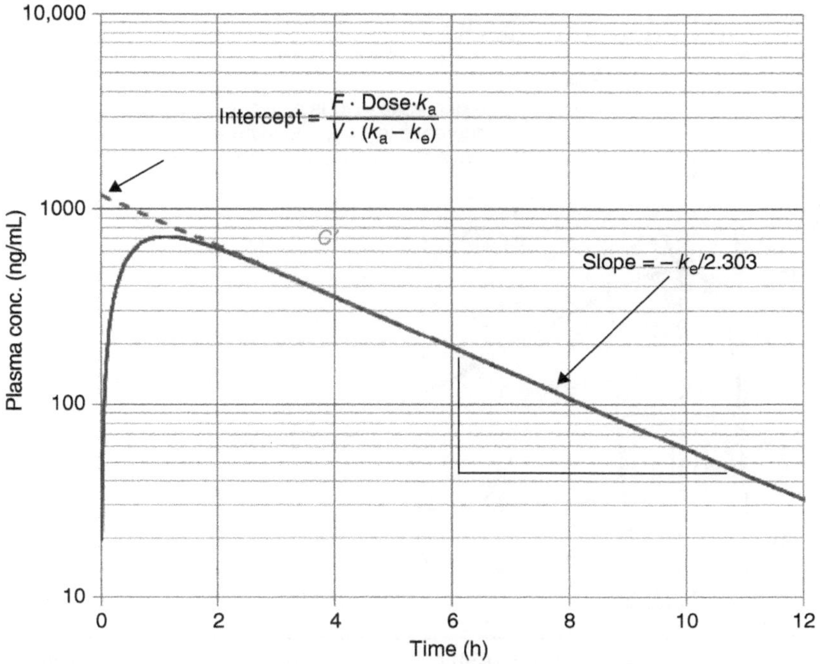

FIGURE 1.8 The semilog plot of plasma profile versus time of a compound that follows one compartment model with first-order input and first-order output.

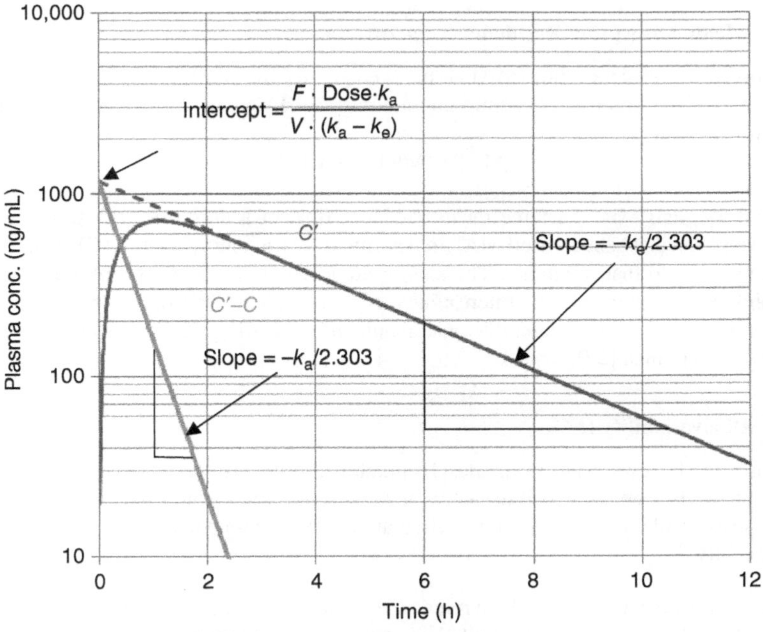

FIGURE 1.9 The semilog plot of residual versus time.

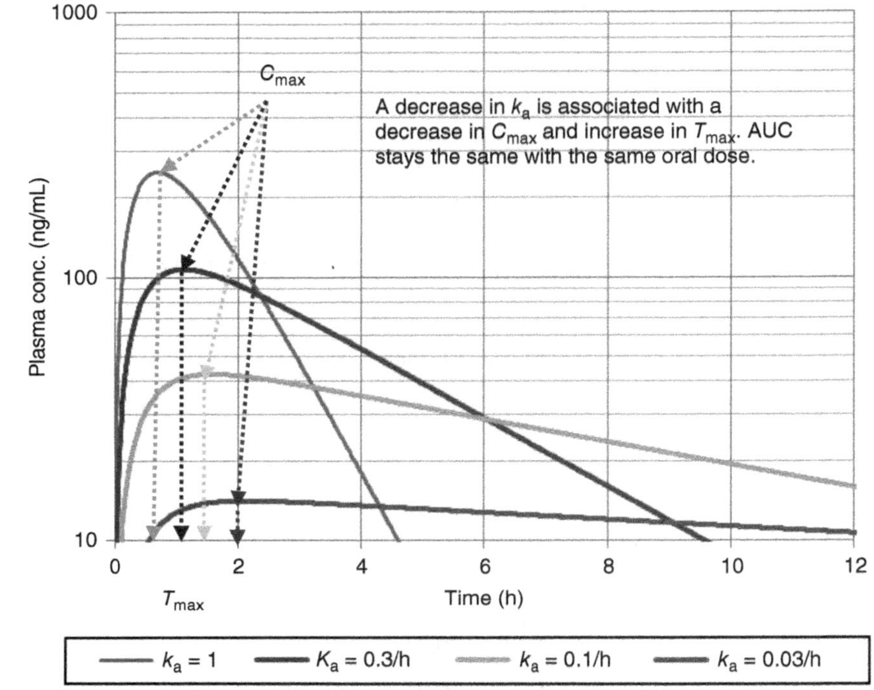

FIGURE 1.10 The impact of changes in k_a values on the oral plasma profile of a compound.

1.3.10 Mean Absorption Residence Time (MAT)

Mean absorption residence time (MAT) is the average time for a molecule to cross the intestinal membrane and arrive at the systemic circulation [26]. It is calculated using the following equation:

$$\text{MAT} = \text{MRT}_{\text{po}} - \text{MRT}_{\text{iv}} \tag{1.24}$$

where MRT_{PO} and MRT_{iv} are the mean residence time of a drug after PO and IV dosing, respectively. Takahashi et al. used MAT to determine gastric emptying rate (GER) in monkey and compared it to that in human. The team used acetaminophen as a probe substrate since it has high passive permeability. Interestingly, acetaminophen MAT of 1.02 h in cynomolgus monkeys was only slightly longer than that in humans suggesting that monkey GER is comparable to that in human [27].

1.3.11 Bioavailability (*F*%)

According to the European Medicines Evaluation Agency (EMEA), bioavailability (*F*%) is "the rate and extent to which an active moiety is absorbed from a pharmaceutical form, and becomes available in the systemic circulation." As a parameter, there are two types of bioavailability:

1. Absolute bioavailability, which refers to the fraction of the extravascular (e.g., oral) dose that reaches the systemic circulation unchanged in reference to an intravenous dose.

It is usually determined by calculating the respective AUC after oral and intravenous administration as depicted in Equation 1.24. This calculation assumes that the drug complies with linear kinetics following dosing by both routes. Therefore, to avoid the effect of nonlinearity, the plasma concentrations following both intravenous and oral dosing should be similar.

$$\text{Absolute bioavailability} = \frac{\text{AUC}_{\text{PO}}}{\text{AUC}_{\text{IV}}} \times \frac{\text{Dose}_{\text{IV}}}{\text{Dose}_{\text{PO}}} \tag{1.25}$$

2. Relative bioavailability, which refers to the fraction of a dose of drug reaching the systemic circulation relative to a reference product, is usually calculated as

$$\text{Relative bioavailability} = \frac{\text{AUC}_{\text{test}}}{\text{AUC}_{\text{ref}}} \times \frac{\text{Dose}_{\text{ref}}}{\text{Dose}_{\text{test}}} \tag{1.26}$$

Oral bioavailability is determined by the fraction of dose absorbed (f_a) in the gastrointestinal tract and fraction of dose that does not undergo metabolism in the intestinal tract (f_g) and liver (f_h) (Figure 1.11). Oral bioavailability is mathematically expressed by the following equation:

$$F = f_a \cdot f_g \cdot f_h \tag{1.27}$$

Furthermore, oral bioavailability is a multiplicity parameter due to the anatomical sequential nature of the overall process (Figure 1.11).

f_h is calculated using the following equation:

$$f_h = 1 - E_h = 1 - \frac{\text{CL}_h}{Q} \tag{1.28}$$

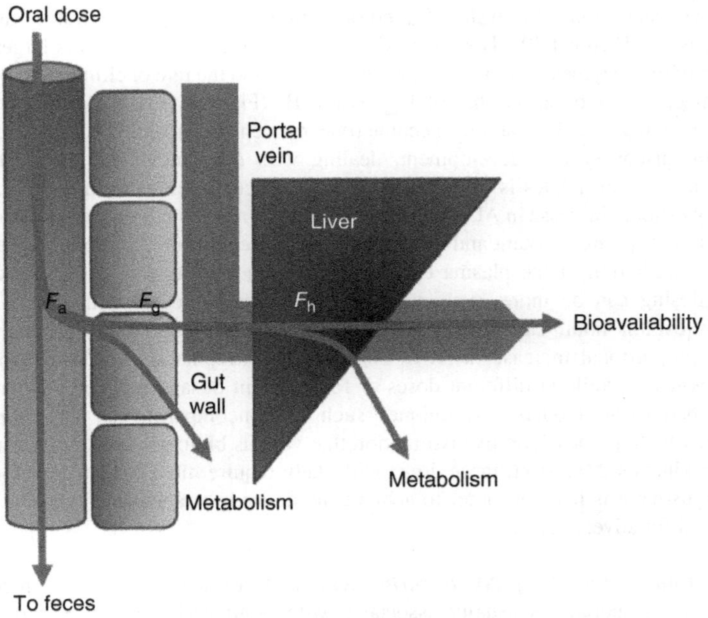

FIGURE 1.11 Oral bioavailability is a multiplicity parameter and a product of f_a, f_g, and f_h. *(See color plate section for the color representation of this figure.)*

Thus, if a drug has a high hepatic extraction ($E_h > 0.7$), then its bioavailability will be low when it is given orally ($F \leq 0.3$). On the other hand, if a drug has low hepatic extraction ($E_h < 0.3$), then the extent of bioavailability will be high provided that it is completely absorbed and not significantly metabolized by the intestine.

TIP:

Always keep the values of the hepatic blood flow in preclinical species and human in mind. Develop a habit of calculating extraction ratio and f_h from CL using Equation 1.28. Based on which, you would be able to determine if hepatic first pass is a major contributing factor for your compound poor bioavailability, if observed following oral dosing.

1.3.12 Linear and Nonlinear Pharmacokinetics

Drug metabolism, renal tubular secretion, biliary secretion, and other active processes are usually mediated by metabolizing enzymes or transporter proteins. These systems usually have good substrate selectivity and are capacity limited. They are usually described by Michaelis–Menten kinetics:

$$v = -\frac{dc}{dt} = \frac{V_{max} \cdot C}{K_m + C} \tag{1.29}$$

where v is equal to $-dc/dt$, which is the differential rate of change in free drug concentration with time, C the free drug concentration that can undergo the change, V_{max} the maximum elimination or transport rate, and K_m the Michaelis constant, which is equal to free drug concentration that results in $V = V_{max}/2$. The values of V_{max} and K_m are dependent on the nature of the drug and enzymatic/transporter process involved. This equation implies that when the free drug concentration is lower than K_m, no saturation of the enzymes or transporters protein occurs (Case I) (Figure 1.12). However, when the free drug concentration is larger than K_m, saturation of the enzymes or transporter proteins occurs and the rate of elimination or transport is maximized and approaches that of V_{max} (Case II) (Figure 1.12). Here, pharmacokinetic parameters such as CL, V_{dss}, and $t_{1/2}$ become time, concentration, and dose dependent.

In drug discovery and development, dealing with drug molecules that follow linear pharmacokinetics principles is a desirable property. Since, an increase in dose is associated with a proportional increase in AUC and all relevant pharmacokinetics parameters such as CL, V_{dss}, MRT, and $t_{1/2}$ are constant and dose, time, and concentration independent (Figure 1.13). Therefore, prediction of the plasma exposure following various dosing regimens and over multiple dosing can be more easily achieved. However, there are situations where drug molecules exhibit nonlinearity, where increasing doses is associated with more than or less than proportional increase in AUC. As a result, extrapolation and projection of drug pharmacokinetic profile at different doses or for different dosage regimen cannot be easily accomplished using modeling techniques such as noncompartmental or compartmental modeling, which are based on the assumption that various biological processes comply with first-order kinetics. Moreover, these drugs will likely require more careful monitoring when dosage adjustment is made in order to achieve the desired therapeutic effects and minimize the potential for adverse effects.

1.3.12.1 Causes for Drug Nonlinearity Saturation of active processes mediated by enzymes or transporters is usually associated with nonlinearity that affects drug overall absorption, distribution, metabolism, or elimination (ADME). The causes of nonlinear kinetics are depicted in Table 1.2.

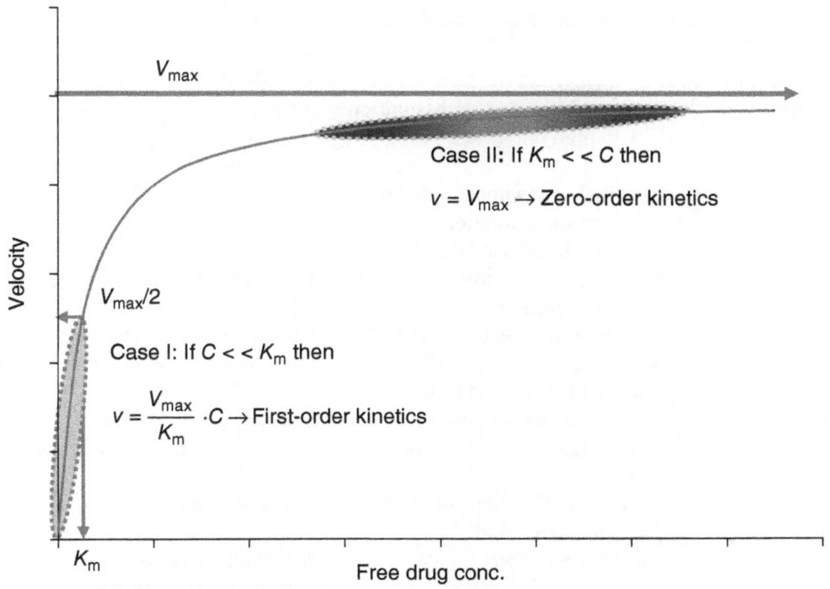

FIGURE 1.12 The relationship between drug elimination/transport rate and free drug concentration for a Michaelis–Menten kinetics complying biological process.

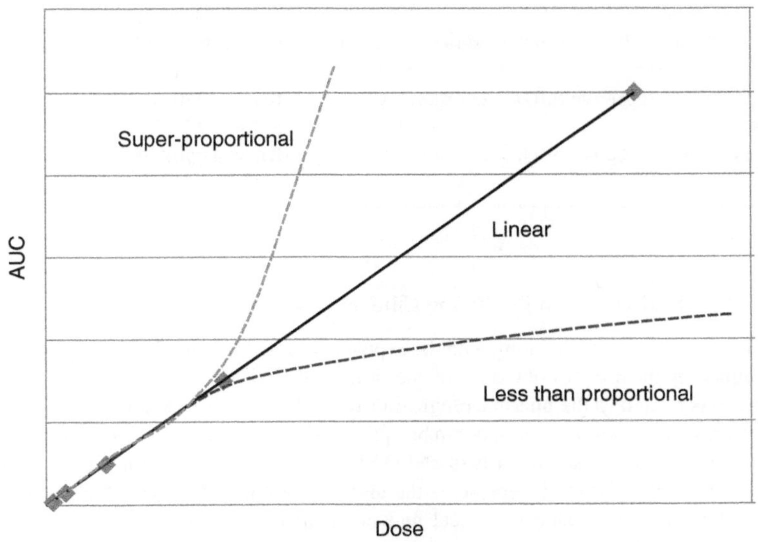

FIGURE 1.13 The relationship between dose and AUC is indicator of presence or absence of drug linearity.

TABLE 1.2 The Factors that may Contribute to Drug Nonlinear Kinetics

Process	Factor	Example
Absorption	Poor solubility and/or dissolution with high oral dose	Amphotericin, griseofulvin, danazol, and chorothiazide [25, 28, 29]
	Intestinal active uptake-mediated absorption saturation e.g., system L and OATP2B1	Gabapentin and celiprolol [30, 31]
	Intestinal P-glycoprotein-mediated efflux saturation	Talinolol and UK343,664 [32, 33]
Distribution	Plasma protein binding saturation	Ceftriaxone and warfarin [34–36]
	Transport into and out of tissues saturation e.g., OATP1B1	Methotrexate and bosentan [37, 38]
Metabolism	Intestinal and/or hepatic-mediated first-pass effect saturation	Salicylic acid, lorcainide, theophylline, and simepravir [39–41]
	Cofactor and/or metabolizing enzyme depletion	Acetaminophen [42]
	Enzyme induction	Avasimibe, carbamazepine, phenobarbital, phenytoin, ritonavir, efavirenz, and bosentan [38, 43–48]
	Metabolite-dependent inhibition	Paroxetine, erythromycin, clarithromycin, and diltiazem [49–51]
Renal excretion	Active renal secretion	Clofarabine, adefovir, cidofovir, and tenofovir [52, 53]
Active hepatic uptake	Active biliary secretion	Rosuvastatin and pravastatin [54–56]
Biliary excretion	Active biliary secretion	Rosuvastatin and pravastatin [54–56]

1.3.13 PK/TK Modeling in Predicting Clinical Dose

PK/TK is an area of science dealing with the exposure of test compound and metabolites, which is determined by the kinetics of exposure and drug ADME. Generally, the extent and duration of exposure is related to the pharmacological or toxicological effects, and thus, the change of the occurrence of the observed effects can be optimized by altering the dose or exposure period.

A basis for toxicity assessment is the NOAEL or no observed effect level (NOEL) or low observed effect level. NOAEL represents the highest dose at and below which no significant adverse effects are seen. Due to ethical and practical reasons, assessment of the NOAEL is derived only from animal toxicity data and extrapolated to identify a clinical dose that is significantly lower than the NOAEL. The same strain and species used in the toxicology studies should be used in the TK studies. Initial studies may involve one sex of each species; however, the use of multiple species is relevant to build confidence in models predicting human effects. Using dose–response relationship, the statistical confidence limits of dose at which the incidence or frequency of a toxic effect is established.

Developing mathematical models is of value only when the mechanisms of toxicity are understood and/or the parameters that are being used in model building have established relationships to the observed toxicity effect. While it is important to understand whether the parent molecule or its metabolite is responsible for toxicity, an estimate of whether humans are of lesser, equal, or greater sensitivity in comparison to the test species is needed for translating the preclinical data to predict human effects. To establish relationship between exposure and dosimetry, a PK/TK model should incorporate the rate and extent of absorption, compound and/or metabolites distributed in the body, metabolism and kinetics of metabolites (if appropriate), elimination rate and elimination route(s) and the influence of dose on all the above processes (dose-dependency).

A range of modeling approaches is used to simulate and project plasma exposure in preclinical species and human. Below is a summary of these approaches with emphasis on their advantages and limitations.

1.3.14 Noncompartmental Pharmacokinetics

Various pharmacokinetic parameters such as CL, V_d, $t_{1/2}$, MRT, and $F\%$, can be determined using noncompartmental methods. These methods are based on the empirical determination of AUC and AUMC described above. Unlike compartmental models (see below), these calculation methods can be applied to any other models, provided that the drug follows linear pharmacokinetics. However, the main limitations of the noncompartmental method are it is not robust and lacks any mechanistic interpretation of the data since derived pharmacokinetic parameters have no meaningful physiological relevance. Furthermore, the method does not provide insight into the mechanism of drug–drug interaction. The method cannot be used for the simulation of different plasma concentration–time profiles when there are alterations in dosing regimen or multiple dosing regimens are used.

1.3.15 Compartmental Pharmacokinetics

Compartmental models of pharmacokinetic analysis are widely used to describe drug distribution and disposition. In these models, the body is assumed to be composed of one compartment or more and the drug kinetics can be defined by differential equations generally of first-order process. These compartments are virtual and do not have any physiological significance. However, they may represent a group of tissues or organs with similar distribution characteristics. For example, highly blood perfused body organs such as liver, lungs, and kidney often have different drug distribution than fat tissue. Compartmental models are usually arranged in a mammillary format, such that there is a one or more compartment that feeds from a central compartment.

1.3.15.1 One-Compartment Open Model In the one-compartment model, the body is assumed to be a homogenous unit where the drug is rapidly distributing throughout the body and once eliminated it follows a monoexponential decline (Figure 1.14). Following intravenous dosing, the plasma drug concentration can be calculated as

$$C = C^{\circ} \cdot e^{-K_e t} \tag{1.30}$$

where C° is the plasma drug concentration immediately after intravenous dosing. C° is also calculated as

$$C^{\circ} = D/V_c \tag{1.31}$$

Unlike other compartmental models, there is only one V_d, where $V_c = V_{dss}$.

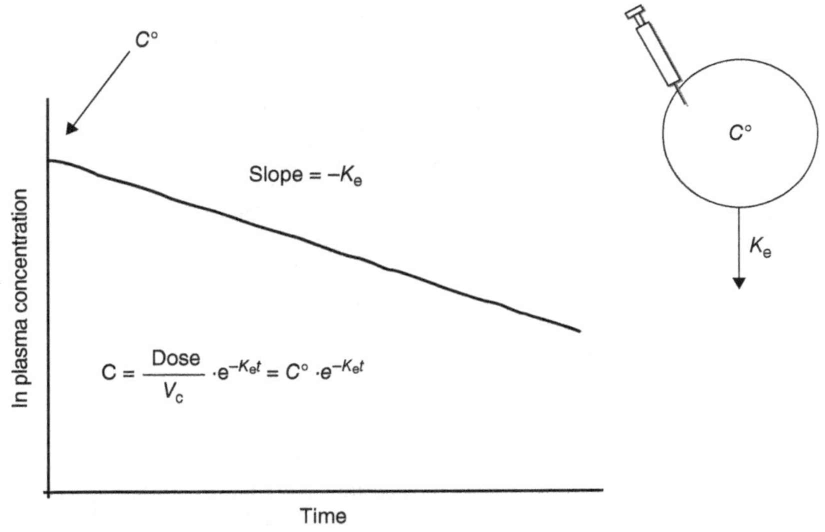

FIGURE 1.14 One-compartment model.

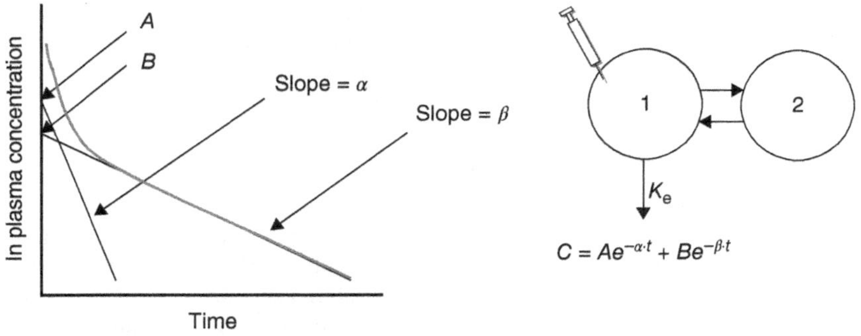

FIGURE 1.15 Two-compartment model.

1.3.15.2 Two-Compartment Open Model When the drug concentration versus time profile demonstrates a biexponential decline following intravenous dosing, a two-compartment model that is the sum of two first-order processes (distribution and elimination) will better describe the data (Figure 1.15). A drug that follows the pharmacokinetics of a two-compartment model does not rapidly distribute throughout the body as evident in the one-compartment model. In the two-compartment model, the drug is assumed to distribute into two compartments, the central and tissue compartments. The central compartment represents the highly perfused body organs where the drug distributes rapidly and uniformly. On the other hand, in the tissue compartment, the drug distributes more slowly.

For a drug that follows the two-compartment model, the rate of drug plasma concentration change following intravenous dose can be determined as

$$C = Ae^{-\alpha \cdot t} + Be^{-\beta \cdot t} \tag{1.32}$$

where A and B are functions of the administered dose and α and β the first-order constants for the distribution and elimination phase, respectively.

In this chapter, only the one- and two-compartment models following intravenous dosing were described. Other models with extravascular dosing have an additional compartment with an absorption rate constant describing input into the central compartment. Models with three or greater compartments may be used if the drug concentration versus time are described with additional exponential terms. However, these models are more complex.

1.3.16 Physiology-Based Pharmacokinetic (PBPK) Modeling

PBPK modeling is quantitative description of pharmacokinetics using both drug-specific *in vitro* and physiochemical properties as well as the species-specific anatomical and physiological information. Mathematically, the model constitutes multiple compartments, each corresponding to different body organs or tissues and linked together based on their anatomical placement with respect to blood flow (Figure 1.16). The vital physiological information in building these models are tissue volumes, blood flow to the tissue and tissue composition, which are considered system parameters that are drug independent (Figure 1.17). The compound-specific physicochemical and *in vitro* information (drug parameters) such as

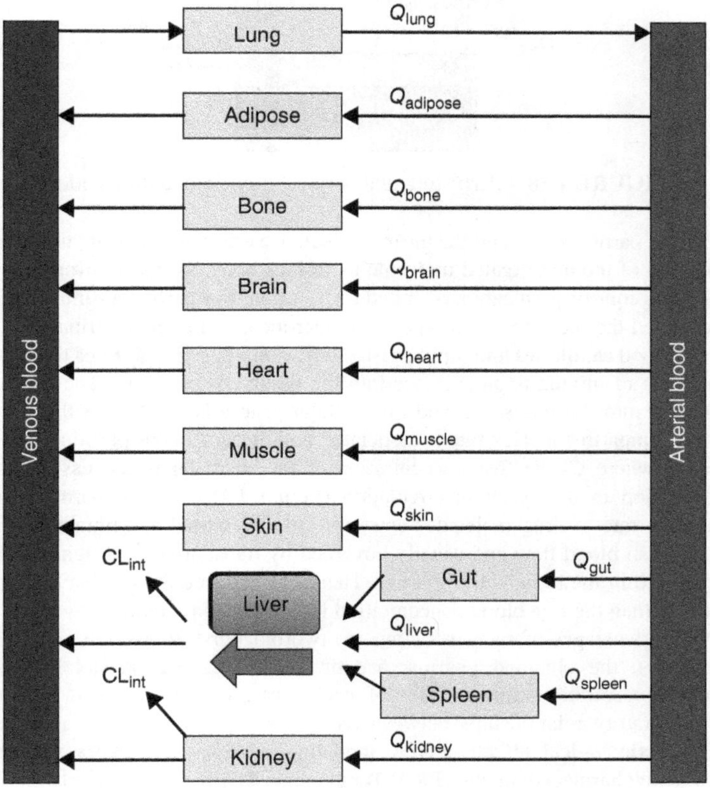

FIGURE 1.16 Physiologically based pharmacokinetic (PBPK) model incorporating physiological compartments depending on the drug's distribution.

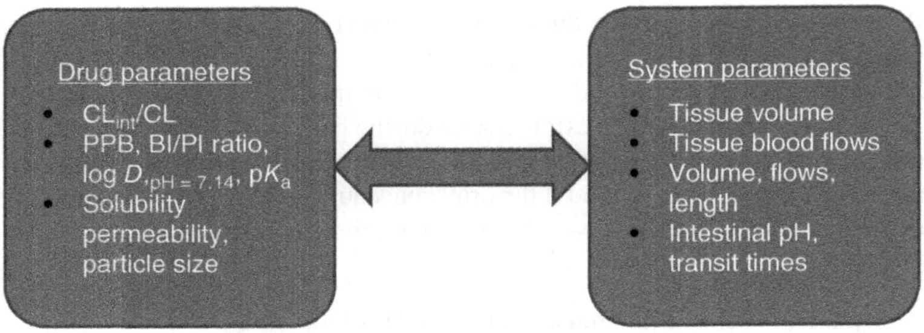

FIGURE 1.17 Known drug and system parameters used to build PBPK models.

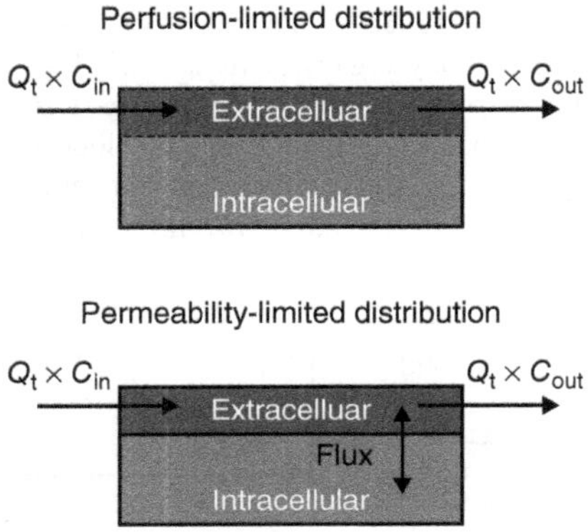

FIGURE 1.18 Perfusion- and permeability-limited distribution.

tissue to plasma partition (K_p) and the intrinsic clearance are provided as inputs (Figure 1.17). The distribution of the investigated molecule in various body organs is governed by whether it is either perfusion- or permeability-limited distribution. For perfusion-limited distribution, this is considered the most common type of compartment and mainly attributed to the leaky nature of the blood capillaries and rapid passive diffusion of drug molecules that is associated with high extent of mixing upon entrance into the tissue compartment. The rate of entry of drug molecules into the interstitial and intracellular spaces is faster than the rate of blood flow into the compartment. This rate is associated with instantaneous partition into the tissue and $C_{out} = C_f$, where C_f the free concentration of the chemical in the tissue and C_{out} the free concentration in the systemic circulation (Figure 1.18). As for permeability-limited distribution, the rate of drug molecules entry into the interstitial and intracellular spaces is slower than organ blood flow and usually governed by transporters and tight junctions such as those reported in the blood–brain barrier. Therefore, the free drug tissue concentration is typically lower than the free blood concentration in systemic circulation.

The major advantages of such modeling are twofold. First, unlike the empirical models (e.g., allometry), the obtained pharmacokinetic parameters are physiologically relevant and second, concentration–time profiles of each tissue can be obtained simultaneously. With the exploratory relationships between tissue concentration profiles and the pharmacological or toxicological effects, PBPK modeling provides framework for mechanistic pharmacokinetic/pharmacodynamic (PK/PD) modeling. PBPK models are also most reliable in dose–response and tissue exposure assessments under various physiological conditions (e.g., age, disease condition) [57].

1.3.17 Modeling to Predict Single and Multiple Dose Pharmacokinetic Profile

As previously discussed, compartmental models can be effectively used to project plasma concentrations that would be achieved following different dosage regimen and/or multiple dosing. However, for these projections to be accurate, the drug pharmacokinetic profile should follow first-order kinetics where various pharmacokinetic parameters such as CL, V_d, $t_{1/2}$, and $F\%$ do not change with dose.

1.4 TOXICOGENOMICS AND BIOMARKERS

Toxicology safety biomarkers are functional or structural measurements that correlate with a morphologic histopathologic or clinical pathology change in an organ system such as the liver or heart. A translatable biomarker is critical in drug discovery and development. Toxicogenomics is the integration of the *omics* technologies (genomics, proteomics, and metabonomics), bioinformatics, and toxicology to better understand drug- or toxicant-induced alterations in biochemical networks (gene, protein, and metabolite) of drug candidate development in a pharmaceutical setting. Therefore, toxicogenomics data can be used as drug toxicity/exposure biomarkers or signatures that provide insights into the toxic mechanism of action of a drug candidate and support safety risk assessment. Such an approach can be used in high-throughput screening in discovery research. In fact, gene expression is used in the clinic to predict pathologic conditions (e.g., breast cancer) prognosis and response to therapy. The issues with toxicogenomics experiments are variability in experimental design, strain, gender, age of experimental animals, husbandry, and nutrition; interanimal variation and clinical health effects of test compounds; and organs heterogeneity. For example, although the liver appears heterogeneous, it is composed of approximately 58% hepatocytes, 19% endothelial cells, 14% Kupffer cells, 4% biliary epithelium, and 5% stellate cells.

A good example of the application of toxicogenomics was its use to identify a transcriptional biomarker of the histopathological liver change of oval cell-mediated bile duct hyperplasia (BDH). BDH is a histopathologic finding that occurs in both rodent and nonrodent species. BDH can progress to cholangiocarcinoma with low margins of safety, which can lead to costly, late stage compound terminations and increased risk to patient safety. Thus, interpretation of the significance of BDH requires consideration of a number of variables, including duration of drug exposure, margins of safety, intended drug indication, and availability of biomarkers to monitor for patient safety. The specificity and sensitivity of the discovered candidate biomarker, called deleted in malignant brain tumor (DMBT1), was evaluated in livers of rats treated with more than 30 different compounds comprising hepatotoxicities that ranged from BDH and hepatocyte proliferation, to phospholipidosis, hepatocellular vacuolation, apoptosis, and inflammation [58]. Multidisciplinary collaboration among toxicologic pathologists, toxicologists, biologists, toxicokinetics, and statisticians is needed for successful toxicogenomic efforts.

1.5 SPECIES DIFFERENCE IN DRUG DISPOSITION

Oral exposure of drug molecules is a product of their absorption and hepatic and intestinal first pass. Species differences were reported in the oral exposure of various drug molecules. Several investigators attributed these findings mainly to differences in anatomical and physiological factors such as hepatic blood flow, metabolizing enzyme type, and expression or extent of protein binding [59]. There are significant species differences in bile flow rate and hepatic blood flow (Table 1.1), which may explain some of this variation. In addition, bile composition (acid, ions, electrolytes) also varies between species and may further explain reported species differences in drug biliary excretion rate and thus differences in pharmacokinetics disposition of various drugs [60].

Nelson et al. reported that so far 14 CYP gene families have been identified in mammals with significant variations in the primary sequence of amino acids across species. However, these members of the superfamily had highly conserved regions of amino acid sequence [61]. Similar findings were also reported with uridine diphosphoglucose transferases and carboxylesterases [19, 62]. Overall, these small differences in the amino acid sequences can lead to significant differences in substrate affinity and specificity, which translates into differences in the metabolism rate and metabolism profiles. As a general rule, compounds with good passive absorption, high rat hepatic extraction ratio, and poor oral bioavailability tend to have better oral bioavailability in higher species such as dogs, monkeys, and humans. There are many cited examples that are consistent with this trend. For example, atomoxetine is a CYP2D6 substrate with an absolute human oral bioavailability of 94% and 63% in poor and extensive metabolizers of CYP2D6, respectively [63]. The moderate to high human oral bioavailability suggests nearly complete oral absorption of atomoxetine. However, preclinical evaluations showed that the absolute oral bioavailability of atomoxetine in rat was only 4% [64], but was 74% in dog [64]. Overall, the disposition of atomoxetine is similar in rats, dogs, and humans with a primary oxidative metabolite of 4-hydroxyatomoxetine that is subsequently conjugated to form 4-hydroxyatomoxetine-*O*-glucuronide. In a radiolabeled study in rats administered ^{14}C-atomoxetine, atomoxetine AUC following oral administration accounted for only 2% of the total ^{14}C AUC as compared to 30% of the ^{14}C AUC following intravenous administration, indicating extensive first-pass metabolism in rats [64]. In a corresponding radiolabeled study in dogs, atomoxetine AUC following oral administration accounted for 33% of the total ^{14}C AUC as compared to 39% of the ^{14}C AUC following intravenous administration, indicating considerably less pronounced first-pass metabolism [64]. This example clearly illustrates the importance of understanding not only the species differences in a drug's metabolic fate but also the extent of species differences in the first-pass metabolism when utilizing preclinical data to project human oral bioavailability.

Indinavir, a CYP3A4 substrate, is an HIV protease inhibitor for which variable oral bioavailability has been observed in preclinical species, ranging from 72% in dogs to 19% in monkeys, and was 24% in rats [65]. This variability was mainly attributed to species differences in the extent of hepatic first-pass metabolism. Chemical and immunochemical inhibition studies indicated the potential involvement of CYP3A isoforms in the metabolism of indinavir in rats, dogs, and monkeys [65], which is consistent with the observation that CYP3A4 is the main isoform responsible for the oxidative metabolism of indinavir in human liver microsomes [66]. The *in vitro* profile of indinavir metabolism was qualitatively similar across species [65]. In addition, an *in vitro–in vivo* correlation was established in rats and dogs using the *in vivo* hepatic clearance and hepatic first-pass extraction ratio obtained from *in vitro* rat and dog metabolic data, respectively. Based on the *in vitro–in vivo* correlation established in rats and dogs, the *in vitro* intrinsic clearance of indinavir in human liver microsomes projected a small first-pass metabolism in humans ($E_h = 0.25$), which was consistent with indinavir's high oral bioavailability (60–65%) observed in humans at clinically relevant doses [65] [67]. This example depicts the importance of establishing an *in vitro–in vivo* correlation in tested preclinical species so as to use it as a basis to project human clearance and oral bioavailability. Overall, these successful medications would not be on the market if the discovery team solely depended on rat oral bioavailability to evaluate their metabolism in humans.

Species differences in the extent of protein binding of various xenobiotics were also reported. It is interesting to note that albeit their structural and functional homologies, there are minor differences in the amino acid sequence in plasma protein molecules such as albumin among various mammals. Thus, protein binding may be another contributing factor to the differences in both the binding affinity and sites of drugs in protein molecules among different species [68]. For more information on the impact of species differences of metabolizing enzymes and transporters on drug disposition following oral dosing, see Chapter 11.

1.6 MIST (METABOLITES IN SAFETY TESTING)

Guidance for Industry on Safety Testing of Drug Metabolites, MIST, published in 2008 from the Center for Drug Evaluation and Research (CDER) defining the threshold for human metabolites that raises safety concerns as 10% of total drug exposure. Additional guidelines issued from the International Conference on Harmonization Guideline (ICH) on Nonclinical Safety Studies for the Conduct of Human Clinical Trials and Marketing Authorization for Pharmaceuticals, ICH M3[R2], defines the threshold as 10% of total drug related exposure, which supersedes when in conflict with the CDER guidance. Others have suggested that total metabolite burden, which account for dose is a more relevant threshold [69, 70]. Early clinical assessment of human metabolites in plasma or urine enables identification of significant human metabolites and whether they are present at equal or greater exposure levels (AUC; area-under-the-curve) in any one of the preclinical safety species is critical for addressing MIST implications. While rare, this early assessment is exceedingly important if the metabolite is formed in human only. Traditionally, preclinical measurement of circulation concentrations of metabolites only occurs to address specialized cases such as when extensive metabolism of parent drug is expected. When the human metabolite profile is expected to be qualitatively similar to at least one of the preclinical safety testing species, measurement of circulating parent in plasma is adequate in preclinical pharmacokinetic and toxicokinetic studies. Drug development history has taught us that human drug metabolites can play significant roles in primary and secondary pharmacology, general toxicity, and potentially idiosyncratic drug toxicity (Table 1.3). Therefore, one must develop fit-for-purpose metabolite assessment strategy and use the knowledge of the chemical space of the drug candidate in question, structural activity relationship for both primary and related secondary pharmacology, metabolic clearance pathways and potential target organ toxicity to evaluate if measuring metabolites in preclinical pharmacokinetic/toxicokinetic samples is warranted and would provide the discovery team with knowledge to design a best-in-class compound. Once a compound has progressed through early development, usually post-POC, definitive radiolabeled ADME studies are conducted to identify metabolites and assess the need to monitor these in the clinic or qualify in toxicology studies. Realistically, successful drug discovery teams know well in advance the outcome of these studies.

TABLE 1.3 Examples of Human Drug Metabolites that are known to be Responsible for General Toxicity and Potentially Idiosyncratic Drug Toxicity

Parent Drug	Metabolite	Suspected Metabolite Activity
Acetaminophen	*N*-Acetyl-*p*-benzoquinone imine (NAPQI)	Toxicity
Troglitazone	GSH conjugation/ring scission	Idiosyncratic toxicity
Losartan	5-Carboxylic acid (10–40× potency)	Pharmacologically active
Dexfenfluramine	D-Norfenfluramine	Pharmacologically active

1.7 PHARMACOLOGICALLY ACTIVE METABOLITES

Metabolism of drug candidates can occur at sites of the molecule that do not alter the active pharmacophore, and sometimes, result in metabolites that have greater pharmacological activity or longer pharmacodynamic duration. This result is usually first suspected when existing PK/PD relationships fail to predict the accurate dose required for the desired effect and thus may serve as a trigger for metabolite scouting activities. Prodrug strategies are the intentional design of active pharmacological species from an inactive precursor. Where prodrug approaches achieve quantitative conversion of inactive precursor to the active species, circulating metabolite analysis is the only manner to confirm that dose was administered.

1.8 REACTIVE METABOLITES

The MIST guidance addresses a majority of direct or indirect conjugates; the latter class of metabolites considered harmless but toxicity continues to be a major challenge leading to drug development failures. Reactive metabolites are frequently suspected when hepatotoxicity or idiosyncratic toxicity is observed. By the very nature of these species, circulating concentrations of reactive metabolites often occur at trace levels making them difficult to detect and quantitate. Recent strategies to identify reactive metabolites as conjugates to biomolecules such as glutathione or cysteine offer an approach to confirm the formation of these species *in vivo* [71]. Not only is the confirmation of suspected reactive metabolites informative, but the design of enzymology studies to understand pathways that lead to their formation allow for an understanding of the potential for species-specific reactions.

1.9 ENABLING TECHNOLOGIES

Plasma pooling methods were compared to the traditional approaches of obtaining quantitative information on the levels of circulating metabolites in preclinical species. The exposure values obtained via sample pooling are comparable to those obtained by traditional methods of analyzing samples individually. When kinetic information is not needed, this approach allows metabolite identification and exposure burden estimates with a modest resource investment. While metabolite analysis without authentic standards can be accomplished using high-resolution or accurate mass LCMS techniques, new techniques using chromatographic measures such as retention time shifting or background subtraction of vehicle-dosed samples offer methods to identify metabolites [72].

There are various approaches to assess the metabolite exposure margin between toxicology species and humans: either by direct or indirect comparison. The preferences in when and how to pursue metabolite assessment are based on the overall development strategy. Therefore, it is important to understand the utility and limitations of analytical instruments in order to apply an appropriate analytical tool to address specific questions posed at different stages of drug development. The urgency of metabolite monitoring depends on the intrinsic nature of the compound, therapeutic intent and objective of the clinical development; there is not a unified approach that provides efficient resource utilization.

1.10 CONCLUSION

PK and TK are important surrogates of drug exposure, efficacy, and safety in early and late drug discovery and development. Determinants of a drug PK/TK include absorption, distribution, metabolism, and elimination. Several toxicity assessment studies (e.g., acute, single, repeat dose studies, GLP) in both rodent and nonrodent species are used to better refine safety margins, PK/PD modeling, and set appropriate dosages before starting FIH clinical trials. Critical parameters that define PK/TK profile are AUC, C_{max}, t_{max}, V_d, V_{dss}, $t_{1/2}$, and F. Mathematical modeling can be of great value in the PK/TK evaluation. There is an emerging and increased utility of toxicogenomics to better define drug toxicity/exposure to support compound selection and safety risk assessment. There are significant species differences in the oral exposure to various drugs that must be taken into consideration. Such differences are related to various anatomical and physiological factors including hepatic and renal blood flow, metabolizing enzymes (i.e., CYPs), and protein binding. A valid strategy should be a fit-for-purpose metabolite assessment strategy that uses the knowledge of the chemical space of a drug candidate being developed, structural activity relationship to pharmacology and toxicology, and metabolism clearance pathways. Ultimately, a well-designed PK/TK evaluation would provide the discovery team with knowledge to design and advance a best-in-class compound.

1.11 CHAPTER 1 TUTORIAL

1. Which of the following statements are true (**T**) or false (**F**) with regard to first-order or zero-order processes?
   - A constant $t_{1/2}$ is characteristic for a first-order process.
   - The rate constant for a zero-order process has the unit [concentration per time].
   - A drug that is eliminated by a first-order process with the rate constant $0.1\,h^{-1}$ has a longer half-life than a drug eliminated by a first-order process with the rate constant $0.3\,h^{-1}$.
   - If a drug follows linear pharmacokinetics, then the transfer of drug between different compartments in a compartmental pharmacokinetic model always follows zero-order kinetics.
2. Define the following concepts:
   - Area under the curve
   - Clearance
3. What is the impact of increasing clearance on volume of distribution and half-life according to Equation 1.16?
4. A patient received an intravenous bolus injection of the Amikacin. Amikacin elimination follows first-order kinetics. The plasma concentration 2 h after the dose is 31 µg/mL. Six hours after the dose was given, it is 2.3 µg/mL. Calculate the elimination rate constant, the half-life of Amikacin, the initial concentration of the drug, and the area under the curve using integration method?

5. The data summarized in the following table depict the average plasma exposure of a new medication obtained from 10 healthy volunteers (average weight 70 kg).

Time (h)	Mean Plasma Conc. (ng/mL) IV Dose 1 mg/kg	Oral Solution 10 mg/kg	Oral Tablet 10 mg/kg
0	10,000	0	0
0.25	7,650	495	100
0.5	6,260	1,330	250
1	4,630	3,540	600
2	3,710	4,650	800
3	2,110	3,500	1,100
4	1,940	2,730	1,200
6	1,020	1,450	800
8	390	807	500
12	138	288	150
24	9	68	100

a. From the above data, calculate the following:
- Total body clearance (CL)
- Volume of distribution (V_d)
- Half-life ($t_{1/2}$)
- First-order elimination rate constant (K).

b. What is the relative oral bioavailability of the oral tablet to oral solution dosing? Why it is not 100%?

c. Which of the oral formulations tested, demonstrated flip–flop kinetics? What is the basis for your observation?

1.12 CHAPTER 1 TUTORIAL ANSWERS KEY

1. Which of the following statements are true (**T**) or false (**F**) with regard to first-order or zero-order processes?
- A constant $t_{1/2}$ is characteristic for a first-order process. **(T)**
- The rate constant for a zero-order process has the unit [concentration per time]. **(F)**
- A drug that is eliminated by a first-order process with the rate constant 0.1 h^{-1} has a longer half-life than a drug eliminated by a first-order process with the rate constant 0.3 h^{-1}. **(T)**
- If a drug follows linear pharmacokinetics, then the transfer of drug between different compartments in a compartmental pharmacokinetic model always follows zero-order kinetics. **(F)**

2. Define the following concepts:
- Area under the curve:

It is the total amount of unchanged drug that reaches the systemic circulation following intravenous or extravascular administration. The unit for AUC is concentration per unit time (e.g., ng*h/mL)

- Clearance:

  It is a primary pharmacokinetic parameter that describes the process of irreversible elimination of a drug from the systemic circulation. It is defined as the volume of blood or plasma that is totally cleared of its content of drug per unit time.

3. What is the impact of increasing clearance on volume of distribution and half-life according to Equation 1.16?

Clearance and volume of distribution are independent primary parameters where changes in either won't affect the other. Any change in either volume of distribution of clearance will only affect the half-life.

4. A patient received an intravenous bolus injection of the Amikacin. Amikacin elimination follows first-order kinetics. The plasma concentration 2 h after the dose is 31 μg/mL. Six hours after the dose was given, it is 2.3 μg/mL. Calculate the elimination rate constant, the half-life of Amikacin, the initial concentration of the drug, and the area under the curve using integration method?

$$-K/2.3 = [\log 2.3 - \log 31]/6 = -0.188, \quad K = 0.434\,\text{h}^{-1}$$

$$t_{1/2} = 0.693/K = 0.693/0.434 = 1.6\,\text{h}$$

$$\text{AUC}(0 \rightarrow 6\,\text{h}) = (C_2 + C_1)/2 * (t_2 - t_1) = (31 + 2.3)/2 * 6 = 99.9\,\mu\text{g}\,\text{h/mL}$$

5. The data summarized in the following table depict the average plasma exposure of a new medication obtained from 10 healthy volunteers (average weight 70 kg).

	Mean Plasma Conc. (ng/mL)		
Time (h)	IV Dose 1 mg/kg	Oral Solution 10 mg/kg	Oral Tablet 10 mg/kg
0	10,000	0	0
0.25	7,650	495	100
0.5	6,260	1,330	250
1	4,630	3,540	600
2	3,710	4,650	800
3	2,110	3,500	1,100
4	1,940	2,730	1,200
6	1,020	1,450	800
8	390	807	500
12	138	288	150
24	9	68	100

a. From the above data, calculate the following:

- Total body clearance (CL)

$$= \text{Dose/AUC} = 1{,}000{,}000\,\text{ng/kg}/22{,}120\,\text{ng}\,\text{h/mL}/60\,\text{min/h} = 0.75\,\text{mL/min/kg}$$

- Volume of distribution (V_d)

$$V_{dss} = \text{CL} * \text{MRT} = \text{CL} * \text{AUMC/AUC}$$

$$= 0.75\,\text{mL/min/kg} * 60\,\text{min/h} * 3.493\,\text{h} * /1000\,\text{mL} * 1\,\text{L}$$

$$= 157\,\text{mL/kg}$$

- Half-life ($t_{1/2}$)

$$= 0.693 * V_{dss}/CL = 0.693 * 157\,mL/kg/0.75\,mL/min/kg/60\,min * h = 2.42\,h$$

- First-order elimination rate constant (K)

$$K = 0.693/t_{1/2} = 0.693/2.42 = 0.286$$

b. What is the relative oral bioavailability of the oral tablet to oral solution dosing? Why is not 100%?

The relative oral bioavailability is 50% and the cause of this is the potential poor solubility and dissolution of the tablet formulation in comparison to oral solution.

c. Which of the oral formulations tested, demonstrated flip–flop kinetics? What is the basis for your observation?

The oral tablet appears to have a flip–flop kinetics since its half-life following PO dosing is longer than that observed in the IV formulation.

REFERENCES

1. Meibohm, B. and Derendorf, H., *Basic concepts of pharmacokinetic/pharmacodynamic (PK/PD) modelling*. Int J Clin Pharmacol Ther, 1997. 35(10): p. 401–413.
2. Roberts, M.S., Rumble, R.H., Wanwimolruk, S., Thomas, D. and Brooks, P.M., *Pharmacokinetics of aspirin and salicylate in elderly subjects and in patients with alcoholic liver disease*. Eur J Clin Pharmacol, 1983. 25(2): p. 253–261.
3. Lin, J., Sahakian, D.C., de Morais, S.M., Xu, J.J., Polzer, R.J. and Winter, S.M., *The role of absorption, distribution, metabolism, excretion and toxicity in drug discovery*. Current Topics in Medicinal Chemistry, 2003. 3(10): p. 1125–1154.
4. DiMasi, J.A., Hansen, R.W. and Grabowski, H.G., *The price of innovation: new estimates of drug development costs*. J. Health Economics, 2003. 22(2): p. 151–185.
5. van De Waterbeemd, H., Smith, D.A., Beaumont, K. and Walker, D.K., *Property-based design: optimization of drug absorption and pharmacokinetics*. J Med Chem, 2001. 44(9): p. 1313–1333.
6. Kola, I. and Landis, J., *Can the pharmaceutical industry reduce attrition rates?* Nat Rev Drug Discov, 2004. 3(8): p. 711–715.
7. Hellriegel, E.T., Bjornsson, T.D. and Hauck, W.W., *Interpatient variability in bioavailability is related to the extent of absorption: implications for bioavailability and bioequivalence studies*. Clin Pharmacol Ther, 1996. 60(6): p. 601–607.
8. Varma, M.V., Obach, R.S., Rotter, C., Miller, H.R., Chang, G., Steyn, S.J., El-Kattan, A. and Troutman, M.D., *Physicochemical space for optimum oral bioavailability: contribution of human intestinal absorption and first-pass elimination*. J Med Chem, 2010. 53(3): p. 1098–1108.
9. Di, L., *The role of drug metabolizing enzymes in clearance*. Expert Opin Drug Metab Toxicol, 2014. 10(3): p. 379–393.
10. Marathe, P.H. and Rodrigues, A.D., In vivo *animal models for investigating potential CYP3A- and Pgp-mediated drug–drug interactions*. Curr Drug Metab, 2006. 7(7): p. 687–704.
11. Chu, X., Bleasby, K. and Evers, R., *Species differences in drug transporters and implications for translating preclinical findings to humans*. Expert Opin Drug Metab Toxicol, 2013. 9(3): p. 237–252.
12. Raunio, H., Taavitsainen, P., Honkakoski, P., Juvonen, R. and Pelkonen, O., In vitro *methods in the prediction of kinetics of drugs: focus on drug metabolism*. Altern Lab Anim, 2004. 32(4): p. 425–430.

13. Miners, J.O., Smith, P.A., Sorich, M.J., McKinnon, R.A. and Mackenzie, P.I., *Predicting human drug glucuronidation parameters: application of* in vitro *and* in silico *modeling approaches*. Annu Rev Pharmacol Toxicol, 2004. 44: p. 1–25.

14. Poggesi, I., *Predicting human pharmacokinetics from preclinical data*. Curr Opin Drug Discov Devel, 2004. 7(1): p. 100–111.

15. Obach, R.S., Baxter, J.G., Liston, T.E., Silber, B.M., Jones, B.C., MacIntyre, F., Rance, D.J. and Wastall, P., *The prediction of human pharmacokinetic parameters from preclinical and* in vitro *metabolism data*. J Pharmacol Exp Ther, 1997. 283(1): p. 46–58.

16. Rowland, M. and Tozer, T., *Clinical Pharmacokinetics*. 3rd ed. 1995, Lippincott Williams & Wilkins: Philadelphia.

17. Gibaldi, M. and Perrier, D., *Pharmacokinetics*. 2nd ed.. 1982, Marcel Dekker, Inc: New York.

18. Arrowsmith, J., *Trial watch: phase II failures: 2008–2010*. Nat Rev Drug Discov, 2011. 10(5): p. 328–329.

19. Hosokawa, M., Maki, T. and Satoh, T., *Characterization of molecular species of liver microsomal carboxylesterases of several animal species and humans*. Arch Biochem Biophys, 1990. 277(2): p. 219–227.

20. Williams, J.A., Hurst, S.I., Bauman, J., Jones, B.C., Hyland, R., Gibbs, J.P., Obach, R.S. and Ball, S.E., *Reaction phenotyping in drug discovery: moving forward with confidence?* Curr Drug Metab, 2003. 4(6): p. 527–534.

21. Wells, T.G., Mortensen, M.E., Dietrich, A., Walson, P.D., Blasier, D. and Kearns, G.L., *Comparison of the pharmacokinetics of naproxen tablets and suspension in children*. J Clin Pharmacol, 1994. 34(1): p. 30–33.

22. Brunton, L., Lazo, J. and Parker, K., eds. *Goodman & Gilman's The Pharmacological Basis of Therapeutics*. 10th ed.. 2001, Mc Graw Hill: New York. p. 2021.

23. Schentag, J.J., *Antimicrobial action and pharmacokinetics/pharmacodynamics: the use of AUIC to improve efficacy and avoid resistance*. J Chemother, 1999. 11(6): p. 426–439.

24. MacGowan, A.P., *Role of pharmacokinetics and pharmacodynamics: does the dose matter?* Clin Infect Dis, 2001. 33(Suppl 3): p. S238–S239.

25. Robbie, G., Wu, T.C. and Chiou, W.L., *Poor and unusually prolonged oral absorption of amphotericin B in rats*. Pharm Res, 1999. 16(3): p. 455–458.

26. Veng-Pedersen, P., *Mean time parameters in pharmacokinetics. Definition, computation and clinical implications (Part I)*. Clin Pharmacokinet, 1989. 17(5): p. 345–366.

27. Takahashi, M., Washio, T., Suzuki, N., Igeta, K., Fujii, Y., Hayashi, M., Shirasaka, Y. and Yamashita, S., *Characterization of gastrointestinal drug absorption in cynomolgus monkeys*. Mol Pharm, 2008. 5(2): p. 340–348.

28. Lennernas, H., *Modeling gastrointestinal drug absorption requires more* in vivo *biopharmaceutical data: experience from* in vivo *dissolution and permeability studies in humans*. Curr Drug Metab, 2007. 8(7): p. 645–657.

29. Dressman, J.B., Fleisher, D. and Amidon, G.L., *Physicochemical model for dose-dependent drug absorption*. J Pharm Sci, 1984. 73(9): p. 1274–1279.

30. Chen, C., Cowles, V.E. and Sweeney, M., *The intestinal absorption mechanism of gabapentin makes it appropriate for gastroretentive delivery*. Curr Clin Pharmacol, 2013. 8(1): p. 67–72.

31. Ieiri, I., et al., *Microdosing clinical study: pharmacokinetic, pharmacogenomic (SLCO2B1), and interaction (grapefruit juice) profiles of celiprolol following the oral microdose and therapeutic dose*. J Clin Pharmacol, 2012. 52(7): p. 1078–1089.

32. Tubic, M., Wagner, D., Spahn-Langguth, H., Bolger, M.B. and Langguth, P., In silico *modeling of non-linear drug absorption for the P-gp substrate talinolol and of consequences for the resulting pharmacodynamic effect*. Pharm Res, 2006. 23(8): p. 1712–1720.

33. Abuasal, B.S., Bolger, M.B., Walker, D.K. and Kaddoumi, A., In silico *modeling for the nonlinear absorption kinetics of UK-343,664: a P-gp and CYP3A4 substrate*. Mol Pharm, 2012. 9(3): p. 492–504.

34. Popick, A.C., Crouthamel, W.G. and Bekersky, I., *Plasma protein binding of ceftriaxone*. Xenobiotica, 1987. 17(10): p. 1139–1145.
35. Di Bari, L., Ripoli, S., Pradhan, S. and Salvadori, P., *Interactions between quercetin and warfarin for albumin binding: a new eye on food/drug interference*. Chirality, 2010. 22(6): p. 593–596.
36. Covell, D.G., Abbrecht, P.H. and Berman, M., *The effect of hepatic uptake on the disappearance of warfarin from the plasma of rats: a kinetic analysis*. J Pharmacokinet Biopharm, 1983. 11(2): p. 127–145.
37. Scheufler, E., *Evidence of nonlinear pharmacokinetics of methotrexate in the rat*. Pharmacology, 1982. 25(1): p. 51–56.
38. Dingemanse, J. and van Giersbergen, P.L., *Clinical pharmacology of bosentan, a dual endothelin receptor antagonist*. Clin Pharmacokinet, 2004. 43(15): p. 1089–1115.
39. Klotz, U., Muller-Seydlitz, P. and Heimburg, P., *Pharmacokinetics of lorcainide in man: a new antiarrhythmic agent*. Clin Pharmacokinet, 1978. 3(5): p. 407–418.
40. Ludden, T.M., *Nonlinear pharmacokinetics: clinical implications*. Clin Pharmacokinet, 1991. 20(6): p. 429–446.
41. Ouwerkerk-Mahadevan, S., Beumont-Mauviel, M., Mortier, S., Peeters, M., Verloes, R., Truyers, C., Mannens, G., Wynant, I. and Simion, A., *Evaluation of the pharmacokinetics and renal excretion of simeprevir in subjects with renal impairment*. Drugs R D, 2015. 15(3): p. 261–270.
42. Lee, W.M., *Review article: drug-induced hepatotoxicity*. Aliment Pharmacol Ther, 1993. 7(5): p. 477–485.
43. Vlase, L., Popa, A., Neag, M., Muntean, D., Baldea, I. and Leucuta, S.E., *Pharmacokinetic interaction between zolpidem and carbamazepine in healthy volunteers*. J Clin Pharmacol, 2011. 51(8): p. 1233–1236.
44. Fleishaker, J.C., Pearson, L.K. and Peters, G.R., *Gender does not affect the degree of induction of tirilazad clearance by phenobarbital*. Eur J Clin Pharmacol, 1996. 50(1–2): p. 139–145.
45. Spaans, E., van den Heuvel, M.W., Schnabel, P.G., Peeters, P.A., Chin-Kon-Sung, U.G., Colbers, E.P. and Sitsen, J.M., *Concomitant use of mirtazapine and phenytoin: a drug–drug interaction study in healthy male subjects*. Eur J Clin Pharmacol, 2002. 58(6): p. 423–429.
46. Lim, M.L., Min, S.S., Eron, J.J., Bertz, R.J., Robinson, M., Gaedigk, A. and Kashuba, A.D., *Coadministration of lopinavir/ritonavir and phenytoin results in two-way drug interaction through cytochrome P-450 induction*. J Acquir Immune Defic Syndr, 2004. 36(5): p. 1034–1040.
47. Zhu, M., Kaul, S., Nandy, P., Grasela, D.M. and Pfister, M., *Model-based approach to characterize efavirenz autoinduction and concurrent enzyme induction with carbamazepine*. Antimicrob Agents Chemother, 2009. 53(6): p. 2346–2353.
48. Sahi, J., et al., *Avasimibe induces CYP3A4 and multiple drug resistance protein 1 gene expression through activation of the pregnane X receptor*. J Pharmacol Exp Ther, 2003. 306(3): p. 1027–1034.
49. Zhang, X., Jones, D.R. and Hall, S.D., *Prediction of the effect of erythromycin, diltiazem, and their metabolites, alone and in combination, on CYP3A4 inhibition*. Drug Metab Dispos, 2009. 37(1): p. 150–160.
50. Rowland Yeo, K., Jamei, M., Yang, J., Tucker, G.T. and Rostami-Hodjegan, A., *Physiologically based mechanistic modelling to predict complex drug–drug interactions involving simultaneous competitive and time-dependent enzyme inhibition by parent compound and its metabolite in both liver and gut – the effect of diltiazem on the time-course of exposure to triazolam*. Eur J Pharm Sci, 2010. 39(5): p. 298–309.
51. Hemeryck, A. and Belpaire, F.M., *Selective serotonin reuptake inhibitors and cytochrome P-450 mediated drug–drug interactions: an update*. Curr Drug Metab, 2002. 3(1): p. 13–37.
52. Ajavon, A.D., Bonate, P.L. and Taft, D.R., *Renal excretion of clofarabine: assessment of dose-linearity and role of renal transport systems on drug excretion*. Eur J Pharm Sci, 2010. 40(3): p. 209–216.
53. Uwai, Y., Ida, H., Tsuji, Y., Katsura, T. and Inui, K., *Renal transport of adefovir, cidofovir, and tenofovir by SLC22A family members (hOAT1, hOAT3, and hOCT2)*. Pharm Res, 2007. 24(4): p. 811–815.

54. Pfeifer, N.D., Yang, K. and Brouwer, K.L., *Hepatic basolateral efflux contributes significantly to rosuvastatin disposition I: characterization of basolateral versus biliary clearance using a novel protocol in sandwich-cultured hepatocytes*. J Pharmacol Exp Ther, 2013. 347(3): p. 727–736.

55. Kitamura, S., Maeda, K., Wang, Y. and Sugiyama, Y., *Involvement of multiple transporters in the hepatobiliary transport of rosuvastatin*. Drug Metab Dispos, 2008. 36(10): p. 2014–2023.

56. Tokui, T., Nakai, D., Nakagomi, R., Yawo, H., Abe, T. and Sugiyama, Y., *Pravastatin, an HMG-CoA reductase inhibitor, is transported by rat organic anion transporting polypeptide, oatp2*. Pharm Res, 1999. 16(6): p. 904–908.

57. Peters, S.A., Ungell, A.L. and Dolgos, H., *Physiologically based pharmacokinetic (PBPK) modeling and simulation: applications in lead optimization*. Curr Opin Drug Discov Devel, 2009. 12(4): p. 509–518.

58. Watson, D.E., Timothy, P.R. and Stevens, J.L., Case History: Toxicology Biomarker Development Using Toxicogenomics, in *Optimizing the "Drug-Like" Properties of Leads in Drug Discovery*. 2006, Springer: New York. p. 256–270.

59. Lin, J.H. and Lu, A.Y.H., *Role of pharmacokinetics and metabolism in drug discovery and development*. Pharmacol. Rev., 1997. 49(4): p. 403–449.

60. Hurst, S., Loi, C.M., Brodfuehrer, J. and El-Kattan, A., *Impact of physiological, physicochemical and biopharmaceutical factors in absorption and metabolism mechanisms on the drug oral bioavailability of rats and humans*. Expert Opin. Drug Metab. Toxicol., 2007. 3(4): p. 469–489.

61. Nelson, D.R., et al., *The P450 superfamily: update on new sequences, gene mapping, accession numbers, early trivial names of enzymes, and nomenclature*. DNA Cell. Biol., 1993. 12(1): p. 1–51.

62. Burchell, B., Ebner, T., Baird, S., Bin Senafi, S., Clarke, D., Brierley, C. and Sutherland, L., *Use of cloned and expressed human liver UDP-glucuronosyltransferases for analysis of drug glucuronide formation and assessment of drug toxicity*. Environ Health Perspect, 1994. 102(Suppl 9): p. 19–23.

63. Sauer, J.M., Ring, B.J. and Witcher, J.W., *Clinical pharmacokinetics of atomoxetine*. Clin Pharmacokinet, 2005. 44(6): p. 571–590.

64. Mattiuz, E.L., et al., *Disposition and metabolic fate of atomoxetine hydrochloride: pharmacokinetics, metabolism, and excretion in the Fischer 344 rat and beagle dog*. Drug Metab Dispos, 2003. 31(1): p. 88–97.

65. Lin, J.H., Chiba, M., Balani, S.K., Chen, I.W., Kwei, G.Y., Vastag, K.J. and Nishime, J.A., *Species differences in the pharmacokinetics and metabolism of indinavir, a potent human immunodeficiency virus protease inhibitor*. Drug Metab Dispos, 1996. 24(10): p. 1111–1120.

66. Chiba, M., Hensleigh, M., Nishime, J.A., Balani, S.K. and Lin, J.H., *Role of cytochrome P450 3A4 in human metabolism of MK-639, a potent human immunodeficiency virus protease inhibitor*. Drug Metab Dispos, 1996. 24(3): p. 307–314.

67. Yeh, K.C., Stone, J.A., Carides, A.D., Rolan, P., Woolf, E. and Ju, W.D., *Simultaneous investigation of indinavir nonlinear pharmacokinetics and bioavailability in healthy volunteers using stable isotope labeling technique: study design and model-independent data analysis*. J Pharm Sci, 1999. 88(5): p. 568–573.

68. Walton, K., Dorne, J.L. and Renwick, A.G., *Species-specific uncertainty factors for compounds eliminated principally by renal excretion in humans*. Food Chem Toxicol, 2004. 42(2): p. 261–274.

69. Frederick, C.B. and Obach, R.S., *Metabolites in safety testing: "MIST" for the clinical pharmacologist*. Clin Pharmacol Ther, 2010. 87(3): p. 345–350.

70. Smith, D.A., Obach, R.S., Williams, D.P. and Park, B.K., *Clearing the MIST (metabolites in safety testing) of time: The impact of duration of administration on drug metabolite toxicity*. Chem Biol Interact, 2009. 179(1): p. 60–67.

71. Jian, W., Yao, M., Zhang, D. and Zhu, M., *Rapid detection and characterization of in vitro and urinary N-acetyl-L-cysteine conjugates using quadrupole-linear ion trap mass spectrometry and polarity switching*. Chem Res Toxicol, 2009. 22(7): p. 1246–1255.

72. Zhu, P., Ding, W., Tong, W., Ghosal, A., Alton, K. and Chowdhury, S., *A retention-time-shift-tolerant background subtraction and noise reduction algorithm (BgS-NoRA) for extraction of drug metabolites in liquid chromatography/mass spectrometry data from biological matrices*. Rapid Commun Mass Spectrom, 2009. 23(11): p. 1563–1572.

2

GIT ANATOMY AND PHYSIOLOGY AND DRUG ORAL BIOAVAILABILITY: IMPACT OF SPECIES DIFFERENCES

2.1 INTRODUCTION

The discovery and development of therapeutically efficacious drugs is a time-consuming, resource-intensive, and complicated process. A new drug requires 10–15 years to reach the marketplace in the United States [1]. Based on the data from Tufts Center for the Study of Drug Development, the average capitalized cost to bring one new biopharmaceutical product to market, including the cost of failures, is $1.24 billion in 2005 dollars [2]. Most of the drugs that are available in the marketplace are administered via the oral route, which is a convenient and cost-effective route of administration [3]. Thus, oral bioavailability is one of the key parameters frequently evaluated in drug discovery when considering a new chemical entity (NCE) for further development, as poor oral bioavailability is one of the major contributing factors to therapeutic variability [4]. Hellriegel et al. reported a significant inverse relationship between the oral bioavailability of drugs from several therapeutic classes and the coefficient of interindividual variability in their oral bioavailability [5]. The increased variability associated with poor oral bioavailability may predispose the patient to a higher risk of exposure to toxic or subtherapeutic plasma drug concentration. This variability is particularly important for drugs with narrow therapeutic window or potential for resistance development such as cytotoxic drugs and antibiotics [4].

Evaluation of an NCE's pharmacokinetic profile including oral bioavailability determination is routinely performed in animal species such as rat, dog, and nonhuman primate (e.g., monkey) during drug discovery and early preclinical development. Typically, the *in vivo* pharmacokinetic evaluations are performed in rodents during early discovery since NCEs are synthesized in small quantities, which are insufficient to perform such studies in larger animal species [6]. This data, in conjunction with that obtained from *in silico* and *in vitro* studies, are commonly used to project the pharmacokinetic profile in humans in order to provide guidance

Oral Bioavailability Assessment: Basics and Strategies for Drug Discovery and Development, First Edition.
Ayman F. El-Kattan.

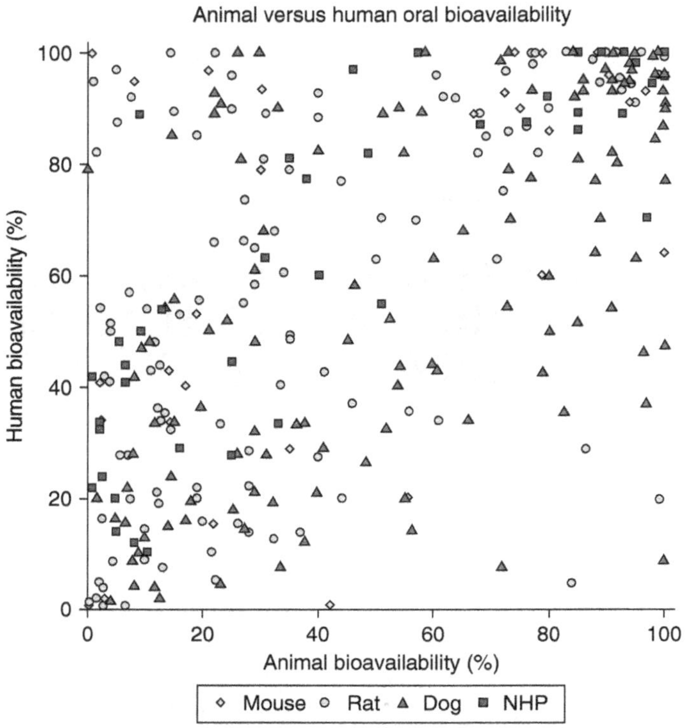

FIGURE 2.1 Plot of oral bioavailability (*F*) in animal species versus oral bioavailability in humans (in percentage). Diamonds are for mouse, circles for rats, triangles for dogs, and squares for nonhuman primates (NHP). Musther et al. 2013 [7]. Reproduced with permission of Elsevier.

for chemical synthesis to optimize the ADME properties of a chemical series. When more promising candidates are identified, synthesis of larger quantity can be made for pharmacokinetic assessment in a nonrodent species such as dog or monkey to refine the projection of human pharmacokinetic profiles in support of further development of the NCE. However, the approach of predicting oral bioavailability in humans from animal and *in vitro* data still presents many challenges due to species-dependent physiological, transporters, and metabolic differences. These challenges are illustrated in Figure 2.1 and demonstrate the poor relationship between human oral bioavailability and that measured in various preclinical species [7].

The purpose of this chapter is to provide a critical review of the species similarities and differences in the absorption and metabolism mechanisms that affect oral bioavailability in humans and three commonly used preclinical species: rats, dogs, and monkeys.

2.2 PHYSIOLOGICAL FACTORS THAT IMPACT ORAL DRUG ABSORPTION

An extensive review of the anatomical and physiological differences in gastrointestinal tract among preclinical animal species and human has been described previously and the following is a brief overview [8–11].

2.2.1 Gastrointestinal Anatomy and Physiology

In rats, dogs, monkeys, and humans, the gastrointestinal tract consists mainly of the stomach, small intestine (duodenum, jejunum, and ileum), and large intestine (cecum, colon, and rectum). The dimensions and surface areas of these components differ among these species. Table 2.1 illustrates a comparison among the radius, anatomical lengths, and relative contributions of the major subdivisions of the gastrointestinal tract in rats, dogs, monkeys, and humans [8–11].

Although the body weight of human (70 kg) is larger than that of dog (10 kg) and rat (0.25 kg), the human gastrointestinal tract (8.35 m) is only 2.0- and 5.6-fold longer than the dog and rat gastrointestinal tracts, respectively (Tables 2.1 and 2.2). However, the relative size of the small intestine, which is considered the primary site of drug absorption, to the total length of the gastrointestinal tract is 83% in rats, 86% in dogs, and 81% in humans. As for the large intestine, its relative size in rats, dogs, and humans is 17%, 14%, and 19%, respectively [12, 13].

The radius of the small intestine in humans is 1.5–2 cm, while in rats, dogs, and monkeys it is 0.2, 0.5, and 0.4 cm, respectively (Table 2.1). Interestingly, the radius of the human small intestine decreases as one progresses distally from the upper to the end of the small intestine (1.7–1.0 cm, respectively).

Table 2.2 shows a comparison of the absolute surface areas of the gastrointestinal tracts in rats, dogs, monkeys, and humans. The human small intestine surface area is 158- and 11-fold larger than that of the rat and dog, respectively. Yet, the human small intestine is only fourfold larger than the monkey small intestine surface area. The large difference in the surface area between rat and dog versus human and monkey small intestine is attributed to the fact that human and monkey small intestines have three anatomical modifications that significantly increase the surface area of the small intestine [14, 15]. Indeed, rat and dog small intestines have only two of these. The human and monkey small intestines have additional grossly observable folds of mucosa (plicae circulares or folds of kerckring) that increase the surface

TABLE 2.1 Comparison of the Anatomical Lengths of the Intestinal Tract and Its Major Subdivision in Rats, Dogs, Monkeys, and Humans [9–12]

Species	Region of Intestine	Radius (cm)	Length (m)	Total (%)
Rat	Small intestine	0.2	1.25	83
	Large intestine		0.25	17
	Total intestinal tract		1.50	
Dog	Small intestine	0.5	4.14	86
	Large intestine		0.68	14
	Total intestinal tract		4.82	
Monkey	Small intestine	0.4[a]/0.8[b]	ND	ND
	Large intestine		ND	ND
	Total intestinal tract		ND	ND
Human	Small intestine	1.5–2	6.80	81
	Large intestine		1.55	19
	Total intestinal tract		8.35	

[a]Rhesus monkey.
[b]Cynomolgus monkey.
ND, Not determined.

TABLE 2.2 Comparison of the Absolute Surface Areas and Surface Area of the Gastrointestinal Tract Relative to Total Body Surface Area in Rats, Dogs, Monkeys, and Humans [8–11]

	Human		Rat		Dog		Monkey	
Body weight (kg)	70		0.3		10.7–12.6		2.7–3.0	
Body surface area (BSA; m^2)	1.8		0.03–0.06		0.39–0.78		0.3	
Region of intestinal tract	Absolute SA (m^2)[a]	SA relative to BSA	Absolute SA (m^2)[a]	SA relative to BSA	Absolute SA (m^2)[a]	SA relative to BSA	Absolute SA (m^2)[a]	SA relative to BSA
Stomach	0.053	0.029	0.00062	0.01–0.02	0.0344–0.0426	0.04–0.11	0.0143–0.0306	0.048–0.1
Small intestine	252	140	1.6	26.67–53.33	24.75–35	31.73–89.74	54.6–66	182–220
Large intestine	0.35	0.19	0.034	0.57–1.13	0.023–0.245	0.029–0.63	0.23–0.31	0.23–0.31

[a] Surface area.

area by threefold. From the plicae circulares project microscopic finger-like pieces of tissue called villi that increase the surface area by 5- to 10-fold depending on the species. Each villus is covered in microvilli, which increases the surface area by 20- to 25-fold. When the small intestine is normalized to the total body surface area, the relative surface area of the human small intestine is 4- and 2.3-fold larger than that of the rat and dog, respectively. However, it is only 0.7-fold the relative surface area of the monkey small intestine [8].

Unlike the small intestine, the large intestine surface area does not have villi and is divided into geographical areas by transverse furrows. In addition, the large intestine enterocytes differ slightly from that of the small intestine and its microvilli are less packed [9]. Overall, this contributes to the smaller surface area of the large intestine in all species investigated (Table 2.2) and is consistent with the fact that small intestine is the major site of drug absorption in these species.

2.2.2 Unstirred Water Layer

Adjacent to the absorptive membrane of the enterocyte is an unstirred water layer (UWL) that functions as a potential barrier for the permeation of various drug molecules across the intestinal membrane. The UWL is comprised of strings of water lamella that become more stirred toward the lumen bulk and away from the apical surface of the enterocyte. The thickness of this UWL in rats and dogs is comparable to that in humans and about 300 μm [16]. In the literature, monkey UWL was not determined. However, based on antipyrine effective intestinal permeability calculated in monkeys and considering that monkey plicate expansion is smaller compared to that of humans, it is suggested that the thickness of monkey UWL is similar to that of humans [17].

Chiou evaluated the impact of UWL adjacent to the intestinal membrane on the rate and extent of absorption for passively absorbed drugs with different membrane absorption half-lives (10–300 min) in humans, dogs, rabbits, rats, and mice. Results suggested that the presence of the UWL is generally expected to have a relatively minor, if any, effect on the rate and extent of absorption of drugs across the gastrointestinal tract [18].

2.2.3 Gastrointestinal Transit Times

The absorption rate of a drug molecule is generally a function of drug absorption through the gastrointestinal tract. Absorption rate is determined by the residence time and absorption in each gastrointestinal segment [19]. Gastric emptying time, also known as gastric transit time, depends on the content of the stomach. The gastric transit time describes the time required for the stomach to empty its content in the small intestine [20]. In humans, the half-life of the stomach content for small particles and liquids is calculated to be 10 min and 1 h in fasted and fed conditions, respectively. Table 2.3 illustrates the impact of meal volume and caloric density on gastric emptying time in humans. Overall, higher caloric density and smaller meal volume are associated with longer gastric transit time [20]. This mainly impacts the rate of absorption of rapidly dissolved and well-absorbed drugs (BCS Class 1) and leads to a decrease in drug C_{max} and a corresponding longer T_{max} and/or lag time in the fed conditions. However, the extent of absorption and oral bioavailability are typically not affected [20–22]. It should be emphasized that in the fasted conditions, the gastric emptying rate of a solution for dogs, monkeys, and humans is similar [23], while in the fed conditions, gastric emptying time in dogs is longer than that reported in humans [10, 24].

Intestinal transit time influences the absorption of drugs with limited mucosal permeability, substrates for carrier-mediated uptake, subject to intestinal degradation, or products whose dissolution is the rate-limiting step for systemic absorption [25]. In contrast to gastric transit

TABLE 2.3 The Impact of Food Caloric Density and Meal Volume on the Gastric Emptying Rate in Humans

	Caloric Density (kcal)		
Meal Volume (mL)	200	300	400
200	56	74	92
400	41	59	77
600	42	60	78
800	38	56	74

Source: Mudie et al. 2010 [20]. Reproduced with permission of American Chemical Society.

time, intestinal transit time is independent of the feeding conditions and the physical composition of the intestinal contents [26]. In rats, dogs, monkeys, and humans, intestinal transit time is around 3–4 h [10, 23, 27]. In rats, the transit time in the large intestine is around 15 h and the transit time in the large intestine of humans and dogs can vary from 8 to 72 h [9, 28]. Overall, the intestinal content and formulation size has minimal impact on the transit time in the human large intestine [29].

2.2.4 The Gastrointestinal pH

Most of the drug molecules are either acidic or basic in nature. The extent of ionization plays a pivotal role in determining their dissolution rate and passive permeability across the gastrointestinal tract. Therefore, pH at the absorption site is a critical factor in facilitating or decreasing the dissolution rate and absorption of various ionizable drug molecules [1]. Figure 2.2 shows the pH values of the gastrointestinal tract of rats, dogs, monkeys, and humans. In rats and dogs, the pH of luminal content (chyme) in the stomach is around 3.3, while in humans and monkeys, the pH of chyme is more acidic and around 2.5 [25]. When the chyme arrives in the duodenum, it is quickly neutralized by the secretion of the pancreatic bicarbonate and bile. The pH values of chyme become progressively more alkaline in the distal portion of the small intestine in rats, dogs, monkeys, and humans. However, the pH of chyme in the large intestine is generally more acidic than the pH observed in the small intestine in these species, possibly due to fermentation mediated by the microbial flora [31, 32].

TIP:

Food may have a significant impact on the absorption of compounds with poor solubility such as BDDCS Class II and IV compounds. To project food effect on NCE oral pharmacokinetics profile in humans and assess the feasibility of formulation in mitigating this liability, Lentz et al. recommended the use of pentagastrin-treated beagle dog as a preclinical model to project food effect in humans. Pentagastrin is a synthetic polypeptide that lowers dog basal gastric pH and makes it comparable to that observed in humans. This approach was validated using a set of compounds and proven to be effective as a screen prior to clinical formulation efforts to mitigate potential food effect liability for NCEs with poor solubility and dissolution [31]. Recent progress in modeling and simulation tools such as GastroPlus and SimCYP provided higher confidence in the prediction of food effects on NCE systemic profiles following oral dosing.

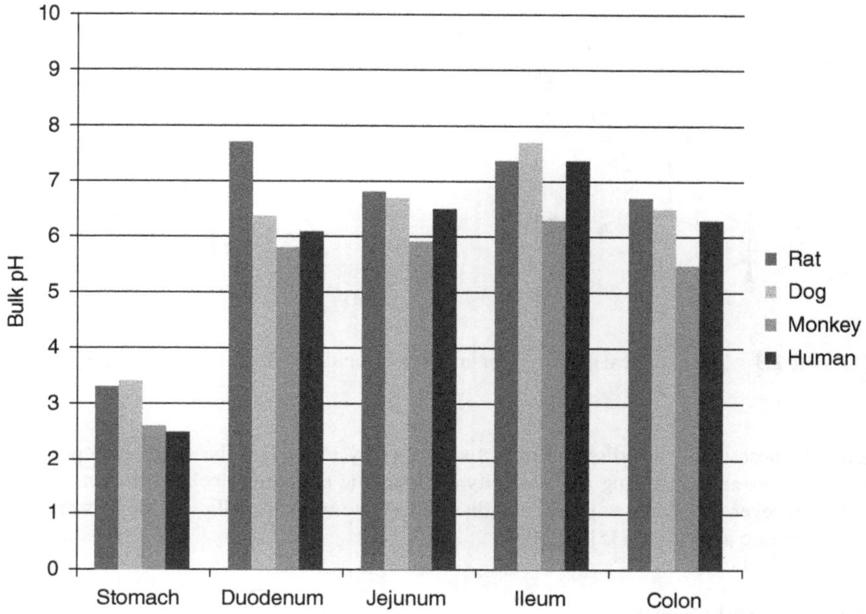

FIGURE 2.2 Comparative bulk pH across the gastrointestinal tract of rats, dogs, monkeys, and humans [25, 30]. Martinez et al. 2002 [25]. Reproduced with permission of Elsevier.

2.2.5 Bile Fluid

Bile is produced by hepatocytes and is drained through the bile ducts that penetrate the liver [8]. During this process, the epithelial cells add a watery solution that is rich in bicarbonates, which increases the alkalinity of the solution. In humans and dogs, bile is stored and concentrated up to five times its original potency in the gall bladder by water reabsorption. Gall bladder content is emptied following food intake. Emptying process is mediated by secretin and cholecystokinin, which stimulates the gallbladder to contract and secrete bile as the acidic chyme enters the duodenum. Bile secretion rate in humans and dogs is 2–22 and 19–36 mL/kg/day, respectively. However, rats lack a gall bladder and bile is secreted continuously at an estimated rate of 48 mL/kg as a dilute solution into the duodenum each day [9, 25]. In monkeys, the bile secretion rate is 25 mL/kg [30].

Bile acts as a detergent that emulsifies fats by increasing the surface area to help digestive enzymes action, and thus, aid in their absorption in the small intestine. Bile is a complex mixture of bicarbonate solution, cholesterol, phospholipids such as lecithin and bilirubin, and bile salts such as those of taurocholic acid and deoxycholic acid derivatives (Figure 2.3). Bile salts are also amphipathic water-soluble molecules that are derived from cholesterol and ionized at physiological conditions [33]. There are major differences in the chemical structure of various bile salts identified. These are shown in stereochemistry of the hydroxyl groups and their numbers. Overall, the bile salt is around 20 Å long, with an average radius of about 3.5 Å [34]. Bile salts are combined with phospholipids to break down fat globules via the process of emulsification. Emulsified droplets then are organized into many micelles, which increase their absorption. Since bile increases the absorption of fats, it also plays a pivotal role in the absorption of the fat-soluble vitamins, steroids, and lipophilic xenobiotics [35]. Indeed, bile

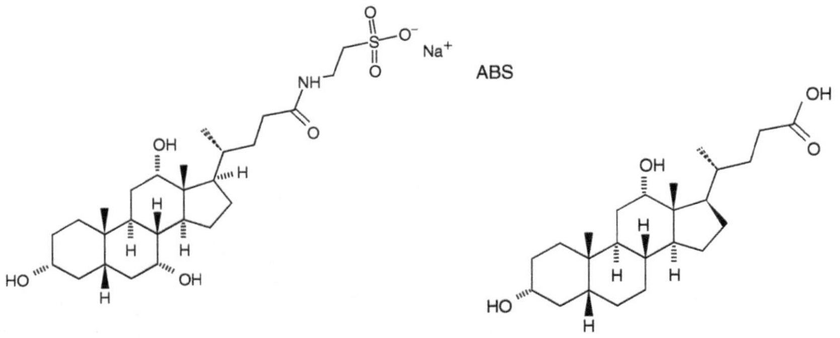

FIGURE 2.3 The chemical structures of taurocholic acid and deoxycholic acid, respectively.

increases these molecules dissolution and solubility by improving the wetting properties of solid particles and enhancing the solubility of lipophilic drugs in micellar solutions, respectively. However, bile tends to reduce the dissolution rate of hydrophilic compounds and their absorption rate and extent [35].

2.2.6 Bacterial Microflora

In rats, dogs, monkeys, and humans, bacterial microflora exist in most of the gastrointestinal tract and is an important component of the luminal content. In rats and dogs, there is significant number of bacterial microflora in the stomach as well as small and large intestines. In humans, there is no bacterial microflora in the stomach and upper small intestine. This lack of bacterial microflora is mainly attributed to the low pH of the gastric content of human stomach. However, large numbers of bacterial microflora populate the human's distal small and large intestines [36].

Intestinal bacterial microflora play a crucial role in the metabolism of various chemicals, endogenous substrates, and xenobiotics through hydrolysis of glycosides, sulfate esters, amide, esters, dehydroxylation, deamidation, decarboxylation, dealkylation, acetylation, esterification, and reduction of double bonds, nitro and diazo groups [37].

Indeed, hydrolysis of glucuronide conjugates is a pivotal metabolic reaction, which is mediated by the β-glucuronidase enzyme that is produced by the intestinal bacterial microflora. The abundance of this enzyme appears to be higher in both rat and dog gastrointestinal tracts compared to humans [9]. However, the β-glucuronidase activity in the human gastrointestinal tract is still substantial and generates 0.02 and 0.9 μmol of product formed per hour per gram of the content for the proximal and distal regions of the small intestine, respectively [38]. Interestingly, this process sometimes contributes to the enterohepatic circulation of several drugs. Where following biliary excretion, drug conjugates are broken down by β-glucuronidase, which is then followed by intestinal reabsorption of parent molecule. Enterohepatic circulation typically leads to multiple-peak phenomenon and a longer terminal half-life.

With regard to *in vivo* models for human intestinal microflora metabolism, conventional laboratory animals are limited. A germ-free rat model with associated human colonic bacteria has been used as an *in vivo* model for human projections [37]. Since there are significant species differences in the distribution, amount, strains, and metabolic activities of microflora that populate the gastrointestinal tract of preclinical species and human, more studies are needed to enable a meaningful interpretation of the data observed in preclinical models and further emphasize the need to exercise caution in the projection of compounds profiles in humans based on preclinical data [38, 39].

2.2.7 Lymphatic Absorption

The intestinal lymphatic route plays a key role in the absorption of drugs that are highly lipophilic and may provide an opportunity for development of controlled release formulation via this route of delivery [40]. Similarities and differences in drug lymphatic uptake are determined by the lymph flow to various absorption sites, mechanisms of drug lymphatic absorption, mechanism of lipid digestion, and bile composition and secretion pattern [25]. However, little is known about species differences in lymphatic uptake of drugs and its impact on species differences in drug absorption and bioavailability [41].

2.3 MECHANISM OF ORAL ABSORPTION

Following oral dosing, drug molecules can cross the luminal membrane through various mechanisms such as passive diffusion or active transport. Passive diffusion is comprised of two pathways: (i) the paracellular pathway in which a drug diffuses through the aqueous pores at the tight junctions between the intestinal enterocytes and (ii) the transcellular (lipophilic) pathway, which requires drug diffusion across the lipid cell membrane of the enterocyte (Figure 2.4). Active transport pathway is mediated by transporters and is functionally divided into active drug influx and efflux. The relevance of each route to the absorption of a compound is determined by the compound's physicochemical properties and its affinity for various transport proteins (Figure 2.4).

2.3.1 Passive Diffusion

In paracellular diffusion, drug molecules are absorbed by diffusion and convective volume flow through the intercellular spaces governed by the tight junctions or zonula occludens, which formulate the major rate-limiting barrier [42]. Drug molecules absorbed through this pathway are typically small in size (e.g., molecular weight $[M_{wt}] < 350$ Da) and hydrophilic ($\log P < 0$). Because the junctional complex has a net negative charge, positively charged molecules pass more readily, whereas negatively charged molecules are repelled. There are significant species differences in drug absorption via this mechanism among rats, dogs, and humans. For example, He et al. conducted cross-species comparison of the absolute bioavailability of several hydrophilic and metabolically stable drugs (acyclovir, atenolol, nedocromil, dazmegrel, sumatriptan, TRH, and nadolol) (Table 2.4). The authors reported a good correlation between the bioavailability of these drugs in humans and rats, with rats slightly underpredicting the bioavailability in humans. In contrast, there was no apparent correlation in bioavailability between dogs and humans, with dogs overpredicting the bioavailability and oral absorption in humans. For example, acyclovir and nadolol oral bioavailability was 75% and 90% in dogs, but was only 28% and 30%, respectively, in humans (Table 2.4). The authors concluded that rats are a better model than dogs for predicting human bioavailability of paracellularly absorbed compounds [44]. Furthermore, the paracellular pathway offers a limited window for absorption since it accounts for ~0.01–0.1% of the total surface area of intestinal membrane. In addition, the tight junctions between cells become tighter traveling from the jejunum toward the colon. Therefore, compounds that are absorbed via paracellular pathway are not optimum for traditional controlled release formulation that target

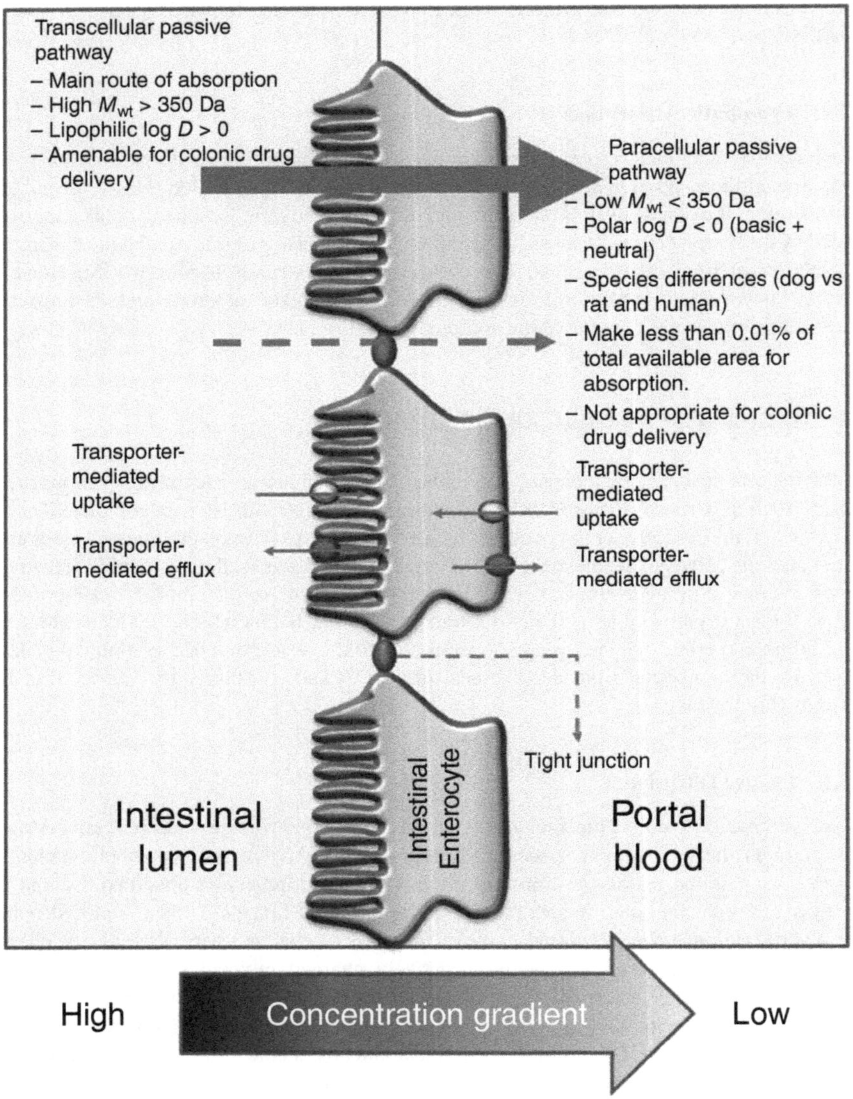

FIGURE 2.4 Possible routes of drug absorption across intestinal enterocytes. *(See color plate section for the color representation of this figure.)*

delivery through the colon. For example, metformin is a compound that is predominantly absorbed via paracellular pathway and its absorption is confined to the small intestine. As a result, metformin is not a good candidate for controlled release formulation that targets the colon. Instead, it is formulated using sustained release gastric retentive technology, in which the tablet swells into a gel-like mass. This would limit its passage through the stomach pylorus and extends its gastric residence. Once the gel-like mass is formed, metformin slowly dissolves and diffuses from the polymer containing tablet matrix, leading to a longer half-life [45].

TABLE 2.4 The Absorption of a Set of Hydrophilic Drugs in Rats, Dogs, and Humans

	Acyclovir	Atenolol	Nedocromil	Dazmegrel	Sumatriptan	TRH	Nadolol
Oral bioavailability (%) rat	23	56	2	64	54	2	19
Oral bioavailability (%) dog	75	100	80	68	92	13	90
Oral bioavailability (%) human	28	55	2	98	57	1.5	30

Source: He et al. 1998. [43]. Reproduced with permission of Elsevier.

TIP:

Compounds absorbed via the paracellular pathway tend to have erratic and poor oral absorption especially if they are not cationic in nature. In addition, they are not appropriate candidates for controlled release formulation that targets the colon. Other technologies to extend half-life should be considered such as gastric retentive technology [45]. It should be kept in mind that dog is not a good model for predicting the absorption of paracellular absorbed molecules in humans and tends to overpredict human absorption fraction. This overprediction is attributed to the bigger radius of the dog tight junction relative to that of humans, 9 Å versus 6 Å, respectively [43].

The transcellular pathway is the main route of absorption for most drug molecules. In general, the rate of passive transcellular permeability is mainly determined by the rate of transport across the apical cell membrane. This is controlled by the physicochemical properties of the absorbed molecules. Unlike the paracellular pathway, compounds that are absorbed through the transcellular pathway are unionized, with $\log P > 0$ and $M_{wt} > 350$ Da. In addition, the hydrogen-bonding capacity as determined by the number of hydrogen bond donors and hydrogen bond acceptors is less than 10 and 5, respectively [46].

2.3.2 Active Transport

Intestinal transporters have a significant impact on the absorption of biopharmaceutics drug disposition classification system (BDDCS) Class II, III, and IV molecules [47]. These are membrane-bound proteins that are either located on the apical or basolateral side of the enterocyte. Depending on the direction in which they translocate substrates across the cell membrane, intestinal transporters are classified as efflux or influx transporters. Intestinal efflux transporters are part of the ATP-binding cassette superfamily (ABC superfamily). These are primary transporters that use ATP hydrolysis to mediate the active export of xenobiotics from the cytoplasm to intestinal lumen or blood depending on their location at the enterocyte surface. Intestinal influx transporters are part of the solute carrier (SLC) superfamilies and they mainly employ a proton and Na^+ gradient created by primary active carriers (Na^+/K^+-ATPase, Na^+/H^+-ATPase) to drive the transport of endogenous substrates and xenobiotics into the enterocyte and called secondary transporters.

Recent studies suggest that a plethora of transporter genes are expressed in the gastrointestinal tracts of rats, dogs, monkeys, and humans [48–50]. The following section is an overview of the widely studied intestinal efflux and influx transporters and their reported species differences.

2.3.2.1 Efflux Transporters The most investigated efflux transporters are P-glycoprotein (P-gp), breast cancer resistance protein (BCRP), and multidrug resistance-associated proteins 2 (MRP2) [51]. They are located at the apical surface of the enterocyte and limit the oral absorption of widely prescribed drugs such as statins, antibiotics, HIV protease inhibitors, immunosuppressants, anticancer, and cardiac drugs. These substrates are usually transported against their concentration gradient using ATP hydrolysis as a source of energy. Expression of some of these transporters in the gastrointestinal tract has been quantified, which displays regional distribution patterns [52–54].

P-gp (ABCB1) P-gp (*ABCB1*) is considered the best-characterized member of the apical efflux transporters [51]. It is a product of the multidrug resistance (MDR1) gene and consists of two homologous halves with a total of 12 transmembrane helices and 1 ABC.

This membrane-bound protein plays a critical role in limiting the intestinal absorption of a large number of lipophilic drugs that are either neutral or basic such as anthracyclines (doxorubicin, daunorubicin), alkaloids (reserpine, vincristine, vinblastine), specific peptides (valinomycin, cyclosporine), steroid hormones (aldosterone, hydrocortisone), and local anesthetics (dibucaine) [47, 55–57]. This membrane-bound protein also mediates the active secretion of substrates, for example, digoxin from the systemic circulation into the intestinal lumen [58]. Dogs, monkeys, and humans only have 1 MDR1 gene, while mice and rats contain two paralogous genes (Mdr1a and Mdr1b) that encode for P-gp [59]. Separate research groups studied the relative mRNA expression of MDR1 in the gastrointestinal tract in rats, dogs, monkeys, and humans using semiquantitative reverse-transcription polymerase chain reaction (RTC-PCR). Interestingly, small intestines of dogs and monkeys appear to have higher relative mRNA expression of MDR1 compared to that of humans [60, 61]. However, duodenum of rats had lower mRNA expression of MDR1 compared to that of humans [52].

At the protein level, P-gp in rats, dogs, and rhesus monkeys are 86.6%, 90.9%, and 96.3% identical to P-gp amino-acid sequence in humans, respectively [62]. In humans, relative protein expression of P-gp is not homogenous but rather has a regional dependence that progressively increases from the proximal to distal region of the small intestine [54, 63]. Therefore, the absorption and oral bioavailability of a P-gp substrate may depend on the site of absorption and delivery relative to this regional distribution pattern [64, 65]. Therefore, potential species differences in regional distribution and function of P-gp should be considered when predicting absorption of an NCE in humans from preclinical species [37].

In vitro, species differences were reported in the P-gp-mediated transport activity, which was also shown to be substrate dependent [66]. For example, Xia et al. used human MDR1, rhesus monkey MDR1, and beagle dog Mdr1 transfected insect cell membranes and a set of 21 compounds with different physicochemical properties to investigate species difference in the P-gp binding affinity (αKa) via P-gp ATPase assays. The compounds list included ritonavir, saquinavir, vinblastine, quercetin, verapamil, digoxin, progesterone, rhodamine 123, propranolol, cyclosporine, etoposide, amiodarone, tamoxifen, thioridazine, physostigmine, paclitaxel, erythromycin, chloroquine, daunomycin, Hoechst 33342, miconazole, terfenadine, diltiazem, and dexamethasone. Consistent with the closer similarity in the gene and amino acid sequence between human and monkey MDR1, the authors reported a good correlation coefficient ($R^2 = 0.83$) of αKa between the two species (Figure 2.5a). However, the correlation coefficient (R^2) of αKa between human and beagle dog P-gp was only 0.36 [67] (Figure 2.5b). Another research group compared efflux ratio for 12 compounds generated in 7 multidrug resistance 1 (MDR1) transfected LLCK-PK1 cell lines, in which the cDNA was from human, monkey, dog, rat (Mdr1a and Mdr1b), and mouse (Mdr1a and Mdr1b). The group conducted a correlation analysis of the efflux ratio between two animal species and reported only a good correlation between human and monkey MDR1 ($R^2 = 0.954$). Other combinations had poor correlation [68].

BCRP (ABCG2) BCRP (*ABCG2*) is a 72 kDa membrane protein that is composed of 655 amino acids and arranged in six transmembrane segments (TM) and has one ABC at the amino terminus [69]. BCRP is also believed to be a half transporter that may function as a homodimer or homotetramer. BCRP limits the oral absorption of substrate compounds and/or facilitates their active secretion back into the gastrointestinal tract. BCRP exhibits broad substrate specificity and diverse chemical space. Substrates to BCRP include chemotherapy agents (mitoxantrone, camptothecins, tyrosine kinase inhibitors), antivirals (zidovudine, lamivudine), HMG-CoA reductase inhibitors (statins), benzimidazoles, and antibiotics (ciprofloxacin, rifampicin) [47]. Research groups studied the relative intestinal mRNA distribution of BCRP in various preclinical species and human small intestine. Similar to MDR1, BCRP relative mRNA expression in the small intestine of dogs and monkeys is higher

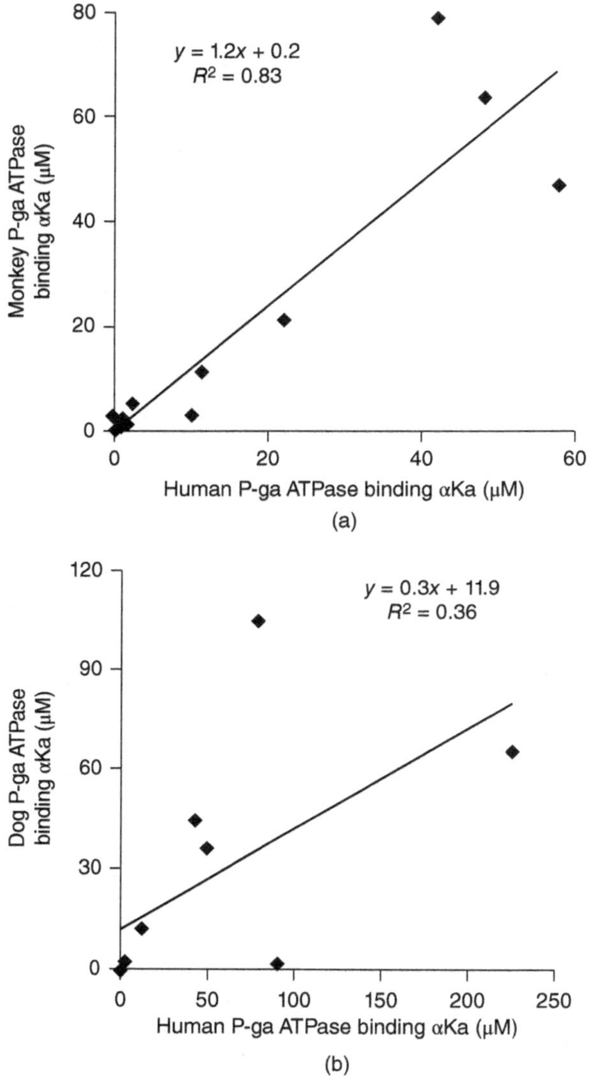

FIGURE 2.5 Correlations of P-gp binding affinity (αKa) between human MDR1 and rhesus monkey Mdr1 (a) and between human MDR1 and beagle dog Mdr1 (b). Xia et al. 2006 [67]. Reproduced with permission of Springer.

compared to that of humans [61, 70, 71]. It is important to realize that BCRP intestinal relative mRNA expression tends to be lower compared to overall ABCB1 (MDR1) expression along the dog intestine [70]. This expression is dissimilar to the findings observed in human jejunum where ABCG2 mRNA expression is higher than ABCB1 [71].

MRP2 (ABCC2) MRP2 (*ABCC2*) is another member of ABC transporter superfamily, which is colocated with BCRP and P-gp transporters at the apical membrane of the enterocytes. MRP2 has an apparent molecular mass of 190–200 kDa and is composed of 1545 amino acids [72]. MRP2 is composed of two TM and two ABC with an N-terminal TM domain that is composed of five TM segments [72]. Rat, dog, and monkey Mrp2 share 78%, 84%, and 96% identity with the amino acid sequence of human MRP2, respectively [72–75]. Known MRP2 substrates are lipophilic and usually acidic in nature. They are either therapeutic drugs (e.g., rosuvastatin, bosentan, methotrexate, cyclosporine, carbamazepine, rifampin, pravastatin, cisplatin, etoposide, doxorubicin, epirubicin, and mitoxantrone) or phase II metabolites (anionic drug conjugates of glucuronides, sulfates, or glutathiones), and endogenous compounds [76–79]. Different research groups investigated the relative mRNA expression of MRP2 in rat, dog, monkey, and human small intestine and colon using RTC-PCR. Overall, MRP2 mRNA expression in human small intestine and colon is lower compared to that shown in preclinical species tested [52, 61, 70, 71]. Ninomiya et al. studied the intrinsic transport activity of leukotriene C4, 17β-estradiol 17-(β-D-glucuronide), bromosulfophthalein, and cholecystokinin octapeptide. Vesicles studies were conducted using recombinant mouse, rat, dog, and monkey Mrp2 cDNA transfected insect cells. It is interesting to note that the substrate affinity (K_m) for the compounds investigated were comparable. However, intrinsic transport activity (V_{max}/K_m) was only different for leukotriene C4 and bromsulphthalein. The group attributed the findings to differences in the K_m or V_{max} values as well as to major differences in substrate and modulator recognition among species studied. Furthermore, preclinical studies demonstrated the central role of MRP2 in limiting the oral absorption and active intestinal secretion of its substrates following intravenous dosing [80, 81].

Taken together, compared to rat and dog, monkey major efflux transporters (MDR1, BCRP, MRP2) have close similarity to these expressed in human gastrointestinal tract in terms of gene and amino acid sequence, function, and regional distribution. However, significantly higher expression of efflux transporters along the monkey gastrointestinal tract relative to human was reported (Table 2.5). This higher expression may lead to a considerably lower intestinal membrane permeability and fraction absorbed in monkeys relative to humans. Indeed, this factor has been postulated for rationalizing furosemide (a known BCRP and MRP2 substrate) lower intestinal permeability and fraction absorbed in monkey ($f_a = 0.38$) relative to human ($f_a = 0.73$) [61, 82]. Further studies to understand the impact of higher efflux transporter expression along the monkey gastrointestinal tract on intestinal permeability and f_a using a larger set of efflux transporters substrates are warranted. These studies should also be accompanied with both a quantification of efflux transporter protein amount along the monkey gastrointestinal tract, similar to the work recently published by Oswald and coworkers. In addition, integration of these findings in physiologically based pharmacokinetics models (PBPK) to better characterize absorption in preclinical models in comparison to that of humans is pivotal [54]. Overall, this modeling allows better understanding of the potential factors that lead to species differences in oral bioavailability determination for NCE that are substrates for efflux transporters. Also, these models help to increase confidence in the decision-making process to advance an NCE into first in human studies based on preclinical pharmacokinetic studies.

2.3.2.2 Influx Transporters The role of influx transporters is critical for enabling the intestinal absorption of BDDCS Class III and IV drug molecules [83]. These compounds have low passive permeability and tend to be hydrophilic [47]. Examples of intestinal influx transporters include PEPT1 transporter, organic cation transporter (OCT), organic anion transporter (OAT), plasma membrane monoamine transporter (PMAT), large neutral amino acids (system L), bile acids, nucleosides, monocarboxylic acid transporters, and organic anion-transporting polypeptides (OATP) [84]. This section focuses on the most investigated intestinal influx transporters with a wealth of clinical data, which includes PEPT1 and OATP2B1 transporters.

TABLE 2.5 Relative Expression of mRNA for MDR1, MRP2, and BCRP in Small Intestine and Colon of Cynomolgus Monkeys and Humans Determined using RT-PCR

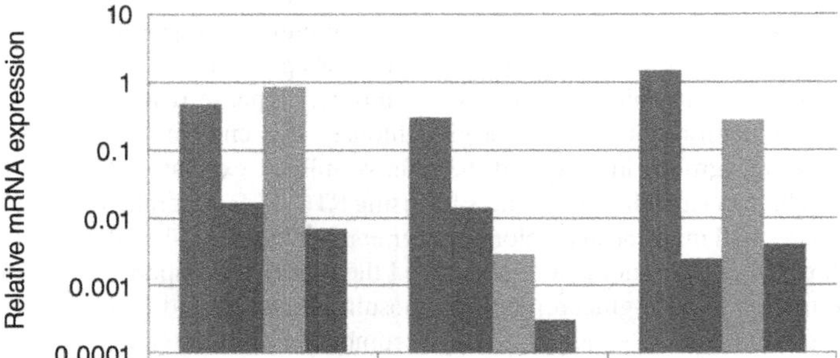

	MDR1	MRP2	BCRP
■ Monkey small intestine	0.479	0.302	1.45
■ Human small intestine	0.017	0.014	0.0024
■ Monkey colon	0.846	0.0029	0.278
■ Human colon	0.007	0.0003	0.004

GAPDH was used as an endogenous RNA to normalize for differences in the amount of total RNA.
Source: Takahashi et al. 2008 [61]. Reproduced with permission of American Chemical Society.

PEPT1 (SLC15A1) Human PEPT1 (*SLC15A1*) is a low-affinity and high-capacity transporter protein that is electrogenic and pH gradient dependent. The cloned human PEPT1 cDNA sequence encodes a 708 amino acid protein (M_{wt} 79 kDa) with several putative glycosylation and phosphorylation sites. There are 12 TM domains and a large extracellular loop between the IX and X TM domains, which possess intracellularly localized N- and C-termini [47]. PEPT1 is a well-characterized influx transporter that plays a pivotal role in the oral absorption of hydrophilic peptide-based drugs that include β-lactam antibiotics such as cephalosporins and penicillins, angiotensin converting enzyme inhibitors such as zofenopril, fosinopril, benazepril, quinapril, trandolapril, spirapril, cilazapril, ramipril, moexipril, quinaprilat, enalapril maleate, and perindopril, selected rennin inhibitors, antitumor agents such as bestatin, and dopamine receptor antagonists such as sulpiride [85–89].

The gene sequences of PEPT1 transporter appear to be highly conserved in rats, dogs, monkeys, and humans, with the rat, dog, and monkey gene sequences being 82%, 85%, and 94%, respectively, identical to that of humans [90–92]. Regional intestinal distribution pattern of PEPT1 also appears to be comparable among rats, dogs, and humans where higher expression was reported in small intestine relative to colon in the three species tested [54, 70, 93, 94]. Interestingly, RTC-PCR studies indicated PEPT1 mRNA expression in monkey gastrointestinal tract is limited to the small intestine [92].

Based on reported absorption fraction (f_a) calculated in rats, dogs, monkeys, and humans following oral absorption, PEPT1 substrates tend to have comparable absorption profiles (Table 2.6). This species similarity in oral absorption is consistent with their close similarity in their gene sequence and mRNA regional distribution in the gastrointestinal tract in species tested [95–97]. Overall, preclinical species such as rats, dogs, and monkeys are viable models to project human intestinal absorption of compounds absorbed via active uptake by PEPT1 transporter.

TABLE 2.6 The Absorption of a Set of PEPT1 Transporter Substrates in Rats, Dogs, Monkeys, and Humans [95–97]

Compound	M_{wt} (Da)	Human F_a%	Monkey F_a%	Dog F_a%	Rat F_a%
Valacyclovir	361	100	100		
Bepridil	421	99	83		
Captopril	217	68	79		
Benazepril	425	30	32	39	50
Enalapril maleate	376	60		61	34
Fosinopril	564	30		25	
Ramipril	417	60		43	56
Cefadroxil	381	100			95

OATP2B1 (SLCO2B1) Human OATP2B1 (*SLCO2B1*) is part of the OATP superfamily, which is another important class of influx transporters expressed in the small intestine [98–100]. Indeed, human OATP2B1 protein is immunolocalized at the apical surface of the enterocytes [101] and Caco-2 cell monolayers [99]. It is worth noting that mRNA expression and protein quantification studies indicated minimal or no expression of other members of OATP superfamily (e.g., OATP1A1, OATP1B1, and OATP1B3) in human gastrointestinal tract [54]. Heterologous expression of OATP2B1 produces a sodium-independent and pH-gradient-dependent transporter [102]. The isolated human OATP2B1 cDNA comprises 4108 base pair (bp) that encodes a protein with 709 amino acids. It consists of 12 TM domains and its structure showed relatively shorter helices X and XII compared to the OATP1 family. While it has narrow substrate specificity relative to other OATPs, OATP2B1 plays a key role in the intestinal absorption of certain xenobiotics. For example, it transports sulfate-conjugated steroids such as estrone-3-sulfate, dehydroepiandrosterone-sulfate, 3-hydroxy-3-methylglutaryl-coenzyme A (HMG-CoA) reductase inhibitors, which include rosuvastatin, fluvastatin, pravastatin, atorvastatin, cerivastatin, pitavastatin, lovastatin, and simvastatin, the antidiabetic agent glibenclamide, bosentan, and fexofenadine [47, 103–105].

Human OATP2B1 direct rat ortholog transporter, rOatp2b1 also known as Oatp9 (Slc21a9), has 77% identical amino acid sequence [106]. Notwithstanding the similarity in the function, rat Oatp2b1 has wider substrate specificity relative to human OATP2B1 [107].

Cynomolgus monkey OATP2B1 amino acid sequence is 96.6% identical to human OATP2B1. Shen et al. studied the uptake of three known OATP2B1 substrates (estradiol-17β-D-glucuronide, cholecystokinin octapeptide, and estrone-3-sulfate) in HEK-293 cells stably expressing cynomolgus monkey and human OATP2B1. Consistent with the close similarity in amino acid sequence, the group reported comparable uptake profiles in monkey and human OATP2B1 expressing cells for the three model drugs tested under similar assay conditions (Figure 2.6) [108].

2.3.3 Summary of Intestinal Transporters That Impact Oral Drug Absorption

Marked species differences were reported in the anatomical and physiological factors that influence drug absorption in rats, dogs, monkeys, and humans. In a series of papers, Chiou et al. reported that the fraction of oral dose absorbed in rats and monkeys for a set of drugs with diverse physicochemical properties correlate well with that observed in human ($R^2 = 0.971$ and 0.974, respectively). However, a less robust correlation was observed, with several drugs showing much better absorption in dogs than in humans ($R^2 = 0.5123$) (Figure 2.7). Thus, caution

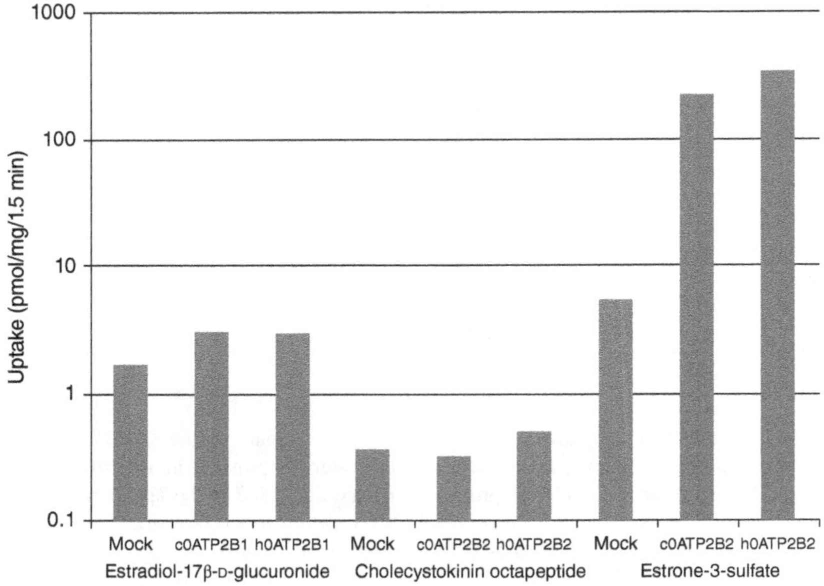

FIGURE 2.6 Uptake of [^{3}H]estradiol-17β-D-glucuronide (1 μM), [^{3}H]cholecystokinin octapeptide (0.1 μM) and [^{3}H]estrone-3-sulfate (1 μM) into mock, cynomolgus OATP2B1, and human OATP2B1 expressing HEK-293 cells was determined following 1.5 min of incubation. Adapted from Shen 2013 [108].

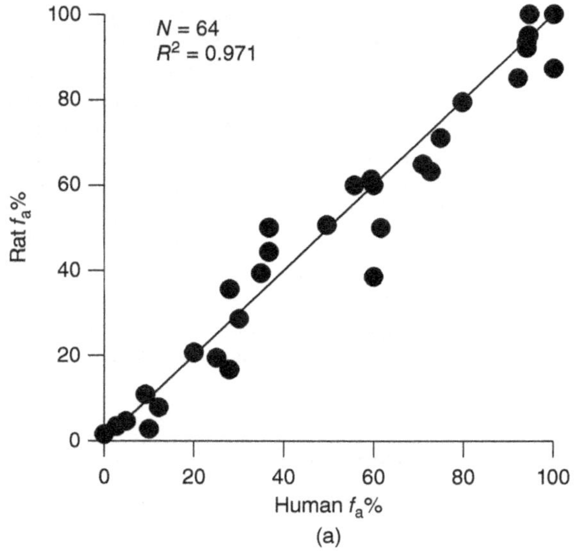

FIGURE 2.7 Correlation of percent of oral dose absorbed (f_a) between humans versus rats (a), monkeys (b), and dogs (c). Chiou et al. 1998 [95–97]. Reproduced with permission of Springer.

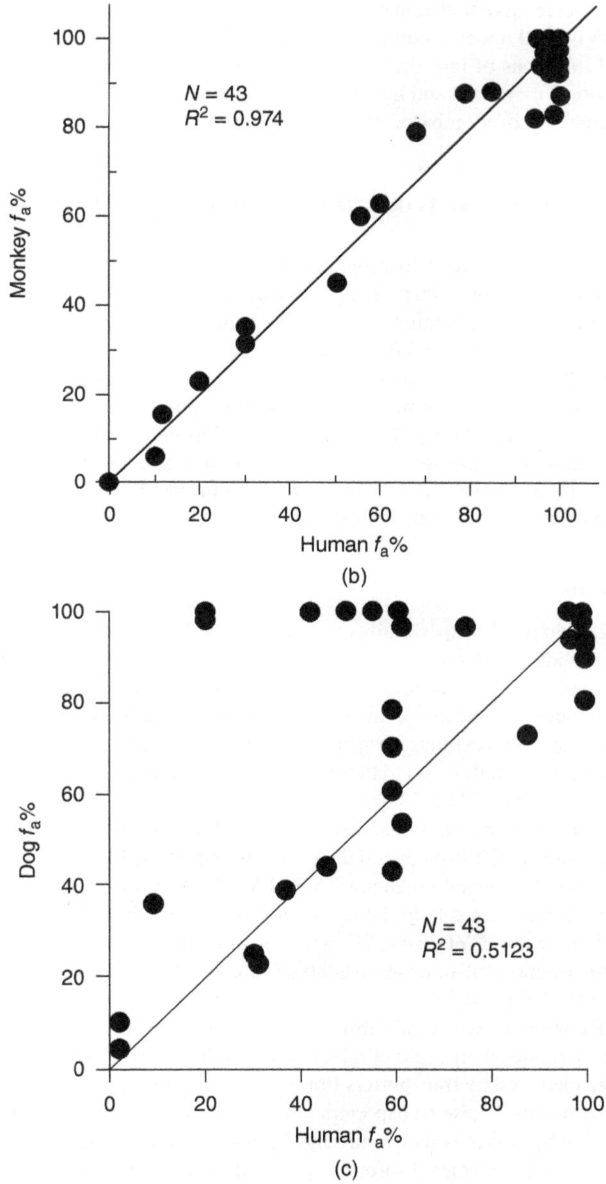

FIGURE 2.7 (*Continued*)

should be exercised when extrapolating bioavailability data from dogs to humans [95–97]. It should be stressed that an additional factor should be considered and involves the role that intestinal efflux and influx transporters plays in oral drug absorption. Little is known regarding the expression, regional distribution, substrate specificity, and functional characterization of efflux and influx transporters particularly in upper preclinical species such as dog and monkey gastrointestinal tract relative to that in humans. Indeed, investigating these areas will be pivotal in improving our overall understanding of the impact of potential species differences in tissue localization and functions of intestinal transporters on drug absorption and oral bioavailability among rats, dogs, monkeys, and humans. This knowledge will also improve the quality of prediction of human absorption based on available preclinical data.

2.4 PHYSIOLOGICAL FACTORS THAT IMPACT DRUG METABOLISM

Extrapolation of preclinical data to project human clearance and bioavailability for NCE predominantly eliminated by metabolizing enzymes requires a thorough understanding of the metabolic fate and identification of rate determining step for the compound clearance [109]. Appropriate *in vitro* and *in vivo* models of these species are essential to establish *in vitro* and *in vivo* (IVIVC) correlation between the NCE intrinsic metabolic clearance and its clearance observed *in vivo*. These models are also prerequisite for the successful prediction of human pharmacokinetics [110]. The following section will focus on cross-species comparison of the major drug metabolizing enzyme families that are involved in the hepatic and intestinal first-pass metabolism, which are cytochrome P450, uridine 5′-diphosphate (UDP)-glucuronosyltranferase, and aldehyde oxidase metabolizing enzymes.

2.4.1 Cytochrome P450s

The cytochrome P450 (CYP) superfamily of enzymes is divided into multiple families and subfamilies, with individual isoform identified and grouped into each family or subfamily based on similarity in amino acid sequences [111, 112]. In mammals, CYPs are membrane-bound enzymes that are located in the endoplasmic reticulum of the cells. As a result, CYP enzymes substrates need to have high cellular permeability to be metabolized. Compounds with low permeability would have limited access to these metabolizing enzymes, unless they are substrates to active uptake transporters [113, 114].

CYP enzymes have been found in almost all human body organs; however, liver and intestine are the major sites of CYP-mediated drug metabolism [115, 116]. The main human CYP enzymes involved in drug metabolism are CYPs 1A2, 2C9, 2C19, 2D6, and 3A4 with CYPs 2B6, 2C8, and 3A5 playing a role in the metabolism of certain therapeutic agents [117–124]. It is interesting to note that CYPs 3A4, 2C, and 2D6 comprise nearly 70% of the total hepatic CYP protein with around 80% of drugs available in the market metabolized predominantly by these enzymes [116, 125].

Overall, CYP enzymes have broad substrate specificity, which renders their substrates more hydrophilic and facilitates their rapid elimination through kidney and/or bile. The rapid elimination is usually mediated by transporters lining the kidney proximal tubules and canalicular side of liver hepatocytes. These transporters also make some molecules more amenable for phase II conjugation by enzymes such as uridine diphosphate glucuronosyltransferases (UGTs) and sulfonyl transferase enzymes. Profound species differences in CYP enzymes amino acid sequence, expression, substrate specificity, and catalytic activity are well documented in the literature [126]. This leads to major species differences in the metabolism and pharmacokinetic profiles for CYPs substrates.

2.4.1.1 CYP3A In humans, CYP3A is the major drug metabolizing enzymes subfamily (Table 2.7). While its liver content is only 30% of the CYPs in human liver, CYP3A is responsible for the metabolism of around 50% of the drugs available on the market [127]. It consists of four isozymes (CYP3A4, CYP3A5, CYP3A7, and CYP3A43). The most abundant

TABLE 2.7 CYP Enzymes of Major Drug Metabolizing CYP Family in Human, Mouse, Rat, Dog, and Monkey

Family	Subfamily	Human	Mouse	Rat	Dog	Monkey
CYP1	A	1A1, 1A2	1A1, 1A2	1A1, 1A2	1A1, 1A2	1A1, 1A2
	B	1B1	1B1	1B1	1B1	1B1
CYP2	A	2A6, 2A7, 2A13	2A4, 2A5, 2A12, 2A22	2A1, 2A2, 2A3	2A13, 2A25	2A23, 2A24
	B	2B6, 2B7	2B9, 2B10	2B1, 2B2, 2B3	2B11	2B17
	C	2C8, 2C9, 2C18, 2C19	2C29, 2C37, 2C38, 2C39, 2C40, 2C44, 2C50, 2C54, 2C55	6, 2C7*, 2C11*, 2C12*, 2C13*, 2C22, 2C23	2C21, 2C41	2C20, 2C43
	D	2D6, 2D7, 2D8	9, 2D10, 2D11, 2D12, 2D13, 2D22, 2D26, 2D34, 2D40	1, 2D2, 2D3, 2D4, 2D5, 2D18	2D15	2D17, 2D19, 2D29, 2D30
	E	2E1	2E1	2E1	2E1	2E1
CYP3	A	3A4, 3A5, 3A7, 3A43	3A11, 3A13, 3A16, 3A25, 3A41, 3A44	1/3A23, 3A2*, 3A9*, 3A18*, 3A62	3A12, 3A26	3A8

Source: Martignoni et al. 2006 [6]. Reproduced with permission of Taylor & Francis.

liver CYP protein level is CYP3A4 with a median value of 68.1 pmol/mg microsomal protein [128]. CYP3A5 is a polymorphically expressed enzyme with more than 50% of the total CYP3A in some individuals [116, 124, 129, 130]. CYP3A enzymes are expressed in the intestine and located at the apex of enterocytes with the greatest enzymatic activity in the duodenum, followed by jejunum and ileum [131]. CYP3A7 is considered a fetus-specific form of CYP3A; however, the CYP3A7*1C polymorphism causes the persistence of enzymatic activity of CYP3A7 during adult life [132]. CYP3A43 is minimally expressed in liver with low catalytic activity compared to CYP3A4 [133]. Although gender differences in CYP3A4 activity utilizing the probe substrate midazolam have been observed, the overall impact on the systemic exposure to midazolam was of negligible clinical importance [134].

In rats, CYP3A consists of several isoforms including CYP3A1, CYP3A2, CYP3A9, CYP3A18, CYP3A23, and CYP3A62. It is interesting to note that the expression of these isoforms in rats appears to be gender specific. For example, CYP3A2 and CYP3A18 are male-specific isoforms, whereas CYP3A9 is a female-dominant isoform [135, 136]. CYP3A1 and CYP3A2 were only detected in the liver, whereas CYP3A62 is the predominant form in the intestine and its expression is similar to that of human CYP3A4. CYP3A9 and CYP3A18 were detected in both liver and intestines [137]. GeneChip analysis of both rat and human intestinal samples showed no correlation between the enzyme expression level of CYP3A4 and CYP3A9 in either duodenum or colon [52]. As for the enzymatic activity of CYP3A enzymes, nifedipine and sildenafil are known probe substrates for human CYP3A4 that are not metabolized by rat CYP3A1 but by rat CYP2C [138, 139]. However, CYP3A is involved in the metabolism of midazolam in both rats and humans [140]. Species differences have also been reported in CYP3A enzymes induction, where rifampin is a stronger inducer for humans than rats [141], whereas dexamethasone is a stronger inducer in rats than humans CYP3A [141].

In dogs, CYP3A consists of two isoforms (CYP3A12 and CYP3A26) [142]. These enzymes showed marked differences in the rates of metabolism for several steroids (hydroxylation reactions) [142]. 6-β-Hydroxytestosterone activity (human CYP3A probe metabolic pathway) has been observed in dog intestines [143]. Ketoconazole is a potent inhibitor of 6-β-hydroxytestosterone activity in both dog and human liver microsomes [144]. The effect of ketoconazole in rat liver microsomes is dependent on sex [144]. Ketoconazole is also a moderate inhibitor of 1′-hydroxy bufuralol metabolism in dog (CYP2D) but not in human liver microsomes. While troleandomycin is a potent *in vitro* inhibitor of 6-β-hydroxytestosterone activity in both dogs and humans, its specificity against other CYP enzymes has not been fully evaluated [145].

In cynomolgus monkey, there is one CYP3A isoform, which is CYP3A8. CYP3A8 comprises around 20% of the total CYPs expressed in the liver [6]. While CYP3A8 shares 93% identity with the amino acid sequence of the human CYP3A4, there are major species differences in terms of protein levels and catalytic activity between the two isoforms. For example, CYP3A8 has around four- to fivefold higher protein levels compared with CYP3A4. Furthermore, typical CYP3A4 probe substrates such as midazolam and erythromycin have significantly higher enzymatic rates in monkey than in human liver microsomes. The higher enzymatic rate is partially attributed to the higher levels of CYP3A protein in monkey relative to human liver.

In summary, there are major species differences in the CYP3A isoforms expression: inducibility, substrate specificity, and catalytic activity. This renders extrapolation of clearance for CYP3A substrates from animals to humans quite perilous.

2.4.1.2 CYP2C In humans, the CYP2C superfamily consists of CYP2C8, CYP2C9, CYP2C18, and CYP2C19, which display race-related differences and polymorphic expression [112, 146, 147]. Around 16% of drugs available in the market are partially/completely metabolized by this superfamily [6]. CYP2C superfamily makes ≈20% of the total CYP located in the liver. The abundance of CYP2C protein is greater in human liver compared to small intestine [148]. The order of relative abundance of CYP2C protein was CYP2C9 > CYP2C8 > CYP2C19. It should be emphasized that CYP2C18 is not detected in either human liver or small intestine and only expressed in human skin epidermis [6]. Sex differences in CYP2C metabolism have not been observed in humans [149, 150].

In rats, the CYP2C family consists of several isoforms, including CYP2C6, CYP2C7, CYP2C11, CYP2C12, CYP2C13, CYP2C22, and CYP2C23 (Table 2.7). There are sex and developmental stage-dependent differences in the expression of the CYP2C family in rats [151, 152]. For example, CYP2C11 is a male-specific enzyme that is expressed in adult rat liver and only induced at puberty, where it comprises 50% of the total CYP content in adult male rat liver. Similarly, CYP2C13 is a male-specific enzyme that is expressed in rat liver and brain. Furthermore, CYP2C12 has higher protein expression in the liver of adult female rats than male rats. Interestingly, these differences do not exist in either immature or adult rats. Unlike these CYP2C isoforms, CYP2C7 enzyme is female specific, while CYP2C6 expression is gender independent.

In dogs, the CYP2C family consists of several isoforms, including CYP2C21 and CYP2C41 [6]. They share 70% identity with amino acid and gene sequences. Canine CYP2C21 and CYP2C41 display 74–83% gene and 67–76% amino acid similarity with human CYP2Cs. However, canine CYP2C21 is less homologous to the human CYP2Cs compared to CYP2C41. CYP2C21 is expressed in all dogs, while CYP2C41 expression is rather polymorphic. The *in vitro* metabolism of several human CYP2C9 probes (tolbutamide, 7-hydroxylation of S-warfarin) is impaired in dogs. In contrast, the *in vitro* metabolism of the human CYP2C9 probe R-mephenytoin (4-hydroxylation) was much greater in dog than human microsomes [153]. These examples illustrate important species differences between human and dog CYP2C.

The CYP2C family consists of two isoforms (CYP2C20 and 2C43) that are expressed in the liver of monkey. They share 83% and 77% identity with amino acid and gene sequences, respectively. Consistent with the high similarity in gene and amino acid sequences of CYP2C43 with human CYP2C19, this enzyme metabolized (*S*)-mephenytoin, a known probe substrate of human CYP2C19. However, CYP2C43 did not metabolize tolbutamide, a positive probe substrate of human CYP2C9. Therefore, it is widely accepted that CYP2C43 is functionally related to human CYP2C19 rather CYP2C9 [6].

2.4.1.3 CYP2D In humans, CYP2D subfamily consists of one isoform (CYP2D6), which is expressed in various tissues including liver and small intestine [116, 154, 155]. Although CYP2D6 only accounts for 4% of the total CYP content, it is involved in the metabolism of around 30% of the drugs available in the market at present [156]. CYP2D6 exhibits polymorphism with more than 80 identified alleles within the human population [155]. This leads to significant interindividual variability in CYP2D6 substrates disposition and pharmacokinetics.

In rats, there are six CYP2D isoforms (CYP2D1, CYP2D2, CYP2D3, CYP2D4, CYP2D5, and CYP2D18) [112] (Table 2.7). These isoforms share more than 70% identity with human CYP2D6 gene sequence [157]. *In vitro* evaluation showed that expressed CYP2D isozymes, including rat CYP2D1, CYP2D2, CYP2D3, and CYP2D4 and human recombinant CYP2D6 isoforms, possessed catalytic activity for the 1′-hydroxylation of bufuralol [158]. However, the 1′,2′-ethenylation of bufuralol was catalyzed only by rat CYP2D4 and human CYP2D6, indicating a difference between the rat and human CYP2D catalytic properties. The inhibition profile is also different for rat CYP2D compared to human as evidenced by the fact that human (but not rat) CYP2D is inhibited by quinidine, while rat CYP2D is weakly inhibited by quinine [159, 160].

In dogs, CYP2D15 is the major CYP2D enzyme with several identified variants namely CYP2D15 WT2, CYP2D15 V1, and CYP2D15 V2 [161]. *In vitro*, bufuralol 1′-hydroxylation enzyme kinetics were similar and quinidine inhibited its metabolic reaction to a similar degree in both dogs and humans. However, in contrast to human microsomal metabolism, quinine also showed moderate inhibition in dogs. In addition, antihuman CYP2D did not inhibit 1′-hydroxylation of bufuralol in dogs but did moderately inhibit 1′-hydroxylation of bufuralol in rat and human liver microsomes [144].

In monkeys, significant strain differences in the expression of various CYP2D isoforms have been reported [6]. For example, CYP2D17 is expressed in the liver of cynomolgus monkey. It shares 93% identity with human CYP2D6, while CYP2D42 is expressed in rhesus monkey. Similar trends have been observed with other monkey strains [6].

2.4.1.4 CYP1A The CYP1A subfamily consists of CYP1A1 and CYP1A2 isoforms that are expressed in rats, dogs, monkeys, and humans. CYP1A2 is constitutively expressed in the livers of both humans [162] and rats [163] but is expressed at low levels in the livers of untreated dogs and monkeys [164] (Table 2.7). CYP1A1 is constitutively expressed at very low levels in the livers of rats, dogs [165], monkeys, and humans and present in the small intestine [154, 166, 167], lung [168], and placenta [169, 170] of rats and humans. The presence of extrahepatic CYP1A1 for dogs has not been established. CYP1A1 expression levels are variable in the small intestine of humans and have been shown to be inducible by cigarette smoking and ingestion of chargrilled meats [171, 172]. Furafylline inhibits both rat and human CYP1A2, but the potency is 1000-fold less for rats than humans. This species difference in furafylline inhibition between rats and humans suggests a major difference in the active site topography between rat and human orthologs of CYP1A2 [173, 174]. Cross-species differences are also apparent between dog and human orthologs of CYP1A. In a cross-species comparison study, antihuman 1A antibody and human CYP1A inhibitors (α-naphthoflavone and furafylline) inhibited the metabolism of chlorzoxazone (human CYP2E1 probe substrate) in dog microsomes in addition to the human CYP1A2 probe substrate (7-ethoxy-resorufin O-dealkylation) [144].

2.4.2 UGT

Glucuronosyltransferases (UGT) superfamily of enzymes consists of two families (UGT1 and UGT2). These enzymes primarily catalyze the linkage of glucuronic acid component of UDP-glucuronic acid, to a substrate with acceptor functional group such as carboxylic acid moiety, hydroxyl, and amine groups [175]. This glucuronidation process is effective in rendering lipophilic xenobiotics and toxins water soluble, which facilitates their renal and/or biliary excretion. Known substrates of UGT enzymes are amitriptyline, acetaminophen, carvedilol, clofibric acid, fenofibrate, furosemide, naloxone, naproxen, olanzapine, propofol, raloxifene, and valproic acid [156].

In rats and humans, UGTs are located in various tissues with tissue specificity dependent on the isoform. In humans, UGT1A1, 1A3, 1A4, 1A6, 1A9, 2B4, 2B7, 2B15, and 2B17 have been identified in the liver based on mRNA and protein levels [113, 176]. UGT2B7 and UGT2B15 are the most abundant UGT enzymes in the liver with average protein levels of 82.9 ± 36.1 and 62.1 ± 31.5 pmol/mg protein, respectively [128]. UGT1A1, 1A3, 1A6, 1A8, and 2B7 have been identified in human small intestine based on protein levels determined using proteomic approaches (Figure 2.8). In rats, the UGTs identified in the liver based on mRNA levels include UGT1A1, 1A5, 1A6, 2B1, 2B2, 2B3, 2B6, and 2B12 [177]. UGT1A1, 1A2, 1A3, 1A6, 1A7, 2B3, 2B8, and 2B12 have been identified in rat intestines based on mRNA levels [177].

In a recent publication, Heikkinen et al. quantified protein levels of UGT and CYP metabolizing enzymes in beagle dog intestine and liver using mass-spectrometry-based quantitative proteomics [178]. The group quantified seven UGT enzymes in both liver and intestine. Interestingly, UGT1A11 and UGT1A9 were only present in the intestine, while UGT1A7 and UGT2B31 were only reported in dog liver. The most abundant UGTs in dog intestine were UGT1A11 and UGT1A2, whereas UGT2B31 was the most abundant UGT enzyme in the liver. When compared to the overall abundance of CYP enzymes, the UGT enzymes had similar abundance to the sum of CYP enzymes in the dog liver. However, UGT enzymes overall abundance was four times higher than that reported with CYP enzymes in dog intestine.

The mRNA expression levels of eight UGT isoforms in large intestine, liver, and small intestine of cynomolgus monkey were assessed using real-time reverse transcription PCR

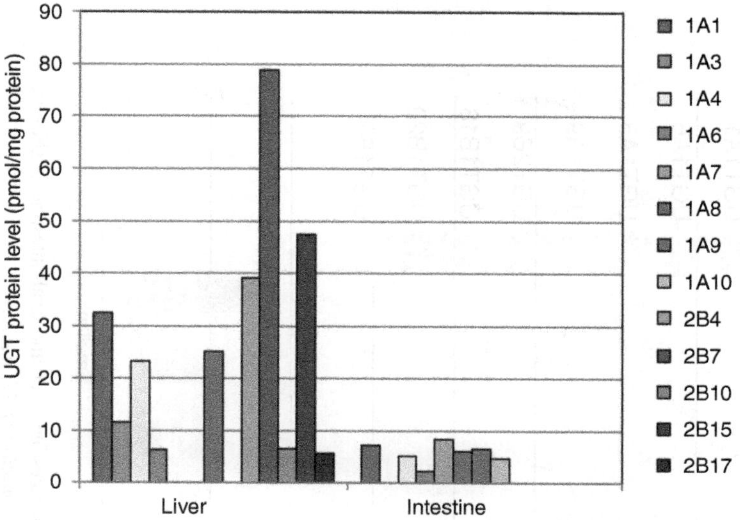

FIGURE 2.8 Major UDP-glucuronosyltransferases expressed in human liver and intestine based on protein level using proteomic approaches. Di 2014. [113]. Reproduced with permission of Taylor & Francis.

by Nishimura et al. [179]. As illustrated in Figure 2.9, UGT enzymes expression was tissue specific. The highest expression of six UGT enzymes isoforms, namely, UGT1A6, UGT1A9, UGT2B9, UGT1B18, UGT1B20, and UGT1B3 were reported in the monkey liver (Figure 2.9), whereas two isoforms (UG1A1 and UGT1A8) had the highest expression in monkey large intestine. High expression was also noted in small intestine (Figure 2.9). It should be emphasized that UGT1A8 mRNA has a selective expression in large and small intestines of cynomolgus monkeys.

UGT-mediated intestinal metabolism has a significant impact on the disposition and pharmacokinetics of UGT substrates. Furukawa et al. reported species differences in the intestinal glucuronidation activities among humans, rats, dogs, and monkeys [180]. While correlated, human *in vitro* intrinsic clearance ($CL_{int,u,UGT}$) calculated using intestinal microsomes for known UGT substrates was typically lower than that reported in other species investigated. This result is not unexpected and consistent with a general trend of lower clearance in human relative to other species that is partially ascribed to the lower human hepatic blood flow in comparison with preclinical species [30]. An inverse relationship between $CL_{int,u,UGT}$ and $f_a f_g$ of UGT substrates for human, rat, dog, and monkey was observed. In dogs, the $CL_{int,u,UGT}$ values relating to $f_a f_g$ values were 3–4 times higher than other species investigated. These species differences are ascribed to different UGT isoforms mediating the metabolism of these compounds in each species.

2.4.3 Aldehyde Oxidase

Human aldehyde oxidase (AO) is a soluble complex molybdoflavoprotein that unlike CYP and UGT enzymes is located in the cytosolic compartment of cells. Molybdenum is an integral part of the enzyme and required along with flavin adenine dinucleotide (FAD) and iron–sulfur center for AO enzyme activity [181]. AO catalyzes the oxidation of organic aldehyde into carboxylic

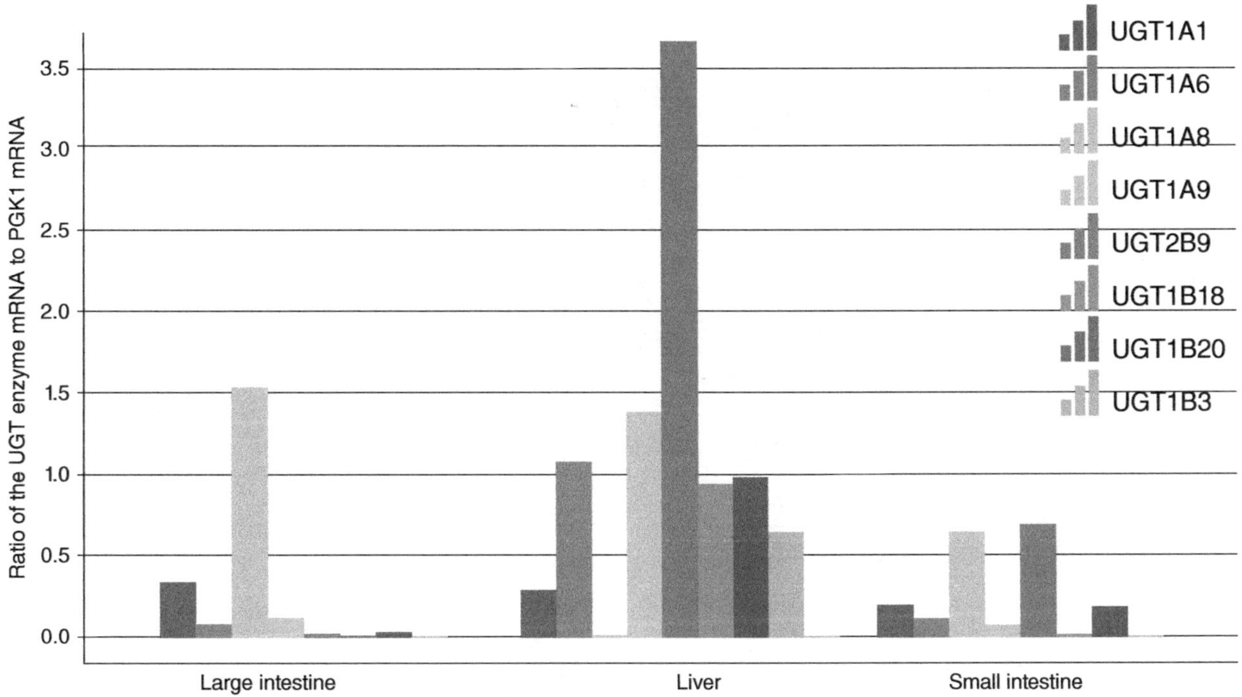

FIGURE 2.9 Expression of UDP-glucuronosyltransferase mRNAs in cynomolgus monkey large intestine, liver, and small intestine. Nishimura et al. 2009 [179]. Reproduced with permission of Elsevier. *(See color plate section for the color representation of this figure.)*

acid and nucleophilic oxidation of nitrogenous heterocyclic compounds [113]. Examples on xenobiotics that are substrates for AO include 6-deoxypenciclovir, 6-*O*-benzylguanine, acyclovir, BIBX1382, carbazeran, famciclovir, methotrexate, phenanthridine, phthalazine, zaleplon, ziprasidone, zoniporide, PF-4217903, and PF-945863 [182–186].

Moriwaki and coworkers demonstrated the widespread distribution of AO in various excretory organs of the human body, using immunohistochemistry staining [187]. For example, it was detected in lung (epithelial cells and alveolar cells), gastrointestinal tract (small intestine and colon), and kidney with its highest expression in the liver [181]. Similar trends were also observed in preclinical species [188].

There are significant species and strain (rat and mice) differences of AO. For example, both humans and monkeys have one single active AO gene (*AOX1*) that codes one functional enzyme. On the other hand, mice and rats have four separate *Aox* genes, which are *Aox1*, *Aox3*, *Aox4*, *and Aox3l*. They encode an equivalent number of structurally conserved enzymes. Interestingly, *Aox3* is the most expressed *Aox* form in rats and mice strains and contributes to the majority of the metabolic activity in the liver cytosol of these preclinical species. This leads to species differences in the metabolic profiles for AO substrates assessed in humans and rats or mice [189].

From enzymatic activity perspective, marked species differences were observed. In general, AO enzymatic activity tends to be high in both monkeys and humans followed by rats [190]. Interestingly, no AO enzymatic activity is reported in dogs. This is attributed to the inactivation of the *AOX1* gene and suppression/pseudogenization of *AOX3* counterpart [188].

In summary, caution should be exercised when projecting the human clearance of AO substrate. Instead of using allometric scaling from preclinical species with different AO genetic profiles, for example, mice, rats, and dogs, discovery DMPK scientists are recommended to scale human clearance using *in vitro* tools that are based on human hepatocytes and S9 fraction (Chapter 12) [191].

Failing to adhere to these guidelines may lead to clinical trials attrition due to compound poor human oral pharmacokinetic profile, which has devastating economic consequences. More importantly, a poor human oral pharmacokinetic profile impedes our ability to develop novel therapeutic agents for the treatment of devastating diseases in a timely manner. For example, FK3453 is a novel drug that was developed by Astellas for the treatment of Parkinson (Figure 2.8). The compound showed promising pharmacokinetic profiles in rats and dogs with high metabolic stability in liver microsomes. This result suggests minor contribution of CYP enzymes to FK3453 overall metabolism in species tested. Allometric scaling from rats and dogs projected a favorable human intrinsic clearance of 1.3 mL/min/kg. When tested in humans, the compound had poor oral plasma exposure with the oxidative metabolite of the aminopyrimidine moiety (M4) identified as a major metabolite (Figure 2.10). Furthermore, FK3453 observed human intrinsic clearance was 3000 mL/min/kg. Follow-up *in vitro* studies confirmed AO critical role in the metabolism of FK3453 and contribution to its high hepatic extraction ratio and poor oral bioavailability in humans. Overall, the results from this study led to FK3453 clinical trial termination. Similar trends were also reported with the development of a selective c-MET inhibitor, SGX523 and a novel p38 MAP kinase inhibitor. Their clinical trials were discontinued since their observed clearance in humans was significantly higher than that projected based on allometric scaling from rats and dogs, leading to high hepatic extraction and poor oral pharmacokinetics profile in humans [193, 194].

2.4.4 Species Differences in the First-Pass Metabolism and Its Impact on Oral Bioavailability

In addition to characterizing the species differences in the metabolic profile of an NCE (including the specific enzymatic pathways), an understanding of the species differences in the rate and

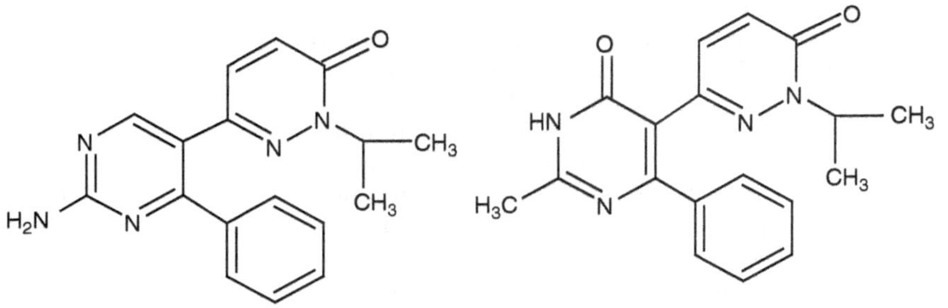

FIGURE 2.10 Chemical structures of FK3453 and M4. Zwisler et al. 2010. [192]. Reproduced with permission of John Wiley & Sons.

extent of first-pass metabolism is crucial for the appropriate projection of human bioavailability from preclinical data. The importance of this understanding is illustrated using the examples of atomoxetine and indinavir as follows.

Atomoxetine is a CYP2D6 substrate with an absolute human oral bioavailability of 94% and 63% in poor and extensive metabolizers of CYP2D6, respectively [195]. The moderate-to-high human oral bioavailability suggests nearly complete oral absorption of atomoxetine. However, preclinical evaluations indicated that the absolute oral bioavailability of atomoxetine in rats was only 4% [196] but was 74% in dogs [196]. Overall, the disposition of atomoxetine is similar in rats, dogs, and humans with a primary oxidative metabolite of 4-hydroxyatomoxetine that is subsequently conjugated to form 4-hydroxyatomoxetine-*O*-glucuronide. In a radiolabeled study in rats administered ^{14}C-atomoxetine, atomoxetine AUC following oral administration accounted for only 2% of the total ^{14}C AUC as compared to 30% of the ^{14}C AUC following intravenous administration, indicating extensive first-pass metabolism in rats [196]. In a corresponding radiolabeled study in dogs, atomoxetine AUC following oral administration accounted for 33% of the total ^{14}C AUC compared to 39% of the ^{14}C AUC following intravenous administration, indicating considerably less pronounced first-pass metabolism [196]. This example clearly illustrates the importance of understanding not only the species differences in a drug's metabolic fate but also the extent of species differences in the first-pass metabolism when utilizing preclinical data to project human oral bioavailability.

Indinavir, a CYP3A4 substrate, is an HIV protease inhibitor for which variable oral bioavailability has been observed in preclinical species, ranging from 72% in dogs to 19% in monkeys, and was 24% in rats [197]. This variability was mainly attributed to species differences in the extent of hepatic first-pass metabolism. Chemical and immunochemical inhibition studies indicated the potential involvement of CYP3A isoforms in the metabolism of indinavir in rats, dogs, and monkeys [197], which is consistent with the observation that CYP3A4 is the main isoform responsible for the oxidative metabolism of indinavir in human liver microsomes [198]. The *in vitro* profile of indinavir metabolism was qualitatively similar across species [197]. In addition, an *in vitro–in vivo* correlation was established in rats and dogs using the *in vivo* hepatic clearance and hepatic first-pass extraction ratio obtained from *in vitro* rat and dog metabolic data, respectively. Based on the *in vitro–in vivo* correlation established in rats and dogs, the *in vitro* intrinsic clearance of indinavir in human liver microsomes projected a small first-pass metabolism in humans ($E_h = 0.25$), which was consistent with indinavir's high oral bioavailability (60–65%) observed in humans at clinically relevant doses [197, 199]. This example depicts the importance of establishing an *in vitro–in vivo* correlation in tested preclinical species so as to use it as a basis to project human clearance and oral bioavailability.

2.4.5 Summary of Physiological Factors That Impact Drug Metabolism

Species differences in metabolic enzymes among rats, dogs, monkeys, and humans have been well characterized with the CYP enzymes and to a lesser extent with the UGT superfamily of enzymes. An important implication of this species differences is that the projection of hepatic and intestinal extraction and oral bioavailability from rats, dogs, and monkey data to human needs be made with caution, taking into account potential species differences in first-pass metabolism. Despite this limitation, there are approaches that can be utilized to improve the robustness of the projection of human oral bioavailability as illustrated by the examples from atomoxetine and indinavir. Knowledge of the validity (or lack thereof) regarding the *in vitro–in vivo* relationship for metabolic clearance in the preclinical species, along with an understanding of the similarities and differences in the metabolic profile and clearance rate between rat/dog and human *in vitro*, will provide important insights into the appropriateness of utilizing preclinical and/or *in vitro* data to predict human oral bioavailability.

2.5 CONCLUSION

Oral bioavailability is a pivotal factor in the successful discovery and development of NCEs. As discussed in this chapter, the key determinants of oral bioavailability are driven by complex interplay of the physicochemical property of a compound, its absorption (solubility and permeability including passive permeability and active transport), and first-pass intestinal and hepatic metabolism characteristics. Many of the physiological factors that influence these processes show distinct species differences among rats, dogs, monkeys, and humans. Considerable advances have been made to understand these critical factors, which in turn have led to the development of various *in silico* and *in vitro* models to assess a drug candidate's potential to be orally bioavailable early in the discovery program. While passive permeability can be reasonably predicted based on the physicochemical properties of a compound such as lipophilicity and polarity, key gaps in our understanding on the species differences in active transport and metabolism still exist. The lack of a complete or comprehensive understanding renders extrapolation of the bioavailability data from preclinical models such as rats, dogs, and monkeys *in vivo* studies to humans challenging. Future advances in these areas undoubtedly will improve our ability to predict oral bioavailability in humans and further increase the success and expedite the process of developing successful drugs.

2.6 CHAPTER 2 TUTORIALS

1. You are working on a compound in a small pharmaceutical company. It has the following physicochemical properties and preclinical *in vivo* profiles:
   - $M_{wt} = 225$ Da
   - $\mathrm{Bp}K_a = 8.5$ [Basic]
   - Intestinal permeability = Low consistent with paracellular absorbed molecule. In rats and humans, the compound has poor absorption [$f_a = 10\%$] and has almost complete absorption in dogs
   - Metabolically stable in preclinical and human liver microsomes and hepatocyte

- Completely eliminated in urine as unchanged drug in all preclinical species tested
- The compound has a short half-life that is not appropriate for once daily dosing.

The formulation colleague reached out for the team to propose once daily dosing controlled release formulation that targets colon where he/she will use dog as a preclinical species to test his/her theory. What is your position and would you support his/her proposal? Why?

2. Monkey is generally perceived to be a good model for assessing oral absorption and intestinal first pass of new molecular entities (NMEs) due to similarities in metabolizing enzymes and transporters sequence homology. What do you think about this notion and what are the things you need to be mindful of when you dose NMEs in monkey?

3. You are working on a promising compound that appears to have good metabolic stability in all preclinical and human microsomal preparation [what does that mean?]. Interestingly, it appears to have poor stability in human and monkey hepatocytes but not dog hepatocytes. The team is debating moving forward with this molecule. Some propose doing allometric scaling from dog and use it as a basis to predict human clearance, which appears to be favorable rendering the compound to be appropriate for QD dosing, since the projected human clearance is low [1 mL/min/kg] and the volume of distribution is moderate [2 L/kg]. What would you do?

2.7 CHAPTER 2 TUTORIALS KEY

1. You are working on a compound in a small pharmaceutical company. It has the following physicochemical properties and preclinical *in vivo* profiles:
   - $M_{wt} = 225$ Da
   - Bp$K_a = 8.5$ [Basic]
   - Intestinal permeability = Low consistent with paracellular absorbed molecule. In rats and humans the compound has poor absorption [$f_a = 10\%$] and has almost complete absorption in dog
   - Metabolically stable in preclinical and human liver microsomes and hepatocyte
   - Completely eliminated in urine as unchanged drug in all preclinical species tested
   - The compound has a short half-life that is not appropriate for once daily dosing.

   The formulation colleague reached out for the team to propose once daily dosing controlled release formulation that targets colon where he/she will use dog as a preclinical species to test his/her theory. What is your position and would you support his/her proposal? Why?

   Answer: Since the compound is absorbed through paracellular pathway, dog is predicted to overestimate the absorption of this molecule and it is not a viable model to assess the utility of formulation to predict profile in humans. Furthermore, colon targeting for compounds absorbed via paracellular pathway is not an appropriate approach and should not be considered. Instead, other technologies such as gastric retention technology may be a better option to consider.

2. Monkey is generally perceived to be a good model for assessing oral absorption and intestinal first pass of new molecular entities (NMEs) due to similarities in metabolizing enzymes and transporters sequence homology. What do you think about this notion and what are the things you need to be mindful of when you dose NMEs in monkey?

Answer: As discussed in this chapter, major species differences are noted between human and monkey metabolizing enzymes and efflux transporter. In particular, in terms of their functional activity and expression in the intestine and liver. Therefore, it is pivotal to understand the properties of a tested article in terms of its affinity to various enzymes and transporters. This would provide better perspective on the cause of potential species difference. This allows a better prediction of human absorption and first-pass effect.

3. You are working on a promising compound that appears to have good metabolic stability in all preclinical and human microsomal preparation [what does it mean?]. Interestingly, it appears to have poor stability in human and monkey hepatocytes but not dog hepatocytes. The team is debating moving forward with this molecule. Some propose doing allometric scaling from dogs and use it as a basis to predict human clearance, which appears to be favorable rendering the compound to be appropriate for QD dosing, since the projected human clearance is low [1 mL/min/kg] and the volume of distribution is moderate [2 L/kg]. What would you do?

Answer: The first thing is to investigate the cause of the difference in the metabolic stability between microsomes and hepatocytes reagents. As noted in the aldehyde oxidase enzyme section, this is a cytosolic enzyme; microsomal preparation does not express this enzyme. The differences observed between the metabolic stability observed between the two preparations combined with high metabolic stability in dog hepatocytes suggest that the compound is mainly metabolized by aldehyde oxidase enzyme that has no expression in dog liver. Therefore, using dog as a preclinical model to predict the clearance in human would most likely be associated with underestimation of the human clearance. This typically would increase the odds of attrition. Always determine the driver for your compound metabolism and elimination before you predict the profile of the tested article.

REFERENCES

1. Thomas, V.H., Bhattachar, S., Hitchingham, L., Zocharski, P., Naath, M., Surendran, N., Stoner, C.L. and El-Kattan, A., *The road map to oral bioavailability: an industrial perspective*. Expert Opin Drug Metab Toxicol, 2006. 2(4): p. 591–608.
2. Kaitin, K.I., *Obstacles and opportunities in new drug development*. Clin Pharmacol Ther, 2008. 83(2): p. 210–212.
3. Lipinski, C.A., Lombardo, F., Dominy, B.W. and Feeney, P.J., *Experimental and computational approaches to estimate solubility and permeability in drug discovery and development settings*. Adv Drug Deliv Rev, 2001. 46(1–3): p. 3–26.
4. Bardelmeijer, H.A., van Tellingen, O., Schellens, J.H. and Beijnen, J.H., *The oral route for the administration of cytotoxic drugs: strategies to increase the efficiency and consistency of drug delivery*. Invest New Drugs, 2000. 18(3): p. 231–241.
5. Hellriegel, E.T., Bjornsson, T.D. and Hauck, W.W., *Interpatient variability in bioavailability is related to the extent of absorption: implications for bioavailability and bioequivalence studies*. Clin Pharmacol Ther, 1996. 60(6): p. 601–607.
6. Martignoni, M., Groothuis, G.M. and de Kanter, R., *Species differences between mouse, rat, dog, monkey and human CYP-mediated drug metabolism, inhibition and induction*. Expert Opin Drug Metab Toxicol, 2006. 2(6): p. 875–894.
7. Musther, H., Olivares-Morales, A., Hatley, O.J., Liu, B. and Rostami Hodjegan, A., *Animal versus human oral drug bioavailability: do they correlate?* Eur J Pharm Sci, 2014. 57: p. 280–291.
8. DeSesso, J.M. and Jacobson, C.F., *Anatomical and physiological parameters affecting gastrointestinal absorption in humans and rats*. Food Chem Toxicol, 2001. 39(3): p. 209–228.

9. Kararli, T.T., *Comparison of the gastrointestinal anatomy, physiology, and biochemistry of humans and commonly used laboratory animals*. Biopharm Drug Dispos, 1995. 16(5): p. 351–380.
10. Dressman, J.B., *Comparison of canine and human gastrointestinal physiology*. Pharm Res, 1986. 3(3): p. 123–131.
11. DeSesso, J.M. and Williams, A.L., *Contrasting the gastrointestinal tracts of mammals: factors that influence absorption*. Annu Rep Med Chem, 2008. 43: p. 353–371.
12. Hebel, R. and Stromberg, M.W., *Anatomy and Embryology of the Laboratory Rat*. 1986, BioMed Verlag: Worthsee, Germany.
13. Snyder, W.S., Cook, M.J., Nasset, E.S., Karhausen, L.R., Howells, G.P. and Tipton, I.H., *Report of the Task Group on Reference Man*. 1975, New York.
14. Rowland, M. and Tozer, T.N., *Clinical Pharmacokinetics Concepts and Application*. 3rd ed. 1995, Lippincott Williams & Wilkins: Philadelphia.
15. Shargel, L. and Yu, A., *Applied Biopharmaceutics and Pharmacokinetics*. 4th ed. 1999, Appleton and Lange: Stamford.
16. Pappenheimer, J.R., *Scaling of dimensions of small intestines in non-ruminant eutherian mammals and its significance for absorptive mechanisms*. Comp Biochem Physiol A Mol Integr Physiol, 1998. 121(1): p. 45–58.
17. Takahashi, M., Washio, T., Suzuki, N., Igeta, K. and Yamashita, S., *Investigation of the intestinal permeability and first-pass metabolism of drugs in cynomolgus monkeys using single-pass intestinal perfusion*. Biol Pharm Bull, 2010. 33(1): p. 111–116.
18. Chiou, W.L., *Effect of 'unstirred' water layer in the intestine on the rate and extent of absorption after oral administration*. Biopharm Drug Dispos, 1994. 15(8): p. 709–717.
19. Kimura, T. and Higaki, K., *Gastrointestinal transit and drug absorption*. Biol Pharm Bull, 2002. 25(2): p. 149–164.
20. Mudie, D.M., Amidon, G.L. and Amidon, G.E., *Physiological parameters for oral delivery and* in vitro *testing*. Mol Pharm, 2010. 7(5): p. 1388–1405.
21. Custodio, J.M., Wu, C.Y. and Benet, L.Z., *Predicting drug disposition, absorption/elimination/transporter interplay and the role of food on drug absorption*. Adv Drug Deliv Rev, 2008. 60(6): p. 717–733.
22. Singh, B.N., *Effects of food on clinical pharmacokinetics*. Clin Pharmacokinet, 1999. 37(3): p. 213–255.
23. Ikegami, K., Tagawa, K., Narisawa, S. and Osawa, T., *Suitability of the cynomolgus monkey as an animal model for drug absorption studies of oral dosage forms from the viewpoint of gastrointestinal physiology*. Biol Pharm Bull, 2003. 26(10): p. 1442–1447.
24. Lui, C.Y., Amidon, G.L., Berardi, R.R., Fleisher, D., Youngberg, C. and Dressman, J.B., *Comparison of gastrointestinal pH in dogs and humans: implications on the use of the beagle dog as a model for oral absorption in humans*. J Pharm Sci, 1986. 75(3): p. 271–274.
25. Martinez, M., Amidon, G., Clarke, L., Jones, W.W., Mitra, A. and Riviere, J., *Applying the biopharmaceutics classification system to veterinary pharmaceutical products. Part II. Physiological considerations*. Adv Drug Deliv Rev, 2002. 54(6): p. 825–850.
26. Garanero, G., Ramanathan, S. and Amidon, G., Gastrointestinal Dissolution and Absorption of Drugs, in *Drug Bioavailability: Estimation of Solubility, Permeability, Absorption and Bioavailability*, Waterbeemd, H.v.d., Lennernas, H. and Artursson, P., Editors. 2005, Wiley-VCH: Zurich. p. 191–210.
27. Stenberg, P., Bergstrom, C.A.S., Luthman, K. and Artursson, P., *Theoretical predictions of drug absorption in drug discovery and development*. Clin Pharmacokinet, 2002. 41(11): p. 877–899.
28. Enck, P., Merlin, V., Erckenbrecht, J.F. and Wienbeck, M., *Stress effects on gastrointestinal transit in the rat*. Gut, 1989. 30(4): p. 455–459.
29. Barrow, L., Steed, K.P., Spiller, R.C., Watts, P.J., Melia, C.D., Davies, M.C. and Wilson, C.G., *Scintigraphic demonstration of lactulose-induced accelerated proximal colon transit*. Gastroenterology, 1992. 103(4): p. 1167–1173.

30. Davies, B. and Morris, T., *Physiological parameters in laboratory animals and humans*. Pharm Res, 1993. 10(7): p. 1093–1095.

31. Lentz, K.A., *Current methods for predicting human food effect*. AAPS J, 2008. 10(2): p. 282–288.

32. Lentz, K.A., Quitko, M., Morgan, D.G., Grace, J.E., Jr., Gleason, C. and Marathe, P.H., *Development and validation of a preclinical food effect model*. J Pharm Sci, 2007. 96(2): p. 459–472.

33. Hofmann, A.F. and Hagey, L.R., *Bile acids: chemistry, pathochemistry, biology, pathobiology, and therapeutics*. Cell Mol Life Sci, 2008. 65(16): p. 2461–2483.

34. Monte, M.J., Marin, J.J., Antelo, A. and Vazquez-Tato, J., *Bile acids: chemistry, physiology, and pathophysiology*. World J Gastroenterol, 2009. 15(7): p. 804–816.

35. Charman, W.N., Porter, C.J., Mithani, S. and Dressman, J.B., *Physiochemical and physiological mechanisms for the effects of food on drug absorption: the role of lipids and pH*. J Pharm Sci, 1997. 86(3): p. 269–282.

36. Borriello, S.P., *Bacteria and gastrointestinal secretion and motility*. Scand J Gastroenterol Suppl, 1984. 93: p. 115–121.

37. Hurst, S., Loi, C.M., Brodfuehrer, J. and El-Kattan, A., *Impact of physiological, physicochemical and biopharmaceutical factors in absorption and metabolism mechanisms on the drug oral bioavailability of rats and humans*. Expert Opin Drug Metab Toxicol, 2007. 3(4): p. 469–489.

38. Roberts, M.S., Magnusson, B.M., Burczynski, F.J. and Weiss, M., *Enterohepatic circulation: physiological, pharmacokinetic and clinical implications*. Clin Pharmacokinet, 2002. 41(10): p. 751–790.

39. Drasar, B.S., Montgomery, F. and Tomkins, A.M., *Diet and faecal flora in three dietary groups in rural northern Nigeria*. J Hyg (Lond), 1986. 96(1): p. 59–65.

40. Trevaskis, N.L., Porter, C.J. and Charman, W.N., *The lymph lipid precursor pool is a key determinant of intestinal lymphatic drug transport*. J Pharmacol Exp Ther, 2006. 316(2): p. 881–891.

41. Edwards, G.A., Porter, C.J., Caliph, S.M., Khoo, S.M. and Charman, W.N., *Animal models for the study of intestinal lymphatic drug transport*. Adv Drug Deliv Rev, 2001. 50(1–2): p. 45–60.

42. Lennernas, H., *Does fluid flow across the intestinal mucosa affect quantitative oral drug absorption? Is it time for a reevaluation?* Pharm Res, 1995. 12(11): p. 1573–1582.

43. He, Y.L., Murby, S., Warhurst, G., Gifford, L., Walker, D., Ayrton, J., Eastmond, R. and Rowland, M., *Species differences in size discrimination in the paracellular pathway reflected by oral bioavailability of poly(ethylene glycol) and D-peptides*. J Pharm Sci, 1998. 87(5): p. 626–633.

44. Bader, T.R., Herneth, A.M., Blaicher, W., Steininger, R., Mühlbacher, F., Lechner, G. and Grabenwöger, F., *Hepatic perfusion after liver transplantation: noninvasive measurement with dynamic single-section CT*. Radiology, 1998. 209(1): p. 129–134.

45. Graham, G.G., et al., *Clinical pharmacokinetics of metformin*. Clin Pharmacokinet, 2011. 50(2): p. 81–98.

46. Avdeef, A., *Physicochemical profiling (solubility, permeability and charge state)*. Curr Top Med Chem, 2001. 1(4): p. 277–351.

47. Varma, M.V., Ambler, C.M., Ullah, M., Rotter, C.J., Sun, H., Litchfield, J., Fenner, K.S. and El-Kattan, A.F., *Targeting intestinal transporters for optimizing oral drug absorption*. Curr Drug Metab, 2010. 11(9): p. 730–742.

48. Hilgendorf, C., Ahlin, G., Seithel, A., Artursson, P., Ungell, A.L. and Karlsson, J., *Expression of thirty-six drug transporter genes in human intestine, liver, kidney, and organotypic cell lines*. Drug Metab Dispos, 2007. 35(8): p. 1333–1340.

49. Bleasby, K., et al., *Expression profiles of 50 xenobiotic transporter genes in humans and pre-clinical species: a resource for investigations into drug disposition*. Xenobiotica, 2006. 36(10–11): p. 963–988.

50. Shimoyama, Y., et al., *Expression of monocarboxylate transporter 1 (MCT1) in the dog intestine*. J Vet Med Sci, 2007. 69(6): p. 599–604.

51. International Transporter Consortium, et al., *Membrane transporters in drug development*. Nat Rev Drug Discov, 2010. 9(3): p. 215–236.

52. Cao, X., et al., *Why is it challenging to predict intestinal drug absorption and oral bioavailability in human using rat model*. Pharm Res, 2006. 23(8): p. 1675–1686.

53. Groer, C., Bruck, S., Lai, Y., Paulick, A., Busemann, A., Heidecke, C.D., Siegmund, W. and Oswald, S., *LC-MS/MS-based quantification of clinically relevant intestinal uptake and efflux transporter proteins*. J Pharm Biomed Anal, 2013. 85: p. 253–261.

54. Oswald, S., Groer, C., Drozdzik, M. and Siegmund, W., *Mass spectrometry-based targeted proteomics as a tool to elucidate the expression and function of intestinal drug transporters*. AAPS J, 2013. 15(4): p. 1128–1140.

55. Abel, S., et al., *Potential role for P-glycoprotein in the non-proportional pharmacokinetics of UK-343,664 in man*. Xenobiotica, 2001. 31(8–9): p. 665–676.

56. El Ela, A.A., Hartter, S., Schmitt, U., Hiemke, C., Spahn-Langguth, H. and Langguth, P., *Identification of P-glycoprotein substrates and inhibitors among psychoactive compounds – implications for pharmacokinetics of selected substrates*. J Pharm Pharmacol, 2004. 56(8): p. 967–975.

57. Salphati, L. and Benet, L.Z., *Effects of ketoconazole on digoxin absorption and disposition in rat*. Pharmacolology, 1998. 56(6): p. 308–313.

58. Drescher, S., Glaeser, H., Murdter, T., Hitzl, M., Eichelbaum, M. and Fromm, M.F., *P-glycoprotein-mediated intestinal and biliary digoxin transport in humans*. Clin Pharmacol Ther, 2003. 73(3): p. 223–231.

59. Devault, A. and Gros, P., *Two members of the mouse mdr gene family confer multidrug resistance with overlapping but distinct drug specificities*. Mol Cell Biol, 1990. 10(4): p. 1652–1663.

60. Nishimura, M. and Naito, S., *Tissue-specific mRNA expression profiles of human ATP-binding cassette and solute carrier transporter superfamilies*. Drug Metab Pharmacokinet, 2005. 20(6): p. 452–477.

61. Takahashi, M., Washio, T., Suzuki, N., Igeta, K., Fujii, Y., Hayashi, M., Shirasaka, Y. and Yamashita, S., *Characterization of gastrointestinal drug absorption in cynomolgus monkeys*. Mol Pharm, 2008. 5(2): p. 340–348.

62. Kim, I.W., Booth-Genthe, C. and Ambudkar, S.V., *Relationship between drugs and functional activity of various mammalian P-glycoproteins (ABCB1)*. Mini Rev Med Chem, 2008. 8(3): p. 193–200.

63. Mouly, S. and Paine, M.F., *P-glycoprotein increases from proximal to distal regions of human small intestine*. Pharm Res, 2003. 20(10): p. 1595–1599.

64. Lown, K.S., et al., *Role of intestinal P-glycoprotein (mdr1) in interpatient variation in the oral bioavailability of cyclosporine*. Clin Pharmacol Ther, 1997. 62(3): p. 248–260.

65. Tamura, S., Ohike, A., Ibuki, R., Amidon, G.L. and Yamashita, S., *Tacrolimus is a class II low-solubility high-permeability drug: the effect of P-glycoprotein efflux on regional permeability of tacrolimus in rats*. J Pharm Sci, 2002. 91(3): p. 719–729.

66. Katoh, M., Suzuyama, N., Takeuchi, T., Yoshitomi, S., Asahi, S. and Yokoi, T., *Kinetic analyses for species differences in P-glycoprotein-mediated drug transport*. J Pharm Sci, 2006. 95(12): p. 2673–2683.

67. Xia, C.Q., Xiao, G., Liu, N., Pimprale, S., Fox, L., Patten, C.J., Crespi, C.L., Miwa, G. and Gan, L.S., *Comparison of species differences of P-glycoproteins in beagle dog, rhesus monkey, and human using Atpase activity assays*. Mol Pharm, 2006. 3(1): p. 78–86.

68. Takeuchi, T., Yoshitomi, S., Higuchi, T., Ikemoto, K., Niwa, S., Ebihara, T., Katoh, M., Yokoi, T. and Asahi, S., *Establishment and characterization of the transformants stably-expressing MDR1 derived from various animal species in LLC-PK1*. Pharm Res, 2006. 23(7): p. 1460–1472.

69. Han, B. and Zhang, J.T., *Multidrug resistance in cancer chemotherapy and xenobiotic protection mediated by the half ATP-binding cassette transporter ABCG2*. Curr Med Chem Anticancer Agents, 2004. 4(1): p. 31–42.

70. Haller, S., Schuler, F., Lazic, S.E., Bachir-Cherif, D., Kramer, S.D., Parrott, N.J., Steiner, G. and Belli, S., *Expression profiles of metabolic enzymes and drug transporters in the liver and along the intestine of beagle dogs*. Drug Metab Dispos, 2012. 40(8): p. 1603–1610.

71. Seithel, A., Karlsson, J., Hilgendorf, C., Bjorquist, A. and Ungell, A.L., *Variability in mRNA expression of ABC- and SLC-transporters in human intestinal cells: comparison between human segments and Caco-2 cells*. Eur J Pharm Sci, 2006. 28(4): p. 291–299.
72. Nies, A.T. and D. Keppler, *The apical conjugate efflux pump ABCC2 (MRP2)*. Pflugers Arch, 2007. 453(5): p. 643–659.
73. Buchler, M., Konig, J., Brom, M., Kartenbeck, J., Spring, H., Horie, T. and Keppler, D., *cDNA cloning of the hepatocyte canalicular isoform of the multidrug resistance protein, cMrp, reveals a novel conjugate export pump deficient in hyperbilirubinemic mutant rats*. J Biol Chem, 1996. 271(25): p. 15091–15098.
74. Conrad, S., Viertelhaus, A., Orzechowski, A., Hoogstraate, J., Gjellan, K., Schrenk, D. and Kauffmann, H.M., *Sequencing and tissue distribution of the canine MRP2 gene compared with MRP1 and MDR1*. Toxicology, 2001. 156(2–3): p. 81–91.
75. Yasunaga, M., Takemura, M., Fujita, K., Yabuuchi, H. and Wada, M., *Molecular cloning and functional characterization of cynomolgus monkey multidrug resistance-associated protein 2 (MRP2)*. Eur J Pharm Sci, 2008. 35(4): p. 326–334.
76. Pfeifer, N.D., Yang, K. and Brouwer, K.L., *Hepatic basolateral efflux contributes significantly to rosuvastatin disposition: I. Characterization of basolateral vs. biliary clearance using a novel protocol in sandwich-cultured hepatocytes*. J Pharmacol Exp Ther, 2013. 347(3): p. 727–736.
77. Misaka, S., Muller, F. and Fromm, M.F., *Clinical relevance of drug efflux pumps in the gut*. Curr Opin Pharmacol, 2013. 13(6): p. 847–852.
78. Kato, R., Nishide, M., Kozu, C., Iwamoto, A., Urashima, K., Suzuki, K., Ijiri, Y., Hayashi, T. and Tanaka, K., *Is cyclosporine A transport inhibited by pravastatin via multidrug resistant protein 2?* Eur J Clin Pharmacol, 2010. 66(2): p. 153–158.
79. Borst, P., Evers, R., Kool, M. and Wijnholds, J., *A family of drug transporters: the multidrug resistance-associated proteins*. J Natl Cancer Inst, 2000. 92(16): p. 1295–1302.
80. Dietrich, C.G., de Waart, D.R., Ottenhoff, R., Bootsma, A.H., van Gennip, A.H. and Elferink, R.P., *Mrp2-deficiency in the rat impairs biliary and intestinal excretion and influences metabolism and disposition of the food-derived carcinogen 2-amino-1-methyl-6-phenylimidazo*. Carcinogenesis, 2001. 22(5): p. 805–811.
81. Dietrich, C.G., de Waart, D.R., Ottenhoff, R., Schoots, I.G. and Elferink, R.P., *Increased bioavailability of the food-derived carcinogen 2-amino-1-methyl-6-phenylimidazo[4,5-b]pyridine in MRP2-deficient rats*. Mol Pharmacol, 2001. 59(5): p. 974–980.
82. Lin, X., Skolnik, S., Chen, X. and Wang, J., *Attenuation of intestinal absorption by major efflux transporters: quantitative tools and strategies using a Caco-2 model*. Drug Metab Dispos, 2011. 39(2): p. 265–274.
83. Benet, L.Z., Amidon, G.L., Barends, D.M., Lennernas, H., Polli, J.E., Shah, V.P., Stavchansky, S.A. and Yu, L.X., *The use of BDDCS in classifying the permeability of marketed drugs*. Pharm Res, 2008. 25(3): p. 483–488.
84. Estudante, M., Morais, J.G., Soveral, G. and Benet, L.Z., *Intestinal drug transporters: an overview*. Adv Drug Deliv Rev, 2013. 65(10): p. 1340–1356.
85. Bretschneider, B., Brandsch, M. and Neubert, R., *Intestinal transport of beta-lactam antibiotics: analysis of the affinity at the H+/peptide symporter (PEPT1), the uptake into Caco-2 cell monolayers and the transepithelial flux*. Pharm Res, 1999. 16(1): p. 55–61.
86. Knutter, I., Wollesky, C., Kottra, G., Hahn, M.G., Fischer, W., Zebisch, K., Neubert, R.H., Daniel, H. and Brandsch, M., *Transport of angiotensin-converting enzyme inhibitors by H+/peptide transporters revisited*. J Pharmacol Exp Ther, 2008. 327(2): p. 432–441.
87. Terada, T., Saito, H., Mukai, M. and Inui, K., *Characterization of stably transfected kidney epithelial cell line expressing rat H+/peptide cotransporter PEPT1: localization of PEPT1 and transport of beta-lactam antibiotics*. J Pharmacol Exp Ther, 1997. 281(3): p. 1415–1421.
88. Watanabe, K., Sawano, T., Endo, T., Sakata, M. and Sato, J., *Studies on intestinal absorption of sulpiride (2): transepithelial transport of sulpiride across the human intestinal cell line Caco-2*. Biol Pharm Bull, 2002. 25(10): p. 1345–1350.

89. Watanabe, K., Sawano, T., Jinriki, T. and Sato, J., *Studies on intestinal absorption of sulpiride (3): intestinal absorption of sulpiride in rats*. Biol Pharm Bull, 2004. 27(1): p. 77–81.

90. Perkins, E.J. and Abraham, T., *Pharmacokinetics, metabolism, and excretion of the intestinal peptide transporter 1 (SLC15A1)-targeted prodrug (1S,2S,5R,6S)-2-[(2'S)-(2-amino)propionyl]aminobicyclo[3.1.0.]hexen-2,6-di carboxylic acid (LY544344) in rats and dogs: assessment of first-pass bioactivation and dose linearity*. Drug Metab Dispos, 2007. 35(10): p. 1903–1909.

91. Miyamoto, K., et al., *Sequence, tissue distribution and developmental changes in rat intestinal oligopeptide transporter*. Biochim Biophys Acta, 1996. 1305(1–2): p. 34–38.

92. Zhang, E.Y., Emerick, R.M., Pak, Y.A., Wrighton, S.A. and Hillgren, K.M., *Comparison of human and monkey peptide transporters: PEPT1 and PEPT2*. Mol Pharm, 2004. 1(3): p. 201–210.

93. Englund, G., Rorsman, F., Ronnblom, A., Karlbom, U., Lazorova, L., Grasjo, J., Kindmark, A. and Artursson, P., *Regional levels of drug transporters along the human intestinal tract: co-expression of ABC and SLC transporters and comparison with Caco-2 cells*. Eur J Pharm Sci, 2006. 29(3–4): p. 269–277.

94. Lu, H. and Klaassen, C., *Tissue distribution and thyroid hormone regulation of Pept1 and Pept2 mRNA in rodents*. Peptides, 2006. 27(4): p. 850–857.

95. Chiou, W.L. and Barve, A., *Linear correlation of the fraction of oral dose absorbed of 64 drugs between humans and rats*. Pharm Res, 1998. 15(11): p. 1792–1795.

96. Chiou, W.L., Jeong, H.Y., Chung, S.M. and Wu, T.C., *Evaluation of using dog as an animal model to study the fraction of oral dose absorbed of 43 drugs in humans*. Pharm Res, 2000. 17(2): p. 135–140.

97. Chiou, W.L. and Buehler, P.W., *Comparison of oral absorption and bioavailablity of drugs between monkey and human*. Pharm Res, 2002. 19(6): p. 868–874.

98. Kullak-Ublick, G.A., Ismair, M.G., Stieger, B., Landmann, L., Huber, R., Pizzagalli, F., Fattinger, K., Meier, P.J. and Hagenbuch, B., *Organic anion-transporting polypeptide B (OATP-B) and its functional comparison with three other OATPs of human liver*. Gastroenterology, 2001. 120(2): p. 525–533.

99. Sai, Y., Kaneko, Y., Ito, S., Mitsuoka, K., Kato, Y., Tamai, I., Artursson, P. and Tsuji, A., *Predominant contribution of organic anion transporting polypeptide OATP-B (OATP2B1) to apical uptake of estrone-3-sulfate by human intestinal Caco-2 cells*. Drug Metab Dispos, 2006. 34(8): p. 1423–1431.

100. Tamai, I., Nezu, J., Uchino, H., Sai, Y., Oku, A., Shimane, M. and Tsuji, A., *Molecular identification and characterization of novel members of the human organic anion transporter (OATP) family*. Biochem Biophys Res Commun, 2000. 273(1): p. 251–260.

101. Kobayashi, D., Nozawa, T., Imai, K., Nezu, J., Tsuji, A. and Tamai, I., *Involvement of human organic anion transporting polypeptide OATP-B (SLC21A9) in pH-dependent transport across intestinal apical membrane*. J Pharmacol Exp Ther, 2003. 306(2): p. 703–708.

102. Nozawa, T., Imai, K., Nezu, J., Tsuji, A. and Tamai, I., *Functional characterization of pH-sensitive organic anion transporting polypeptide OATP-B in human*. J Pharmacol Exp Ther, 2004. 308(2): p. 438–445.

103. Varma, M.V., Rotter, C.J., Chupka, J., Whalen, K.M., Duignan, D.B., Feng, B., Litchfield, J., Goosen, T.C. and El-Kattan, A.F., *pH-sensitive interaction of HMG-CoA reductase inhibitors (statins) with organic anion transporting polypeptide 2B1*. Mol Pharm, 2011. 8(4): p. 1303–1313.

104. Treiber, A., Schneiter, R., Hausler, S. and Stieger, B., *Bosentan is a substrate of human OATP1B1 and OATP1B3: inhibition of hepatic uptake as the common mechanism of its interactions with cyclosporin A, rifampicin, and sildenafil*. Drug Metab Dispos, 2007. 35(8): p. 1400–1407.

105. Imanaga, J., et al., *The effects of the SLCO2B1 c.1457C > T polymorphism and apple juice on the pharmacokinetics of fexofenadine and midazolam in humans*. Pharmacogenet Genomics, 2011. 21(2): p. 84–93.

106. Chu, X., Bleasby, K. and Evers, R., *Species differences in drug transporters and implications for translating preclinical findings to humans*. Expert Opin Drug Metab Toxicol, 2013. 9(3): p. 237–252.

107. Hagenbuch, B. and Meier, P.J., *The superfamily of organic anion transporting polypeptides*. Biochim Biophys Acta, 2003. 1609(1): p. 1–18.

108. Shen, H., et al., *Cynomolgus monkey as a potential model to assess drug interactions involving hepatic organic anion transporting polypeptides:* in vitro, in vivo, *and* in vitro-*to*-in vivo *extrapolation*. J Pharmacol Exp Ther, 2013. 344(3): p. 673–685.

109. Yoshida, K., Maeda, K. and Sugiyama, Y., *Hepatic and intestinal drug transporters: prediction of pharmacokinetic effects caused by drug–drug interactions and genetic polymorphisms*. Annu Rev Pharmacol Toxicol, 2013. 53: p. 581–612.

110. Hosea, N.A., et al., *Prediction of human pharmacokinetics from preclinical information: comparative accuracy of quantitative prediction approaches*. J Clin Pharmacol, 2009. 49(5): p. 513–533.

111. Gonzalez, F.J., *The molecular biology of cytochrome P450s*. Pharmacol Rev, 1988. 40(4): p. 243–288.

112. Nelson, D.R., et al., *P450 superfamily: update on new sequences, gene mapping, accession numbers and nomenclature*. Pharmacogenetics, 1996. 6(1): p. 1–42.

113. Di, L., *The role of drug metabolizing enzymes in clearance*. Expert Opin Drug Metab Toxicol, 2014. 10(3): p. 379–393.

114. Varma, M.V., Steyn, S.J., Allerton, C. and El-Kattan, A.F., *Predicting clearance mechanism in drug discovery: extended clearance classification system (ECCS)*. Pharm Res, 2015. 32(12): p. 3785–3802.

115. Peters, W. and Kremers, P., *Cytochromes P-450 in the intestinal mucosa of man*. Biochem Pharmacol, 1989. 38(9): p. 1535–1538.

116. Shimada, T., Yamazaki, H., Mimura, M., Inui, Y. and Guengerich, F.P., *Interindividual variations in human liver cytochrome P-450 enzymes involved in the oxidation of drugs, carcinogens and toxic chemicals: studies with liver microsomes of 30 Japanese and 30 Caucasians*. J Pharmacol Exp Ther, 1994. 270(1): p. 414–423.

117. Spatzenegger, M. and Jaeger, W., *Clinical importance of hepatic cytochrome P450 in drug metabolism*. Drug Metab Rev, 1995. 27(3): p. 397–417.

118. McGinnity, D.F., Parker, A.J., Soars, M. and Riley, R.J., *Automated definition of the enzymology of drug oxidation by the major human drug metabolizing cytochrome P450s*. Drug Metab Dispos, 2000. 28(11): p. 1327–1334.

119. Dai, D., Zeldin, D.C., Blaisdell, J.A., Chanas, B., Coulter, S.J., Ghanayem, B.I. and Goldstein, J.A., *Polymorphisms in human CYP2C8 decrease metabolism of the anticancer drug paclitaxel and arachidonic acid*. Pharmacogenetics, 2001. 11(7): p. 597–607.

120. Desai, P.B., Duan, J.Z., Zhu, Y.W. and Kouzi, S., *Human liver microsomal metabolism of paclitaxel and drug interactions*. Eur J Drug Metab Pharmacokinet, 1998. 23(3): p. 417–424.

121. O'Donnell, C.J., Grime, K., Courtney, P., Slee, D. and Riley, R.J., *The development of a cocktail CYP2B6, CYP2C8, and CYP3A5 inhibition assay and a preliminary assessment of utility in a drug discovery setting*. Drug Metab Dispos, 2007. 35(3): p. 381–385.

122. Casabar, R.C.T., Wallace, A.D., Hodgson, E. and Rose, R.L., *Metabolism of endosulfan-alpha by human liver microsomes and its utility as a simultaneous* in vitro *probe for CYP2B6 and CYP3A4*. Drug Metab Dispos, 2006. 34(10): p. 1779–1785.

123. Turpeinen, M., Raunio, H. and Pelkonen, O., *The functional role of CYP2B6 in human drug metabolism: substrates and inhibitors* in vitro, in vivo *and* in silico. Curr Drug Metab, 2006. 7(7): p. 705–714.

124. Mirghani, R.A., et al., *CYP3A5 genotype has significant effect on quinine 3-hydroxylation in Tanzanians, who have lower total CYP3A activity than a Swedish population*. Pharmacogenet Genomics, 2006. 16(9): p. 637–645.

125. Smith, D.A., Abel, S.M., Hyland, R. and Jones, B.C., *Human cytochrome P450s: selectivity and measurement* in vivo. Xenobiotica, 1998. 28(12): p. 1095–1128.

126. Stevens, J.C., Shipley, L.A., Cashman, J.R., Vandenbranden, M. and Wrighton, S.A., *Comparison of human and rhesus monkey* in vitro *phase I and phase II hepatic drug metabolism activities*. Drug Metab Dispos, 1993. 21(5): p. 753–760.

127. Williams, J.A., Hurst, S.I., Bauman, J., Jones, B.C., Hyland, R., Gibbs, J.P., Obach, R.S. and Ball, S.E., *Reaction phenotyping in drug discovery: moving forward with confidence?* Curr Drug Metab, 2003. 4(6): p. 527–534.
128. Achour, B., Russell, M.R., Barber, J. and Rostami-Hodjegan, A., *Simultaneous quantification of the abundance of several cytochrome P450 and uridine 5'-diphospho-glucuronosyltransferase enzymes in human liver microsomes using multiplexed targeted proteomics.* Drug Metab Dispos, 2014. 42(4): p. 500–510.
129. Wrighton, S.A., Brian, W.R., Sari, M.A., Iwasaki, M., Guengerich, F.P., Raucy, J.L., Molowa, D.T. and Vandenbranden, M., *Studies on the expression and metabolic capabilities of human liver cytochrome P450IIIA5 (HLp3).* Mol Pharmacol, 1990. 38(2): p. 207–213.
130. Kuehl, P., et al., *Sequence diversity in CYP3A promoters and characterization of the genetic basis of polymorphic CYP3A5 expression.* Nat Genet, 2001. 27(4): p. 383–391.
131. Paine, M.F., Khalighi, M., Fisher, J.M., Shen, D.D., Kunze, K.L., Marsh, C.L., Perkins, J.D. and Thummel, K.E., *Characterization of interintestinal and intraintestinal variations in human CYP3A-dependent metabolism.* J Pharmacol Exp Ther, 1997. 283(3): p. 1552–1562.
132. Smit, P., et al., *A common polymorphism in the CYP3A7 gene is associated with a nearly 50% reduction in serum dehydroepiandrosterone sulfate levels.* J Clin Endocrinol Metab, 2005. 90(9): p. 5313–5316.
133. Gellner, K., et al., *Genomic organization of the human CYP3A locus: identification of a new, inducible CYP3A gene.* Pharmacogenetics, 2001. 11(2): p. 111–121.
134. Chen, M., Ma, L., Drusano, G.L., Bertino, Jr., J.S. and Nafziger, A.N., *Sex differences in CYP3A activity using intravenous and oral midazolam.* Clin Pharmacol Ther, 2006. 80(5): p. 531–538.
135. Robertson, G.R., Farrell, G.C. and Liddle, C., *Sexually dimorphic expression of rat CYP3A9 and CYP3A18 genes is regulated by growth hormone.* Biochem Biophys Res Commun, 1998. 242(1): p. 57–60.
136. Yamazoe, Y., Murayama, N., Shimada, M., Yamauchi, K., Nagata, K., Imaoka, S., Funae, Y. and Kato, R., *A sex-specific form of cytochrome P-450 catalyzing propoxycoumarin O-depropylation and its identity with testosterone 6 beta-hydroxylase in untreated rat livers: reconstitution of the activity with microsomal lipids.* J Biochem, 1988. 104(5): p. 785–790.
137. Matsubara, T., Kim, H.J., Miyata, M., Shimada, M., Nagata, K. and Yamazoe, Y., *Isolation and characterization of a new major intestinal CYP3A form, CYP3A62, in the rat.* J Pharmacol Exp Ther, 2004. 309(3): p. 1282–1290.
138. Smith, D.A., *Species differences in metabolism and pharmacokinetics: are we close to an understanding?* Drug Metab Rev, 1991. 23(3–4): p. 355–373.
139. Warrington, J.S., Von Moltke, L.L., Shader, R.I. and Greenblatt, D.J., In vitro *biotransformation of sildenafil (Viagra) in the male rat: the role of CYP2C11.* Drug Metab Dispos, 2002. 30(6): p. 655–657.
140. Ghosal, A., Satoh, H., Thomas, P.E., Bush, E. and Moore, D., *Inhibition and kinetics of cytochrome P4503A activity in microsomes from rat, human, and cDNA-expressed human cytochrome P450.* Drug Metab Dispos, 1996. 24(9): p. 940–947.
141. Lu, C. and Li, A.P., *Species comparison in P450 induction: effects of dexamethasone, omeprazole, and rifampin on P450 isoforms 1A and 3A in primary cultured hepatocytes from man, Sprague-Dawley rat, minipig, and beagle dog.* Chem Biol Interact, 2001. 134(3): p. 271–281.
142. Fraser, D.J., Feyereisen, R., Harlow, G.R. and Halpert, J.R., *Isolation, heterologous expression and functional characterization of a novel cytochrome P450 3A enzyme from a canine liver cDNA library.* J Pharmacol Exp Ther, 1997. 283(3): p. 1425–1432.
143. Prueksaritanont, T., Gorham, L.M., Hochman, J.H., Tran, L.O. and Vyas, K.P., *Comparative studies of drug-metabolizing enzymes in dog, monkey, and human small intestines, and in Caco-2 cells.* Drug Metab Dispos, 1996. 24(6): p. 634–642.

144. Bogaards, J.J., Bertrand, M., Jackson, P., Oudshoorn, M.J., Weaver, R.J., van Bladeren, P.J. and Walther, B., *Determining the best animal model for human cytochrome P450 activities: a comparison of mouse, rat, rabbit, dog, micropig, monkey and man*. Xenobiotica, 2000. 30(12): p. 1131–1152.

145. Chauret, N., Gauthier, A., Martin, J. and Nicoll-Griffith, D.A., In vitro *comparison of cytochrome P450-mediated metabolic activities in human, dog, cat, and horse*. Drug Metab Dispos, 1997. 25(10): p. 1130–1136.

146. Goldstein, J.A., *Clinical relevance of genetic polymorphisms in the human CYP2C subfamily*. Br J Clin Pharmacol, 2001. 52(4): p. 349–355.

147. Lapple, F., von Richter, O., Fromm, M.F., Richter, T., Thon, K.P., Wisser, H., Griese, E.U., Eichelbaum, M. and Kivisto, K.T., *Differential expression and function of CYP2C isoforms in human intestine and liver*. Pharmacogenetics, 2003. 13(9): p. 565–575.

148. Blaisdell, J., Goldstein, J.A. and Bai, S.A., *Isolation of a new canine cytochrome P450 CDNA from the cytochrome P450 2C subfamily (CYP2C41) and evidence for polymorphic differences in its expression*. Drug Metab Dispos, 1998. 26(3): p. 278–283.

149. Ohhira, S., Enomoto, M. and Matsui, H., In vitro *metabolism of tributyltin and triphenyltin by human cytochrome P-450 isoforms*. Toxicology, 2006. 228(2–3): p. 171–177.

150. Mugford, C.A. and Kedderis, G.L., *Sex-dependent metabolism of xenobiotics*. Drug Metab Rev, 1998. 30(3): p. 441–498.

151. Mode, A., Ahlgren, R., Lahuna, O. and Gustafsson, J.A., *Gender differences in rat hepatic CYP2C gene expression – regulation by growth hormone*. Growth Horm IGF Res, 1998. 8(Suppl B): p. 61–67.

152. Rich, K.J. and Boobis, A.R., *Expression and inducibility of P450 enzymes during liver ontogeny*. Microsc Res Tech, 1997. 39(5): p. 424–435.

153. Graham, M.J., Bell, A.R., Crewe, H.K., Moorcraft, C.L., Walker, L., Whittaker, E.F. and Lennard, M.S., *mRNA and protein expression of dog liver cytochromes P450 in relation to the metabolism of human CYP2C substrates*. Xenobiotica, 2003. 33(3): p. 225–237.

154. Paine, M.F., Hart, H.L., Ludington, S.S., Haining, R.L., Rettie, A.E. and Zeldin, D.C., *The human intestinal cytochrome P450 "pie"*. Drug Metab Dispos, 2006. 34(5): p. 880–886.

155. Zanger, U.M., Raimundo, S. and Eichelbaum, M., *Cytochrome P450 2D6: overview and update on pharmacology, genetics, biochemistry*. Naunyn Schmiedebergs Arch Pharmacol, 2004. 369(1): p. 23–37.

156. Williams, J.A., Hyland, R., Jones, B.C., Smith, D.A., Hurst, S., Goosen, T.C., Peterkin, V., Koup, J.R. and Ball, S.E., *Drug–drug interactions for UDP-glucuronosyltransferase substrates: a pharmacokinetic explanation for typically observed low exposure (AUCi/AUC) ratios*. Drug Metab Dispos, 2004. 32(11): p. 1201–1208.

157. Venhorst, J., ter Laak, A.M., Commandeur, J.N., Funae, Y., Hiroi, T. and Vermeulen, N.P., *Homology modeling of rat and human cytochrome P450 2D (CYP2D) isoforms and computational rationalization of experimental ligand-binding specificities*. J Med Chem, 2003. 46(1): p. 74–86.

158. Hiroi, T., Chow, T., Imaoka, S. and Funae, Y., *Catalytic specificity of CYP2D isoforms in rat and human*. Drug Metab Dispos, 2002. 30(9): p. 970–976.

159. Muralidharan, G., Hawes, E.M., McKay, G., Korchinski, E.D. and Midha, K.K., *Quinidine but not quinine inhibits in man the oxidative metabolic routes of methoxyphenamine which involve debrisoquine 4-hydroxylase*. Eur J Clin Pharmacol, 1991. 41(5): p. 471–474.

160. Muralidharan, G., Hawes, E.M., McKay, G. and Midha, K.K., *Quinine is a more potent inhibitor than quinidine in rat of the oxidative metabolic routes of methoxyphenamine which involve debrisoquine 4-hydroxylase*. Xenobiotica, 1991. 21(11): p. 1441–1450.

161. Paulson, S.K., Engel, L., Reitz, B., Bolten, S., Burton, E.G., Maziasz, T.J., Yan, B. and Schoenhard, G.L., *Evidence for polymorphism in the canine metabolism of the cyclooxygenase 2 inhibitor, celecoxib*. Drug Metab Dispos, 1999. 27(10): p. 1133–1142.

162. Shimada, T., Martin, M.V., Pruess-Schwartz, D., Marnett, L.J. and Guengerich, F.P., *Roles of individual human cytochrome P-450 enzymes in the bioactivation of benzo(a)pyrene, 7,8-dihydroxy-7,8-dihydrobenzo(a)pyrene, and other dihydrodiol derivatives of polycyclic aromatic hydrocarbons*. Cancer Res, 1989. 49(22): p. 6304–6312.

163. Iba, M.M., Storch, A., Ghosal, A., Bennett, S., Reuhl, K.R. and Lowndes, H.E., *Constitutive and inducible levels of CYP1A1 and CYP1A2 in rat cerebral cortex and cerebellum*. Arch Toxicol, 2003. 77(10): p. 547–554.

164. Graham, R.A., Downey, A., Mudra, D., Krueger, L., Carroll, K., Chengelis, C., Madan, A. and Parkinson, A., In vivo *and* in vitro *induction of cytochrome P450 enzymes in beagle dogs*. Drug Metab Dispos, 2002. 30(11): p. 1206–1213.

165. Uchida, T., Komori, M., Kitada, M. and Kamataki, T., *Isolation of cDNAs coding for three different forms of liver microsomal cytochrome P-450 from polychlorinated biphenyl-treated beagle dogs*. Mol Pharmacol, 1990. 38(5): p. 644–651.

166. Guengerich, F.P., *Comparisons of catalytic selectivity of cytochrome P450 subfamily enzymes from different species*. Chem Biol Interact, 1997. 106(3): p. 161–182.

167. Lindell, M., Lang, M. and Lennernas, H., *Expression of genes encoding for drug metabolising cytochrome P450 enzymes and P-glycoprotein in the rat small intestine; comparison to the liver*. Eur J Drug Metab Pharmacokinet, 2003. 28(1): p. 41–48.

168. Shimada, T., Yamazaki, H., Mimura, M., Wakamiya, N., Ueng, Y.F., Guengerich, F.P. and Inui, Y., *Characterization of microsomal cytochrome P450 enzymes involved in the oxidation of xenobiotic chemicals in human fetal liver and adult lungs*. Drug Metab Dispos, 1996. 24(5): p. 515–522.

169. Hakkola, J., Pasanen, M., Hukkanen, J., Pelkonen, O., Maenpaa, J., Edwards, R.J., Boobis, A.R. and Raunio, H., *Expression of xenobiotic-metabolizing cytochrome P450 forms in human full-term placenta*. Biochem Pharmacol, 1996. 51(4): p. 403–411.

170. Hakkola, J., Raunio, H., Purkunen, R., Pelkonen, O., Saarikoski, S., Cresteil, T. and Pasanen, M., *Detection of cytochrome P450 gene expression in human placenta in first trimester of pregnancy*. Biochem Pharmacol, 1996. 52(2): p. 379–383.

171. Benet, L.Z., Izumi, T., Zhang, Y., Silverman, J.A. and Wacher, V.J., *Intestinal MDR transport proteins and P-450 enzymes as barriers to oral drug delivery*. J Control Release, 1999. 62(1–2): p. 25–31.

172. Fisher, J.M., Wrighton, S.A., Watkins, P.B., Schmiedlin-Ren, P., Calamia, J.C., Shen, D.D., Kunze, K.L. and Thummel, K.E., *First-pass midazolam metabolism catalyzed by 1alpha,25-dihydroxy vitamin D3-modified Caco-2 cell monolayers*. J Pharmacol Exp Ther, 1999. 289(2): p. 1134–1142.

173. Sesardic, D., Boobis, A.R., Murray, B.P., Murray, S., Segura, J., de la Torre, R. and Davies, D.S., *Furafylline is a potent and selective inhibitor of cytochrome P450IA2 in man*. Br J Clin Pharmacol, 1990. 29(6): p. 651–663.

174. Sesardic, D., Edwards, R.J., Davies, D.S., Thomas, P.E., Levin, W. and Boobis, A.R., *High affinity phenacetin O-deethylase is catalysed specifically by cytochrome P450d (P450IA2) in the liver of the rat*. Biochem Pharmacol, 1990. 39(3): p. 489–498.

175. Wu, B., Kulkarni, K., Basu, S., Zhang, S. and Hu, M., *First-pass metabolism via UDP-glucuronosyltransferase: a barrier to oral bioavailability of phenolics*. J Pharm Sci, 2011. 100(9): p. 3655–3681.

176. King, C.D., Rios, G.R., Green, M.D. and Tephly, T.R., *UDP-glucuronosyltransferases*. Curr Drug Metab, 2000. 1(2): p. 143–161.

177. Shelby, M.K., Cherrington, N.J., Vansell, N.R. and Klaassen, C.D., *Tissue mRNA expression of the rat UDP-glucuronosyltransferase gene family*. Drug Metab Dispos, 2003. 31(3): p. 326–333.

178. Heikkinen, A.T., et al., *Quantitative ADME proteomics – CYP and UGT enzymes in the beagle dog liver and intestine*. Pharm Res, 2015. 32(1): p. 74–90.

179. Nishimura, M., Koeda, A., Morikawa, H., Satoh, T., Narimatsu, S. and Naito, S., *Tissue-specific mRNA expression profiles of drug-metabolizing enzymes and transporters in the cynomolgus monkey*. Drug Metab Pharmacokinet, 2009. 24(2): p. 139–144.

180. Furukawa, T., Naritomi, Y., Tetsuka, K., Nakamori, F., Moriguchi, H., Yamano, K., Terashita, S., Tabata, K. and Teramura, T., *Species differences in intestinal glucuronidation activities between humans, rats, dogs and monkeys*. Xenobiotica, 2014. 44(3): p. 205–216.

181. Kitamura, S., Sugihara, K., Nakatani, K., Ohta, S., Ohhara, T., Ninomiya, S., Green, C.E. and Tyson, C.A., *Variation of hepatic methotrexate 7-hydroxylase activity in animals and humans*. IUBMB Life, 1999. 48(6): p. 607–611.

182. Zientek, M., Jiang, Y., Youdim, K. and Obach, R.S., In vitro–in vivo *correlation for intrinsic clearance for drugs metabolized by human aldehyde oxidase*. Drug Metab Dispos, 2010. 38(8): p. 1322–1327.

183. Rashidi, M.R., Smith, J.A., Clarke, S.E. and Beedham, C., In vitro *oxidation of famciclovir and 6-deoxypenciclovir by aldehyde oxidase from human, guinea pig, rabbit, and rat liver*. Drug Metab Dispos, 1997. 25(7): p. 805–813.

184. Jordan, C.G., Rashidi, M.R., Laljee, H., Clarke, S.E., Brown, J.E. and Beedham, C., *Aldehyde oxidase-catalysed oxidation of methotrexate in the liver of guinea-pig, rabbit and man*. J Pharm Pharmacol, 1999. 51(4): p. 411–418.

185. Clarke, S.E., Harrell, A.W. and Chenery, R.J., *Role of aldehyde oxidase in the* in vitro *conversion of famciclovir to penciclovir in human liver*. Drug Metab Dispos, 1995. 23(2): p. 251–254.

186. Beedham, C., Miceli, J.J. and Obach, R.S., *Ziprasidone metabolism, aldehyde oxidase, and clinical implications*. J Clin Psychopharmacol, 2003. 23(3): p. 229–232.

187. Moriwaki, Y., Yamamoto, T., Takahashi, S., Tsutsumi, Z. and Hada, T., *Widespread cellular distribution of aldehyde oxidase in human tissues found by immunohistochemistry staining*. Histol Histopathol, 2001. 16(3): p. 745–753.

188. Garattini, E. and Terao, M., *The role of aldehyde oxidase in drug metabolism*. Expert Opin Drug Metab Toxicol, 2012. 8(4): p. 487–503.

189. Garattini, E. and Terao, M., *Aldehyde oxidase and its importance in novel drug discovery: present and future challenges*. Expert Opin Drug Discov, 2013. 8(6): p. 641–654.

190. Beedham, C., Peet, C.F., Panoutsopoulos, G.I., Carter, H. and Smith, J.A., *Role of aldehyde oxidase in biogenic amine metabolism*. Prog Brain Res, 1995. 106: p. 345–353.

191. Akabane, T., Tanaka, K., Irie, M., Terashita, S. and Teramura, T., *Case report of extensive metabolism by aldehyde oxidase in humans: pharmacokinetics and metabolite profile of FK3453 in rats, dogs, and humans*. Xenobiotica, 2011. 41(5): p. 372–384.

192. Zwisler, S.T., Enggaard, T.P., Noehr-Jensen, L., Mikkelsen, S., Verstuyft, C., Becquemont, L., Sindrup, S.H. and Brosen, K., *The antinociceptive effect and adverse drug reactions of oxycodone in human experimental pain in relation to genetic variations in the OPRM1 and ABCB1 genes*. Fundam Clin Pharmacol, 2010. 24(4): p. 517–524.

193. Diamond, S., Boer, J., Maduskuie, Jr., T.P., Falahatpisheh, N., Li, Y. and Yeleswaram, S., *Species-specific metabolism of SGX523 by aldehyde oxidase and the toxicological implications*. Drug Metab Dispos, 2010. 38(8): p. 1277–1285.

194. Zhang, X., Liu, H.H., Weller, P., Zheng, M., Tao, W., Wang, J., Liao, G., Monshouwer, M. and Peltz, G., In silico *and* in vitro *pharmacogenetics: aldehyde oxidase rapidly metabolizes a p38 kinase inhibitor*. Pharmacogenomics J, 2011. 11(1): p. 15–24.

195. Sauer, J.M., Ring, B.J. and Witcher, J.W., *Clinical pharmacokinetics of atomoxetine*. Clin Pharmacokinet, 2005. 44(6): p. 571–590.

196. Mattiuz, E.L., et al., *Disposition and metabolic fate of atomoxetine hydrochloride: pharmacokinetics, metabolism, and excretion in the Fischer 344 rat and beagle dog*. Drug Metab Dispos, 2003. 31(1): p. 88–97.

197. Lin, J.H., Chiba, M., Balani, S.K., Chen, I.W., Kwei, G.Y., Vastag, K.J. and Nishime, J.A., *Species differences in the pharmacokinetics and metabolism of indinavir, a potent human immunodeficiency virus protease inhibitor*. Drug Metab Dispos, 1996. 24(10): p. 1111–1120.

198. Chiba, M., Hensleigh, M., Nishime, J.A., Balani, S.K. and Lin, J.H., *Role of cytochrome P450 3A4 in human metabolism of MK-639, a potent human immunodeficiency virus protease inhibitor*. Drug Metab Dispos, 1996. 24(3): p. 307–314.

199. Yeh, K.C., Stone, J.A., Carides, A.D., Rolan, P., Woolf, E. and Ju, W.D., *Simultaneous investigation of indinavir nonlinear pharmacokinetics and bioavailability in healthy volunteers using stable isotope labeling technique: study design and model-independent data analysis*. J Pharm Sci, 1999. 88(5): p. 568–573.

3

DRUG ROUTES OF EXCRETION

3.1 INTRODUCTION

Understanding route of elimination and determining clearance drivers for new molecular entities (NMEs) is one of the key deliverables of drug metabolism and pharmacokinetics (DMPK) scientists. Other than metabolism, renal elimination and active hepatic uptake followed by biliary elimination are key mechanisms for drug excretion [1]. Our group proposed a tool, the extended clearance classification system (ECCS), to be utilized early in drug discovery to enable prospective prediction of clearance rate-determining step based on *in vitro* and calculated inputs available in early drug discovery and include permeability, ionization, and molecular weight. This classification system classifies molecules into six categories:

1. ECCS Class 1a: This class of molecules are either acid/zwitterion with high permeability and low molecular weight. Their clearance rate-determining step is metabolism.
2. ECCS Class 1b: This class of molecules are either acid/zwitterion with high permeability and molecular weight (>400 Da). Active hepatic uptake mediated by organic anion transporting polypeptides is the clearance rate-determining step that once the molecule is in the liver is metabolized.
3. ECCS Class 2: These are base/neutral molecules that have high permeability and metabolism is their clearance rate-determining step.
4. ECCS Class 3a: Similar to Class 1a, they are acid/zwitterion with low molecular weight (<400 Da) and permeability. Renal elimination is their clearance rate-determining step.

Oral Bioavailability Assessment: Basics and Strategies for Drug Discovery and Development, First Edition.
Ayman F. El-Kattan.

5. ECCS Class 3b: These are acid/zwitterion with high molecular weight and low permeability. Their clearance rate-determining step is either renal elimination and/or active hepatic uptake that is followed by biliary elimination.
6. ECCS Class 4: These are base/neutral molecules that have low permeability and renal elimination is their rate-determining step.

In this chapter, we focus on the drivers of renal elimination, active hepatic uptake, and biliary excretion. This includes review of the transporters that drive these elimination mechanism and the physicochemical properties that characterize drug molecules that are eliminated via these routes, that is, ECCS Class 1b, 3a, 3b, and 4.

3.2 RENAL ELIMINATION

3.2.1 Mathematical Principles Governing Renal Elimination

Renal clearance of a drug molecule is an outcome of the glomerular filtration, tubular secretion, and reabsorption processes. It is mathematically described as

$$CL_{renal} = (f_b.GFR + CL_{sec}) \cdot (1 - f_{reabs}) \quad (3.1)$$

where f_b is the unbound drug fraction in blood, GFR is glomerular filtration rate, CL_{sec} is renal secretory clearance, and f_{reabs} is the fraction of filtered and secreted drug that is reabsorbed in renal proximal tubules. Drug filtration through the glomerulus and nephron is governed by the glomerular filtration rate and the drug f_b. Overall, the contribution of GFR to the total renal clearance is small, considering its low value reported in human to be 1.78 mL/min/kg. It is interesting to note that compounds with high permeability, that is, ECCS Class 1 and Class 2 have low renal clearance and recovery (Chapter 8). This is primarily driven by a proportional relationship between f_{reabs} and permeability, where compounds with high permeability would have high f_{reabs} resulting in low renal clearance and recovery and vice versa (Equation 3.1) [1].

Active tubular secretion is a major driver of low permeability compound excretion in the urine mediated by their uptake from blood to proximal tubular lumen. This process is determined by uptake transporters that are lining the basolateral side of the proximal tubular cells, blood flow, and drug f_b as shown in Equation 3.2 [1–6]. As shown in Equation 3.2, CL_{sec} is calculated as

$$CL_{sec} = Q_r \cdot \frac{f_b \cdot CL_{int,sec}}{Q_r + f_b \cdot CL_{int,sec}} \quad (3.2)$$

This assumes that renal secretion is transporter mediated that obeys well-stirred model principles, where Q_r is the renal blood flow and $CL_{int,sec}$ is the intrinsic secretory clearance, which is calculated using Equation 3.3.

$$CL_{int,sec} = \frac{PS_{influx,b} \cdot PS_{efflux,a}}{(PS_{efflux,b} + PS_{efflux,a})} \quad (3.3)$$

$PS_{influx,b}$, $PS_{efflux,b}$, $PS_{influx,a}$, and $PS_{efflux,a}$ are influx and efflux intrinsic transport clearances across the basolateral and apical membranes of proximal tubule cells, respectively. Renal secretion is an active process that is mediated by uptake transporters of the organic ion transporter family (*SLC22*) that line the basolateral membrane of proximal tubule cells, for example, organic cation transporter 2 (OCT2) and organic anion transporters 1, 2, and 3 (OAT1, OAT2, and OAT3) [7–9]. Typically, these transporter substrates have low permeability, that is, ECCS Class 3 and Class 4 [10]. They also have a high counter concentration-gradient created by water reabsorption. This leads to a minimum contribution of passive permeability to overall tubular secretion. It should be emphasized that the rate-determining step of active secretion across the proximal tubule is predominantly driven by active basolateral uptake from blood compartment [11]. However, the efflux transporters at the urine side would mainly determine the kidney tissue concentrations.

3.2.2 Uptake Renal Transporters

3.2.2.1 OAT1 and OAT3 Organic anion transporter 1 (OAT1, SLC22A6) and organic anion transporter 3 (OAT3, SLC22A8) are major uptake renal transporters that line the basolateral membrane of the human proximal tubules. Typically, compounds that are substrates for OAT1 and OAT3 are hydrophilic small organic acids [1, 6]. Examples of compounds that are substrates for OAT1 transporters are shown in Table 3.1. Their K_m values ranges from 0.06830 to 895.5 μM. OAT1 substrates belong to angiotensin converting enzyme (ACE) inhibitors, angiotensin II inhibitors (ARBs), antibiotics, anticoagulants, antimetabolites, antiviral agents, diagnostics, dipeptidyl peptidase-4 inhibitors, diuretics, fibric acid derivatives, food products, H2-receptor antagonists, neuroprotective, hypnotics, NSAIDs, nucleoside reverse transcriptase inhibitors (NRTIs), and stroke treatments therapeutic classes. Table 3.2 provides a partial list of known OAT3 substrates with K_m range of 0.12–1882 μM. Similar to OAT 1, OAT2 substrates are part of ACE inhibitors, antibiotics, anticoagulants, antidotes, antimetabolites, antivirals, beta-3 adrenoreceptor agonists, calcium channel blockers, anti-cancer agents, corticosteroids, diagnostics agents, vitamins, dipeptidyl peptidase-4 inhibitors, diuretics, estrogens, fibric acid derivatives, food products, gastrointestinal agents, H1 receptor antagonists, herbal medications, statins, immunosuppressants, neuroprotectors, NSAIDs, NRTIs, SGLT2 inhibitors, stroke treatments, and topoisomerase inhibitors therapeutic classes. It is clear that there is an overlap in the substrate specificity between the two transporters. This is not unexpected considering the similarity in the amino acid sequences between the two transmembrane proteins.

3.2.2.2 OCT2 Encoded by SLC22A2 gene, OCT2 transporter is a major uptake transporter that is kidney specific and located at the basolateral membrane of human proximal tubule cells. It plays a key role in the renal secretion of organic bases including drug molecules that belong to therapeutic classes such as alkylating agents, α–β adrenergic antagonists, anesthetics, antiarrhythmics, antibiotics, anticholinergics, anticonvulsants, antidotes, antiparasitics, antipsychotics, biguanides, bronchodilators, calcium channel blockers, cancer treatments, cardiovascular drugs, cholinergic agonists, corticosteroids, dietary supplements and vitamins, dipeptidyl peptidase-4 inhibitors, diuretics, dopamine–norepinephrine reuptake inhibitors, dopaminergic agonists, H2 receptor antagonists, herbal medications, inotropic agents, muscarinic antagonists, neurotransmitters, NMDA receptor antagonists, nucleoside reverse transcriptase inhibitors, antiparkinsonians, platinums, and triptans. The K_m range of compounds investigated is 0.61–4000 μM (Table 3.3).

3.2.3 Renal Efflux Transporters

3.2.3.1 MATE1 and MATE2K Proton/organic cation antiporters multidrug and toxin extrusion, MATE1/SLC47A1 and MATE2-K/SLC47A2, are located in the apical membrane of the renal epithelium. It should be emphasized that MATE1 is also expressed at the canalicular membrane of liver hepatocytes. MATE1 and MATE2-K function as efflux transporter,

TABLE 3.1 Partial List of Known OAT1 Transporter Substrates

Object	Object Therapeutic Class	K_m (μM)	V_{max} (pmol/min/mg)	V_{max}/K_m (μL/min/mg)	References
2,4-Dichlorophenoxyacetate	None	0.727	0.469	0.645	[12]
6-Carboxyfluorescein	None	12.7			[13]
Captopril	Angiotensin converting enzyme (ACE) inhibitors				[14]
Quinaprilat	Angiotensin converting enzyme (ACE) inhibitors				[14]
Cefdinir	Antibiotics				[14]
Ceftibuten	Antibiotics				[14]
Ceftizoxime	Antibiotics				[14]
Cephaloridine	Antibiotics				[14]
Edaravone sulfate	Neuroprotectors	10.8	323	29.9	[15]
Tetracycline	Antibiotics				[16]
Ganciclovir	Antivirals	895.5	1165.5	1.3	[16]
Oseltamivir carboxylate (Ro 64-0802)	Antivirals				[16]
Acetylsalicylic acid	NSAIDS				[16]
Zidovudine	Nucleoside reverse transcriptase inhibitors (NRTIs)	45.9	254.6	5.53	[16]
Acyclovir	Antivirals	839	434	0.5	[17]
Adefovir	Antivirals	23.8	46 pmol/min/Mil cells	1.93 μL/min/Mill cells	[17]
Ibuprofen	NSAIDS				[17]
Indomethacin	NSAIDS				[17]
Ketoprofen	NSAIDS				[17]
Salicylic acid	NSAIDS				[17]
Aminopterin	Antimetabolites	226	36.25 pmol/h/oocyte	0.16 μL/h/oocyte	[18]
Zonampanel	Stroke treatments	1.4	41.8	29.9	[18]
Amoxicillin	Antibiotics				[16]
Methotrexate	Antimetabolites	724	217.1 pmol/h/oocyte	0.3 μL/h/oocyte	[16]

Anagliptin	Dipeptidyl peptidase-4 inhibitors				[19]
Ranitidine	H-2 receptor antagonists				[19]
Avibactam	Antibiotics				[14]
Leucovorin (folinic acid)	Antidotes				[14]
Cimetidine	H-2 receptor antagonists				[14]
Cidofovir	Antivirals	63.36	43.5 pmol/h/oocyte	0.011 µL/min/oocyte	[20]
Bezafibrate	Fibric acid derivatives	290	768	2.65	[20]
Zalcitabine	Nucleoside reverse transcriptase inhibitors (NRTIs)				[20]
Olmesartan	Angiotensin II inhibitors (angiotensin receptor blockers or ARBs)	0.0683	0.195	2.86	[21]
Bumetanide	Diuretics				[21]
Caffeic acid	Food products	28.6	183	6.4	[22]
Caffeic acid-3-*O*-glucuronide	Food products				[22]
Caffeic acid-4-*O*-sulfate	Food products				[22]
Dihydrocaffeic acid	Food products	21.4	218	10.2	[22]
Dihydrocaffeic acid-3-*O*-glucuronide	Food products				[22]
Dihydrocaffeic acid-3-*O*-sulfate	Food products	76.7	2070 nmol/min/mg	27	[22]
Dihydrocaffeic acid-4-*O*-glucuronide	Food products				[22]
Dihydrocaffeic acid-4-*O*-sulfate	Food products				[22]
Dihydroferulic acid	Food products				[22]
Ferulic acid	Food products				[22]
Ferulic acid-4-*O*-sulfate	Food products	54.2	454	8.38	[22]
Isoferulic acid-3-*O*-glucuronide	Food products				[22]

(*continued*)

TABLE 3.1 (*Continued*)

Object	Object Therapeutic Class	K_m (μM)	V_{max} (pmol/min/mg)	V_{max}/K_m (μL/min/mg)	References
Isoferulic acid-3-*O*-sulfate	Food products	32.7	1290	39.4	[22]
Gaboxadol	Hypnotics–sedatives	115	2 pmol/min/Mill cells	0.19 mL/min/Mill cells	[22]
p-Aminohippuric acid (PAH)	None	3.6	7.1 pmol/h/oocyte	1.97 μL/h/oocyte	[18]
Raltegravir	HIV-integrase strand transfer inhibitors	150	36 pmol/h/oocyte	0.2 μL/h/oocyte	[23]
Tenofovir	Nucleoside reverse transcriptase inhibitors (NRTIs)	25	129 pmol/h/oocyte	5.2 μL/h/oocyte	[22]
Genistein-4-*O*-sulfate	Food products	5.07	28.3	5.58	[22]
Quercetin-3-glucuronide	Food products				[22]
Quercetin-3-*O*-sulfate	Food products	1.73	105	60.7	[22]
Quercetin-7-glucuronide	Food products				[22]
Dimesna	Agents [24]	636			[25]
Ubenimex (Bestatin)	Other antineoplastics	679	1614	2.4	[25]
Glycyrrhetinic acid glucuronide	Dietary supplements and vitamins	49			[26]
Chlorothiazide	Diuretics	14.5	108	7.4	[27]
Tc-99m-mercaptoacetyl-triglycine (MAG3)	Diagnostic agents				[28]
Penicillin G (benzylpenicillin)	Antibiotics	884			[29]
Pemetrexed	Antimetabolites				[30]
P-cresyl sulfate	None	127.8	9800	77	[31]
Clopidogrel carboxylate	Anticoagulants and antiplatelets	5.61	41.7	7.43	[32]
P-Aminobenzoic acid	None	12.7	480	37.8	[33]
Fisetin	Food products				[34]
Cefixime	Antibiotics	181	1646	9.09	[35]
Tanshinol	Herbal medications	121	1414	35.4	[36]
Furosemide	Diuretics	38.9			[37]

TABLE 3.2 Partial List of Known OAT3 Transporter Substrates

Object	Object Therapeutic Class	K_m (μM)	V_{max} (pmol/min/mg)	V_{max}/K_m (μL/min/mg)	References
Penicillin G (benzylpenicillin)	Antibiotics	13.9	1.53	0.11	[12]
Edaravone sulfate	Neuroprotectors	15.1	359	23.8	[15]
Estrone sulfate	Estrogens	3.1	5.7 pmol/h/oocyte	1.8 μL/h/oocyte	[38]
Cimetidine	H-2 receptor antagonists	57.4	92.2 pmol/h/oocyte	1.6 μL/h/oocyte	[38]
P-Aminohippuric acid (PAH)	None	87.2	20.4 pmol/h/oocyte	0.2 μL/h/oocyte	[38]
Tetracycline	Antibiotics	566.2			[39]
Zidovudine	Nucleoside reverse transcriptase inhibitors (NRTIs)	145.1	757.7	5.22	[40]
Valacyclovir	Antivirals	57.9			[40]
Ibuprofen	NSAIDS				[41]
Indomethacin	NSAIDS				[41]
Ketoprofen	NSAIDS				[41]
Salicylic acid	NSAIDS				[41]
Bumetanide	Diuretics	1.586			[42]
Zonampanel	Stroke treatments	7.7	44	5.7	[43]
Famotidine	H-2 receptor antagonists	124	448	3.61	[44]
Ranitidine	H-2 receptor antagonists	234	551 nmol/cm^2/h	2.35 μL/cm^2/min	[44]
Captopril	Angiotensin converting enzyme (ACE) inhibitors				[14]
Cefaclor	Antibiotics				[14]
Cefazolin	Antibiotics				[14]
Cefdinir	Antibiotics				[14]
Cefoselis	Antibiotics				[14]
Cefotiam	Antibiotics				[14]
Ceftibuten	Antibiotics				[14]
Ceftizoxime	Antibiotics				[14]

(*continued*)

TABLE 3.2 (*Continued*)

Object	Object Therapeutic Class	K_m (μM)	V_{max} (pmol/min/mg)	V_{max}/K_m (μL/min/mg)	References
Cephaloridine	Antibiotics				[14]
Leucovorin (folinic acid)	Antidotes				[14]
Pitavastatin acid	HMG CoA reductase inhibitors (statins)	3.3	0.76 pmol/h/oocyte	0.23 μL/h/oocyte	[45]
Fexofenadine (terfenadine carboxylate)	H-1 receptor antagonists	70.2	120	0.584	[46]
Sitagliptin	Dipeptidyl peptidase-4 inhibitors	162	38.5 pmol/min/Mill cells	0.25 μL/min/Mill cells	[47]
Adefovir	Antivirals				[48]
Cidofovir	Antivirals				[48]
Tenofovir	Nucleoside reverse transcriptase inhibitors (NRTIs)				[48]
Pravastatin	HMG CoA reductase inhibitors (statins)	27.7			[49]
Mycophenolic acid 7-*O*-glucuronide	Immunosuppressants				[48, 50]
Topotecan hydroxyl acid	Topoisomerase inhibitors	56.5	0.02 pmol/min/oocyte	0.0004 μL/min/oocyte	[51]
Rosuvastatin	HMG CoA reductase inhibitors (statins)	7.4			[52]
Olmesartan	Angiotensin II inhibitors (angiotensin receptor blockers or ARBs)	0.12	1.28	10.7	[21]
Oseltamivir carboxylate (Ro 64-0802)	Antivirals				[21]
Quinaprilat	Angiotensin converting enzyme (ACE) inhibitors	13.4			[53]

Dapagliflozin	Sodium-dependent glucose cotransporter 2 (SGLT2) inhibitors				[54]
Dapagliflozin 3-*O*-glucuronide	Sodium-dependent glucose cotransporter 2 (SGLT2) inhibitors	115			[54]
Ethinyl estradiol sulfate	Estrogens	21.1	67.5	3.2	[55]
Aminopterin	Antimetabolites				[18, 56]
Pnu-288034	Antibiotics	44	1,000	20	[57]
Tazobactam	Antibiotics				[58]
Eluxadoline	Gastrointestinal agents				[59]
Daidzein-7,4-*O*-disulfate	Food products				[22]
Daidzein-7-*O*-glucuronide	Food products	19.1	420	22	[22]
Genistein-4-*O*-sulfate	Food products				[22]
Genistein-7-*O*-glucuronide	Food products	7.94	210	26.4	[22]
Glycitein-7-*O*-glucuronide	Food products	17.6	171	9.72	[22]
Quercetin-3-glucuronide	Food products	5.25	73	13.9	[22]
Dimesna	Cancer chemopreventive agents	390			[25]
Caffeic acid	Food products				[22]
Caffeic acid-3-*O*-glucuronide	Food products				[22]
Caffeic acid-3-*O*-sulfate	Food products				[22]
Caffeic acid-4-*O*-glucuronide	Food products				[22]
Caffeic acid-4-*O*-sulfate	Food products				[22]
Dihydrocaffeic acid	Food products				[22]
Dihydrocaffeic acid-3-*O*-glucuronide	Food products				[22]
Dihydrocaffeic acid-3-*O*-sulfate	Food products				[22]
Dihydrocaffeic acid-4-*O*-glucuronide	Food products				[22]

(continued)

TABLE 3.2 (*Continued*)

Object	Object Therapeutic Class	K_m (μM)	V_{max} (pmol/min/mg)	V_{max}/K_m (μL/min/mg)	References
Dihydrocaffeic acid-4-*O*-sulfate	Food products				[22]
Dihydroferulic acid-4-*O*-glucuronide	Food products				[22]
Dihydroferulic acid-4-*O*-sulfate	Food products				[22]
Ferulic acid-4-*O*-glucuronide	Food products				[22]
Ferulic acid-4-*O*-sulfate	Food products				[22]
Isoferulic acid-3-*O*-glucuronide	Food products				[22]
Isoferulic acid-3-*O*-sulfate	Food products				[22]
Acyclovir	Antivirals	772	797	1	[17]
Ubenimex (Bestatin)	Other antineoplastics	632	2,606	4.1	[60]
Glycyrrhetinic acid glucuronide	Dietary supplements and vitamins	30.1			[26]
Bilastine	H-1 receptor antagonists				[61]
6-Carboxyfluorescein	None	7.24			[61]
6-Beta-hydroxycortisol	Corticosteroids	183	782	4.3	[62]
Anagliptin	Dipeptidyl peptidase-4 inhibitors				[19]
Anagliptin carboxylate metabolite (M1)	Dipeptidyl peptidase-4 inhibitors				[19]
(*R*)-Fexofenadine	H-1 receptor antagonists				[63]
(*S*)-Fexofenadine	H-1 receptor antagonists				[63]
Entecavir	Antivirals	23	2,200	100	[24]
Chlorothiazide	Diuretics	37.6	1,421	37,800	[24]
Cefotaxime	Antibiotics	717	305,000	0.43	[28]
Tc-99m-mercaptoacetyl-triglycine (MAG3)	Diagnostic agents				[28]

3-(4-((5-Acetamide-2-phenylbenzo(D)oxazol-7-Yl)methoxy)-3,5-dichlorophenyl)-2-aminopropanoic acid (Nac-JPH203)	Cancer treatments				[64]
JPH203 (KYT-0353, 3-(4-((5-amino-2-phenylbenzo(D)oxazol-7-Yl)methoxy)-3,5-dichlorophenyl)-2-aminopropanoic acid)	Cancer treatments	32.5	2,030	62.3	[64]
Methotrexate	Antimetabolites	76.6	32.3	0.42	[30]
Pemetrexed	Antimetabolites	28.2	136	4.8	[30]
P-Cresyl sulfate	None				[31]
Avibactam	Antibiotics				[65]
Cilostazol	Anticoagulants and Antiplatelets	23.6	0.726	0.031	[65]
Morinidazole sulfate (M7)	Antibiotics	54	864	16	[66]
R-Morinidazole N+ glucuronide (M8-2)	Antibiotics	26.2	44.5	1.7	[66]
S-Morinidazole N+ glucuronide (M8-1)	Antibiotics	209	218	1	[66]
Cefixime	Antibiotics	81	1,841	22.7	[35]
Catalposide	Herbal medications	41.5			[67]
Tanshinol	Herbal medications	1,888	182	0.0963	[36]
Kuc-7322	Beta3-adrenoreceptor agonist	153.8	97.3	0.633	[68]
Bezafibrate	Fibric acid derivatives	132	410	3.11	[69]
Furosemide	Diuretics	21.5			[70]
CI-1045	Calcium channel blockers	810			[4]

TABLE 3.3 Partial List of Known OCT2 Transporter Substrates

Object	Object Therapeutic Class	K_m (μM)	V_{max} (pmol/min/mg)	V_{max}/K_m (μL/min/mg)	References
Pentamidine	Antiparasitics				[71]
Lamivudine	Nucleoside reverse transcriptase inhibitors (NRTIs)	248	1,100	4.4	[71]
Zalcitabine	Nucleoside reverse transcriptase inhibitors (NRTIs)	232	600	2.6	[71]
Atenolol	Alpha/beta adrenergic antagonists				[72]
Bupivacaine	Anesthetics				[72]
Amiodarone	Antiarrhythmics				[72]
Procainamide	Antiarrhythmics				[72]
Trimethoprim	Antibiotics				[72]
Atropine	Antidotes				[72]
Chlorprothixene	Antipsychotics				[72]
Albuterol (salbutamol)	Bronchodilators				[72]
Terbutaline	Bronchodilators				[72]
(*S*)-Xamoterol	Cardiovascular drugs				[72]
Varenicline	Cholinergic agonists				[72]
Amiloride	Diuretics				[72]
Triamterene	Diuretics				[72]
Bupropion	Dopamine–norepinephrine reuptake inhibitors				[72]
Apomorphine	Dopaminergic agonists				[72]
Berberine	Herbal medications	1.01	92.7	91.4	[72]
Selegiline	MAO-B inhibitors				[72]
Glycopyrronium	Muscarinic antagonists				[72]
Asymmetric dimethylarginine	None	967	6.3 nmol/min/mg	6.5	[72]
Guanfacine	Other antihypertensives	96.19	13,030	135	[72]
Sumatriptan	Triptans				[72]

1-Methyl-4-phenylpyridinium (MPP+)	None	24.5	2.5 nmol/min/mg	102	[73]
Tetraethylammonium (TEA)	None	500	23,000	46	[73]
4-(4-(Dimethylamino)styryl)-*N*-methylpyridinium iodide (ASP+)	None				[74]
4-(4-Chlorophenyl)-1-[4-(4-fluorophenyl)-4-oxobutyl]-pyridinium (HPP+)	Antipsychotics	2.79	256.1 pmol/min/μg	91.8	[75]
6-Beta-hydroxycortisol	Corticosteroids				[62]
Ranitidine	H-2 receptor antagonists	65.2	265	4.06	[76]
Histamine	Neurotransmitters	1,300	300 pmol/h/oocyte	0.2 μL/h/oocyte	[76]
Noradrenaline	Neurotransmitters	1,900	420 pmol/h/oocyte	0.2 μL/h/oocyte	[76]
Serotonin (5-hydroxytryptamine)	Neurotransmitters	80	360 pmol/h/oocyte	4.5 μL/h/oocyte	[76]
Memantine	NMDA receptor antagonists				[76]
Amantadine	Other antiparkinsonians				[76]
Cimetidine	H-2 receptor antagonists	72.6	2,170	29.9	[44]
Famotidine	H-2 receptor antagonists	56.1	204	3.63	[44]
cis-Diammine(pyridine) chloroplatinum II (cDPCP)	Platinums				[77]
Cisplatin	Platinums	11.4	13.7	1.2	[78]
Cobicistat	None				[79]
Creatinine	None	4,000	23,500	10	[80]
Furamidine	Antiparasitics				[81]
Linagliptin	Dipeptidyl peptidase-4 inhibitors				[82]
Ormaplatin	Platinums				[83]
Pt[Dach]Cl(2)	Platinums				[83]
Transplatin (*trans*-diamminedichloroplatinum)	Platinums				[83]
Picoplatin	Platinums				[84]

(continued)

TABLE 3.3 (*Continued*)

Object	Object Therapeutic Class	K_m (μM)	V_{max} (pmol/min/mg)	V_{max}/K_m (μL/min/mg)	References
Pramipexole	Dopaminergic agonists	15.4	0.476 pmol/cm^2/s	0.031 μL/cm^2/s	[85]
Ipratropium	Anticholinergics				[86]
Tiotropium	Anticholinergics				[86]
Ifosfamide	Alkylating agents				[87]
Oxybutynin	Muscarinic antagonists				[88]
Trospium	Muscarinic antagonists	8.03	91.7	11.4	[88]
Rhodamine 123	Other	0.61	39.9	65.4	[89]
Phenformin	Biguanides	37.4	7.83 pmol/min/oocyte	0.209 μL/min/oocyte	[90]
N,N,N-Trimethyl-2-[methyl(7-nitrobenzo[*C*][L, 2,5] oxadiazol-4-Yl)amino] ethanaminium iodide (NBD-MTMA)	None	9	6.8 pmol/cm^2/min	0.76 μL/cm^2/min	[91]
Oxaliplatin	Platinums				[91]
L-Arginine	None				[92]
Propranolol	Alpha/beta adrenergic antagonists	11.8	201	17	[93]
Metformin	Biguanides	3,171	10,700	3.4	[93]
Sepantronium bromide (YM155)	Cancer treatments	2.67	324	12.3	[93]
Dopamine	Inotropic agents	932	7,559	8.1	[93]
Thiamine (Vitamin B1)	Dietary supplements and vitamins	59.9	545	910	[94]
N-Methylnicotinamide	Dietary supplements and vitamins				[95]
Gabapentin	Anticonvulsants				[4]
Pregabalin	Anticonvulsants				[4]
Pd-0299685	Calcium channel blockers	569			[4]

yet they do not belong to ABC Efflux transporter family and play a critical role in the tubular secretion of cationic drugs such as metformin and cimetidine, chloroquine [96–99]. Examples of compounds that are substrates for MATE1 and MATE2K transporters are listed in Tables 3.4 and 3.5, respectively. Compounds that are MATE1 substrates belong to α–β adrenergic antagonists, antiarrhythmics, antibiotics, anticoagulants, antimalarials, antivirals, biguanides, cancer chemopreventive agents, cholinergic agonists, dietary supplements, and vitamins, dopaminergic agonists, estrogens, H1/2 receptor antagonists, herbal medications, nucleoside reverse transcriptase inhibitors, platinums, and topoisomerase inhibitors therapeutic classes with a K_m range of 0.39–5120 μM. Compounds that are MATE2K substrates belong to α–β adrenergic antagonists, antiarrhythmics, antimalarials, antivirals, biguanides, calcium channel blockers, cholinergic agonists, corticosteroids, dietary supplements and vitamins, dopaminergic agonists, estrogens, H1/2 receptor antagonists, nucleoside reverse transcriptase inhibitors, platinums, and topoisomerase with a K_m range of 3.94–4320 μM.

3.2.3.2 P-glycoprotein P-glycoprotein (P-gp) is an ATP-dependent transmembrane efflux pump encoded by ABCB1 gene belonging to ABC superfamily and is the most characterized drug transporter [113, 114]. P-gp is ubiquitous in nature and expressed in various body organs such as gastrointestinal tract, liver, brain, kidney, and testis [115]. In kidney, P-gp is expressed at the apical side of the brush border membrane of proximal tubule. Compounds that are substrates for P-gp tend to be neutral/cationic in nature. Similar to MATE1, P-gp is expressed at the canalicular membrane of liver hepatocytes and plays a key role in the biliary elimination of drug molecules that are typically basic, lipophilic with large molecular weight. Known compounds include statins, beta blockers, diuretics, immunosuppressants, and anticancer agents [113–118]. A list of compounds that are substrates for P-gp is shown in Table 9.1.

3.2.3.3 BCRP Breast cancer-resistant protein (BCRP) is an efflux transporter that is encoded by ABCG2 and a member of the ABC family. Similar to P-gp, it is ubiquitous in nature and expressed in placenta, kidney, liver, testis, brain, mammary tissue, and intestine [119]. BCRP has broad substrate specificity, with compounds typically acidic in nature. They include anticancer agents (e.g., mitoxantrone, camptothecins, tyrosine kinase inhibitors), antiviral agents, statins, antibiotics, diuretics, and antimalarial agents [70, 119–122]. A list of compounds that are substrates for BCRP is shown in Table 6.5.

3.2.4 Physicochemical Factors That Affect Renal Drug Excretion

Renal clearance is determined by the interplay of three processes, which are bidirectional passive, active uptake, and efflux transport mechanisms across the nephron units [10, 123]. Typically, compounds with high passive permeability are more prone to be reabsorbed from urine into blood along the nephron. In part, this is attributed to the high concentration gradient that is driven by water reabsorption process [6, 10, 123, 124]. It should be emphasized that other than the compound passive permeability, the extent and rate of renal reabsorption is affected by urine pH and flow rate. Unlike tubular reabsorption, apparent low passive permeability across basolateral membrane compared to apical membrane [10], and a high counter concentration-gradient created by water reabsorption makes contribution of passive tubular secretion negligible, if any. Due to involvement of active transporters, the secretion process is saturable and may be inhibited by coadministrated drugs, suggesting the possibility of clinical DDIs [2, 3, 10, 125]. It is also influenced by transporter pharmacogenomics [81, 126, 127].

Drugs that are ECCS Class 1 and Class 2 with high passive permeability are reabsorbed throughout the whole length of nephron. However, renal secretion is limited to proximal

TABLE 3.4 Partial List of Known Compounds that are Substrates for MATE1 Transporters

Object	Object Therapeutic Class	K_m (μM)	V_{max} (pmol/min/mg)	V_{max}/K_m (μL/min/mg)	References
Cisplatin	Platinums				[100]
Oxaliplatin	Platinums				[100]
1-Methyl-4-phenylpyridinium (MPP+)	None	100	0.74 nmol/min/mg	7.4	[101]
Acyclovir	Antivirals	2640	0.62 nmol/min/mg	0.23	[101]
Cephalexin	Antibiotics				[101]
Cephradine	Antibiotics				[101]
Cimetidine	H-2 receptor antagonists	170	0.13 nmol/min/mg	0.76	[101]
Creatinine	None				[101]
Estrone sulfate	Estrogens	470	0.26 nmol/min/mg	0.55	[101]
Ganciclovir	Antivirals	5120	1.06 nmol/min/mg	0.21	[101]
Guanidine	Other	2100	0.89 nmol/min/mg	0.42	[101]
Levofloxacin	Antibiotics				[101]
Procainamide	Antiarrhythmics	1230	3.78 nmol/min/mg	3.07	[101]
Quinine	Antimalarials				[101]
Tetraethylammonium (TEA)	None	380	1.19 nmol/min/mg	3.13	[101]
Topotecan	Topoisomerase inhibitors	70	0.21 nmol/min/mg	3	[101]
Metformin	Biguanides	179.1	1.75 nmol/min/mg	9.6819	[101]
Pnu-288034	Antibiotics	340	3.8 nmol/min/mg	0.01	[57]
Dimesna	Cancer chemopreventive agents				[25]
Mesna	Cancer chemopreventive agents				[25]
Chloroquine	Antimalarials				[98]

Varenicline	Cholinergic agonists				[102]
6-Beta-hydroxycortisol	Corticosteroids	200			[62]
(*R*)-Fexofenadine	H-1 receptor antagonists				[63]
(*S*)-Fexofenadine	H-1 receptor antagonists				[63]
4-(4-(Dimethylamino)styryl)-*N*-methylpyridinium	None	34.1	34.5 pmol/min/μg	1.01 μL/min/μg	[103]
Asymmetric dimethylarginine	None				[92]
L-Arginine	None				[92]
Lamivudine	Nucleoside reverse transcriptase inhibitors (NRTIs)				[104]
Atecegatran metoxil metabolite AR-H067637 (Active)	Anticoagulants and antiplatelets	0.39	25	64	[105]
Nitidine chloride	Herbal medications	0.897	18.5	20.6	[106]
Thiamine (vitamin B1)	Dietary supplements and vitamins	830	2.94 nmol/min/mg	3.54	[107]
N-Methylnicotinamide	Dietary supplements and vitamins				[95]
Pramipexole	Dopaminergic agonists				[108]
Atenolol	Alpha/beta adrenergic antagonists	32	323	10	[109]

TABLE 3.5 Partial List of Known Compounds that are Substrates for MATE2K Transporters

Object	Object Therapeutic Class	K_m (μM)	V_{max} (pmol/min/mg)	V_{max}/K_m (μL/min/mg)	References
Quinidine	Antiarrhythmics				[110]
Quinine	Antimalarials				[110]
Verapamil	Calcium channel blockers				[110]
Cisplatin	Platinums				[100]
Oxaliplatin	Platinums				[100]
1-Methyl-4-phenylpyridinium (MPP+)	None	110	0.57 nmol/min/mg	5.18	[101]
Acyclovir	Antivirals	4320	0.94 nmol/min/mg	0.22	[101]
Cimetidine	H-2 receptor antagonists	120	0.11 nmol/min/mg	0.92	[101]
Creatinine	None				[101]
Estrone sulfate	Estrogens	850	0.42 nmol/min/mg	0.49	[101]
Ganciclovir	Antivirals	4280	0.8 nmol/min/mg	0.19	[101]
Guanidine	Other	4200	0.58 nmol/min/mg	0.14	[101]
Metformin	Biguanides	1980	0.84 nmol/min/mg	0.42	[101]
Procainamide	Antiarrhythmics	1580	3.38 nmol/min/mg	2.14	[101]
Tetraethylammonium (TEA)	None	760	0.88 nmol/min/mg	1.16	[101]
Topotecan	Topoisomerase inhibitors	60	0.13 nmol/min/mg	2.17	[101]
Fexofenadine (terfenadine carboxylate)	H-1 receptor antagonists				[111]
Varenicline	Cholinergic agonists				[102]
N-Methylnicotinamide	Dietary supplements and vitamins	422			[112]
6-Beta-hydroxycortisol	Corticosteroids				[62]
Lamivudine	Nucleoside reverse transcriptase inhibitors (NRTIs)				[104]
Thiamine (vitamin B1)	Dietary supplements and vitamins	3.94	25.4	6.45	[94]
Pramipexole	Dopaminergic agonists				[108]
Atenolol	Alpha/beta adrenergic antagonists	76	500	6.6	[109]

tubules. Indeed, physicochemical properties of drug molecules that determine passive permeability are key determinants of drug renal clearance [6]. For example, increasing lipophilicity would increase passive permeability and is associated with a decrease in renal clearance. However, polar descriptors such as polar surface area or hydrogen bond donors or acceptors are positively correlated to renal clearance [128, 129].

As discussed earlier, polyspecific members of the organic ion transporter family (*SLC22*) that are localized at the proximal tubule play a critical role in the renal secretion process of low permeability molecules, that is, ECCS Class 3 and Class 4. The members consist of OCT2, OAT1, and OAT3 uptake transporters [7–9]. Hydrophobicity and basicity are indicated to be the principal determinants of substrate interaction with OCTs, while hydrophobicity and acidity are associated with OATs interaction [130]. However, most of the high-affinity substrates to these transporters are relatively hydrophilic ($c\log P < 0$) [7, 131]. Furthermore, hydrogen bonding ability seems to be an advantageous mechanism to stabilize the substrate-transporter complex [130]. Collectively, hydrophilic and ionized compounds with hydrogen bonding ability are most likely to be secreted in the clinic, mainly because they (i) can interact with the renal transporter systems at the proximal tubule and (ii) do not possess appropriate physicochemical descriptors to undergo reabsorption process along the length of the nephron [6, 124].

The discriminating physicochemical profiles between net secreted and reabsorbed drugs can be helpful in predicting renal DDI potential. For example, two hydrophilic bases ($c\log D < 0$) with predominately renal elimination are more likely to result in clinical interaction as discussed subsequently.

3.3 HEPATOBILIARY ELIMINATION

3.3.1 Mathematical Principles Governing Hepatobiliary Elimination

We were taught that hepatic clearance is an additive process that is the sum of active hepatic uptake (PS_{influx}), passive hepatic diffusion (PS_{pd}), and metabolic and/or biliary intrinsic clearance (CL_{int}). In the last decade, a better understanding of hepatic clearance emerged that is not consistent with the proposed additive nature. For example, the systemic plasma exposure of anionic compounds that have active hepatic uptake that is mediated by organic anion transporting polypeptide and followed by metabolism by CYP or UGT enzymes, for example, statins and sartans, were not influenced by CYP or UGT inhibitors. This observation was not similar to the trend observed with compounds that were taken up by the liver hepatocyte via passive diffusion and metabolized by the same enzymes. Equation 3.4 describes the interplay of the various processes that govern hepatic elimination and is shown as follows [132]:

$$CL_{int,h} = \frac{(PS_{influx} + PS_{pd}) \cdot CL_{int}}{(PS_{efflux} + PS_{pd} + CL_{int})} \tag{3.4}$$

Assuming well-stirred conditions, the total hepatic clearance (CL_h) is calculated by Equation 3.5 [133]:

$$CL_h = Q_h \cdot \frac{f_b \cdot (PS_{influx} + PS_{pd}) \cdot CL_{int}}{Q_h \cdot (PS_{efflux} + PS_{pd} + CL_{int}) + f_b \cdot (PS_{influx} + PS_{pd}) \cdot CL_{int}} = Q_h \cdot E_h \tag{3.5}$$

where PS_{influx} and PS_{efflux} are the active (transporter-mediated) sinusoidal influx and basolateral efflux clearances. PS_{pd} represents passive diffusion clearance. CL_{int} is the sum of the metabolic and biliary intrinsic clearance ($CL_{int,met} + CL_{int,bile}$), Q_h is the liver blood flow, E_h is the hepatic extraction ratio, and $f_{u,b}$ is the unbound fraction in blood and it is calculated using Equation 3.6.

$$f_{u,b} = \frac{f_{u,plasma}}{B/P} \tag{3.6}$$

where $f_{u,plasma}$ is the unbound fraction in plasma and B/P is the blood to plasma ratio.

It should be emphasized that if only one process is considered, the hepatic clearance value would be biased [132, 134–138]. With that in mind, Equation 3.4 can still be simplified if instantaneous equilibrium, also known as rapid equilibrium, is achieved between liver and blood compartments. This phenomenon is observed when the compound is not a good substrate for hepatic OATP transporters, and it has PS_{pd} that is significantly higher than intrinsic clearance mediated by metabolism and/or biliary elimination. Equation 3.5 is reduced to Equation 3.7 [133]:

$$CL_h = Q_h \cdot \frac{f_b \cdot CL_{int}}{Q_h + f_b \cdot CL_{int}} = Q_h \cdot E_h \tag{3.7}$$

This is a key equation that is widely used in drug discovery to predict hepatic clearance for compounds that are predominantly metabolized by CYP enzymes/UGTs and have a rapid equilibrium between liver and blood compartments [139–141]. Rapid equilibrium compounds, ECCS Class 1A and 2, hepatic clearance is predicted by either HLM or hepatocyte depending on the driving rate-determining step of the compound clearance [142–146].

If PS_{pass} is significantly lower than CL_{int} and the compound is a substrate for uptake transporters (e.g., OATP1B1 and 1B3 transporters), that is, $PS_{influx} \ggg PS_{pass}$, hepatic uptake can be the rate-determining step of the systemic clearance of the drug molecule. Indeed, Equation 3.5 can be reduced to Equation 3.8.

$$CL_h = Q_h \cdot \frac{f_b \cdot PS_{influx}}{Q_h + f_b \cdot PS_{influx}} = Q_h \cdot E_h \tag{3.8}$$

Examples of drug molecules that have hepatic uptake as the rate-determining step for their systemic clearance include HMG-CoA reductase inhibitors (statins) and angiotensin II antagonists (sartans) [136, 147–150]. Indeed, many studies suggested that active uptake mediated by OATP1B1 and 1B3 transporters expressed at the basolateral membrane of liver hepatocytes could be the rate-determining step in hepatic clearance, not only for compounds that are metabolically stable such as rosuvastatin [151] and pravastatin [152–154] but also for compounds that are extensively metabolized, for example, atorvastatin [155, 156], glyburide [157, 158], and repaglinide [159, 160]. This is substantiated by clinically relevant drug–drug interactions (DDIs) leading to changes in systemic exposure of these drugs that are primarily attributed to the inhibition of hepatic uptake mediated by members of the OATP family rather than the inhibition of hepatic metabolism or biliary efflux [135, 155, 161].

3.3.2 Active Hepatic Uptake Transporters

Organic anion transporting polypeptide (OATP1B1, OATP1B3, OATP2B1) and organic cation transporter 1 (OCT1) on the hepatocyte sinusoidal membrane are involved in the hepatic uptake of a wide variety of drugs [135, 162, 163]. The following section is a discussion of these transporters, their expression, and known substrates.

TIP:

A prerequisite for putting transporters' role in drug disposition in perspective is to understand their substrate chemical space and determine the compound permeability. It is pivotal to emphasize that for transporters to influence drug disposition the substrate should not always have poor permeability, a perception that some drug discovery scientists may have. This is clearly shown with compounds with active hepatic uptake that is mediated by OATP transporters and that have high passive permeability, for example, atorvastatin.

3.3.2.1 OATP1B1/OATP1B3/OATP2B1 There are 11 functional human proteins that belong to the organic anion transporter polypeptide family. It is composed of 12 putative transmembrane domains that are expressed in various body organs that include liver, kidney, and brain [37, 164]. They play a key role in the absorption, distribution, and elimination of endogenous substances and xenobiotics. The three main isoforms that are expressed at the sinusoidal membrane of hepatocytes are OATP1B1, OATP1B3, and OATP2B1. These transporters are involved in sodium and ATP-independent transport. Typically, compounds that are substrates for these transporters are amphiphilic organic anions. It should be emphasized that OATP1B1 and 1B3 are exclusively expressed in the liver, unlike OATP2B1 that is ubiquitous [1, 148]. Table 3.6 lists known substrates for OATP1B1 transporters. Compounds that are substrates for OATP1B1 belong to androgens, angiotensin II inhibitors, antibiotics, anticoagulants, antidotes, antimetabolites, beta-3 agonists, antivirals, calcium channel blockers, cancer chemopreventive agents, cancer treatment, corticosteroids, diagnostic agents, dietary supplements and vitamins, dipeptidyl peptidase 4-inhibitors, diuretics, estrogens, fibric acid derivatives, food products, gastrointestinal agents, H1 receptor antagonists, statins, immunosuppressants, neuroprotectives, nucleoside reverse transcriptase inhibitors, sodium-dependent glucose cotransporter 2, stroke treatments and topoisomerase inhibitors therapeutic classes. Their K_m values range is 0.089–5870 μM.

Table 3.7 lists known substrates for OATP1B3 transporters. Compounds that are substrates for OATP1B3 belong to androgens, angiotensin II inhibitors, antiarrhythmics, antibiotics, antimetabolites, antivirals, cancer treatment, cardiovascular drugs, diagnostic agents, dipeptidyl peptidase-4 inhibitors, diuretics, endothelin receptor antagonists, estrogens, food products, H-1 antagonists, herbal medications, statins, hormone replacements, hormones, immunosuppressants, kinase inhibitors, meglitinides, NSAIDs, antilipemics, antineoplastics, protease inhibitors, taxanes, thiazolidinediones, topoisomerase inhibitors, and vasodilators therapeutic classes with a K_m range of 0.089–5870 μM.

3.3.2.2 OCT1 OCT1 (SLC22A1) is a facilitative, electrogenic, and, similar to OATP transporters, sodium-independent carrier protein. It is mainly expressed at the sinusoidal membrane of hepatocyte. OCT1 plays a key role in mediating the transport of a variety of organic cations across the plasma membrane in hepatocytes. Compounds that are substrates for OCT1 transporter are low molecular weight, hydrophilic, organic cations. Table 3.8 lists known OCT1 substrates and they are part of α–β adrenergic antagonists, antiarrhythmics, antibiotics, anticholinergics, anticoagulants, anticonvulsants, antidotes, antiparasitics, antipsychotics, antivirals, biguanides, bronchodilators, calcium channel blockers, cancer treatments, cardiovascular drugs, cholinergic agonists, dietary supplements and vitamins, diuretics, estrogens, gastrointestinal agents, H-1 receptor antagonists, herbal medications, inotropic agents, kinase inhibitors, miscellaneous agents, muscarinic antagonists, kinase inhibitors, neurotransmitters, nucleoside reverse transcriptase inhibitors, opioids, analgesics, antiemetics, antihypertensives, platinums, protease inhibitors, serotonin HT3 receptor antagonists, and triptan therapeutic classes. They have a K_m range of 0.54–2600 μM.

TABLE 3.6 Partial List of Known Compounds that are Substrates for OATP1B1 Transporters

Object	Object Therapeutic Class	K_m (μM)	V_{max} (pmol/min/mg)	V_{max}/K_m (μL/min/mg)	References
Dehydroepiandrosterone (DHEA)	OATP1B1				[165]
3,3,5-Triiodothyronine T(3)	OATP1B1				[166]
BQ-123 (Cyclo(Trp-Asp-Pro-Val-Leu))	OATP1B1				[166]
Dehydroepiandrosterone sulfate (DHEAS)	OATP1B1				[166]
Prostaglandin E2	OATP1B1				[166]
Thyroxine	OATP1B1				[166]
Pravastatin	OATP1B1	11.5	10.2 pmol/min/106 cells	0.9 μL/min/106 cells	[167]
Taurocholic acid	OATP1B1	25.5	77 pmol/min/106 cells	3 μL/min/106 cells	[167]
Methotrexate	OATP1B1				[168]
Rifampin	OATP1B1	1.5	9.3	6.2	[169]
Phalloidine	OATP1B1	39	51	1.3	[170]
Troglitazone sulfate	OATP1B1				[171]
Pitavastatin	OATP1B1	3	230	76.6	[172]
PKI 166 *O*-glucuronide (ACU 154)	OATP1B1				[173]
Estradiol-17-beta-glucuronide	OATP1B1	23.72	36.12 pmol/min/106 cells	1.52 μL/min/106 cells	[174]
7-Ethyl-10-hydroxycamptothecin (SN-38)	OATP1B1				[175]
Caspofungin	OATP1B1				[176]
Atorvastatin	OATP1B1	12.4	70.4	5.65	[177]
Pitavastatin acid	OATP1B1	3.6	0.9 pmol/h/oocyte	0.25 μL/h/oocyte	[45]

Atrasentan	OATP1B1				[178]
Valsartan	OATP1B1	1.39	3.85	2.77	[179]
Enalapril	OATP1B1	262	78	0.3	[180]
Temocaprilat	OATP1B1				[181]
2-Hydroxyatorvastatin acid	OATP1B1				[182]
4-Hydroxyatorvastatin acid	OATP1B1				[182]
Atorvastatin acid	OATP1B1	18.9	105	5.6	[182]
Fluvastatin	OATP1B1	2.45	362	147.8	[183]
Bosentan	OATP1B1	44	1.32 pmol/min/μg	0.03 μL/min/μg	[184]
Olmesartan	OATP1B1	12.8	18.2	1.42	[21]
Rosuvastatin	OATP1B1	4.26	0.159 pmol/min/oocyte	0.037 μL/min/oocyte	[185]
Ym758	OATP1B1	87.9	2590 pmol/min/106 cells	29.5 μL/min/106 cells	[186]
S 8921 glucuronide	OATP1B1	1.93	67.8	51.8	[187]
Ezetimibe glucuronide	OATP1B1				[188]
Cefazolin	OATP1B1		31.7 pmol/h/oocyte	0.001 μL/h/oocyte	[189]
Cefditoren	OATP1B1	3,450	65.3 pmol/h/oocyte	0.0189 μL/h/oocyte	[189]
Cefoperazone	OATP1B1	4,840	19.1 pmol/h/oocyte	0.004 μL/h/oocyte	[189]
Nafcillin	OATP1B1	11,000	141 pmol/h/oocyte	0.01 μL/h/oocyte	[189]
Gimatecan	OATP1B1				[190]
Karenitecin	OATP1B1				[190]
Cp-724,714	OATP1B1				[191]
Mycophenolic acid 7-*O*-glucuronide	OATP1B1				[192]
Darunavir	OATP1B1				[193]
Lopinavir	OATP1B1				[193]
Saquinavir	OATP1B1				[193]
Cholyl-glycylamido-fluorescein (CGamF)	OATP1B1	7.9	62.3	7.8	[194]

(*continued*)

TABLE 3.6 (*Continued*)

Object	Object Therapeutic Class	K_m (μM)	V_{max} (pmol/min/mg)	V_{max}/K_m (μL/min/mg)	References
Sirolimus	OATP1B1				[195]
Ethinyl estradiol sulfate	OATP1B1	0.087	2.42	27.8	[196]
Gadolinium-ethoxybenzyl-diethylenetriamine pentaacetic acid (Gd-EOB-DTPA)	OATP1B1	700	10.5	0.015	[197]
Simeprevir	OATP1B1				[198]
Flavopiridol glucuronide	OATP1B1				[199]
99mtc-Mebrofenin	OATP1B1	1,680	385.9 nmol/min/mg	0.23	[200]
Maraviroc	OATP1B1	33.9	0.188 pmol/min/oocyte	0.0055 μL/min/oocyte	[201]
Penicillin	OATP1B1				[202]
Eltrombopag	OATP1B1				[203]
Mesalamine (5-ASA)	OATP1B1	55.1	28.3	0.51	[204]
Fluorescein sodium	OATP1B1	4.2	30.9	7.3	[205]
Etoposide	OATP1B1	42.2			[206]
Simvastatin acid	OATP1B1				[207]
Cinaciguat (BAY 58-2667)	OATP1B1				[208]
Norursodeoxycholic acid	OATP1B1				[209]
(15R)-11C-TIC (PET Probe)	OATP1B1	0.96			[210]
Docetaxel	OATP1B1	7.6	30.7	4	[211]
Paclitaxel	OATP1B1	0.408	22.1	54.2	[211]
Gadoxetic acid	OATP1B1	1170	6.32	0.005	[212]
Erythromycin	OATP1B1	13.2	11.6	0.8	[213]
Crizotinib	OATP1B1				[214]
Pazopanib	OATP1B1				[214]
Sorafenib	OATP1B1	23.5	128	5.4	[214]
Sorafenib glucuronide	OATP1B1				[214]
Sunitinib	OATP1B1				[214]

7-*tert*-Butyldimethylsilyl-10-hydroxycamptothecin (AR-67) carboxylate	OATP1B1	2.58			[215]
3-(4-((5-Acetamide-2-phenylbenzo(D)oxazol-7-Yl)methoxy)-3,5-dichlorophenyl)-2-aminopropanoic acid (Nac-JPH203)	OATP1B1				[64]
JPH203 (KYT-0353, 3-(4-((5-amino-2-phenylbenzo(D)oxazol-7-Yl)methoxy)-3,5-dichlorophenyl)-2-aminopropanoic acid)	OATP1B1	2.63	0.513 nmol/min/mg	194	[64]
Axitinib	OATP1B1				[216]
Estrone-3-sulfate	OATP1B1	1.93	42.1	21.8	[217]
Bilirubin	OATP1B1				[218]
Bromosulfophthalein (BSP)	OATP1B1	0.28	20.8	74.3	[219]
Tm-25659	OATP1B1	62.8	103.4 pmol/min/oocyte	1.65 μL/min/oocyte	[220]
Faldaprevir	OATP1B1				[221]
M-4 edoxaban metabolite	OATP1B1				[222]
99mtc-*N*-Pyridoxyl-5-methyltryptophan (99mtc-PMT)	OATP1B1				[223]
Nilotinib	OATP1B1	10.14	6.95 nmol/min/mg	0.68	[224]
Vandetanib	OATP1B1	2.72	75.95 nmol/min/mg	27.92	[224]
Tauro-Nor-THCA-24-DBD	OATP1B1	0.2	156.4	782	[225]
Eprosartan	OATP1B1				[226]
Lovastatin	OATP1B1			8.9	[227]
Simvastatin	OATP1B1			14.4	[227]

(continued)

TABLE 3.6 (*Continued*)

Object	Object Therapeutic Class	K_m (μM)	V_{max} (pmol/min/mg)	V_{max}/K_m (μL/min/mg)	References
Pf-049	OATP1B1				[228]
Pf-051	OATP1B1				[228]
Cerivastatin	OATP1B1			16.8 μL/min/106 cells	[229]
Timosaponin B2	OATP1B1	3.69	12.94	3.51	[230]
Fexofenadine (terfenadine carboxylate)	OATP1B1	61.6	41	0.66	[231]
Glyburide (glibenclamide)	OATP1B1	1.24	38.1	30.7	[231]
Nateglinide	OATP1B1	36.4	174	4.8	[231]
Repaglinide	OATP1B1	1.36	7.04	5.2	[231]
Torsemide	OATP1B1	20.9	154	7.4	[231]
Bortezomib	OATP1B1				[232]
Catalposide	OATP1B1	220.5			[67]
Asunaprevir	OATP1B1				[233]
2-Ethylidene-1,5-dimethyl-3,3-diphenylpyrrolidine (EDDP)	OATP1B1				[234]
Furosemide	OATP1B1				[70]
Misoprostol acid	OATP1B1				[235]

TABLE 3.7 Partial List of Known Compounds that are Substrates for OATP1B3 Transporters

Object	Object Therapeutic Class	K_m (μM)	V_{max} (pmol/min/mg)	V_{max}/K_m (μL/min/mg)	References
BQ-123 (Cyclo(Trp-Asp-Pro-Val-Leu))	Endothelin receptor antagonists				[166]
Bromosulfophthalein (BSP)	None	0.4			[166]
Dehydroepiandrosterone sulfate (DHEAS)	Androgens				[166]
Digoxin	Antiarrhythmics				[166]
Ouabain	Cardiovascular drugs				[166]
Thyroxine	Hormones				[166]
3,3,5-Triiodothyronine T(3)	Hormone replacement therapy	6.4			[168]
Methotrexate	Antimetabolites	24.7			[168]
Troglitazone sulfate	Thiazolidinediones				[171]
Cholecystokinin octapeptide (CCK-8)	None	3.82	102	26.7	[172]
Estradiol-17-beta-glucuronide	Estrogens	24.6	56.8	2.31	[172]
Pitavastatin	HMG CoA reductase inhibitors (statins)	3.25	100	30.6	[172]
Fexofenadine (terfenadine carboxylate)	H-1 receptor antagonists	108	56.7	0.525	[236]
Fluo-3	None	6.8			[237]
Paclitaxel	Taxanes	6.79	0.098 pmol/min/oocyte	0.014 μL/min/oocyte	[238]
Pitavastatin acid	HMG CoA reductase inhibitors (statins)	3.8	2.79 pmol/h/oocyte	0.73 μL/h/oocyte	[45]
Amanitin	None	3.7			[239]
Olmesartan	Angiotensin II inhibitors (angiotensin receptor blockers or ARBs)	71.8			[240]

(continued)

TABLE 3.7 (*Continued*)

Object	Object Therapeutic Class	K_m (μM)	V_{max} (pmol/min/mg)	V_{max}/K_m (μL/min/mg)	References
Atrasentan	Endothelin receptor antagonists				[178]
Telmisartan	Angiotensin II inhibitors (angiotensin receptor blockers or ARBs)	0.81	6.7	8.3	[241]
Valsartan	Angiotensin II inhibitors (angiotensin receptor blockers or ARBs)	18.2	135	7.41	[179]
Bosentan	Endothelin receptor antagonists	141	3.21 pmol/min/μg	0.023 μL/min/μg	[184]
Bosentan hydroxy metabolite (Ro 48-5033)	Endothelin receptor antagonists	166	2.14 pmol/min/μg	0.013 μL/min/μg	[184]
Olmesartan	Angiotensin II inhibitors (angiotensin receptor blockers or ARBs)	44.2	265	5.99	[21]
7-Ethyl-10-hydroxycamptothecin (SN-38)	Topoisomerase inhibitors				[242]
Chenodeoxycholyl-lysine (CDCA-NBD)	None	0.53	279	526.4	[242]
Dactinomycin	Other antineoplastics				[242]
Mitoxantrone	Other antineoplastics				[242]
Telmisartan acylglucuronide	Angiotensin II inhibitors (angiotensin receptor blockers or ARBs)	3.4	124	36.5	[243]
Estrone-3-sulfate	Estrogens	58	1860	32	[243]
S 8921 glucuronide	Other antilipemics	1.88	47.4	28.3	[187]
Erythromycin	Antibiotics				[244]
Testosterone	Androgens				[245]

Cefadroxil	Antibiotics	4150	8.5 pmol/h/oocyte	0.002 μL/h/oocyte	[189]
Cefazolin	Antibiotics	3890	10.5 pmol/h/oocyte	0.003 μL/h/oocyte	[189]
Cefditoren	Antibiotics	5870	35.6 pmol/h/oocyte	0.006 μL/h/oocyte	[189]
Cefmetazole	Antibiotics	706	10.7 pmol/h/oocyte	0.015 μL/h/oocyte	[189]
Cefoperazone	Antibiotics	1950	11.2 pmol/h/oocyte	0.006 μL/h/oocyte	[189]
Cephalexin	Antibiotics	1190	1.71 pmol/h/oocyte	0.001 μL/h/oocyte	[189]
Nafcillin	Antibiotics	73.6	11.9 pmol/h/oocyte	0.2 μL/h/oocyte	[189]
Mycophenolic acid 7-*O*-glucuronide	Immunosuppressants	122	313	2.6	[192]
Saquinavir	Protease inhibitors				[193]
Cholyl-glycylamido-fluorescein (CGamF)	None	2.2	315.4	142	[246]
Linagliptin	Dipeptidyl peptidase-4 inhibitors				[82]
Sirolimus	Immunosuppressants				[195]
Gadolinium-ethoxybenzyl-diethylenetriamine pentaacetic acid (Gd-EOB-DTPA)	Diagnostic agents	4100	22.7	0.005	[197]
99mtc-Mebrofenin	Diagnostic agents	2570	371.5 nmol/min/mg	0.145 ml/min/mg	[200]
Indocyanine green	Diagnostic agents				[200]
(−)-Epicatechin gallate	Food products	34.1			[247]
(−)-Epigallocatechin gallate	Food products	13.2			[247]
Cefmetazole	Antibiotics				[202]
Cefotaxime	Antibiotics				[202]
Ceftriaxone	Antibiotics				[202]
Penicillin	Antibiotics				[202]

(*continued*)

TABLE 3.7 (*Continued*)

Object	Object Therapeutic Class	K_m (μM)	V_{max} (pmol/min/mg)	V_{max}/K_m (μL/min/mg)	References
Rifampin	Antibiotics				[202]
Diclofenac	NSAIDS				[248]
Mesalamine (5-ASA)	NSAIDS	77.4	85.1	1.1	[204]
Fluorescein sodium	None	10.9	135	12.4	[205]
Etoposide	Topoisomerase inhibitors				[206]
Nateglinide	Meglitinides				[249]
Norursodeoxycholic acid	Other				[209]
Ursodeoxycholic acid	Other				[209]
(15R)-11C-TIC (PET probe)	None	1.3			[210]
Docetaxel	Taxanes	2.2	27.2	12	[211]
Gadoxetic acid	Other	532	7.43	0.014	[212]
(*R*)-Fexofenadine	H-1 receptor antagonists				[63]
Crizotinib	Kinase inhibitors				[214]
Dasatinib	Kinase inhibitors				[214]
Gefitinib	Kinase inhibitors				[214]
Imatinib	Kinase inhibitors				[214]
Pazopanib	Kinase inhibitors				[214]
Sorafenib	Kinase inhibitors				[214]
Sorafenib glucuronide	Kinase inhibitors				[214]
Vandetanib	Kinase inhibitors				[214]
Vemurafenib	Kinase inhibitors				[214]
Dioscin	Herbal medications	2.08	730	350.96	[250]
7-*tert*-Butyldimethylsilyl-10-hydroxycamptothecin (AR-67) carboxylate	Topoisomerase inhibitors	0.32			[215]

3-(4-((5-Acetamide-2-phenylbenzo(D)oxazol-7-Yl)methoxy)-3,5-dichlorophenyl)-2-aminopropanoic acid (Nac-JPH203)	Cancer treatments				[64]
JPH203 (KYT-0353, 3-(4-((5-Amino-2-phenylbenzo(D)oxazol-7-Yl)methoxy)-3,5-dichlorophenyl)-2-aminopropanoic acid)	Cancer treatments	8.94	0.762 nmol/min/mg	85.3	[64]
Axitinib	Kinase inhibitors				[216]
Bilirubin	None				[218]
Tm-25659	Obesity treatments	66.8	89.8 pmol/min/oocyte	1.34 μL/min/oocyte	[220]
99mtc-*N*-Pyridoxyl-5-methyltryptophan (99mtc-PMT)	Diagnostic agents				[223]
Canertinib	Kinase inhibitors	12.18	15.34 nmol/min/mg	1.25 ml/min/mg	[224]
Nilotinib	Kinase inhibitors	7.84	6.75 nmol/min/mg	0.86 ml/min/mg	[224]
Taurocholic acid	None	18.8	500	26.6	[225]
Tauro-Nor-THCA-24-DBD	None	0.4	57.9	144.75	[225]
Convallatoxin	Vasodilators				[251]
Dihydroouabain	Herbal medications				[251]
Ouabagenin	Herbal medications				[251]
Cerivastatin	HMG CoA reductase inhibitors (statins)			8.6	[227]
Fluvastatin	HMG CoA reductase inhibitors (statins)			29.8	[227]

(continued)

TABLE 3.7 ***(Continued)***

Object	Object Therapeutic Class	K_m (μM)	V_{max} (pmol/min/mg)	V_{max}/K_m (μL/min/mg)	References
Pravastatin	HMG CoA reductase inhibitors (statins)			2.2	[227]
Simvastatin	HMG CoA reductase inhibitors (statins)			14.3	[227]
Pf-049	Diabetes treatments				[228]
Pf-051	Diabetes treatments				[228]
Rosuvastatin	HMG CoA reductase inhibitors (statins)	16.5	251	15.2	[252]
Ginsenoside Re	Herbal medications	14.3	21.5	1.51	[253]
Ginsenoside Rg1	Herbal medications	45.4	12.3	0.27	[253]
Notoginsenoside R1	Cardiovascular drugs	73.4	12.3	0.168	[253]
Timosaponin B2	Herbal medications				[230]
Catalposide	Herbal medications	502			[67]
Furosemide	Diuretics				[70]
Berberine	Herbal medications				[254]
Atorvastatin	HMG CoA reductase inhibitors (statins)				[255]

TABLE 3.8 Partial List of Known Compounds that are Substrates for OCT1 Transporters

Object	Object Therapeutic Class	K_m (μM)	V_{max} (pmol/min/mg)	V_{max}/K_m (μL/min/mg)	References
N-methylquinine	None	19.5			[256]
Acyclovir	Antivirals	151.2	133.4	0.88	[40]
Ganciclovir	Antivirals	516.2	200.3	0.388	[40]
Ranitidine	H-2 receptor antagonists	70			[257]
4-(4-Chlorophenyl)-1-[4-(4-fluorophenyl)-4-hydroxybutyl]pyridinium (RHPP+)	Antipsychotics	5.15	1230.9 pmol/min/μg	239 μL/min/μg	[75]
Cisplatin	Platinums				[100]
Carboplatin	Platinums				[258]
Oxaliplatin	Platinums				[258]
Fexofenadine (terfenadine carboxylate)	H-1 receptor antagonists				[259]
Metformin	Biguanides	2420	6.74 nmol/min/mg	2.78	[260]
1-Methyl-4-phenylpyridinium (MPP+)	None	25	2.3 nmol/min/mg	92	[73]
Tetraethylammonium (TEA)	None	69.2	0.29 nmol/min/mg	4.19	[73]
Cimetidine	H-2 receptor antagonists	2600	3100 pmol/min/106 cells	1.2 μL/min/106 cells	[261]
Lamivudine	Nucleoside reverse transcriptase inhibitors (NRTIs)	249	2 nmol/min/mg	8	[71]
Zalcitabine	Nucleoside reverse transcriptase inhibitors (NRTIs)	242	1 nmol/min/mg	4.1	[71]
cis-Diammine(pyridine) chloroplatinum II (cDPCP)	Platinums				[77]
Furamidine	Antiparasitics	6.1	27.2	4.5	[262]

(*continued*)

TABLE 3.8 (*Continued*)

Object	Object Therapeutic Class	K_m (μM)	V_{max} (pmol/min/mg)	V_{max}/K_m (μL/min/mg)	References
Pentamidine	Antiparasitics	36.4	156	4.3	[262]
Lamivudine	Nucleoside reverse transcriptase inhibitors (NRTIs)	1250	10.4 nmol/min/mg	8.03	[263]
Phenformin	Biguanides	12.5	192 pmol/min/106 cells	15.36 μL/min106 cells	[264]
Picoplatin	Platinums				[84]
4-(4-(Dimethylamino)styryl)-*N*-methylpyridinium	None	9.21			[265]
Tropisetron	Serotonin HT3 receptor antagonists				[266]
Tiotropium	Anticholinergics				[86]
Oxybutynin	Muscarinic antagonists	8.82	484	54.9	[88]
Trospium	Muscarinic antagonists	17	93.2	5.5	[88]
Eltrombopag	Miscellaneous agents				[203]
4,6-Diamino-2-phenylindol (DAPI)	None	8.94			[203]
O-Desmethyltramadol (M1)	Other analgesics	149			[267]
Lamotrigine	Anticonvulsants	62	385 pmol/min/106 cells	6.2 μL/min/106 cells	[268]
Debrisoquine	Other antihypertensives	5.9	41.9	7.1	[269]
Rhodamine 123	Other	0.54	35.9	66.5	[89]
Sorafenib	Kinase inhibitors	3.8	116	30.5	[270]
Monocrotaline	None	25	266	10.6	[271]
Morphine	Opioids	3.4	29	8.7	[272]
(*S*)-Xamoterol	Cardiovascular drugs				[72]
Albuterol (salbutamol)	Bronchodilators				[72]
Amiloride	Diuretics				[72]

Atropine	Antidotes				[72]
Benzamil	Calcium channel blockers				[72]
Camostat	Protease inhibitors				[72]
Clidinium	Muscarinic antagonists				[72]
Disopyramide	Antiarrhythmics				[72]
Formoterol	Bronchodilators				[72]
Glycopyrronium	Muscarinic antagonists				[72]
Indacaterol	Alpha/beta adrenergic antagonists				[72]
Ipratropium	Anticholinergics	9			[72]
Mepenzolic acid	Anticholinergics				[72]
Metoclopramide	Other antiemetics				[72]
Nizatidine	H-2 receptor antagonists				[72]
Perphenazine	Antipsychotics				[72]
Procainamide	Antiarrhythmics				[72]
Sumatriptan	Triptans				[72]
Terbutaline	Bronchodilators				[72]
Triamterene	Diuretics				[72]
Trimethoprim	Antibiotics				[72]
Varenicline	Cholinergic agonists				[72]
Atecegatran metoxil metabolite AR-H067637 (active)	Anticoagulants and antiplatelets	26	1480	57	[105]
Estrone-3-sulfate	Estrogens				[223]
Dopamine	Inotropic agents				[273]

(*continued*)

TABLE 3.8 (*Continued*)

Object	Object Therapeutic Class	K_m (μM)	V_{max} (pmol/min/mg)	V_{max}/K_m (μL/min/mg)	References
Noradrenaline	Neurotransmitters				[273]
Serotonin (5-hydroxytryptamine)	Neurotransmitters	197	561	2.85	[273]
Nitidine chloride	Herbal medications	0.797	249	312	[106]
Retrorsine	None				[274]
Thiamine (vitamin B1)	Dietary supplements and vitamins	780	2.77 nmol/min/mg	3.55	[107]
Nilotinib	Kinase inhibitors				[275]
2-Ethylidene-1,5-dimethyl-3,3-diphenylpyrrolidine (EDDP)	Opioids	4.5	11.1	2.5	[234]
Imatinib	Kinase inhibitors				[276]
Berberrubine	Herbal medications	1.27	225	177	[277]
Coptisine	Herbal medications	5.8	609	105	[277]
Epiberberine	Herbal medications	4.37	281	64.3	[277]
Jatrorrhizine	Herbal medications	4.46	251	56.3	[277]
Histamine	Neurotransmitters				[278]
Atenolol	Alpha/beta adrenergic antagonists	128	1355	10.6	[109]

3.3.3 Biliary Efflux Transporters

Biliary excretion is a major elimination pathway for many drugs and/or their metabolites. It has a significant impact on exposure, pharmacological, and toxicological effects of drugs [279, 280]. Drug secretion into bile is primarily mediated by membrane transporters. The major transporters expressed on the canalicular membrane are multidrug resistance-associated protein 2 (MRP2), multidrug resistance 1 and 3 (MDR1 and MDR3), BCRP, and multidrug and toxin extrusion 1 (MATE1), the bile salt export pump (BSEP) [281, 282]. The following section is a discussion on the efflux transporters that were not discussed earlier under renal elimination section.

3.3.3.1 MRP2 MRP2 is encoded by ABCC2 gene and expressed at the canalicular membrane of hepatocytes. It plays a key role in the biliary secretion of many endogenous substances and xenobiotics [164]. Table 3.9 lists known substrates for MRP2 and they are part of angiotensin converting enzyme inhibitors, anthracyclines, antibiotics, anticoagulants, antigouts, antimetabolites, antivirals, cancer chemopreventive agents, cardiovascular medications, cholesterol absorption inhibitors, diagnostic agents, diuretics, dipeptidyl peptidase-4-inhibitors, endothelin receptor antagonists, estrogens, erectile dysfunction medication, fibric acid derivatives, food products, gastrointestinal agents, herbal medications, statins, immunosuppressants, NSAIDs, obesity treatment, antiemetics, prostaglandins, protease inhibitors, respiratory agents, selective serotonin reuptake inhibitors, sodium-dependent glucose cotransporter inhibitors, taxanes, thiazolidinediones, topoisomerase inhibitors, and vinca alkaloids therapeutic classes, with a K_m range of 12.8–1540 μM.

3.3.3.2 BSEP BSEP is a hepatobiliary efflux transporter that is encoded by ABCB11 gene and located in the canalicular membrane of hepatocytes. It mediates the biliary secretion of numerous conjugated bile salts into the bile canaliculus. A mounting evidence that suggests that alteration of the activity of BSEP by either inhibition or genetically is associated with a serious pathophysiological role that is typically associated with the development of drug-induced liver injury [164].

3.3.4 Physicochemical Factors That Affect Hepatobiliary Drug Excretion

OATPs are the key transporters involved in active hepatic uptake of compounds that are typically ECCS Class 1B and 3B [324]. The main physicochemical properties for OATP substrates are not fully understood and not conclusive. Our group investigated the physicochemical properties for compounds that are human OATP substrates. It should be emphasized that drug ionization state is a key factor in determining the substrate interaction with OATP transporters [325]. These compounds are typically acids or zwitterions. In addition, these substrates have large molecular weight, that is >400 Da with polar surface area. Interestingly, permeability does not appear to be a rate-determining step since acid/zwitterion compounds that are high and low permeability molecules are still good OATP substrates as shown for atorvastatin and rosuvastatin, respectively [1]. Collectively, our group noted that hepatic active uptake could be the predominant clearance mechanism for acids and zwitterions with a molecular weight >400 Da, irrespective of their membrane permeability [325]. Varma et al. investigated the properties of compounds that are biliary secreted. They reported that there is a significant overlap in the physicochemical properties for compounds that are biliary eliminated with those that are substrate for OATP transporters, where the dominant physicochemical properties of drugs excreted in bile are similar and are ionization state, molecular weight, lipophilicity, and polarity [279, 325–327]. However, it should be emphasized that unlike active hepatic uptake, biliary eliminated compounds have low permeability [325, 328].

TABLE 3.9 Partial List of Known Compounds that are Substrates for MRP2 Transporters

Object	Object Therapeutic Class	K_m (μM)	V_{max} (pmol/min/mg)	V_{max}/K_m (μL/min/mg)	References
Irinotecan	Topoisomerase inhibitors	90.8			[283]
Vinblastine	Vinca alkaloids	71.8	0.54 pmol/cm^2/s	0.008 μL/cm^2/s	[284]
Leucovorin (folinic acid)	Antidotes				[285]
Estrone-3-sulfate	Estrogens				[286]
Atorvastatin	HMG CoA reductase inhibitors (statins)				[287]
Ethacrynic acid glutathione	None				[288]
Valsartan	Angiotensin II inhibitors (angiotensin receptor blockers or ARBs)	30.4	895	29.4	[288]
Methotrexate	Antimetabolites	1540	4.2 nmol/min/mg	2.7	[289]
Cp-671,305	Respiratory agents				[290]
Olmesartan	Angiotensin II inhibitors (angiotensin receptor blockers or ARBs)	14.9	9.06	6.08	[21]
Belotecan	Topoisomerase inhibitors				[291]
Icariside II	Herbal medications				[292]
Belotecan	Topoisomerase inhibitors	267.93	116.8 pmol/cm^2/min	0.43 μL/cm^2/min	[293]
Docetaxel	Taxanes				[294]
Ochratoxin A	None				[294]
E3040 glucuronide	None				[295]
Gemfibrozil glucuronide	Fibric acid derivatives				[295]
Troglitazone glucuronide	Thiazolidinediones				[295]
Estradiol-17-beta-glucuronide	Estrogens	146	8286	56.7	[296]
Dactinomycin	Other antineoplastics				[297]
Daunorubicin	Anthracyclines				[297]
Dipyridamole	Anticoagulants and antiplatelets				[297]

Domperidone	Other antiemetics				[297]
Etoposide	Topoisomerase inhibitors				[297]
Fluvastatin	HMG CoA reductase inhibitors (statins)				[297]
Furosemide	Diuretics				[297]
Paclitaxel	Taxanes				[297]
Ritonavir	Protease inhibitors				[297]
Saquinavir	Protease inhibitors				[297]
Sulfasalazine	NSAIDS				[297]
Sulfinpyrazone	Antigout and uricosuric agents				[297]
Sulindac	NSAIDS				[297]
Teniposide	Topoisomerase inhibitors				[297]
Topotecan	Topoisomerase inhibitors				[297]
Bromosulfophthalein (BSP)	None				[298]
Erythromycin	Antibiotics	64.4	45.6	0.71	[299]
Dimesna	Cancer chemopreventive agents				[25]
Mesna	Cancer chemopreventive agents				[25]
Pravastatin	HMG CoA reductase inhibitors (statins)				[300]
Sepantronium bromide (YM155)	Cancer treatments				[301]
Etoposide glucuronide	Topoisomerase inhibitors				[301]
Ezetimibe glucuronide	Cholesterol absorption inhibitors				[206]
Lopinavir	Protease inhibitors				[302]
Fosinopril	Angiotensin converting enzyme (ACE) inhibitors				[302]
3,5-Di-*O*-caffeoylquinic acid	Food products				[303]
Cefadroxil	Antibiotics				[304]
Baicalein glucuronide	Herbal medications				[305]
Oroxylin A glucuronide	Herbal medications				[305]

(*continued*)

TABLE 3.9 (*Continued*)

Object	Object Therapeutic Class	K_m (μM)	V_{max} (pmol/min/mg)	V_{max}/K_m (μL/min/mg)	References
Wogonin glucuronide	Herbal medications				[305]
7-Ethyl-10-hydroxycamptothecin (SN-38)	Topoisomerase inhibitors				[306]
Val-SN-38	Topoisomerase inhibitors				[306]
Vardenafil	Erectile dysfunction treatments	12.8	34.9 pmol/cm^2/min	2.84 μL/min/cm^2	[307]
Betanin	Herbal medications	275	42 pmol/cm^2/min	0.15 μL/cm^2/min	[308]
Isoscutellarin	Food products				[309]
Scutellarin	Herbal medications				[309]
Mycophenolic acid phenyl-glucuronide	Immunosuppressants	224.2	0.5 nmol/min/mg	448.4 ml/min/mg	[310]
Gemifloxacin	Antibiotics	55.76	35.58 pmol/cm^2/min	0.64 μL/cm^2/min	[310]
Anagliptin	Dipeptidyl peptidase-4 inhibitors				[19]
Paroxetine MI glucuronide	Selective serotonin reuptake inhibitors (SSRIs)				[311]
Bosentan	Endothelin receptor antagonists				[312]
Raltegravir	HIV-integrase strand transfer inhibitors				[313]
Gadolinium-ethoxybenzyl-diethylenetriamine pentaacetic acid (Gd-EOB-DTPA)	Diagnostic agents	1000	86.8	0.0868	[314]
Tm-25659	Obesity treatments	59.2	256 pmol/cm^2/min	4.32 μL/cm^2/min	[220]
Mycophenolic acid	Immunosuppressants	16			[315]
Mycophenolic acid 7-*O*-glucuronide	Immunosuppressants				[315]
5(6)-Carboxy-2,7-dichlorofluorescein (CDCF)	None				[316]

Glutathione methylfluorescein (GSMF)	None				[316]
99mtc-*N*-Pyridoxyl-5-methyltryptophan (99mtc-PMT)	Diagnostic agents				[223]
Cholecystokinin octapeptide (CCK-8)	None	27	269	9.96	[223]
Prostaglandin E2	Prostaglandins	174			[223]
Vasopressin	Hormone replacement therapy	315			[223]
Bms-275183	Taxanes				[317]
Ginsenoside Re	Herbal medications	89.3	39.9	0.447	[253]
Ginsenoside Rg1	Herbal medications	93.5	190	2.03	[253]
Notoginsenoside R1	Cardiovascular drugs	108	119	1.1	[253]
Timosaponin B2	Herbal medications				[230]
Forsythoside A	Herbal medications				[318]
Forsythoside B	Herbal medications				[318]
Isoforsythoside	Herbal medications				[318]
Danoprevir	Antivirals				[319]
Tanshinol	Herbal medications				[36]
Aconitine	Herbal medications				[320]
Benzoylaconine	Herbal medications				[320]
Benzoylhypaconine	Herbal medications				[320]
Hypaconitine	Herbal medications				[320]
Mesaconitine	Herbal medications				[320]
Tiliroside	Herbal medications				[321]
2-*O*-Beta-L-galactopyranosylorientin	Herbal medications				[321]
Isorhamnetin (ginkgo biloba derivative)	Herbal medications				[322]
Baicalin	Food products				[323]

3.4 CONCLUSION

ECCS is an effective approach that enables prospective prediction of the rate-determining step of NMEs. Transporters have a significant impact on the clearance of compounds that belong to ECCS Class 1B, 3A, 3B, and 4. Simple physicochemical properties such as ionization state and molecular weight along with *in vitro* permeability input, which are either calculated or measured, are used to prospectively predict the rate-determining step of NMEs. An understanding of the physicochemical drivers of various renal or hepatobiliary elimination mechanisms is key for predicting the rate-determining step of the NMEs clearance rate.

3.5 CHAPTER 3 TUTORIALS

1. True or false

OAT1 and OAT3 are ubiquitous transporters that are found in kidney and liver as well as in other body organs.

OATP1B1 and OATP1B3 are hepatoselective transporters that play a key role in the disposition of mainly neutral and basic molecules.

Compounds that are OATP1B1/1B3 substrates have low permeability.

There is an overlap in the physicochemical profile of compounds that are substrates for active hepatic OATP transporter uptake and canalicular efflux transporters.

Compounds that have high passive permeability are prone to have high renal clearance relative to their total clearance.

OCT1 is a facilitative transporter that is selectively expressed in the liver.

Transporters such as OATP1B1 and 1B3 play a key role in the clearance of ECCS Class 1A and 2 molecules.

Compounds that are ECCS Class 3A are acidic/zwitterion with low molecular weight ($M_{wt} < 400$) and permeability. The rate-determining step is renal elimination that is typically mediated by uptake renal transporters such as OAT1/3/OCT2.

Compounds that are ECCS Class 3B are acidic/zwitterion with low molecular weight ($M_{wt} \geq 400$) and permeability. The rate-determining step is renal elimination/active hepatic uptake that is followed by metabolism.

Compounds that are ECCS Class 4 are basic/neutral with low molecular weight and permeability. The rate-determining step is renal elimination that is typically mediated by uptake renal transporters such as OAT1/3/OCT2.

3.6 CHAPTER 3 TUTORIAL ANSWERS KEY

1. True or false

OAT1 and OAT3 are ubiquitous transporters that are found in kidney and liver as well as in other body organs. **(F)**

OATP1B1 and OATP1B3 are hepatoselective transporters that play a key role in the disposition of mainly neutral and basic molecules. **(F)**

Compounds that are OATP1B1/1B3 substrates have low permeability. **(F)**

There is an overlap in the physicochemical profile of compounds that are substrates for active hepatic OATP transporter uptake and canalicular efflux transporters. **(T)**

Compounds that have high passive permeability are prone to have high renal clearance relative to their total clearance. **(F)**

OCT1 is a facilitative transporter that is selectively expressed in the liver. **(T)**

Transporters such as OATP1B1 and 1B3 play a key role in the clearance of ECCS Class 1A and 2 molecules. **(F)**

Compounds that are ECCS Class 3A are acidic/zwitterion with low molecular weight ($M_{wt} < 400$) and permeability. The rate-determining step is renal elimination that is typically mediated by uptake renal transporters such as OAT1/3/OCT2. **(T)**

Compounds that are ECCS Class 3B are acidic/zwitterion with low molecular weight ($M_{wt} \geq 400$) and permeability. The rate-determining step is renal elimination/active hepatic uptake that is followed by metabolism. **(F)**

Compounds that are ECCS Class 4 are basic/neutral with low molecular weight and permeability. The rate-determining step is renal elimination that is typically mediated by uptake renal transporters such as OAT1/3/OCT2. **(T)**

REFERENCES

1. Varma, M.V., Steyn, S.J., Allerton, C. and El-Kattan, A.F., *Predicting clearance mechanism in drug discovery: extended clearance classification system (ECCS).* Pharm Res, 2015. 32(12): p. 3785–3802.
2. Ho, R.H. and Kim, R.B., *Transporters and drug therapy: implications for drug disposition and disease.* Clin Pharmacol Ther, 2005. 78(3): p. 260–277.
3. Shitara, Y., Sato, H. and Sugiyama, Y., *Evaluation of drug–drug interaction in the hepatobiliary and renal transport of drugs.* Annu Rev Pharmacol Toxicol 2005. 45: p. 689–723.
4. Feng, B., Hurst, S., Lu, Y., Varma, M.V., Rotter, C.J., El-Kattan, A., Lockwood, P. and Corrigan, B., *Quantitative prediction of renal transporter-mediated clinical drug–drug interactions.* Mol Pharm, 2013. 10(11): p. 4207–4215.
5. Feng, B., El-Kattan, A.F. and Radi, Z.A., *Renal transporters in drug disposition, drug–drug interactions, and nephrotoxicity.* Curr Protoc Toxicol, 2012. Chapter 23: Unit 23.3.1–23.3.15.
6. Varma, M.V., Feng, B., Obach, R.S., Troutman, M.D., Chupka, J., Miller, H.R. and El-Kattan, A., *Physicochemical determinants of human renal clearance.* J Med Chem, 2009. 52(15): p. 4844–4852.
7. Ahlin, G., Karlsson, J., Pedersen, J.M., Gustavsson, L., Larsson, R., Matsson, P., Norinder, U., Bergstrom, C.A. and Artursson, P., *Structural requirements for drug inhibition of the liver specific human organic cation transport protein 1.* J Med Chem, 2008. 51(19): p. 5932–5942.
8. Inui, K.I., Masuda, S. and Saito, H., *Cellular and molecular aspects of drug transport in the kidney.* Kidney Int, 2000. 58(3): p. 944–958.
9. Launay-Vacher, V., Izzedine, H., Karie, S., Hulot, J.S., Baumelou, A. and Deray, G., *Renal tubular drug transporters.* Nephron Physiol, 2006. 103(3): p. p97–p106.
10. Masereeuw, R. and Russel, F.G., *Mechanisms and clinical implications of renal drug excretion.* Drug Metab Rev, 2001. 33(3–4): p. 299–351.
11. Sirianni, G.L. and Pang, K.S., *Organ clearance concepts: new perspectives on old principles.* J Pharmacokinet Biopharm, 1997. 25(4): p. 449–470.
12. Nozaki, Y., Kusuhara, H., Kondo, T., Hasegawa, M., Shiroyanagi, Y., Nakazawa, H., Okano, T. and Sugiyama, Y., *Characterization of the uptake of organic anion transporter (OAT) 1 and OAT3 substrates by human kidney slices.* J Pharmacol Exp Ther, 2007. 321(1): p. 362–369.

13. Duan, P., Li, S., Ai, N., Hu, L., Welsh, W.J. and You, G., *Potent inhibitors of human organic anion transporters 1 and 3 from clinical drug libraries: discovery and molecular characterization.* Mol Pharm, 2012. 9(11): p. 3340–3346.
14. Ueo, H., Motohashi, H., Katsura, T. and Inui, K., *Human organic anion transporter hOAT3 is a potent transporter of cephalosporin antibiotics, in comparison with hOAT1.* Biochem Pharmacol, 2005. 70(7): p. 1104–1113.
15. Mizuno, N., Takahashi, T., Iwase, Y., Kusuhara, H., Niwa, T. and Sugiyama, Y., *Human organic anion transporters 1 (hOAT1/SLC22A6) and 3 (hOAT3/SLC22A8) transport edaravone (MCI-186; 3-methyl-1-phenyl-2-pyrazolin-5-one) and its sulfate conjugate.* Drug Metab Dispos, 2007. 35(8): p. 1429–1434.
16. Hill, G., Cihlar, T., Oo, C., Ho, E.S., Prior, K., Wiltshire, H., Barrett, J., Liu, B. and Ward, P., *The anti-influenza drug oseltamivir exhibits low potential to induce pharmacokinetic drug interactions via renal secretion–correlation of* in vivo *and* in vitro *studies.* Drug Metab Dispos, 2002. 30(1): p. 13–19.
17. Cheng, Y., Vapurcuyan, A., Shahidullah, M., Aleksunes, L.M. and Pelis, R.M., *Expression of organic anion transporter 2 in the human kidney and its potential role in the tubular secretion of guanine-containing antiviral drugs.* Drug Metab Dispos, 2012. 40(3): p. 617–624.
18. Uwai, Y. and Iwamoto, K., *Transport of aminopterin by human organic anion transporters hOAT1 and hOAT3: Comparison with methotrexate.* Drug Metab Pharmacokinet, 2010. 25(2): p. 163–169.
19. Furuta, S., Smart, C., Hackett, A., Benning, R. and Warrington, S., *Pharmacokinetics and metabolism of [14C]anagliptin, a novel dipeptidyl peptidase-4 inhibitor, in humans.* Xenobiotica, 2013. 43(5): p. 432–442.
20. Jin, M.J. and Han, H.K., *Interaction of zalcitabine with human organic anion transporter 1.* Pharmazie, 2006. 61(5): p. 491–492.
21. Yamada, A., et al., *Multiple human isoforms of drug transporters contribute to the hepatic and renal transport of olmesartan, a selective antagonist of the angiotensin II AT1-receptor.* Drug Metab Dispos, 2007. 35(12): p. 2166–2176.
22. Wong, C.C., Botting, N.P., Orfila, C., Al-Maharik, N. and Williamson, G., *Flavonoid conjugates interact with organic anion transporters (OATs) and attenuate cytotoxicity of adefovir mediated by organic anion transporter 1 (OAT1/SLC22A6).* Biochem Pharmacol, 2011. 81(7): p. 942–949.
23. Moss, D.M., Kwan, W.S., Liptrott, N.J., Smith, D.L., Siccardi, M., Khoo, S.H., Back, D.J. and Owen, A., *Raltegravir is a substrate for SLC22A6: a putative mechanism for the interaction between raltegravir and tenofovir.* Antimicrob Agents Chemother, 2011. 55(2): p. 879–887.
24. Xu, Q., Wang, C., Meng, Q., Liu, Q., Sun, H., Peng, J., Ma, X., Kaku, T. and Liu, K., *OAT1 and OAT3: targets of drug–drug interaction between entecavir and JBP485.* Eur J Pharm Sci, 2013. 48(4–5): p. 650–657.
25. Cutler, M.J., Urquhart, B.L., Velenosi, T.J., Meyer Zu Schwabedissen, H.E., Dresser, G.K., Leake, B.F., Tirona, R.G., Kim, R.B. and Freeman, D.J., In vitro *and* in vivo *assessment of renal drug transporters in the disposition of mesna and dimesna.* J Clin Pharmacol, 2012. 52(4): p. 530–542.
26. Makino, T., Okajima, K., Uebayashi, R., Ohtake, N., Inoue, K. and Mizukami, H., *3-Monoglucuronyl-glycyrrhretinic acid is a substrate of organic anion transporters expressed in tubular epithelial cells and plays important roles in licorice-induced pseudoaldosteronism by inhibiting 11beta-hydroxysteroid dehydrogenase 2.* J Pharmacol Exp Ther, 2012. 342(2): p. 297–304.
27. Juhasz, V., Beery, E., Nagy, Z., Bui, A., Molnar, E., Zolnerciks, J.K., Magnan, R., Jani, M. and Krajcsi, P., *Chlorothiazide is a substrate for the human uptake transporters OAT1 and OAT3.* J Pharm Sci, 2013. 102(5): p. 1683–1687.
28. Takahara, N., et al., *Drugs interacting with organic anion transporter-1 affect uptake of Tc-99m-mercaptoacetyl-triglycine (MAG3) in the human kidney: therapeutic drug interaction in Tc-99m-MAG3 diagnosis of renal function and possible application of Tc-99m-MAG3 for drug development.* Nucl Med Biol, 2013. 40(5): p. 643–650.
29. Ye, J., Liu, Q., Wang, C., Meng, Q., Sun, H., Peng, J., Ma, X. and Liu, K., *Benzylpenicillin inhibits the renal excretion of acyclovir by OAT1 and OAT3.* Pharmacol Rep, 2013. 65(2): p. 505–512.

30. Kurata, T., Iwamoto, T., Kawahara, Y. and Okuda, M., *Characteristics of pemetrexed transport by renal basolateral organic anion transporter hOAT3*. Drug Metab Pharmacokinet, 2014. 29(2): p. 148–153.

31. Watanabe, H., et al., *Human organic anion transporters function as a high-capacity transporter for p-cresyl sulfate, a uremic toxin*. Clin Exp Nephrol, 2014. 18(5): p. 814–820.

32. Li, L., Song, F., Tu, M., Wang, K., Zhao, L., Wu, X., Zhou, H., Xia, Z. and Jiang, H., *In vitro interaction of clopidogrel and its hydrolysate with OCT1, OCT2 and OAT1*. Int J Pharm, 2014. 465(1–2): p. 5–10.

33. Ingraham, L., Li, M., Renfro, J.L., Parker, S., Vapurcuyan, A., Hanna, I. and Pelis, R.M., *A plasma concentration of alpha-ketoglutarate influences the kinetic interaction of ligands with organic anion transporter 1*. Mol Pharmacol, 2014. 86(1): p. 86–95.

34. An, G., Wang, X. and Morris, M.E., *Flavonoids are inhibitors of human organic anion transporter 1 (OAT1)-mediated transport*. Drug Metab Dispos, 2014. 42(9): p. 1357–1366.

35. Wang, L., et al., *PEPT1- and OAT1/3-mediated drug–drug interactions between bestatin and cefixime* in vivo *and* in vitro *in rats, and* in vitro *in human*. Eur J Pharm Sci, 2014. 63: p. 77–86.

36. Jia, W., et al., *Renal tubular secretion of tanshinol: molecular mechanisms, impact on its systemic exposure, and propensity for dose-related nephrotoxicity and for renal herb–drug interactions*. Drug Metab Dispos, 2015. 43(5): p. 669–678.

37. Ebner, D.C., Bialek, P., El-Kattan, A.F., Ambler, C.M. and Tu, M., *Strategies for skeletal muscle targeting in drug discovery*. Curr Pharm Des, 2015. 21(10): p. 1327–1336.

38. Cha, S.H., Sekine, T., Fukushima, J.I., Kanai, Y., Kobayashi, Y., Goya, T. and Endou, H., *Identification and characterization of human organic anion transporter 3 expressing predominantly in the kidney*. Mol Pharmacol, 2001. 59(5): p. 1277–1286.

39. Babu, E., Takeda, M., Narikawa, S., Kobayashi, Y., Yamamoto, T., Cha, S.H., Sekine, T., Sakthisekaran, D. and Endou, H., *Human organic anion transporters mediate the transport of tetracycline*. Jpn J Pharmacol, 2002. 88(1): p. 69–76.

40. Takeda, M., Khamdang, S., Narikawa, S., Kimura, H., Kobayashi, Y., Yamamoto, T., Cha, S.H., Sekine, T. and Endou, H., *Human organic anion transporters and human organic cation transporters mediate renal antiviral transport*. J Pharmacol Exp Ther, 2002. 300(3): p. 918–924.

41. Khamdang, S., Takeda, M., Noshiro, R., Narikawa, S., Enomoto, A., Anzai, N., Piyachaturawat, P. and Endou, H., *Interactions of human organic anion transporters and human organic cation transporters with nonsteroidal anti-inflammatory drugs*. J Pharmacol Exp Ther, 2002. 303(2): p. 534–539.

42. Hasannejad, H., et al., *Interactions of human organic anion transporters with diuretics*. J Pharmacol Exp Ther, 2004. 308(3): p. 1021–1029.

43. Hashimoto, T., Narikawa, S., Huang, X.L., Minematsu, T., Usui, T., Kamimura, H. and Endou, H., *Characterization of the renal tubular transport of zonampanel, a novel alpha-amino-3-hydroxy-5-methylisoxazole-4-propionic acid receptor antagonist, by human organic anion transporters*. Drug Metab Dispos, 2004. 32(10): p. 1096–1102.

44. Tahara, H., Kusuhara, H., Endou, H., Koepsell, H., Imaoka, T., Fuse, E. and Sugiyama, Y., *A species difference in the transport activities of H2 receptor antagonists by rat and human renal organic anion and cation transporters*. J Pharmacol Exp Ther, 2005. 315(1): p. 337–345.

45. Fujino, H., Saito, T., Ogawa, S. and Kojima, J., *Transporter-mediated influx and efflux mechanisms of pitavastatin, a new inhibitor of HMG-CoA reductase*. J Pharm Pharmacol, 2005. 57(10): p. 1305–1311.

46. Tahara, H., Kusuhara, H., Maeda, K., Koepsell, H., Fuse, E. and Sugiyama, Y., *Inhibition of oat3-mediated renal uptake as a mechanism for drug–drug interaction between fexofenadine and probenecid*. Drug Metab Dispos, 2006. 34(5): p. 743–747.

47. Chu, X.Y., et al., *Transport of the dipeptidyl peptidase-4 inhibitor sitagliptin by human organic anion transporter 3, organic anion transporting polypeptide 4C1, and multidrug resistance P-glycoprotein*. J Pharmacol Exp Ther, 2007. 321(2): p. 673–683.

48. Uwai, Y., Ida, H., Tsuji, Y., Katsura, T. and Inui, K., *Renal transport of adefovir, cidofovir, and tenofovir by SLC22A family members (hOAT1, hOAT3, and hOCT2)*. Pharm Res, 2007. 24(4): p. 811–815.

49. Nakagomi-Hagihara, R., Nakai, D. and Tokui, T., *Inhibition of human organic anion transporter 3 mediated pravastatin transport by gemfibrozil and the metabolites in humans*. Xenobiotica, 2007. 37(4): p. 416–426.

50. Uwai, Y., Motohashi, H., Tsuji, Y., Ueo, H., Katsura, T. and Inui, K., *Interaction and transport characteristics of mycophenolic acid and its glucuronide via human organic anion transporters hOAT1 and hOAT3*. Biochem Pharmacol, 2007. 74(1): p. 161–168.

51. Matsumoto, S., Yoshida, K., Ishiguro, N., Maeda, T. and Tamai, I., *Involvement of rat and human organic anion transporter 3 in the renal tubular secretion of topotecan [(S)-9-dimethylaminomethyl-10-hydroxy-camptothecin hydrochloride]*. J Pharmacol Exp Ther, 2007. 322(3): p. 1246–1252.

52. Windass, A.S., Lowes, S., Wang, Y. and Brown, C.D., *The contribution of organic anion transporters OAT1 and OAT3 to the renal uptake of rosuvastatin*. J Pharmacol Exp Ther, 2007. 322(3): p. 1221–1227.

53. Yuan, H., Feng, B., Yu, Y., Chupka, J., Zheng, J.Y., Heath, T.G. and Bond, B.R., *Renal organic anion transporter-mediated drug–drug interaction between gemcabene and quinapril*. J Pharmacol Exp Ther, 2009. 330(1): p. 191–197.

54. Heller Rouassant, S., Castillo Santos, C.A., Silva Acosta, C., Sanchez Cerdan, C.J. and Ruiz Gomez, J., *Antiviral antibodies in acute laryngotracheobronchitis*. Bol Med Hosp Infant Mex, 1978. 35(1): p. 145–150.

55. Han, Y.H., Busler, D., Hong, Y., Tian, Y., Chen, C. and Rodrigues, A.D., *Transporter studies with the 3-O-sulfate conjugate of 17alpha-ethinylestradiol: assessment of human kidney drug transporters*. Drug Metab Dispos, 2010. 38(7): p. 1064–1071.

56. Uwai, Y., Honjo, H. and Iwamoto, K., *Inhibitory effect of selective cyclooxygenase-2 inhibitor lumiracoxib on human organic anion transporters hOAT1 and hOAT3*. Drug Metab Pharmacokinet, 2010. 25(5): p. 450–455.

57. Lai, Y., Sampson, K.E., Balogh, L.M., Brayman, T.G., Cox, S.R., Adams, W.J., Kumar, V. and Stevens, J.C., *Preclinical and clinical evidence for the collaborative transport and renal secretion of an oxazolidinone antibiotic by organic anion transporter 3 (OAT3/SLC22A8) and multidrug and toxin extrusion protein 1 (MATE1/SLC47A1)*. J Pharmacol Exp Ther, 2010. 334(3): p. 936–944.

58. Thoft, R.A., *Current concepts in ophthalmology: corneal disease*. N Engl J Med, 1978. 298(22): p. 1239–1241.

59. Popescu, C., Gheorghiu, T., Russu, R. and Sepeanu, S., *Asthma and tuberculosis (considerations based on hospital case studies)*. Rev Ig Bacteriol Virusol Parazitol Epidemiol Pneumoftiziol Pneumoftiziol, 1978. 27(1): p. 23–28.

60. Zhu, Y., Meng, Q., Wang, C., Liu, Q., Sun, H., Kaku, T. and Liu, K., *Organic anion transporters involved in the excretion of bestatin in the kidney*. Peptides, 2012. 33(2): p. 265–271.

61. Lucero, M.L., Gonzalo, A., Ganza, A., Leal, N., Soengas, I., Ioja, E., Gedey, S., Jahic, M. and Bednarczyk, D., *Interactions of bilastine, a new oral H(1) antihistamine, with human transporter systems*. Drug Chem Toxicol, 2012. 35(Suppl 1): p. 8–17.

62. Imamura, Y., Murayama, N., Okudaira, N., Kurihara, A., Inoue, K., Yuasa, H., Izumi, T., Kusuhara, H. and Sugiyama, Y., *Effect of the fluoroquinolone antibacterial agent DX-619 on the apparent formation and renal clearances of 6beta-hydroxycortisol, an endogenous probe for CYP3A4 inhibition, in healthy subjects*. Pharm Res, 2013. 30(2): p. 447–457.

63. Kusuhara, H., et al., *Effect of coadministration of single and multiple doses of rifampicin on the pharmacokinetics of fexofenadine enantiomers in healthy subjects*. Drug Metab Dispos, 2013. 41(1): p. 206–213.

64. Toyoshima, J., Kusuhara, H., Wempe, M.F., Endou, H. and Sugiyama, Y., *Investigation of the role of transporters on the hepatic elimination of an LAT1 selective inhibitor JPH203*. J Pharm Sci, 2013. 102(9): p. 3228–3238.

65. Vishwanathan, K., Mair, S., Gupta, A., Atherton, J., Clarkson-Jones, J., Edeki, T. and Das, S., *Assessment of the mass balance recovery and metabolite profile of avibactam in humans and* in vitro *drug–drug interaction potential*. Drug Metab Dispos, 2014. 42(5): p. 932–942.

66. Zhong, K., Li, X., Xie, C., Zhang, Y., Zhong, D. and Chen, X., *Effects of renal impairment on the pharmacokinetics of morinidazole: uptake transporter-mediated renal clearance of the conjugated metabolites*. Antimicrob Agents Chemother, 2014. 58(7): p. 4153–4161.

67. Jeong, H.U., Kwon, M., Lee, Y., Yoo, J.S., Shin, D.H., Song, I.S. and Lee, H.S., *Organic anion transporter 3- and organic anion transporting polypeptides 1B1- and 1B3-mediated transport of catalposide*. Drug Des Devel Ther, 2015. 9: p. 643–653.

68. Abe, Y., Ota, E., Harada, H., Kanbe, H., Kojima, Y., Kanazawa, T., Endo, T., Murakami, M. and Kobayashi, M., *Species differences in the metabolism of ritobegron* in vitro *and assessment of potential interactions with transporters and cytochrome P450 enzymes*. Pharmazie, 2015. 70(1): p. 38–46.

69. Feng, S., Cleary, Y., Parrott, N., Hu, P., Weber, C., Wang, Y., Yin, O.Q. and Shi, J., *Evaluating a physiologically based pharmacokinetic model for prediction of omeprazole clearance and assessing ethnic sensitivity in CYP2C19 metabolic pathway*. Eur J Clin Pharmacol, 2015. 71(5): p. 617–624.

70. Ebner, T., Ishiguro, N. and Taub, M.E., *The use of transporter probe drug cocktails for the assessment of transporter-based drug–drug interactions in a clinical setting–proposal of a four component transporter cocktail*. J Pharm Sci, 2015. 104(9): p. 3220–3228.

71. Jung, N., Lehmann, C., Rubbert, A., Knispel, M., Hartmann, P., van Lunzen, J., Stellbrink, H.J., Faetkenheuer, G. and Taubert, D., *Relevance of the organic cation transporters 1 and 2 for antiretroviral drug therapy in human immunodeficiency virus infection*. Drug Metab Dispos, 2008. 36(8): p. 1616–1623.

72. Hendrickx, R., Johansson, J.G., Lohmann, C., Jenvert, R.M., Blomgren, A., Borjesson, L. and Gustavsson, L., *Identification of novel substrates and structure–activity relationship of cellular uptake mediated by human organic cation transporters 1 and 2*. J Med Chem, 2013. 56(18): p. 7232–7242.

73. Umehara, K.I., Iwatsubo, T., Noguchi, K. and Kamimura, H., *Comparison of the kinetic characteristics of inhibitory effects exerted by biguanides and H2-blockers on human and rat organic cation transporter-mediated transport: Insight into the development of drug candidates*. Xenobiotica, 2007. 37(6): p. 618–634.

74. Duan, H., Hu, T., Foti, R.S., Pan, Y., Swaan, P.W. and Wang, J., *Potent and selective inhibition of plasma membrane monoamine transporter by HIV protease inhibitors*. Drug Metab Dispos, 2015. 43(11): p. 1773–1780.

75. Kang, H.J., Lee, S.S., Lee, C.H., Shim, J.C., Shin, H.J., Liu, K.H., Yoo, M.A. and Shin, J.G., *Neurotoxic pyridinium metabolites of haloperidol are substrates of human organic cation transporters*. Drug Metab Dispos, 2006. 34(7): p. 1145–1151.

76. Busch, A.E., et al., *Human neurons express the polyspecific cation transporter hOCT2, which translocates monoamine neurotransmitters, amantadine, and memantine*. Mol Pharmacol, 1998. 54(2): p. 342–352.

77. Lovejoy, K.S., Todd, R.C., Zhang, S., McCormick, M.S., D'Aquino, J.A., Reardon, J.T., Sancar, A., Giacomini, K.M. and Lippard, S.J., *cis-Diammine(pyridine)chloroplatinum(II), a monofunctional platinum(II) antitumor agent: uptake, structure, function, and prospects*. Proc Natl Acad Sci U S A, 2008. 105(26): p. 8902–8907.

78. Sprowl, J.A., et al., *Oxaliplatin-induced neurotoxicity is dependent on the organic cation transporter OCT2*. Proc Natl Acad Sci U S A, 2013. 110(27): p. 11199–11204.

79. Lepist, E.I., Gillies, H., Smith, W., Hao, J., Hubert, C., St Claire, 3rd, R.L., Brouwer, K.R. and Ray, A.S., *Evaluation of the endothelin receptor antagonists ambrisentan, bosentan, macitentan, and sitaxsentan as hepatobiliary transporter inhibitors and substrates in sandwich-cultured human hepatocytes*. PLoS One, 2014. 9(1): p. e87548.

80. Urakami, Y., Kimura, N., Okuda, M. and Inui, K., *Creatinine transport by basolateral organic cation transporter hOCT2 in the human kidney*. Pharm Res, 2004. 21(6): p. 976–981.

81. Song, I.S., Shin, H.J. and Shin, J.G., *Genetic variants of organic cation transporter 2 (OCT2) significantly reduce metformin uptake in oocytes*. Xenobiotica, 2008. 38(9): p. 1252–1262.

82. Klee, C.B. and Krinks, M.H., *Purification of cyclic 3',5'-nucleotide phosphodiesterase inhibitory protein by affinity chromatography on activator protein coupled to Sepharose*. Biochemistry, 1978. 17(1): p. 120–126.

83. Burger, H., Zoumaro-Djayoon, A., Boersma, A.W., Helleman, J., Berns, E.M., Mathijssen, R.H., Loos, W.J. and Wiemer, E.A., *Differential transport of platinum compounds by the human organic cation transporter hOCT2 (hSLC22A2)*. Br J Pharmacol, 2010. 159(4): p. 898–908.

84. More, S.S., Li, S., Yee, S.W., Chen, L., Xu, Z., Jablons, D.M. and Giacomini, K.M., *Organic cation transporters modulate the uptake and cytotoxicity of picoplatin, a third-generation platinum analogue*. Mol Cancer Ther, 2010. 9(4): p. 1058–1069.

85. Diao, L., Shu, Y. and Polli, J.E., *Uptake of pramipexole by human organic cation transporters*. Mol Pharm, 2010. 7(4): p. 1342–1347.

86. Nakanishi, T., Haruta, T., Shirasaka, Y. and Tamai, I., *Organic cation transporter-mediated renal secretion of ipratropium and tiotropium in rats and humans*. Drug Metab Dispos, 2011. 39(1): p. 117–122.

87. Ciarimboli, G., et al., *New clues for nephrotoxicity induced by ifosfamide: preferential renal uptake via the human organic cation transporter 2*. Mol Pharm, 2011. 8(1): p. 270–279.

88. Wenge, B., Geyer, J. and Bonisch, H., *Oxybutynin and trospium are substrates of the human organic cation transporters*. Naunyn Schmiedebergs Arch Pharmacol, 2011. 383(2): p. 203–208.

89. Jouan, E., Le Vee, M., Denizot, C., Da Violante, G. and Fardel, O., *The mitochondrial fluorescent dye rhodamine 123 is a high-affinity substrate for organic cation transporters (OCTs) 1 and 2*. Fundam Clin Pharmacol, 2014. 28(1): p. 65–77.

90. Sogame, Y., Kitamura, A., Yabuki, M., Komuro, S. and Takano, M., *Transport of biguanides by human organic cation transporter OCT2*. Biomed Pharmacother, 2013. 67(5): p. 425–430.

91. Belzer, M., Morales, M., Jagadish, B., Mash, E.A. and Wright, S.H., *Substrate-dependent ligand inhibition of the human organic cation transporter OCT2*. J Pharmacol Exp Ther, 2013. 346(2): p. 300–310.

92. Strobel, J., Muller, F., Zolk, O., Endress, B., Konig, J., Fromm, M.F. and Maas, R., *Transport of asymmetric dimethylarginine (ADMA) by cationic amino acid transporter 2 (CAT2), organic cation transporter 2 (OCT2) and multidrug and toxin extrusion protein 1 (MATE1)*. Amino Acids, 2013. 45(4): p. 989–1002.

93. Zolk, O., Solbach, T.F., Konig, J. and Fromm, M.F., *Structural determinants of inhibitor interaction with the human organic cation transporter OCT2 (SLC22A2)*. Naunyn Schmiedebergs Arch Pharmacol, 2009. 379(4): p. 337–348.

94. Kato, K., et al., *Investigation of endogenous compounds for assessing the drug interactions in the urinary excretion involving multidrug and toxin extrusion proteins*. Pharm Res, 2014. 31(1): p. 136–147.

95. Muller, F., Pontones, C.A., Renner, B., Mieth, M., Hoier, E., Auge, D., Maas, R., Zolk, O. and Fromm, M.F., *N(1)-methylnicotinamide as an endogenous probe for drug interactions by renal cation transporters: studies on the metformin-trimethoprim interaction*. Eur J Clin Pharmacol, 2015. 71(1): p. 85–94.

96. Ito, S., et al., *Potent and specific inhibition of mMate1-mediated efflux of type I organic cations in the liver and kidney by pyrimethamine*. J Pharmacol Exp Ther, 2010. 333(1): p. 341–350.

97. Ito, S., Kusuhara, H., Yokochi, M., Toyoshima, J., Inoue, K., Yuasa, H. and Sugiyama, Y., *Competitive inhibition of the luminal efflux by multidrug and toxin extrusions, but not basolateral uptake by organic cation transporter 2, is the likely mechanism underlying the pharmacokinetic drug–drug interactions caused by cimetidine in the kidney*. J Pharmacol Exp Ther, 2012. 340(2): p. 393–403.

98. Muller, F., Konig, J., Glaeser, H., Schmidt, I., Zolk, O., Fromm, M.F. and Maas, R., *Molecular mechanism of renal tubular secretion of the antimalarial drug chloroquine*. Antimicrob Agents Chemother, 2011. 55(7): p. 3091–3098.

99. Weiner, I.M. and Roth, L., *Renal excretion of cimetidine*. J Pharmacol Exp Ther, 1981. 216(3): p. 516–520.

100. Yonezawa, A., Masuda, S., Yokoo, S., Katsura, T. and Inui, K., *Cisplatin and oxaliplatin, but not carboplatin and nedaplatin, are substrates for human organic cation transporters (SLC22A1-3 and multidrug and toxin extrusion family)*. J Pharmacol Exp Ther, 2006. 319(2): p. 879–886.

101. Tanihara, Y., Masuda, S., Sato, T., Katsura, T., Ogawa, O. and Inui, K., *Substrate specificity of MATE1 and MATE2-K, human multidrug and toxin extrusions/H(+)-organic cation antiporters*. Biochem Pharmacol, 2007. 74(2): p. 359–371.

102. Kajiwara, M., Masuda, S., Watanabe, S., Terada, T., Katsura, T. and Inui, K., *Renal tubular secretion of varenicline by multidrug and toxin extrusion (MATE) transporters*. Drug Metab Pharmacokinet, 2012. 27(6): p. 563–569.

103. Wittwer, M.B., et al., *Discovery of potent, selective multidrug and toxin extrusion transporter 1 (MATE1, SLC47A1) inhibitors through prescription drug profiling and computational modeling*. J Med Chem, 2013. 56(3): p. 781–795.

104. Muller, F., Konig, J., Hoier, E., Mandery, K. and Fromm, M.F., *Role of organic cation transporter OCT2 and multidrug and toxin extrusion proteins MATE1 and MATE2-K for transport and drug interactions of the antiviral lamivudine*. Biochem Pharmacol, 2013. 86(6): p. 808–815.

105. Matsson, E.M., et al., *Combined* in vitro–in vivo *approach to assess the hepatobiliary disposition of a novel oral thrombin inhibitor*. Mol Pharm, 2013. 10(11): p. 4252–4262.

106. Li, L., Tu, M., Yang, X., Sun, S., Wu, X., Zhou, H., Zeng, S. and Jiang, H., *The contribution of human OCT1, OCT3, and CYP3A4 to nitidine chloride-induced hepatocellular toxicity*. Drug Metab Dispos, 2014. 42(7): p. 1227–1234.

107. Chen, L., et al., *OCT1 is a high-capacity thiamine transporter that regulates hepatic steatosis and is a target of metformin*. Proc Natl Acad Sci U S A, 2014. 111(27): p. 9983–9988.

108. Knop, J., Hoier, E., Ebner, T., Fromm, M.F. and Muller, F., *Renal tubular secretion of pramipexole*. Eur J Pharm Sci, 2015. 79: p. 73–78.

109. Yin, J., Duan, H., Shirasaka, Y., Prasad, B. and Wang, J., *Atenolol renal secretion is mediated by human organic cation transporter 2 and multidrug and toxin extrusion proteins*. Drug Metab Dispos, 2015. 43(12): p. 1872–1881.

110. Masuda, S., Terada, T., Yonezawa, A., Tanihara, Y., Kishimoto, K., Katsura, T., Ogawa, O. and Inui, K., *Identification and functional characterization of a new human kidney-specific H+/organic cation antiporter, kidney-specific multidrug and toxin extrusion 2*. J Am Soc Nephrol, 2006. 17(8): p. 2127–2135.

111. Matsushima, S., et al., *The inhibition of human multidrug and toxin extrusion 1 is involved in the drug–drug interaction caused by cimetidine*. Drug Metab Dispos, 2009. 37(3): p. 555–559.

112. Ito, S., et al., *N-methylnicotinamide is an endogenous probe for evaluation of drug–drug interactions involving multidrug and toxin extrusions (MATE1 and MATE2-K)*. Clin Pharmacol Ther, 2012. 92(5): p. 635–641.

113. Lotsch, J., Skarke, C., Liefhold, J. and Geisslinger, G., *Genetic predictors of the clinical response to opioid analgesics: clinical utility and future perspectives*. Clin Pharmacokinet, 2004. 43(14): p. 983–1013.

114. Kivisto, K.T., Niemi, M. and Fromm, M.F., *Functional interaction of intestinal CYP3A4 and P-glycoprotein*. Fundam Clin Pharmacol, 2004. 18(6): p. 621–626.

115. Jamroziak, K. and Robak, T., *Pharmacogenomics of MDR1/ABCB1 gene: the influence on risk and clinical outcome of haematological malignancies*. Hematology, 2004. 9(2): p. 91–105.

116. Dingemanse, J. and van Giersbergen, P.L., *Clinical pharmacology of bosentan, a dual endothelin receptor antagonist*. Clin Pharmacokinet, 2004. 43(15): p. 1089–1115.

117. Bekaii-Saab, T.S., Perloff, M.D., Weemhoff, J.L., Greenblatt, D.J. and von Moltke, L.L., *Interactions of tamoxifen, N-desmethyltamoxifen and 4-hydroxytamoxifen with P-glycoprotein and CYP3A*. Biopharm Drug Dispos, 2004. 25(7): p. 283–289.

118. Fredericks, S., Holt, D.W. and MacPhee, I.A., *The pharmacogenetics of immunosuppression for organ transplantation: a route to individualization of drug administration*. Am J Pharmacogenomics, 2003. 3(5): p. 291–301.

119. Lee, C.A., et al., *Breast cancer resistance protein (ABCG2) in clinical pharmacokinetics and drug interactions: practical recommendations for clinical victim and perpetrator drug–drug interaction study design*. Drug Metab Dispos, 2015. 43(4): p. 490–509.

120. Bircsak, K.M. and Aleksunes, L.M., *Interaction of isoflavones with the BCRP/ABCG2 drug transporter*. Curr Drug Metab, 2015. 16(2): p. 124–140.

121. Rijpma, S.R., van den Heuvel, J.J., van der Velden, M., Sauerwein, R.W., Russel, F.G. and Koenderink, J.B., *Atovaquone and quinine anti-malarials inhibit ATP binding cassette transporter activity*. Malar J, 2014. 13: p. 359.

122. Noguchi, K., Katayama, K. and Sugimoto, Y., *Human ABC transporter ABCG2/BCRP expression in chemoresistance: basic and clinical perspectives for molecular cancer therapeutics*. Pharmgenomics Pers Med, 2014. 7: p. 53–64.

123. Fagerholm, U., *Prediction of human pharmacokinetics – renal metabolic and excretion clearance*. J Pharm Pharmacol, 2007. 59(11): p. 1463–1471.

124. Feng, B., LaPerle, J.L., Chang, G. and Varma, M.V., *Renal clearance in drug discovery and development: molecular descriptors, drug transporters and disease state*. Expert Opin Drug Metab Toxicol, 2010. 6(8): p. 939–952.

125. Li, M., Anderson, G.D. and Wang, J., *Drug–drug interactions involving membrane transporters in the human kidney*. Expert Opin Drug Metab Toxicol, 2006. 2(4): p. 505–532.

126. Sweet, D.H., Bush, K.T. and Nigam, S.K., *The organic anion transporter family: from physiology to ontogeny and the clinic*. Am J Physiol Renal Physiol, 2001. 281(2): p. F197–F205.

127. Motohashi, H. and Inui, K., *Organic cation transporter OCTs (SLC22) and MATEs (SLC47) in the human kidney*. AAPS J, 2013. 15(2): p. 581–588.

128. Goodwin, J.T., Conradi, R.A., Ho, N.F. and Burton, P.S., *Physicochemical determinants of passive membrane permeability: role of solute hydrogen-bonding potential and volume*. J Med Chem, 2001. 44(22): p. 3721–3729.

129. Varma, M.V., Sateesh, K. and Panchagnula, R., *Functional role of P-glycoprotein in limiting intestinal absorption of drugs: contribution of passive permeability to P-glycoprotein mediated efflux transport*. Mol Pharm, 2005. 2(1): p. 12–21.

130. Ullrich, K.J., *Renal transporters for organic anions and organic cations. Structural requirements for substrates*. J Membr Biol, 1997. 158(2): p. 95–107.

131. Bednarczyk, D., Ekins, S., Wikel, J.H. and Wright, S.H., *Influence of molecular structure on substrate binding to the human organic cation transporter, hOCT1*. Mol Pharmacol, 2003. 63(3): p. 489–498.

132. Liu, L. and Pang, K.S., *The roles of transporters and enzymes in hepatic drug processing*. Drug Metab Dispos, 2005. 33(1): p. 1–9.

133. Pang, K.S. and Rowland, M., *Hepatic clearance of drugs. I. Theoretical considerations of a "well-stirred" model and a "parallel tube" model. Influence of hepatic blood flow, plasma and blood cell binding, and the hepatocellular enzymatic activity on hepatic drug clearance*. J Pharmacokinet Biopharm, 1977. 5(6): p. 625–653.

134. Barton, H.A., Lai, Y., Goosen, T.C., Jones, H.M., El-Kattan, A.F., Gosset, J.R., Lin, J. and Varma, M.V., *Model-based approaches to predict drug–drug interactions associated with hepatic uptake transporters: preclinical, clinical and beyond*. Expert Opin Drug Metab Toxicol, 2013. 9(4): p. 459–472.

135. Shitara, Y., Horie, T. and Sugiyama, Y., *Transporters as a determinant of drug clearance and tissue distribution*. Eur J Pharm Sci, 2006. 27(5): p. 425–446.

136. Shitara, Y. and Sugiyama, Y., *Pharmacokinetic and pharmacodynamic alterations of 3-hydroxy-3-methylglutaryl coenzyme A (HMG-CoA) reductase inhibitors: drug–drug interactions and interindividual differences in transporter and metabolic enzyme functions*. Pharmacol Ther, 2006. 112(1): p. 71–105.

137. Camenisch, G. and Umehara, K., *Predicting human hepatic clearance from* in vitro *drug metabolism and transport data: a scientific and pharmaceutical perspective for assessing drug–drug interactions*. Biopharm Drug Dispos, 2012. 33(4): p. 179–194.

138. Varma, M.V., Lin, J., Bi, Y.A., Rotter, C.J., Fahmi, O.A., Lam, J.L., El-Kattan, A.F., Goosen, T.C. and Lai, Y., *Quantitative prediction of repaglinide-rifampicin complex drug interactions using dynamic and static mechanistic models: delineating differential CYP3A4 induction and OATP1B1 inhibition potential of rifampicin*. Drug Metab Dispos, 2013. 41(5): p. 966–974.

139. De Buck, S.S., Sinha, V.K., Fenu, L.A., Nijsen, M.J., Mackie, C.E. and Gilissen, R.A., *Prediction of human pharmacokinetics using physiologically based modeling: a retrospective analysis of 26 clinically tested drugs*. Drug Metab Dispos, 2007. 35(10): p. 1766–1780.

140. Jones, H.M., et al., *Simulation of human intravenous and oral pharmacokinetics of 21 diverse compounds using physiologically based pharmacokinetic modelling*. Clin Pharmacokinet, 2011. 50(5): p. 331–347.

141. Obach, R.S., Baxter, J.G., Liston, T.E., Silber, B.M., Jones, B.C., Macintyre, F., Rance, D.J. and Wastall, P., *The prediction of human pharmacokinetic parameters from preclinical and* in vitro *metabolism data*. J Pharmacol Exp Ther, 1997. 283(1): p. 46–58.

142. Williams, J.A., Hyland, R., Jones, B.C., Smith, D.A., Hurst, S., Goosen, T.C., Peterkin, V., Koup, J.R. and Ball, S.E., *Drug–drug interactions for UDP-glucuronosyltransferase substrates: a pharmacokinetic explanation for typically observed low exposure (AUCi/AUC) ratios*. Drug Metab Dispos, 2004. 32(11): p. 1201–1208.

143. Di, L., Feng, B., Goosen, T.C., Lai, Y., Steyn, S.J., Varma, M.V. and Obach, R.S., *A perspective on the prediction of drug pharmacokinetics and disposition in drug research and development*. Drug Metab Dispos, 2013. 41(12): p. 1975–1993.

144. Hosea, N.A. and Jones, H.M., *Predicting pharmacokinetic profiles using* in silico *derived parameters*. Mol Pharm, 2013. 10(4): p. 1207–1215.

145. Hosea, N.A., *Drug design tools –* in silico, in vitro *and* in vivo *ADME/PK prediction and interpretation: is PK in monkey an essential part of a good human PK prediction?* Curr Top Med Chem, 2011. 11(4): p. 351–357.

146. Hosea, N.A., et al., *Prediction of human pharmacokinetics from preclinical information: comparative accuracy of quantitative prediction approaches*. J Clin Pharmacol, 2009. 49(5): p. 513–533.

147. Fenner, K.S., Jones, H.M., Ullah, M., Kempshall, S., Dickins, M., Lai, Y., Morgan, P. and Barton, H.A., *The evolution of the OATP hepatic uptake transport protein family in DMPK sciences: from obscure liver transporters to key determinants of hepatobiliary clearance*. Xenobiotica, 2012. 42(1): p. 28–45.

148. International Transporter, C., et al., *Membrane transporters in drug development*. Nat Rev Drug Discov, 2010. 9(3): p. 215–236.

149. Kalliokoski, A. and Niemi, M., *Impact of OATP transporters on pharmacokinetics*. Br J Pharmacol, 2009. 158(3): p. 693–705.

150. Shimizu, K., et al., *Whole-body distribution and radiation dosimetry of [11C]telmisartan as a biomarker for hepatic organic anion transporting polypeptide (OATP) 1B3*. Nucl Med Biol, 2012. 39(6): p. 847–853.

151. Bergman, E., Forsell, P., Tevell, A., Persson, E.M., Hedeland, M., Bondesson, U., Knutson, L. and Lennernas, H., *Biliary secretion of rosuvastatin and bile acids in humans during the absorption phase*. Eur J Pharm Sci, 2006. 29(3–4): p. 205–214.

152. Yamazaki, M., Kobayashi, K. and Sugiyama, Y., *Primary active transport of pravastatin across the liver canalicular membrane in normal and mutant Eisai hyperbilirubinemic rats*. Biopharm Drug Dispos, 1996. 17(7): p. 607–621.

153. Varma, M.V., Lai, Y., Feng, B., Litchfield, J., Goosen, T.C. and Bergman, A., *Physiologically based modeling of pravastatin transporter-mediated hepatobiliary disposition and drug–drug interactions*. Pharm Res, 2012. 29(10): p. 2860–2873.

154. Watanabe, T., Kusuhara, H., Maeda, K., Shitara, Y. and Sugiyama, Y., *Physiologically based pharmacokinetic modeling to predict transporter-mediated clearance and distribution of pravastatin in humans*. J Pharmacol Exp Ther, 2009. 328(2): p. 652–662.

155. Maeda, K., Ikeda, Y., Fujita, T., Yoshida, K., Azuma, Y., Haruyama, Y., Yamane, N., Kumagai, Y. and Sugiyama, Y., *Identification of the rate-determining process in the hepatic clearance of atorvastatin in a clinical cassette microdosing study*. Clin Pharmacol Ther, 2011. 90(4): p. 575–581.

156. Lau, Y., Huang, Y., Frassetto, L. and Benet, L., *Effect of OATP1B transporter inhibition on the pharmacokinetics of atorvastatin in healthy volunteers*. Clin Pharmacol Ther (St Louis, MO, USA), 2006. 81(2): p. 194–204.

157. Varma, M.V., Scialis, R.J., Lin, J., Bi, Y.A., Rotter, C.J., Goosen, T.C. and Yang, X., *Mechanism-based pharmacokinetic modeling to evaluate transporter-enzyme interplay in drug interactions and pharmacogenetics of glyburide*. AAPS J, 2014. 16(4): p. 736–748.

158. Zheng, H., Huang, Y., Frassetto, L. and Benet, L., *Elucidating rifampin's inducing and inhibiting effects on glyburide pharmacokinetics and blood glucose in healthy volunteers: unmasking the differential effects of enzyme induction and transporter inhibition for a drug and its primary metabolite*. Clin Pharmacol Ther (St Louis, MO, USA), 2008. 85(1): p. 78–85.

159. Kalliokoski, A., Backman, J.T., Kurkinen, K.J., Neuvonen, P.J. and Niemi, M., *Effects of gemfibrozil and atorvastatin on the pharmacokinetics of repaglinide in relation to SLCO1B1 polymorphism*. Clin Pharmacol Ther, 2008. 84(4): p. 488–496.

160. Varma, M.V., Lai, Y., Kimoto, E., Goosen, T.C., El-Kattan, A.F. and Kumar, V., *Mechanistic modeling to predict the transporter- and enzyme-mediated drug–drug interactions of repaglinide*. Pharm Res, 2013. 30(4): p. 1188–1199.

161. Prueksaritanont, T., et al., *Pitavastatin is a more sensitive and selective OATP1B clinical probe than rosuvastatin*. Br J Clin Pharmacol, 2014. 78(3): p. 587–598.

162. Fahrmayr, C., Fromm, M.F. and Konig, J., *Hepatic OATP and OCT uptake transporters: their role for drug–drug interactions and pharmacogenetic aspects*. Drug Metab Rev, 2010. 42(3): p. 380–401.

163. Fenner, K.S., Jones, H.M., Ullah, M., Kempshall, S., Dickins, M., Lai, Y., Morgan, P. and Barton, H.A., *The evolution of the OATP hepatic uptake transport protein family in DMPK sciences: from obscure liver transporters to key determinants of hepatobiliary clearance*. Xenobiotica, 2012. 42(1): p. 28–45.

164. Lai, Y., et al., *Impact of drug transporter pharmacogenomics on pharmacokinetic and pharmacodynamic variability – considerations for drug development*. Expert Opin Drug Metab Toxicol, 2012. 8(6): p. 723–743.

165. Hsiang, B., Zhu, Y., Wang, Z., Wu, Y., Sasseville, V., Yang, W.P. and Kirchgessner, T.G., *A novel human hepatic organic anion transporting polypeptide (OATP2). Identification of a liver-specific human organic anion transporting polypeptide and identification of rat and human hydroxymethylglutaryl-CoA reductase inhibitor transporters*. J Biol Chem, 1999. 274(52): p. 37161–37168.

166. Kullak-Ublick, G.A., Ismair, M.G., Stieger, B., Landmann, L., Huber, R., Pizzagalli, F., Fattinger, K., Meier, P.J. and Hagenbuch, B., *Organic anion-transporting polypeptide B (OATP-B) and its functional comparison with three other OATPs of human liver*. Gastroenterology, 2001. 120(2): p. 525–533.

167. Nakai, D., Nakagomi, R., Furuta, Y., Tokui, T., Abe, T., Ikeda, T. and Nishimura, K., *Human liver-specific organic anion transporter, LST-1, mediates uptake of pravastatin by human hepatocytes*. J Pharmacol Exp Ther, 2001. 297(3): p. 861–867.

168. Abe, T., et al., *LST-2, a human liver-specific organic anion transporter, determines methotrexate sensitivity in gastrointestinal cancers*. Gastroenterology, 2001. 120(7): p. 1689–1699.

169. Tirona, R.G., Leake, B.F., Wolkoff, A.W. and Kim, R.B., *Human organic anion transporting polypeptide-C (SLC21A6) is a major determinant of rifampin-mediated pregnane X receptor activation*. J Pharmacol Exp Ther, 2003. 304(1): p. 223–228.

170. Fehrenbach, T., Cui, Y., Faulstich, H. and Keppler, D., *Characterization of the transport of the bicyclic peptide phalloidin by human hepatic transport proteins*. Naunyn Schmiedebergs Arch Pharmacol, 2003. 368(5): p. 415–420.

171. Nozawa, T., Sugiura, S., Nakajima, M., Goto, A., Yokoi, T., Nezu, J., Tsuji, A. and Tamai, I., *Involvement of organic anion transporting polypeptides in the transport of troglitazone sulfate: implications for understanding troglitazone hepatotoxicity*. Drug Metab Dispos, 2004. 32(3): p. 291–294.

172. Hirano, M., Maeda, K., Shitara, Y. and Sugiyama, Y., *Contribution of OATP2 (OATP1B1) and OATP8 (OATP1B3) to the hepatic uptake of pitavastatin in humans*. J Pharmacol Exp Ther, 2004. 311(1): p. 139–146.

173. Takada, T., Weiss, H.M., Kretz, O., Gross, G. and Sugiyama, Y., *Hepatic transport of PKI166, an epidermal growth factor receptor kinase inhibitor of the pyrrolo-pyrimidine class, and its main metabolite, ACU154*. Drug Metab Dispos, 2004. 32(11): p. 1272–1278.

174. Fujino, H., Nakai, D., Nakagomi, R., Saito, M., Tokui, T. and Kojima, J., *Metabolic stability and uptake by human hepatocytes of pitavastatin, a new inhibitor of HMG-CoA reductase*. Arzneimittelforschung, 2004. 54(7): p. 382–388.

175. Nozawa, T., Minami, H., Sugiura, S., Tsuji, A. and Tamai, I., *Role of organic anion transporter OATP1B1 (OATP-C) in hepatic uptake of irinotecan and its active metabolite, 7-ethyl-10-hydroxycamptothecin:* in vitro *evidence and effect of single nucleotide polymorphisms*. Drug Metab Dispos, 2005. 33(3): p. 434–439.

176. Sandhu, P., Lee, W., Xu, X., Leake, B.F., Yamazaki, M., Stone, J.A., Lin, J.H., Pearson, P.G. and Kim, R.B., *Hepatic uptake of the novel antifungal agent caspofungin*. Drug Metab Dispos, 2005. 33(5): p. 676–682.

177. Kameyama, Y., Yamashita, K., Kobayashi, K., Hosokawa, M. and Chiba, K., *Functional characterization of SLCO1B1 (OATP-C) variants, SLCO1B1*5, SLCO1B1*15 and SLCO1B1*15+C1007G, by using transient expression systems of HeLa and HEK293 cells*. Pharmacogenet Genomics, 2005. 15(7): p. 513–522.

178. Katz, D.A., et al., *Organic anion transporting polypeptide 1B1 activity classified by SLCO1B1 genotype influences atrasentan pharmacokinetics*. Clin Pharmacol Ther, 2006. 79(3): p. 186–196.

179. Yamashiro, W., Maeda, K., Hirouchi, M., Adachi, Y., Hu, Z. and Sugiyama, Y., *Involvement of transporters in the hepatic uptake and biliary excretion of valsartan, a selective antagonist of the angiotensin II AT1-receptor, in humans*. Drug Metab Dispos, 2006. 34(7): p. 1247–1254.

180. Liu, L., Cui, Y., Chung, A.Y., Shitara, Y., Sugiyama, Y., Keppler, D. and Pang, K.S., *Vectorial transport of enalapril by Oatp1a1/Mrp2 and OATP1B1 and OATP1B3/MRP2 in rat and human livers*. J Pharmacol Exp Ther, 2006. 318(1): p. 395–402.

181. Maeda, K., et al., *Effects of organic anion transporting polypeptide 1B1 haplotype on pharmacokinetics of pravastatin, valsartan, and temocapril*. Clin Pharmacol Ther, 2006. 79(5): p. 427–439.

182. Lau, Y.Y., Huang, Y., Frassetto, L. and Benet, L.Z., *Effect of OATP1B transporter inhibition on the pharmacokinetics of atorvastatin in healthy volunteers*. Clin Pharmacol Ther, 2007. 81(2): p. 194–204.

183. Noe, J., Portmann, R., Brun, M.E. and Funk, C., *Substrate-dependent drug–drug interactions between gemfibrozil, fluvastatin and other organic anion-transporting peptide (OATP) substrates on OATP1B1, OATP2B1, and OATP1B3*. Drug Metab Dispos, 2007. 35(8): p. 1308–1314.

184. Treiber, A., Schneiter, R., Hausler, S. and Stieger, B., *Bosentan is a substrate of human OATP1B1 and OATP1B3: inhibition of hepatic uptake as the common mechanism of its interactions with cyclosporin A, rifampicin, and sildenafil*. Drug Metab Dispos, 2007. 35(8): p. 1400–1407.

185. Deng, J.W., Shon, J.H., Shin, H.J., Park, S.J., Yeo, C.W., Zhou, H.H., Song, I.S. and Shin, J.G., *Effect of silymarin supplement on the pharmacokinetics of rosuvastatin*. Pharm Res, 2008. 25(8): p. 1807–1814.

186. Umehara, K., Iwai, M., Adachi, Y., Iwatsubo, T., Usui, T. and Kamimura, H., *Hepatic uptake and excretion of (−)-N-{2-[(R)-3-(6,7-dimethoxy-1,2,3,4-tetrahydroisoquinoline-2-carbonyl)piperidino]ethyl}-4-fluorobenzamide (YM758), a novel if channel inhibitor, in rats and humans*. Drug Metab Dispos, 2008. 36(6): p. 1030–1038.

187. Sakamoto, S., Kusuhara, H., Horie, K., Takahashi, K., Baba, T., Ishizaki, J. and Sugiyama, Y., *Identification of the transporters involved in the hepatobiliary transport and intestinal efflux of methyl 1-(3,4-dimethoxyphenyl)-3-(3-ethylvaleryl)-4-hydroxy-6,7,8-trimethoxy-2-naphthoate (S-8921) glucuronide, a pharmacologically active metabolite of S-8921*. Drug Metab Dispos, 2008. 36(8): p. 1553–1561.

188. Oswald, S., Konig, J., Lutjohann, D., Giessmann, T., Kroemer, H.K., Rimmbach, C., Rosskopf, D., Fromm, M.F. and Siegmund, W., *Disposition of ezetimibe is influenced by polymorphisms of the hepatic uptake carrier OATP1B1*. Pharmacogenet Genomics, 2008. 18(7): p. 559–568.

189. Nakakariya, M., Shimada, T., Irokawa, M., Maeda, T. and Tamai, I., *Identification and species similarity of OATP transporters responsible for hepatic uptake of beta-lactam antibiotics*. Drug Metab Pharmacokinet, 2008. 23(5): p. 347–355.

190. Oostendorp, R.L., van de Steeg, E., van der Kruijssen, C.M., Beijnen, J.H., Kenworthy, K.E., Schinkel, A.H. and Schellens, J.H., *Organic anion-transporting polypeptide 1B1 mediates transport of gimatecan and BNP1350 and can be inhibited by several classic ATP-binding cassette (ABC) B1 and/or ABCG2 inhibitors*. Drug Metab Dispos, 2009. 37(4): p. 917–923.

191. Feng, B., et al., *Role of hepatic transporters in the disposition and hepatotoxicity of a HER2 tyrosine kinase inhibitor CP-724,714*. Toxicol Sci, 2009. 108(2): p. 492–500.

192. Picard, N., Yee, S.W., Woillard, J.B., Lebranchu, Y., Le Meur, Y., Giacomini, K.M. and Marquet, P., *The role of organic anion-transporting polypeptides and their common genetic variants in mycophenolic acid pharmacokinetics*. Clin Pharmacol Ther, 2010. 87(1): p. 100–108.

193. Hartkoorn, R.C., et al., *HIV protease inhibitors are substrates for OATP1A2, OATP1B1 and OATP1B3 and lopinavir plasma concentrations are influenced by SLCO1B1 polymorphisms*. Pharmacogenet Genomics, 2010. 20(2): p. 112–120.

194. Annaba, F., Ma, K., Kumar, P., Dudeja, A.K., Kineman, R.D., Shneider, B.L., Saksena, S., Gill, R.K. and Alrefai, W.A., *Ileal apical Na^+-dependent bile acid transporter ASBT is upregulated in rats with diabetes mellitus induced by low doses of streptozotocin*. Am J Physiol Gastrointest Liver Physiol, 2010. 299(4): p. G898–G906.

195. Oswald, S., Nassif, A., Modess, C., Keiser, M., Hanke, U., Engel, A., Lutjohann, D., Weitschies, W. and Siegmund, W., *Pharmacokinetic and pharmacodynamic interactions between the immunosuppressant sirolimus and the lipid-lowering drug ezetimibe in healthy volunteers*. Clin Pharmacol Ther, 2010. 87(6): p. 663–667.

196. Han, Y.H., Busler, D., Hong, Y., Tian, Y., Chen, C. and Rodrigues, A.D., *Transporter studies with the 3-O-sulfate conjugate of 17alpha-ethinylestradiol: assessment of human liver drug transporters*. Drug Metab Dispos, 2010. 38(7): p. 1072–1082.

197. Leonhardt, M., Keiser, M., Oswald, S., Kuhn, J., Jia, J., Grube, M., Kroemer, H.K., Siegmund, W. and Weitschies, W., *Hepatic uptake of the magnetic resonance imaging contrast agent Gd-EOB-DTPA: role of human organic anion transporters*. Drug Metab Dispos, 2010. 38(7): p. 1024–1028.

198. deSantos, L.A., Lukeman, J.M., Wallace, S., Murray, J.A. and Ayala, A.G., *Percutaneous needle biopsy of bone in the cancer patient*. AJR Am J Roentgenol, 1978. 130(4): p. 641–649.

199. Ni, W., et al., *Flavopiridol pharmacogenetics: clinical and functional evidence for the role of SLCO1B1/OATP1B1 in flavopiridol disposition*. PLoS One, 2010. 5(11): p. e13792.

200. de Graaf, W., et al., *Transporters involved in the hepatic uptake of (99m)Tc-mebrofenin and indocyanine green*. J Hepatol, 2011. 54(4): p. 738–745.

201. Siccardi, M., et al., *Maraviroc is a substrate for OATP1B1 in vitro and maraviroc plasma concentrations are influenced by SLCO1B1 521 T>C polymorphism*. Pharmacogenet Genomics, 2010. 20(12): p. 759–765.

202. Yamaguchi, H., et al., *Screening of antibiotics that interact with organic anion-transporting polypeptides 1B1 and 1B3 using fluorescent probes*. Biol Pharm Bull, 2011. 34(3): p. 389–395.

203. Takeuchi, K., Sugiura, T., Umeda, S., Matsubara, K., Horikawa, M., Nakamichi, N., Silver, D.L., Ishiwata, N. and Kato, Y., *Pharmacokinetics and hepatic uptake of eltrombopag, a novel platelet-increasing agent*. Drug Metab Dispos, 2011. 39(6): p. 1088–1096.

204. Konig, J., Glaeser, H., Keiser, M., Mandery, K., Klotz, U. and Fromm, M.F., *Role of organic anion-transporting polypeptides for cellular mesalazine (5-aminosalicylic acid) uptake*. Drug Metab Dispos, 2011. 39(6): p. 1097–1102.

205. De Bruyn, T., Fattah, S., Stieger, B., Augustijns, P. and Annaert, P., *Sodium fluorescein is a probe substrate for hepatic drug transport mediated by OATP1B1 and OATP1B3*. J Pharm Sci, 2011. 100(11): p. 5018–5030.

206. Fahrmayr, C., Konig, J., Auge, D., Mieth, M. and Fromm, M.F., *Identification of drugs and drug metabolites as substrates of multidrug resistance protein 2 (MRP2) using triple-transfected MDCK-OATP1B1-UGT1A1-MRP2 cells*. Br J Pharmacol, 2012. 165(6): p. 1836–1847.

207. Greupink, R., Dillen, L., Monshouwer, M., Huisman, M.T. and Russel, F.G., *Interaction of fluvastatin with the liver-specific Na^+-dependent taurocholate cotransporting polypeptide (NTCP)*. Eur J Pharm Sci, 2011. 44(4): p. 487–496.

208. Frey, R., Scheerans, C., Blunck, M., Muck, W., Gnoth, M.J., Unger, S., Schmidt, A. and Wensing, G., *Pharmacokinetics of the soluble guanylate cyclase activator cinaciguat in individuals with hepatic impairment*. J Clin Pharmacol, 2012. 52(11): p. 1714–1724.

209. Konig, J., Klatt, S., Dilger, K. and Fromm, M.F., *Characterization of ursodeoxycholic and norursodeoxycholic acid as substrates of the hepatic uptake transporters OATP1B1, OATP1B3, OATP2B1 and NTCP*. Basic Clin Pharmacol Toxicol, 2012. 111(2): p. 81–86.

210. Takashima, T., et al., *PET imaging-based evaluation of hepatobiliary transport in humans with (15R)-11C-TIC-Me*. J Nucl Med, 2012. 53(5): p. 741–748.

211. de Graan, A.J., et al., *Influence of polymorphic OATP1B-type carriers on the disposition of docetaxel*. Clin Cancer Res, 2012. 18(16): p. 4433–4440.

212. Nassif, A., et al., *Visualization of hepatic uptake transporter function in healthy subjects by using gadoxetic acid-enhanced MR imaging*. Radiology, 2012. 264(3): p. 741–750.

213. Lancaster, C.S., et al., *OATP1B1 polymorphism as a determinant of erythromycin disposition*. Clin Pharmacol Ther, 2012. 92(5): p. 642–650.

214. Zimmerman, E.I., Hu, S., Roberts, J.L., Gibson, A.A., Orwick, S.J., Li, L., Sparreboom, A. and Baker, S.D., *Contribution of OATP1B1 and OATP1B3 to the disposition of sorafenib and sorafenib-glucuronide*. Clin Cancer Res, 2013. 19(6): p. 1458–1466.

215. Tsakalozou, E., Adane, E.D., Kuo, K.L., Daily, A., Moscow, J.A. and Leggas, M., *The effect of breast cancer resistance protein, multidrug resistant protein 1, and organic anion-transporting polypeptide 1B3 on the antitumor efficacy of the lipophilic camptothecin 7-t-butyldimethylsilyl-10-hydroxycamptothecin (AR-67) in vitro*. Drug Metab Dispos, 2013. 41(7): p. 1404–1413.

216. Reyner, E.L., Sevidal, S., West, M.A., Clouser-Roche, A., Freiwald, S., Fenner, K., Ullah, M., Lee, C.A. and Smith, B.J., In vitro *characterization of axitinib interactions with human efflux and hepatic uptake transporters: implications for disposition and drug interactions*. Drug Metab Dispos, 2013. 41(8): p. 1575–1583.

217. Huang, J., Li, N., Hong, W., Zhan, K., Yu, X., Huang, H. and Hong, M., *Conserved tryptophan residues within putative transmembrane domain 6 affect transport function of organic anion transporting polypeptide 1B1*. Mol Pharmacol, 2013. 84(4): p. 521–527.

218. Chiou, W.J., de Morais, S.M., Kikuchi, R., Voorman, R.L., Li, X. and Bow, D.A., In vitro *OATP1B1 and OATP1B3 inhibition is associated with observations of benign clinical unconjugated hyperbilirubinemia*. Xenobiotica, 2014. 44(3): p. 276–282.

219. Izumi, S., Nozaki, Y., Komori, T., Maeda, K., Takenaka, O., Kusano, K., Yoshimura, T., Kusuhara, H. and Sugiyama, Y., *Substrate-dependent inhibition of organic anion transporting polypeptide 1B1: comparative analysis with prototypical probe substrates estradiol-17beta-glucuronide, estrone-3-sulfate, and sulfobromophthalein*. Drug Metab Dispos, 2013. 41(10): p. 1859–1866.

220. Choi, M.K., Kwon, M., Ahn, J.H., Kim, N.J., Bae, M.A. and Song, I.S., *Transport characteristics and transporter-based drug–drug interactions of TM-25659, a novel TAZ modulator*. Biopharm Drug Dispos, 2014. 35(3): p. 183–194.

221. Ramsden, D., Tweedie, D.J., Chan, T.S., Taub, M.E. and Li, Y., *Bridging* in vitro *and* in vivo *metabolism and transport of faldaprevir in human using a novel cocultured human hepatocyte system, HepatoPac.* Drug Metab Dispos, 2014. 42(3): p. 394–406.

222. Mikkaichi, T., Yoshigae, Y., Masumoto, H., Imaoka, T., Rozehnal, V., Fischer, T., Okudaira, N. and Izumi, T., *Edoxaban transport via P-glycoprotein is a key factor for the drug's disposition.* Drug Metab Dispos, 2014. 42(4): p. 520–528.

223. Kobayashi, M., et al., *Transport mechanisms of hepatic uptake and bile excretion in clinical hepatobiliary scintigraphy with 99mTc-N-pyridoxyl-5-methyltryptophan.* Nucl Med Biol, 2014. 41(4): p. 338–342.

224. Khurana, V., Minocha, M., Pal, D. and Mitra, A.K., *Role of OATP-1B1 and/or OATP-1B3 in hepatic disposition of tyrosine kinase inhibitors.* Drug Metabol Drug Interact, 2014. 29(3): p. 179–190.

225. De Bruyn, T., et al., *Confocal imaging with a fluorescent bile acid analogue closely mimicking hepatic taurocholate disposition.* J Pharm Sci, 2014. 103(6): p. 1872–1881.

226. Sun, P., Wang, C., Liu, Q., Meng, Q., Zhang, A., Huo, X., Sun, H. and Liu, K., *OATP and MRP2-mediated hepatic uptake and biliary excretion of eprosartan in rat and human.* Pharmacol Rep, 2014. 66(2): p. 311–319.

227. Kunze, A., Huwyler, J., Camenisch, G. and Poller, B., *Prediction of organic anion-transporting polypeptide 1B1- and 1B3-mediated hepatic uptake of statins based on transporter protein expression and activity data.* Drug Metab Dispos, 2014. 42(9): p. 1514–1521.

228. Ghosh, A., et al., *Toward a unified model of passive drug permeation II: the physiochemical determinants of unbound tissue distribution with applications to the design of hepatoselective glucokinase activators.* Drug Metab Dispos, 2014. 42(10): p. 1599–1610.

229. Varma, M.V., Bi, Y.A., Kimoto, E. and Lin, J., *Quantitative prediction of transporter- and enzyme-mediated clinical drug–drug interactions of organic anion-transporting polypeptide 1B1 substrates using a mechanistic net-effect model.* J Pharmacol Exp Ther, 2014. 351(1): p. 214–223.

230. Sheng, J., Tian, X., Xu, G., Wu, Z., Chen, C., Wang, L., Pan, L., Huang, C. and Pan, G., *The hepatobiliary disposition of timosaponin b2 is highly dependent on influx/efflux transporters but not metabolism.* Drug Metab Dispos, 2015. 43(1): p. 63–72.

231. Izumi, S., Nozaki, Y., Maeda, K., Komori, T., Takenaka, O., Kusuhara, H. and Sugiyama, Y., *Investigation of the impact of substrate selection on* in vitro *organic anion transporting polypeptide 1B1 inhibition profiles for the prediction of drug–drug interactions.* Drug Metab Dispos, 2015. 43(2): p. 235–247.

232. Clemens, J., Seckinger, A., Hose, D., Theile, D., Longo, M., Haefeli, W.E., Burhenne, J. and Weiss, J., *Cellular uptake kinetics of bortezomib in relation to efficacy in myeloma cells and the influence of drug transporters.* Cancer Chemother Pharmacol, 2015. 75(2): p. 281–291.

233. Eley, T., et al., *Organic anion transporting polypeptide-mediated transport of, and inhibition by, asunaprevir, an inhibitor of hepatitis C virus NS3 protease.* Clin Pharmacol Ther, 2015. 97(2): p. 159–166.

234. Campbell, S.D., Gadel, S., Friedel, C., Crafford, A., Regina, K.J. and Kharasch, E.D., *Influence of HIV antiretrovirals on methadone N-demethylation and transport.* Biochem Pharmacol, 2015. 95(2): p. 115–125.

235. Alfirevic, A., et al., *Misoprostol-induced fever and genetic polymorphisms in drug transporters SLCO1B1 and ABCC4 in women of Latin American and European ancestry.* Pharmacogenomics, 2015. 16(9): p. 919–928.

236. Shimizu, M., Fuse, K., Okudaira, K., Nishigaki, R., Maeda, K., Kusuhara, H. and Sugiyama, Y., *Contribution of OATP (organic anion-transporting polypeptide) family transporters to the hepatic uptake of fexofenadine in humans.* Drug Metab Dispos, 2005. 33(10): p. 1477–1481.

237. Baldes, C., Koenig, P., Neumann, D., Lenhof, H.P., Kohlbacher, O. and Lehr, C.M., *Development of a fluorescence-based assay for screening of modulators of human organic anion transporter 1B3 (OATP1B3).* Eur J Pharm Biopharm, 2006. 62(1): p. 39–43.

238. Smith, N.F., Acharya, M.R., Desai, N., Figg, W.D. and Sparreboom, A., *Identification of OATP1B3 as a high-affinity hepatocellular transporter of paclitaxel.* Cancer Biol Ther, 2005. 4(8): p. 815–818.

239. Letschert, K., Faulstich, H., Keller, D. and Keppler, D., *Molecular characterization and inhibition of amanitin uptake into human hepatocytes*. Toxicol Sci, 2006. 91(1): p. 140–149.

240. Nakagomi-Hagihara, R., Nakai, D., Kawai, K., Yoshigae, Y., Tokui, T., Abe, T. and Ikeda, T., *OATP1B1, OATP1B3, and mrp2 are involved in hepatobiliary transport of olmesartan, a novel angiotensin II blocker*. Drug Metab Dispos, 2006. 34(5): p. 862–869.

241. Ishiguro, N., Maeda, K., Kishimoto, W., Saito, A., Harada, A., Ebner, T., Roth, W., Igarashi, T. and Sugiyama, Y., *Predominant contribution of OATP1B3 to the hepatic uptake of telmisartan, an angiotensin II receptor antagonist, in humans*. Drug Metab Dispos, 2006. 34(7): p. 1109–1115.

242. Yamaguchi, H., Kobayashi, M., Okada, M., Takeuchi, T., Unno, M., Abe, T., Goto, J., Hishinuma, T. and Mano, N., *Rapid screening of antineoplastic candidates for the human organic anion transporter OATP1B3 substrates using fluorescent probes*. Cancer Lett, 2008. 260(1–2): p. 163–169.

243. Ishiguro, N., Maeda, K., Saito, A., Kishimoto, W., Matsushima, S., Ebner, T., Roth, W., Igarashi, T. and Sugiyama, Y., *Establishment of a set of double transfectants coexpressing organic anion transporting polypeptide 1B3 and hepatic efflux transporters for the characterization of the hepatobiliary transport of telmisartan acylglucuronide*. Drug Metab Dispos, 2008. 36(4): p. 796–805.

244. Franke, R.M., Baker, S.D., Mathijssen, R.H., Schuetz, E.G. and Sparreboom, A., *Influence of solute carriers on the pharmacokinetics of CYP3A4 probes*. Clin Pharmacol Ther, 2008. 84(6): p. 704–709.

245. Hamada, A., et al., *Effect of SLCO1B3 haplotype on testosterone transport and clinical outcome in caucasian patients with androgen-independent prostatic cancer*. Clin Cancer Res, 2008. 14(11): p. 3312–3318.

246. Annaert, P., Ye, Z.W., Stieger, B. and Augustijns, P., *Interaction of HIV protease inhibitors with OATP1B1, 1B3, and 2B1*. Xenobiotica, 2010. 40(3): p. 163–176.

247. Roth, M., Timmermann, B.N. and Hagenbuch, B., *Interactions of green tea catechins with organic anion-transporting polypeptides*. Drug Metab Dispos, 2011. 39(5): p. 920–926.

248. Kindla, J., Muller, F., Mieth, M., Fromm, M.F. and Konig, J., *Influence of non-steroidal anti-inflammatory drugs on organic anion transporting polypeptide (OATP) 1B1- and OATP1B3-mediated drug transport*. Drug Metab Dispos, 2011. 39(6): p. 1047–1053.

249. Takanohashi, T., Kubo, S., Arisaka, H., Shinkai, K. and Ubukata, K., *Contribution of organic anion transporting polypeptide (OATP) 1B1 and OATP1B3 to hepatic uptake of nateglinide, and the prediction of drug–drug interactions via these transporters*. J Pharm Pharmacol, 2012. 64(2): p. 199–206.

250. Zhang, A., Wang, C., Liu, Q., Meng, Q., Peng, J., Sun, H., Ma, X., Huo, X. and Liu, K., *Involvement of organic anion-transporting polypeptides in the hepatic uptake of dioscin in rats and humans*. Drug Metab Dispos, 2013. 41(5): p. 994–1003.

251. Gozalpour, E., Greupink, R., Wortelboer, H.M., Bilos, A., Schreurs, M., Russel, F.G. and Koenderink, J.B., *Interaction of digitalis-like compounds with liver uptake transporters NTCP, OATP1B1, and OATP1B3*. Mol Pharm, 2014. 11(6): p. 1844–1855.

252. Bosgra, S., van de Steeg, E., Vlaming, M.L., Verhoeckx, K.C., Huisman, M.T., Verwei, M. and Wortelboer, H.M., *Predicting carrier-mediated hepatic disposition of rosuvastatin in man by scaling from individual transfected cell-lines* in vitro *using absolute transporter protein quantification and PBPK modeling*. Eur J Pharm Sci, 2014. 65: p. 156–166.

253. Jiang, R., et al., *Molecular mechanisms governing different pharmacokinetics of ginsenosides and potential for ginsenoside-perpetrated herb–drug interactions on OATP1B3*. Br J Pharmacol, 2015. 172(4): p. 1059–1073.

254. Chen, C., Wu, Z.T., Ma, L.L., Ni, X., Lin, Y.F., Wang, L., Chen, K.P., Huang, C.G. and Pan, G., *Organic anion-transporting polypeptides contribute to the hepatic uptake of berberine*. Xenobiotica, 2015. 45(12): p. 1138–1146.

255. Knop, J., Misaka, S., Singer, K., Hoier, E., Muller, F., Glaeser, H., Konig, J. and Fromm, M.F., *Inhibitory effects of green tea and (−)-epigallocatechin gallate on transport by OATP1B1, OATP1B3, OCT1, OCT2, MATE1, MATE2-K and P-glycoprotein*. PLoS One, 2015. 10(10): p. e0139370.

256. van Montfoort, J.E., Muller, M., Groothuis, G.M., Meijer, D.K., Koepsell, H. and Meier, P.J., *Comparison of "type I" and "type II" organic cation transport by organic cation transporters and organic anion-transporting polypeptides*. J Pharmacol Exp Ther, 2001. 298(1): p. 110–115.

257. Bourdet, D.L., Pritchard, J.B. and Thakker, D.R., *Differential substrate and inhibitory activities of ranitidine and famotidine toward human organic cation transporter 1 (hOCT1; SLC22A1), hOCT2 (SLC22A2), and hOCT3 (SLC22A3).* J Pharmacol Exp Ther, 2005. 315(3): p. 1288–1297.

258. Zhang, S., et al., *Organic cation transporters are determinants of oxaliplatin cytotoxicity.* Cancer Res, 2006. 66(17): p. 8847–8857.

259. Glaeser, H., et al., *Intestinal drug transporter expression and the impact of grapefruit juice in humans.* Clin Pharmacol Ther, 2007. 81(3): p. 362–370.

260. Shu, Y., et al., *Effect of genetic variation in the organic cation transporter 1 (OCT1) on metformin action.* J Clin Invest, 2007. 117(5): p. 1422–1431.

261. Umehara, K.I., Iwatsubo, T., Noguchi, K. and Kamimura, H., *Functional involvement of organic cation transporter1 (OCT1/Oct1) in the hepatic uptake of organic cations in humans and rats.* Xenobiotica, 2007. 37(8): p. 818–831.

262. Ming, X., Ju, W., Wu, H., Tidwell, R.R., Hall, J.E. and Thakker, D.R., *Transport of dicationic drugs pentamidine and furamidine by human organic cation transporters.* Drug Metab Dispos, 2009. 37(2): p. 424–430.

263. Mitsuoka, K., Shirasaka, Y., Fukushi, A., Sato, M., Nakamura, T., Nakanishi, T. and Tamai, I., *Transport characteristics of L-citrulline in renal apical membrane of proximal tubular cells.* Biopharm Drug Dispos, 2009. 30(3): p. 126–137.

264. Sogame, Y., Kitamura, A., Yabuki, M. and Komuro, S., *A comparison of uptake of metformin and phenformin mediated by hOCT1 in human hepatocytes.* Biopharm Drug Dispos, 2009. 30(8): p. 476–484.

265. Ahlin, G., Chen, L., Lazorova, L., Chen, Y., Ianculescu, A.G., Davis, R.L., Giacomini, K.M. and Artursson, P., *Genotype-dependent effects of inhibitors of the organic cation transporter, OCT1: predictions of metformin interactions.* Pharmacogenomics J, 2011. 11(6): p. 400–411.

266. Tzvetkov, M.V., Saadatmand, A.R., Bokelmann, K., Meineke, I., Kaiser, R. and Brockmoller, J., *Effects of OCT1 polymorphisms on the cellular uptake, plasma concentrations and efficacy of the 5-HT(3) antagonists tropisetron and ondansetron.* Pharmacogenomics J, 2012. 12(1): p. 22–29.

267. Tzvetkov, M.V., Saadatmand, A.R., Lotsch, J., Tegeder, I., Stingl, J.C. and Brockmoller, J., *Genetically polymorphic OCT1: another piece in the puzzle of the variable pharmacokinetics and pharmacodynamics of the opioidergic drug tramadol.* Clin Pharmacol Ther, 2011. 90(1): p. 143–150.

268. Dickens, D., Owen, A., Alfirevic, A., Giannoudis, A., Davies, A., Weksler, B., Romero, I.A., Couraud, P.O. and Pirmohamed, M., *Lamotrigine is a substrate for OCT1 in brain endothelial cells.* Biochem Pharmacol, 2012. 83(6): p. 805–814.

269. Saadatmand, A.R., Tadjerpisheh, S., Brockmoller, J. and Tzvetkov, M.V., *The prototypic pharmacogenetic drug debrisoquine is a substrate of the genetically polymorphic organic cation transporter OCT1.* Biochem Pharmacol, 2012. 83(10): p. 1427–1434.

270. Swift, B., et al., *Sorafenib hepatobiliary disposition: mechanisms of hepatic uptake and disposition of generated metabolites.* Drug Metab Dispos, 2013. 41(6): p. 1179–1186.

271. Tu, M., Sun, S., Wang, K., Peng, X., Wang, R., Li, L., Zeng, S., Zhou, H. and Jiang, H., *Organic cation transporter 1 mediates the uptake of monocrotaline and plays an important role in its hepatotoxicity.* Toxicology, 2013. 311(3): p. 225–230.

272. Tzvetkov, M.V., dos Santos Pereira, J.N., Meineke, I., Saadatmand, A.R., Stingl, J.C. and Brockmoller, J., *Morphine is a substrate of the organic cation transporter OCT1 and polymorphisms in OCT1 gene affect morphine pharmacokinetics after codeine administration.* Biochem Pharmacol, 2013. 86(5): p. 666–678.

273. Boxberger, K.H., Hagenbuch, B. and Lampe, J.N., *Common drugs inhibit human organic cation transporter 1 (OCT1)-mediated neurotransmitter uptake.* Drug Metab Dispos, 2014. 42(6): p. 990–995.

274. Tu, M., et al., *Involvement of organic cation transporter 1 and CYP3A4 in retrorsine-induced toxicity.* Toxicology, 2014. 322: p. 34–42.

275. Yamakawa, Y., Hamada, A., Uchida, T., Sato, D., Yuki, M., Hayashi, M., Kawaguchi, T. and Saito, H., *Distinct interaction of nilotinib and imatinib with P-Glycoprotein in intracellular accumulation and cytotoxicity in CML Cell Line K562 cells.* Biol Pharm Bull, 2014. 37(8): p. 1330–1335.

276. Burger, H., et al., *Lysosomal Sequestration Determines Intracellular Imatinib Levels.* Mol Pharmacol, 2015. 88(3): p. 477–487.

277. Li, L., Sun, S., Weng, Y., Song, F., Zhou, S., Bai, M., Zhou, H., Zeng, S. and Jiang, H., *Interaction of six protoberberine alkaloids with human organic cation transporters 1, 2 and 3.* Xenobiotica, 2016. 46(2): p. 175–183.

278. Yee, S.W., Lin, L., Merski, M., Keiser, M.J., Gupta, A., Zhang, Y., Chien, H.C., Shoichet, B.K. and Giacomini, K.M., *Prediction and validation of enzyme and transporter off-targets for metformin.* J Pharmacokinet Pharmacodyn, 2015. 42(5): p. 463–475.

279. Fagerholm, U., *Prediction of human pharmacokinetics-biliary and intestinal clearance and enterohepatic circulation.* J Pharm Pharmacol, 2008. 60(5): p. 535–542.

280. Gregus, Z. and Klaassen, C.D., *Biliary excretion.* J Clin Pharmacol, 1987. 27(8): p. 537–541.

281. Kusuhara, H. and Sugiyama, Y., *Pharmacokinetic modeling of the hepatobiliary transport mediated by cooperation of uptake and efflux transporters.* Drug Metab Rev, 2010. 42(3): p. 539–550.

282. Lai, Y., *Identification of interspecies difference in hepatobiliary transporters to improve extrapolation of human biliary secretion.* Expert Opin Drug Metab Toxicol, 2009. 5(10): p. 1175–1187.

283. Luo, F.R., Paranjpe, P.V., Guo, A., Rubin, E. and Sinko, P., *Intestinal transport of irinotecan in Caco-2 cells and MDCK II cells overexpressing efflux transporters Pgp, cMOAT, and MRP1.* Drug Metab Dispos, 2002. 30(7): p. 763–770.

284. Tang, F., Horie, K. and Borchardt, R.T., *Are MDCK cells transfected with the human MRP2 gene a good model of the human intestinal mucosa?* Pharm Res, 2002. 19(6): p. 773–779.

285. Chen, Z.S., Robey, R.W., Belinsky, M.G., Shchaveleva, I., Ren, X.Q., Sugimoto, Y., Ross, D.D., Bates, S.E. and Kruh, G.D., *Transport of methotrexate, methotrexate polyglutamates, and 17beta-estradiol 17-(beta-D-glucuronide) by ABCG2: effects of acquired mutations at R482 on methotrexate transport.* Cancer Res, 2003. 63(14): p. 4048–4054.

286. Kopplow, K., Letschert, K., Konig, J., Walter, B. and Keppler, D., *Human hepatobiliary transport of organic anions analyzed by quadruple-transfected cells.* Mol Pharmacol, 2005. 68(4): p. 1031–1038.

287. Lau, Y.Y., Okochi, H., Huang, Y. and Benet, L.Z., *Multiple transporters affect the disposition of atorvastatin and its two active hydroxy metabolites: application of* in vitro *and* ex situ *systems.* J Pharmacol Exp Ther, 2006. 316(2): p. 762–771.

288. Yamazaki, M., et al., *Effects of fibrates on human organic anion-transporting polypeptide 1B1-, multidrug resistance protein 2- and P-glycoprotein-mediated transport.* Xenobiotica, 2005. 35(7): p. 737–753.

289. Nozaki, Y., et al., *Species difference in the inhibitory effect of nonsteroidal anti-inflammatory drugs on the uptake of methotrexate by human kidney slices.* J Pharmacol Exp Ther, 2007. 322(3): p. 1162–1170.

290. Kalgutkar, A.S., et al., *Role of transporters in the disposition of the selective phosphodiesterase-4 inhibitor (+)-2-[4-({[2-(benzo[1,3]dioxol-5-yloxy)-pyridine-3-carbonyl]-amino}-methyl)-3-fluorophenoxy]-propionic acid in rat and human.* Drug Metab Dispos, 2007. 35(11): p. 2111–2118.

291. Namkoong, E.M., Kim, I.W., Kim, D.D., Chung, S.J. and Shim, C.K., *Effect of probenecid on the biliary excretion of belotecan.* Arch Pharm Res, 2007. 30(11): p. 1482–1488.

292. Chen, Y., Zhao, Y.H., Jia, X.B. and Hu, M., *Intestinal absorption mechanisms of prenylated flavonoids present in the heat-processed Epimedium koreanum Nakai (Yin Yanghuo).* Pharm Res, 2008. 25(9): p. 2190–2199.

293. Li, H., Jin, H.E., Kim, W., Han, Y.H., Kim, D.D., Chung, S.J. and Shim, C.K., *Involvement of P-glycoprotein, multidrug resistance protein 2 and breast cancer resistance protein in the transport of belotecan and topotecan in Caco-2 and MDCKII cells.* Pharm Res, 2008. 25(11): p. 2601–2612.

294. Lagas, J.S., van der Kruijssen, C.M., van de Wetering, K., Beijnen, J.H. and Schinkel, A.H., *Transport of diclofenac by breast cancer resistance protein (ABCG2) and stimulation of multidrug resistance*

protein 2 (ABCC2)-mediated drug transport by diclofenac and benzbromarone. Drug Metab Dispos, 2009. 37(1): p. 129–136.

295. Hirouchi, M., Kusuhara, H., Onuki, R., Ogilvie, B.W., Parkinson, A. and Sugiyama, Y., *Construction of triple-transfected cells [organic anion-transporting polypeptide (OATP) 1B1/multidrug resistance-associated protein (MRP) 2/MRP3 and OATP1B1/MRP2/MRP4] for analysis of the sinusoidal function of MRP3 and MRP4*. Drug Metab Dispos, 2009. 37(10): p. 2103–2111.

296. Karlsson, J.E., Heddle, C., Rozkov, A., Rotticci-Mulder, J., Tuvesson, O., Hilgendorf, C. and Andersson, T.B., *High-activity p-glycoprotein, multidrug resistance protein 2, and breast cancer resistance protein membrane vesicles prepared from transiently transfected human embryonic kidney 293-Epstein-Barr virus nuclear antigen cells*. Drug Metab Dispos, 2010. 38(4): p. 705–714.

297. Lin, X., Skolnik, S., Chen, X. and Wang, J., *Attenuation of intestinal absorption by major efflux transporters: quantitative tools and strategies using a Caco-2 model*. Drug Metab Dispos, 2011. 39(2): p. 265–274.

298. Li, J., Volpe, D.A., Wang, Y., Zhang, W., Bode, C., Owen, A. and Hidalgo, I.J., *Use of transporter knockdown Caco-2 cells to investigate the* in vitro *efflux of statin drugs*. Drug Metab Dispos, 2011. 39(7): p. 1196–1202.

299. Franke, R.M., et al., *Effect of ABCC2 (MRP2) transport function on erythromycin metabolism*. Clin Pharmacol Ther, 2011. 89(5): p. 693–701.

300. Elsby, R., Smith, V., Fox, L., Stresser, D., Butters, C., Sharma, P. and Surry, D.D., *Validation of membrane vesicle-based breast cancer resistance protein and multidrug resistance protein 2 assays to assess drug transport and the potential for drug–drug interaction to support regulatory submissions*. Xenobiotica, 2011. 41(9): p. 764–783.

301. Iwai, M., Minematsu, T., Li, Q., Iwatsubo, T. and Usui, T., *Utility of P-glycoprotein and organic cation transporter 1 double-transfected LLC-PK1 cells for studying the interaction of YM155 monobromide, novel small-molecule survivin suppressant, with P-glycoprotein*. Drug Metab Dispos, 2011. 39(12): p. 2314–2320.

302. Elens, L., Tyteca, D., Panin, N., Courtoy, P., Lison, D., Demoulin, J.B. and Haufroid, V., *Functional defect caused by the 4544G>A SNP in ABCC2: potential impact for drug cellular disposition*. Pharmacogenet Genomics, 2011. 21(12): p. 884–893.

303. Farrell, T.L., Poquet, L., Dew, T.P., Barber, S. and Williamson, G., *Predicting phenolic acid absorption in Caco-2 cells: a theoretical permeability model and mechanistic study*. Drug Metab Dispos, 2012. 40(2): p. 397–406.

304. de Waart, D.R., van de Wetering, K., Kunne, C., Duijst, S., Paulusma, C.C. and Oude Elferink, R.P., *Oral availability of cefadroxil depends on ABCC3 and ABCC4*. Drug Metab Dispos, 2012. 40(3): p. 515–521.

305. Li, C., Zhang, L., Zhou, L., Wo, S.K., Lin, G. and Zuo, Z., *Comparison of intestinal absorption and disposition of structurally similar bioactive flavones in Radix Scutellariae*. AAPS J, 2012. 14(1): p. 23–34.

306. Kwak, E.Y., Shim, W.S., Chang, J.E., Chong, S., Kim, D.D., Chung, S.J. and Shim, C.K., *Enhanced intracellular accumulation of a non-nucleoside anti-cancer agent via increased uptake of its valine ester prodrug through amino acid transporters*. Xenobiotica, 2012. 42(7): p. 603–613.

307. Choi, M.K. and Song, I.S., *Characterization of efflux transport of the PDE5 inhibitors, vardenafil and sildenafil*. J Pharm Pharmacol, 2012. 64(8): p. 1074–1083.

308. Tesoriere, L., Gentile, C., Angileri, F., Attanzio, A., Tutone, M., Allegra, M. and Livrea, M.A., *Trans-epithelial transport of the betalain pigments indicaxanthin and betanin across Caco-2 cell monolayers and influence of food matrix*. Eur J Nutr, 2013. 52(3): p. 1077–1087.

309. Gao, C., Zhang, H., Guo, Z., You, T., Chen, X. and Zhong, D., *Mechanistic studies on the absorption and disposition of scutellarin in humans: selective OATP2B1-mediated hepatic uptake is a likely key determinant for its unique pharmacokinetic characteristics*. Drug Metab Dispos, 2012. 40(10): p. 2009–2020.

310. Patel, C.G., Ogasawara, K. and Akhlaghi, F., *Mycophenolic acid glucuronide is transported by multidrug resistance-associated protein 2 and this transport is not inhibited by cyclosporine, tacrolimus or sirolimus*. Xenobiotica, 2013. 43(3): p. 229–235.

311. Matsunaga, N., Nunoya, K., Okada, M., Ogawa, M. and Tamai, I., *Evaluation of hepatic disposition of paroxetine using sandwich-cultured rat and human hepatocytes*. Drug Metab Dispos, 2013. 41(4): p. 735–743.

312. Fahrmayr, C., Konig, J., Auge, D., Mieth, M., Munch, K., Segrestaa, J., Pfeifer, T., Treiber, A. and Fromm, M., *Phase I and II metabolism and MRP2-mediated export of bosentan in a MDCKII-OATP1B1-CYP3A4-UGT1A1-MRP2 quadruple-transfected cell line*. Br J Pharmacol, 2013. 169(1): p. 21–33.

313. Hashiguchi, Y., Hamada, A., Shinohara, T., Tsuchiya, K., Jono, H. and Saito, H., *Role of P-glycoprotein in the efflux of raltegravir from human intestinal cells and CD4+ T-cells as an interaction target for anti-HIV agents*. Biochem Biophys Res Commun, 2013. 439(2): p. 221–227.

314. Jia, J., Puls, D., Oswald, S., Jedlitschky, G., Kuhn, J.P., Weitschies, W., Hosten, N., Siegmund, W. and Keiser, M., *Characterization of the intestinal and hepatic uptake/efflux transport of the magnetic resonance imaging contrast agent gadolinium-ethoxylbenzyl-diethylenetriamine-pentaacetic acid*. Invest Radiol, 2014. 49(2): p. 78–86.

315. El-Sheikh, A.A., Koenderink, J.B., Wouterse, A.C., van den Broek, P.H., Verweij, V.G., Masereeuw, R. and Russel, F.G., *Renal glucuronidation and multidrug resistance protein 2-/multidrug resistance protein 4-mediated efflux of mycophenolic acid: interaction with cyclosporine and tacrolimus*. Transl Res, 2014. 164(1): p. 46–56.

316. Radisch, S., Dickens, D., Lang, T., Bonnett, L., Arlanov, R., Johnson, M.R., Schwab, M., Marson, A.G. and Pirmohamed, M., *A comprehensive functional and clinical analysis of ABCC2 and its impact on treatment response to carbamazepine*. Pharmacogenomics J, 2014. 14(5): p. 481–487.

317. Marchetti, S., Pluim, D., Beijnen, J.H., Mazzanti, R., van Tellingen, O. and Schellens, J.H., *"Effect of the drug transporters ABCB1, ABCC2, and ABCG2 on the disposition and brain accumulation of the taxane analog BMS-275,183"*. Invest New Drugs, 2014. 32(6): p. 1083–1095.

318. Zhou, W., Tan, X., Shan, J., Liu, T., Cai, B. and Di, L., *Effect of chito-oligosaccharide on the intestinal absorptions of phenylethanoid glycosides in* Fructus forsythiae *extract*. Phytomedicine, 2014. 21(12): p. 1549–1558.

319. Brennan, B.J., Poirier, A., Moreira, S., Morcos, P.N., Goelzer, P., Portmann, R., Asthappan, J., Funk, C. and Smith, P.F., *Characterization of the transmembrane transport and absolute bioavailability of the HCV protease inhibitor danoprevir*. Clin Pharmacokinet, 2015. 54(5): p. 537–549.

320. Dai, P., Zhu, L., Yang, X., Zhao, M., Shi, J., Wang, Y., Lu, L. and Liu, Z., *Multidrug resistance-associated protein 2 is involved in the efflux of Aconitum alkaloids determined by MRP2-MDCKII cells*. Life Sci, 2015. 127: p. 66–72.

321. Luo, Z., Morgan, M.R. and Day, A.J., *Transport of trans-tiliroside (kaempferol-3-beta-D-(6"-p-coumaroyl-glucopyranoside) and related flavonoids across Caco-2 cells, as a model of absorption and metabolism in the small intestine*. Xenobiotica, 2015. 45(8): p. 722–730.

322. Xie, Y., Duan, J., Fu, Q., Xia, M., Zhang, L., Li, G., Wu, T. and Ji, G., *Comparison of isorhamnetin absorption properties in total flavones of* Hippophae rhamnoides *L. with its pure form in a Caco-2 cell model mediated by multidrug resistance-associated protein*. Eur J Pharm Sci, 2015. 73: p. 1–8.

323. Kalapos-Kovacs, B., Magda, B., Jani, M., Fekete, Z., Szabo, P.T., Antal, I., Krajcsi, P. and Klebovich, I., *Multiple ABC transporters efflux baicalin*. Phytother Res, 2015. 29(12): p. 1987–1990.

324. Shitara, Y., Maeda, K., Ikejiri, K., Yoshida, K., Horie, T. and Sugiyama, Y., *Clinical significance of organic anion transporting polypeptides (OATPs) in drug disposition: their roles in hepatic clearance and intestinal absorption*. Biopharm Drug Dispos, 2013. 34(1): p. 45–78.

325. Varma, M.V., Chang, G., Lai, Y., Feng, B., El-Kattan, A.F., Litchfield, J. and Goosen, T.C., *Physicochemical property space of hepatobiliary transport and computational models for predicting rat biliary excretion*. Drug Metab Dispos, 2012. 40(8): p. 1527–1537.

326. Yang, X., Gandhi, Y.A., Duignan, D.B. and Morris, M.E., *Prediction of biliary excretion in rats and humans using molecular weight and quantitative structure-pharmacokinetic relationships.* AAPS J, 2009. 11(3): p. 511–525.

327. Luo, G., Johnson, S., Hsueh, M.M., Zheng, J., Cai, H., Xin, B., Chong, S., He, K. and Harper, T.W., In silico *prediction of biliary excretion of drugs in rats based on physicochemical properties.* Drug Metab Dispos, 2010. 38(3): p. 422–430.

328. Varma, M.V., et al., *pH-Dependent solubility and permeability criteria for provisional biopharmaceutics classification (BCS and BDDCS) in early drug discovery.* Mol Pharm, 2012. 9(5): p. 1199–1212.

4

PHYSICOCHEMICAL AND BIOPHARMACEUTICAL PROPERTIES THAT AFFECT DRUG ABSORPTION OF COMPOUNDS ABSORBED BY PASSIVE DIFFUSION

4.1 INTRODUCTION

Absorption is a dynamic process of drug transfer from the site of administration, the gastrointestinal lumen, to the intestinal epithelium. Oral drug absorption can be expressed using Fick's First Law applied to membranes:

$$J_{\text{wall}} = P_{\text{wall}} \cdot \Delta C \cdot \text{SA} \quad (4.1)$$

where J_{wall} is the drug flux across a homogeneous intestinal membrane; SA the intestinal lumen surface area; P_{wall} the effective permeability, which is the rate at which dissolved drug will cross the intestinal wall to reach the portal blood circulation; and ΔC the drug concentration gradient across the intestinal lumen. Therefore, drug flux is a product of drug permeability and solubility [1, 2].

A qualitative understanding of permeability, solubility, and their impact on drug absorption is essential to achieve acceptable oral bioavailability. A proper balance must be established between these two parameters during early SAR optimization. Other factors such as dose and drug stability in the intestinal medium, which can significantly impact oral absorption, must also be considered.

The goals of this chapter are twofold. First, the passive diffusion of an orally absorbed drug molecule is discussed. Second, the physicochemical and biopharmaceutical drug properties that would influence drug permeability, solubility, and dissolution are reviewed.

Oral Bioavailability Assessment: Basics and Strategies for Drug Discovery and Development, First Edition. Ayman F. El-Kattan.

4.2 MECHANISM OF ORAL ABSORPTION VIA PASSIVE DIFFUSION

Following oral dosing, drug molecules can cross the luminal membrane through various mechanisms that involve passive diffusion or active transport. Passive diffusion comprises two pathways: the paracellular pathway, in which drug diffuses through the aqueous pores at the tight junctions between the intestinal enterocytes, and the transcellular (lipophilic) pathway, which requires drug diffusion across the lipid cell membrane of the enterocyte. The active transport pathway is mediated by transporters and is divided into active drug influx and efflux. It is important to note that the relevance of each route is determined by the compound's physicochemical properties and its potential affinity for various transport proteins [1–5].

4.2.1 Paracellular Diffusion

In paracellular diffusion, drug molecules are absorbed by diffusion and convective volume flow through the water-filled intercellular space [6]. In general, drugs that are absorbed through this pathway are small molecules (e.g., molecular weight $[M_{wt}] < 250$ Da) and hydrophilic in nature ($c \log P < 0$). Since the junctional complex has a net negative charge, positively charged molecules pass through more readily, whereas negatively charged molecules are repelled [7]. Furthermore, the paracellular pathway offers a limited window for absorption and accounts for <0.01% of the total surface area of intestinal membrane. In addition, the tight junctions between cells become tighter traveling from the jejunum toward the colon. Therefore, compounds that are absorbed via this pathway are not amenable for traditional controlled release formulation targeting the colon such as metformin and gabapentin. For these two molecules, gastric retentive controlled release formulation technology was effectively used to extend the half-life of these molecules [8–10].

4.2.2 Transcellular Pathway

The transcellular pathway is the major route of absorption for most of the drug molecules. In general, the rate of passive transcellular permeability is mainly determined by the rate of transport across the apical cell membrane, which is controlled by the physicochemical properties of the absorbed compound. Unlike the paracellular pathway, compounds that are absorbed through the transcellular pathway are unionized, with lipophilicity of $\log P > 0$ and $M_{wt} > 300$ g/mole. In addition, the hydrogen-bonding capacity determined by the number of hydrogen-bond donors and hydrogen-bond acceptors are <10 and 5, respectively [11–13].

TIP:

It is highly recommended to investigate the physicochemical properties of NMEs before orally dosing them in preclinical species. For example, compounds that have small molecular weight and hydrophilic tend to be absorbed via the paracellular route. If they will not be actively absorbed via intestinal uptake transporters such as PEPT1 or OATP2B1, these molecules will have high likelihood of having poor oral absorption. In this case, rat will be a better model to use to investigate their f_a relative to the use of dogs, which tend to overestimate the f_a of these molecules and potentially misguide the discovery team.

4.3 PHYSICOCHEMICAL AND BIOPHARMACEUTICAL PROPERTIES THAT INFLUENCE PASSIVE PERMEABILITY

4.3.1 Molecular Size/Volume

Molecular weight is an important physicochemical parameter that should be considered when optimizing intestinal permeability. This parameter has been extensively assessed in the literature. Interestingly, different labels were used to describe molecular weight such as molecular bulkiness, surface area, volume, geometry, conformation, radius, radius of gyration, cross-sectional diameter, hydrodynamic size, and hydrodynamic radius. Herein, for the focus of this chapter, the discussion is limited to molecular radius or diameter, molecular weight, and radius of gyration.

Molecular size plays a key role to determine compound diffusion across intestinal membrane. For example, Pantzar et al. studies suggested that permeability across rat intestinal brush border membrane demonstrates regional size dependence [14]. The first region is for compounds with a molecular diameter that is smaller than 10 Å, where permeability is via aqueous pores of the apical brush border membrane of the enterocyte. The second region is for compounds with a diameter between 10 and 30 Å, where intestinal permeability is ascribed to paracellular transport through the tight junctions. The third region is for compounds with molecular diameters larger than 30 Å, where the permeability is probably due to transcellular transport that involves transcytosis. Similar findings were also reported with Knipp's et al. who demonstrated that permeability via paracellular pathway decreased with increasing molecular radius [15].

In 1998, Camenisch et al. investigated the permeability of a set of molecules ($n = 34$) with a molecular weight range of 150–670 Da using Caco-2 cell-line. Molecules with a molecular weight smaller than 200 Da passed the intestinal mucosa via paracellular and transcellular transport. Molecules with molecular weight from 200 to 500 Da had rather nonrestricted membrane diffusion. However, molecules with a molecular weight larger than 500 Da resulted in a decrease in membrane diffusion. Interestingly, molecules with large molecular weight can permeate brush border membrane via endocytosis/transcytosis.

Hou et al. [16] investigated the relationship between the $\log P_{\text{eff}}$ obtained by Caco-2 and a set of 77 compounds radius of gyration. It should be emphasized that the radius of gyration for compounds investigated considered three pivotal factors, which are mass, molecular radius, and total nuclear mass. Compounds with larger radius of gyration are more difficult to cross Caco-2 monolayers than those with smaller radius of gyration. Interestingly, rigid and larger molecules are associated with lower Caco-2 monolayers permeability.

4.3.2 Hydrogen-Bonding Affinity

To estimate the hydrogen-bonding affinity of a drug molecule, several methods have been reported. Method complexity has ranged from simplistic (e.g., heteroatom (S, O, and N) counts or the consideration of molecules in terms of the number of hydrogen-bond acceptors and donors) to sophisticated (e.g., accounting for parameters such as free-energy and polar surface area (PSA)) [17]. In general, it is difficult to have a reliable estimation of hydrogen-bonding estimation because of these inherent differences in the computational methods. For example, simple counting can be rather problematic and arbitrary since bulky groups or conformational flexibility may shield the hydrogen-bond strength. In addition, simple counting assumes all kinds of H-bonds to be energetically equivalent and, in particular, effects of ionization are ignored. van der Waterbeemd et al. stated that such ad hoc methods (i.e., counting potential hydrogen bonds) are impractical for large series of compounds [18]. Therefore, the use of PSA was shown to be a viable approach that is widely used to project oral drug absorption in human. PSA is assumed to be related to the hydrogen-bonding capacity of the drug molecule and defined as that part of the molecular surface (Van der Waals or solvent accessible) that is due to oxygen and nitrogen atoms or hydrogen atoms attached to them. Some workers also included sulfur and phosphorus and attached hydrogen as polar atoms [1]. PSA has been demonstrated to correlate with passive permeability across biological membranes, which enables the prediction of intestinal transport properties of various drug molecules. In 1996, Palm et al. showed a good correlation between the PSA of drug molecules and their permeability in two widely used *in vitro* intestinal permeability models (Caco-2 monolayers and rat intestinal segments). A homologous series of β-adrenoreceptor antagonists were used as model drug molecules. Excellent correlations were obtained between the PSA of these molecules and their permeabilities in Caco-2 cells and rat ileum ($R^2 = 0.99$ and 0.92, respectively) [19].

In a subsequent study, Palm et al. found an excellent sigmoidal correlation ($R^2 = 0.94$) between PSA and human F_{abs} for 20 drug molecules that had a wide range of human f_a values (0.3–100%) and physicochemical properties. These drug molecules were predominantly absorbed by passive diffusion [20]. Based on the regression analysis, the group concluded that drug molecules with PSA values >140 Å^2 would show poor oral absorption (<10%) and drug molecules with PSA <60 Å^2 could be predicted to show complete oral absorption (>90%). Similar findings were reported by Kelder et al. as the PSA for 2000 oral drugs that had reached Phase II trials were calculated. The group divided the compounds into oral CNS drugs ($n = 776$) and oral non-CNS drugs ($n = 1590$). They reported that the orally active non-CNS drugs that are transported passively through the transcellular route tend to display a PSA value of <120 Å^2. Alternatively, orally active CNS drugs tend to exhibit a PSA value that is <60–70 Å^2 [21].

Our group further assessed the impact of polar descriptors such as PSA on the f_a of a set of 309 molecules. As illustrated in Figure 4.1, there is an inverse relationship between PSA and compounds f_a in human, where compounds with PSA >125 Å^2 tend to have poor absorption in human, which is consistent with earlier findings [4].

4.3.3 Hydrophobicity

Hydrophobicity is a measure of how likely a drug molecule is to be repelled out of an aqueous environment and to favor "lipids" as occurs during permeability across the gastrointestinal tract. Most molecules are believed to be absorbed as a result of diffusion through intestinal membranes in a mucosal layer that is rich in lipids and lipoproteins. Generally, hydrophobicity refers to the degree to which a solute is disruptive to the array of water molecules established by hydrogen bonding. Hydrophobicity is often used synonymously with lipophilicity, although the latter includes the idea of attractive forces between the solute and lipid [22, 23]. $\log P$ and $\log D$ are the most widely used statistics describing hydrophobicity, but are generally considered to be measures of lipophilicity.

4.3.3.1 log P and log D $\log P$ is a partition coefficient and is defined by Equation 4.2:

$$\log P = \log\left(\frac{C_{oct}}{C_{water}}\right) \tag{4.2}$$

where C_{oct} and C_{water} are the concentrations of an unionized solute in layers of n-octanol and water in a two-phase system at equilibrium. The coefficient P can also refer to partitioning in

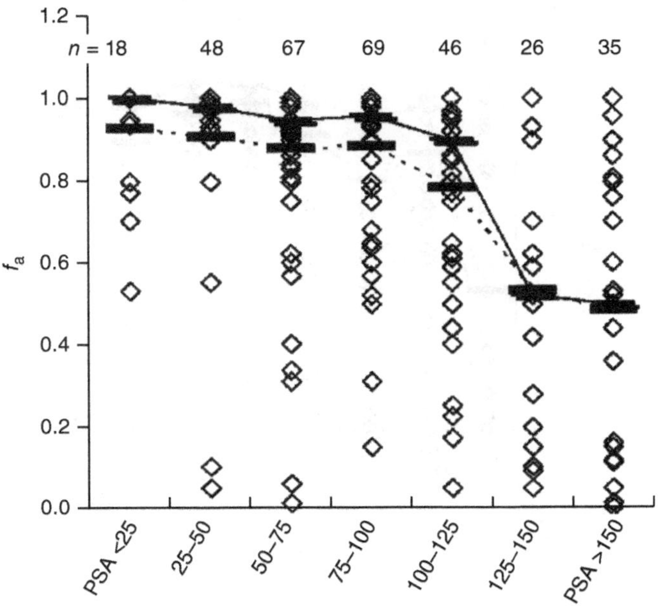

FIGURE 4.1 The relationship between PSA and human f_a for a set of 309 molecules. Varma et al. 2010 [4] with permission.

other organic solvent/water systems. In contrast to octanol, which is an amphiprotic solvent, other solvents that are purely hydrogen bond-donors or -acceptors and aprotic alkanes have been used as alternatives. However, octanol is well established as the organic solvent standard.

$\log P$ assesses the partitioning of only a single form of the solute, generally the neutral form. For ionizable molecules, only one form can be present during testing. The pH of the aqueous phase is rarely mentioned for experimental $\log P$ determinations, but it must be assumed that it was not within ±2 units of the pK_a.

To account for solutes that are fully or partially ionized at the pH of interest, the $\log D$ or distribution coefficient is used (Equation (4.3):

$$\log D \approx \log\left(\frac{C^{\mathrm{N}}_{\mathrm{oct}}}{(C^{\mathrm{I}}_{\mathrm{water}} + C^{\mathrm{N}}_{\mathrm{water}})}\right) \tag{4.3}$$

Here, C^{I} and C^{N} refer to concentrations of the ionized and neutral forms, respectively. It is assumed that only the unionized form partitions into octanol. Equation (4.3) represents an approximation that illustrates the concept; the distribution of all forms between the aqueous and organic layers must be considered [24]. Generally, the organic layer is *n*-octanol and the aqueous solution is buffered to what is defined as "biological pH," 7.4. Neutral compounds and most weak bases, ubiquitous among drug candidates, have $\log D_{7.4} \approx \log P$. However, for molecules with a pK_a 6–9 a substantial fraction of the solute is ionized at 7.4 and the $\log D$ may be significantly different from $\log P$. For this reason, $\log D$ is sometimes substituted for $\log P$ in studies of biologically active molecules.

In 2010, the relationship between lipophilicity descriptor parameters ($c\log P$ and $c\log D_{7.4}$) and oral absorption (f_a) was investigated in our laboratory [4]. As illustrated in Figure 4.2,

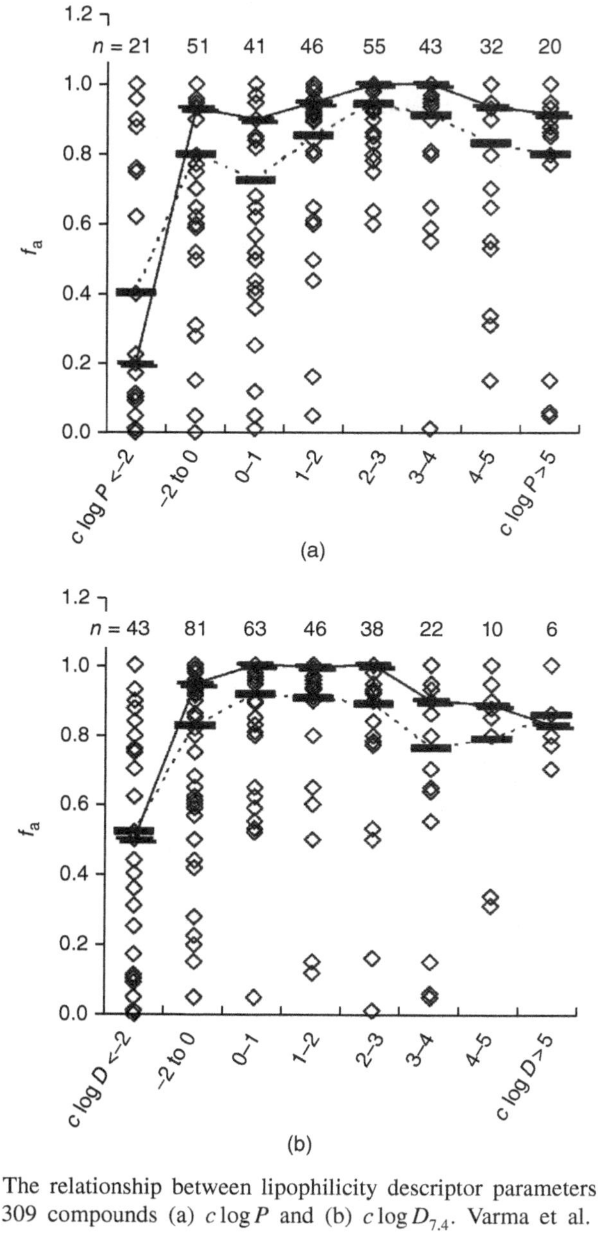

FIGURE 4.2 The relationship between lipophilicity descriptor parameters and human f_a using a set of 309 compounds (a) $c\log P$ and (b) $c\log D_{7.4}$. Varma et al. 2010 [4] with permission.

lipophilicity had a significant impact on f_a. Compounds that are very hydrophilic (i.e., $c \log P$ and $c \log D_{7.4} < -2$) had poor oral absorption, whereas compounds with higher lipophilicity showed improved absorption.

4.3.4 pH Partition Theory and pK_a

Most drug molecules are either weak acids or bases. When given orally, the extent of the ionization of these drugs is dependent on their pK_a and pH gradient inherent to the gastrointestinal tract. The interplay between these two interrelated parameters has a major impact on both drug permeability and solubility, which ultimately affect the rate and extent of oral drug absorption. One of the original concepts that govern the oral drug absorption is the pH partition theory. This theory states that oral drug absorption takes place mainly by a passive diffusion of the unionized form of the drug molecule through the lipophilic intestinal membrane. The Henderson–Hasselbalch equation (Equations (4.4 and (4.5) can be used to calculate the fraction of the unionized form of a basic or acidic drug. Thus, the fraction of the unionized form of monobasic drug can be calculated using

$$\mathrm{pH} = \mathrm{p}K_a + \log_{10}\left(\frac{\text{Unionized concentration}}{\text{Ionized concentration}}\right) \tag{4.4}$$

The fraction of the unionized form of monoacidic drug can be calculated using

$$\mathrm{pH} = \mathrm{p}K_a + \log_{10}\left(\frac{\text{Ionized concentration}}{\text{Unionized concentration}}\right) \tag{4.5}$$

Accordingly, the oral absorption of a weak base will be favored in the intestine rather than the stomach, as the intestine has a higher pH and, as a result, a higher proportion of the unionized drug fraction will be available for oral absorption (Figure 4.3). On the contrary, weak acid absorption will be favored in the stomach owing to its lower pH (Figure 4.3). However, the intestine continues to be the main site of absorption of both weak acids and weak bases because of the larger surface area available to absorption as compared to stomach and colon.

In 1999, van Niel et al. investigated the advantage of reducing the pK_a of the piperazine of the 3-(3-(piperazin-1-yl)propyl)indole series of 5-HT1D receptor using one or two fluorine linkers. The group identified agonists that had high affinity and selectivity to the 5-HT1D receptor *in vitro*. Fluorine addition reduced the basicity of the molecules. The reduced basicity was associated with a significant improvement in the compound oral absorption, which is mainly ascribed to higher intestinal permeability that is predominantly mediated by increase in the unionized permeable fraction [25].

Overall, compounds that are basic in nature tend to have higher f_a in humans compared to those that are acidic as depicted in Figure 4.4 [4].

4.3.5 Rule of Five

Lipinski et al. introduced the rule of five (RO5). The RO5 is one of the first concepts that was implemented to qualitatively predict the oral absorption of drugs in human. A set of 2245 compounds obtained from the World Drug Index (WDI) database were analyzed. These molecules were either considered for or entered into Phase II clinical trials. Data analysis suggested that good oral absorption is observed when drug molecules have <5 hydrogen-bond donors (defined as NH or OH groups), <10 hydrogen-bond acceptors (defined as oxygen or nitrogen atoms, including those that are part of hydrogen-bond donors (HBDs)), M_{wt} <500 and a calculated lipophilicity ($c \log P$) < 5 [12, 13, 26, 27].

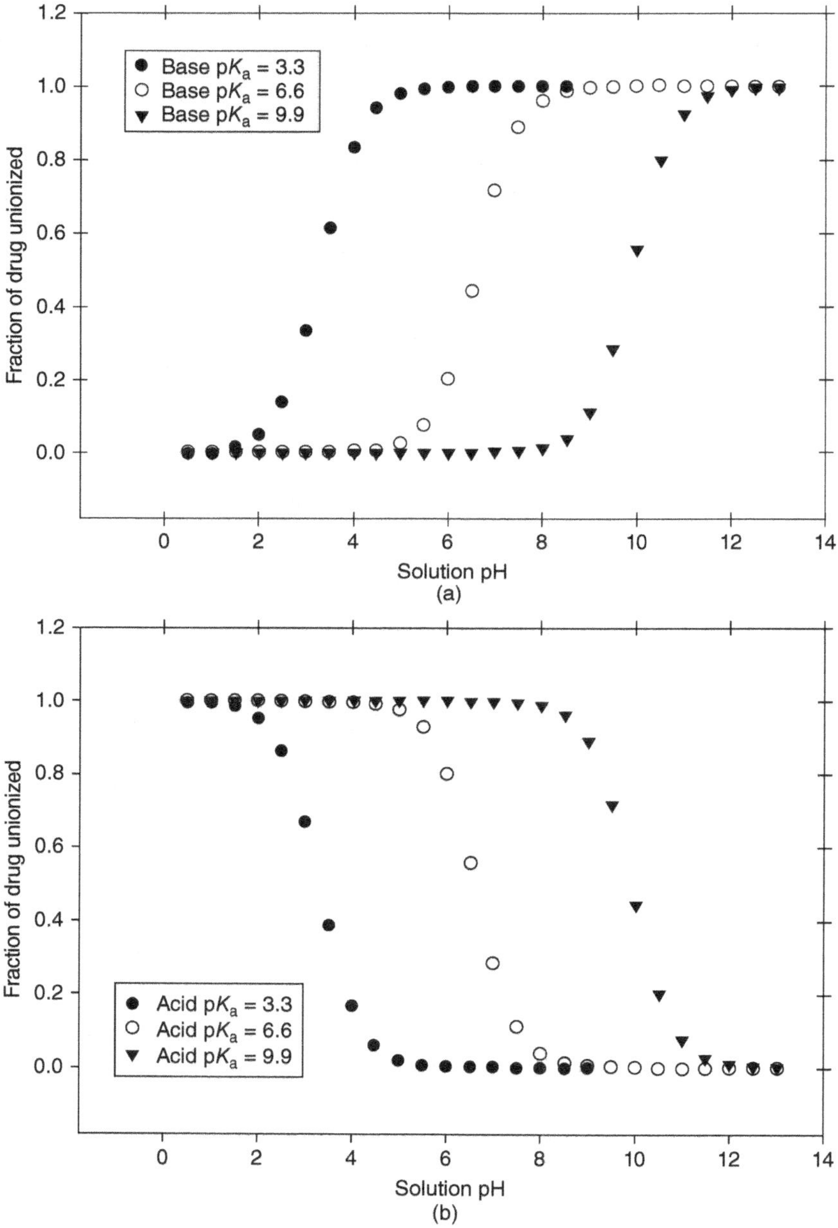

FIGURE 4.3 Schematic diagram to illustrate the relationship between the fraction of drug unionized, and pH and pK_a of (a) bases and (b) acids.

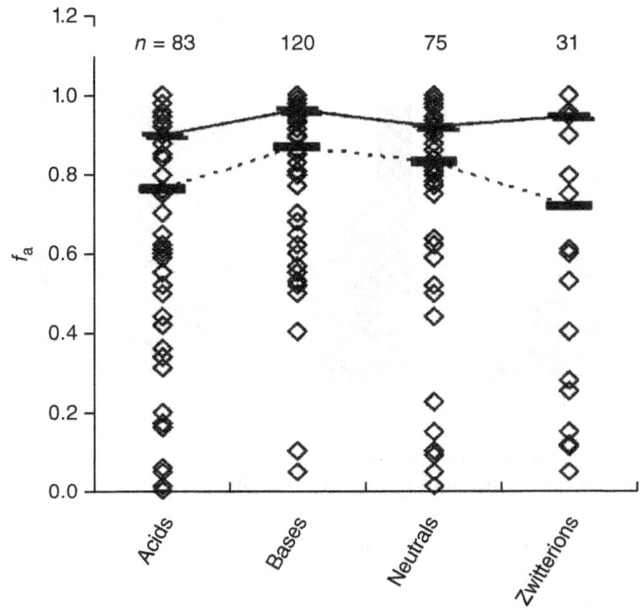

FIGURE 4.4 The relationship between compound ionization state and human f_a ($n = 309$). Varma et al. 2010 [4] with permission.

Varma et al. investigated the relationship between the number of violations of RO5, bioavailability and its components for a set of 309 molecules. As illustrated in Figure 4.5, Varma et al. demonstrated a significant impact of the number of violations of RO5 and oral bioavailability, where the median F decreased from 0.7 to 0.05 when the number of violations increased from 0 to 3. Interestingly, this impact was only observed with f_a. Similar trends were not reported with f_g and f_h, which underscores the relationship of RO5 and intestinal absorption [4].

4.4 SOLUBILITY

Solubility is the concentration of a drug in a saturated solution at a definite temperature and pH. As depicted in Equation (4.1), ensuring a sufficient compound concentration in solution at the primary site of absorption is a requirement for successful oral drug absorption. Different methodologies are used to determine compound solubility at physiological temperature of 37 °C. These are classified as thermodynamic and kinetic solubility methods [12, 13, 26, 27]. Thermodynamic solubility is defined as the drug equilibrium concentration, which is saturated in a given solvent at a definite temperature, typically 37 °C. Many factors influence drug thermodynamic solubility and these include, and not limited to, media pH, temperature, pressure, and drug crystal lattice and melting point [28]. However, kinetic solubility is a determination of a compound solubility in a relevant aqueous medium. The compound is predissolved in organic solvent such as dimethylsulfoxide (DMSO) and then it is titrated with aqueous medium over a period of time until the compound precipitates [29]. In drug discovery and lead development, kinetic solubility measurement is the method of choice since it is pliable to high-throughput screening (HTS) and requires no to minimum equilibration time. The main disadvantage of

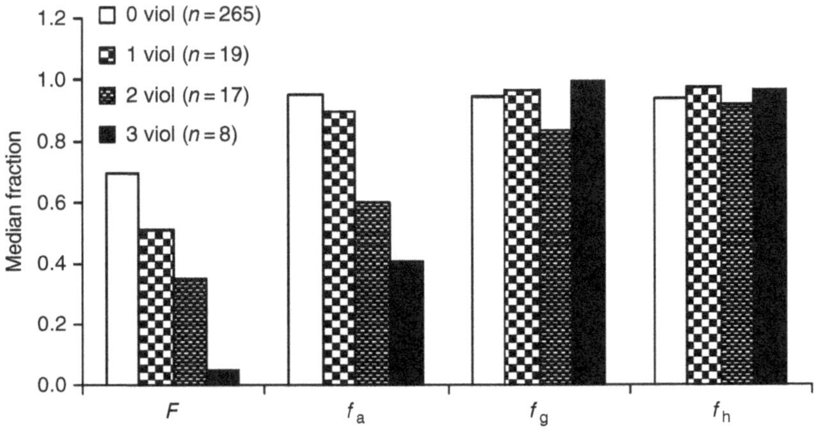

FIGURE 4.5 Relationship between number of violations of RO5 and bioavailability and individual processes. "*n*" is the number of compounds in each bin [4] with permission.

this methodology is the lack of consideration for crystal lattice effect as evident with thermodynamic solubility [27]. Both thermodynamic and kinetic solubility are generally measured in a phosphate-based buffer (pH range 6.5–7) as this medium is considered to be a surrogate for intestinal pH. Sometimes, fasted state simulated intestinal fluid (FASSIF) or fed state simulated intestinal fluid (FESSIF) are also considered as aqueous mediums to assess compounds solubility. In 2001, Lipinski et al. determined the minimum aqueous solubility necessary to achieve adequate oral absorption. The group noted that this value is dependent on both the compound projected human dose and permeability [27]. For example, given a compound series with a projected human dose of ~1 mg/kg and high permeability, determined using the apparent absorption rate constant (K_a), an aqueous solubility of ≥10 µg/mL would be needed to ensure that oral bioavailability is not solubility limited.

4.5 PHYSICOCHEMICAL AND BIOPHARMACEUTICAL PROPERTIES THAT INFLUENCE SOLUBILITY

4.5.1 Polymorphism and Amorphous Form

While being chemically identical, polymorphs of drug substance have different physical properties such as melting temperature, molecular density, hygroscopicity, crystal shape, and enthalpy of fusion, which are ascribed to differences in their molecular packing [30, 31]. Despite these differences, various polymorphs tend to have comparable solubility profiles. In 2005, Pudipeddi and Serajuddin investigated the impact of various polymorphs of drug molecules on their solubility profiles ($n = 81$) [30]. The solubility values of different polymorphs of drug investigated were within twofolds as illustrated in Figure 4.6. This difference in the solubility value is not expected to have a major impact on the compound biopharmaceutical profile depending on the doses used, particle sizes, and solubility values [30]. However, polymorphism may affect the physical and chemical stability of various drug molecules by influencing the rate and mechanism of decay [1]. Specific examples of how polymorphism can affect stability include carbamazepine, indomethacin, and enalapril maleate [32–35].

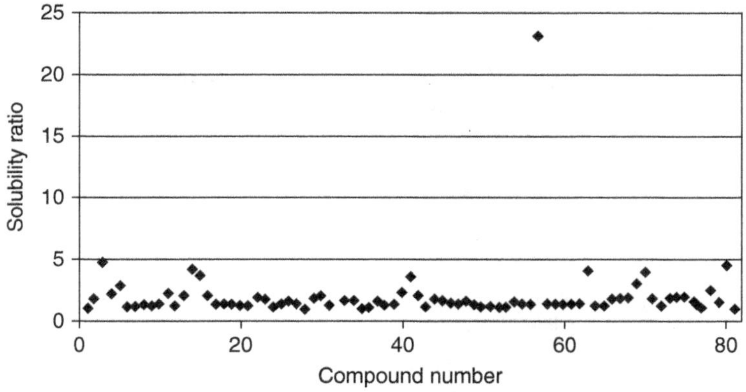

FIGURE 4.6 The solubility ratio for polymorphs ($n = 81$). Pudipeddi and Serajuddin 2005 [30]. Reproduced with permission of Elsevier.

4.5.2 Crystal Lattice Energy and Solubility

A highly ordered crystal structure is typically associated with higher melting point as well as lower aqueous solubility [1]. This highly ordered crystal lattice is primarily a result of two forms of strong cohesive interactions, which are either intermolecular hydrogen bonding and/or short-ranged dispersion forces. These interactions are typically observed with drug molecules that contains one or more of the following properties: significant number of hydrogen-bond acceptor (HBA)/donor (HBD) groups, planar or inflexible conformation, or a high degree of symmetry [36]. Collectively, these factors promote tight packing within the crystal lattice rendering it not readily soluble in aqueous media. In general, a high melting point compound (>250 °C) will have poor thermodynamic solubility, requiring a discovery team to develop a structure activity relationship to reduce its crystal lattice energy leading to higher aqueous solubility. This approach was implemented by Goosen et al. who evaluated the effect of structural modification on the physicochemical properties of thalidomide, which has an aqueous solubility of 52 µg/mL and a melting point of 275 °C [37]. Thalidomide's poor aqueous solubility is ascribed to the acidic imido hydrogen on the glutarimide ring. This group is responsible for the strong hydrogen and dipolar bonding in thalidomide crystalline state. When replaced with an alkyl group, the highly ordered crystal lattice was disrupted (Figure 4.7). As a result, *N*-methyl thalidomide has an aqueous solubility of 276 µg/mL and 159 °C melting point.

TIP:

Flux across the enterocyte is a product of drug passive permeability and solubility. Interestingly, physicochemical properties that influence permeability have also significant impact on solubility, yet in the opposite direction. For example, increasing compound lipophilicity is typically associated with an increase in its permeability. However, it is also accompanied with lower solubility. Therefore, when you modulate physicochemical properties of your lead, be mindful that it is a sword with two edges, only one can help. Be mindful of this balance. It will enable you to design more orally bioavailable molecules.

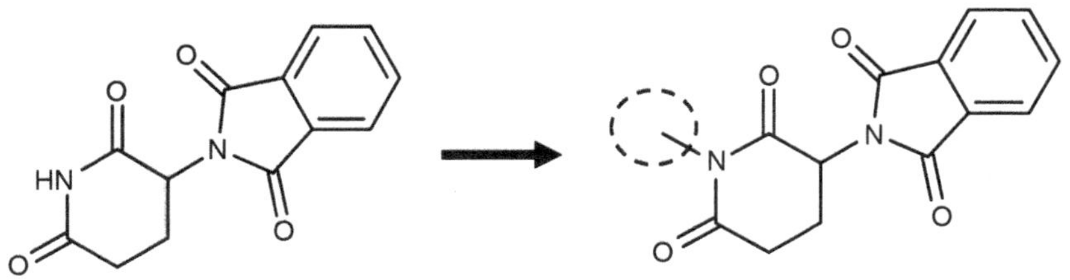

FIGURE 4.7 N-alkylation of thalidomide alters the highly ordered crystal lattice and increases its solubility by fourfold. Adapted from Goosen et al. 2002 [37].

4.5.3 pH and Salt Interplay

Salt is a solid that is composed of a charged drug molecule, typically a weak acid or base, with its oppositely charged counterion. The use of a salt form as pharmaceutical ingredient is an integral part of the pharmaceutical development of many oral dosage forms and is well established in the literature [38, 39]. A salt form of a drug molecule changes the Coulombic attraction between the drug molecule and its counterion and alters the potential energy of the solid state. As depicted in Figure 4.8, a salt of a weak acid in the gastric fluid would alter the pH of the diffusion layer at the surface of the dissolving solid making pH higher (pH 5–6) than the gastric fluid pH of 1–3. As a result, this change in pH significantly increases the solubility of the parent drug molecule, in that layer over its inherent solubility at the pH of the gastric fluid. Once the drug diffuses out, the parent weak acid partially crashes out of solution in a form of fine precipitate of free acid (HA). This is attributed to the fact that the solubility of the dissolved weak acid is significantly higher than its original solubility in the bulk gastric pH. Interestingly, the precipitate has very high surface area that would facilitate its redissolution and absorption across the gastrointestinal tract. Similar trend would be observed with salt of weak base that reaches the small intestine. Overall, if other relevant factors, such as chemical stability, permeability, and intestinal and liver metabolism, remain constant, the dissolution rate of a compound should determine the rate of buildup of blood levels with time and the maximal levels achieved [38, 39]. In summary, the drug salt form usually alters the drug dissolution rate by modifying the diffusion layer pH at the surface of the dissolving solid [38, 39]. Nelson was the first to report this phenomenon in which the salts of acidic theophylline with high diffusion layer pHs had greater *in vitro* dissolution rates than those exhibiting a lower diffusion layer pH. In fact, the rank order of dissolution rates of theophylline was closely correlated with the clinical blood exposure. This report led many additional studies that demonstrated the influence of the salt form on drug dissolution and the benefit of changing nonionized drug to salts [38, 39].

4.5.4 Counterion/Salt Selection

In general, there should be around 2–3 unit separation between the pK_a values of the counterion and the functional group on the drug molecule.

When selecting a salt, scientists should take into consideration the loading of the drug product where high molecular weight counterion may not be acceptable with high loading product.

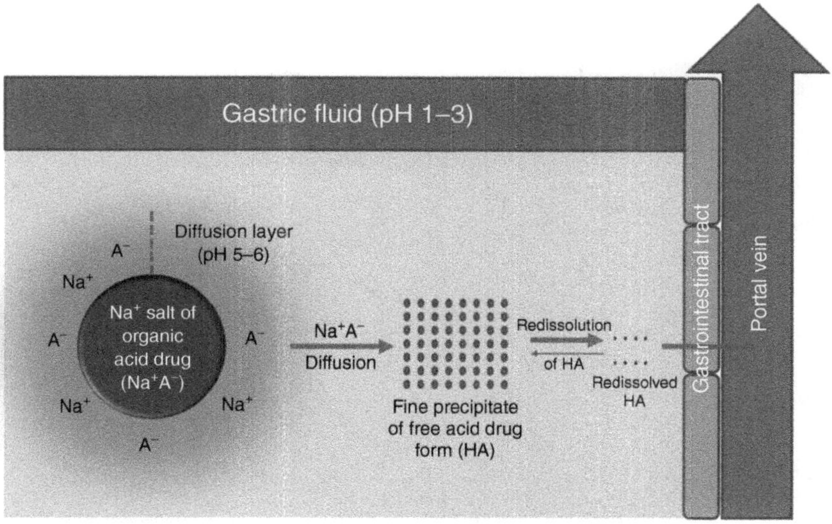

FIGURE 4.8 Dissolution of a salt form of organic acid in gastric fluid. *(See color plate section for the color representation of this figure.)*

Also, dosage form would dictate the counterion selection. For example, counterions for oral formulation would not typically be used in the development of inhalation or ocular formulations. In addition, using counterions that are endogenous ions such as Cl^- is often associated with common ion effect. Counterions will lower the solubility of a hydrochloride salt of a basic drug in gastric fluids, which reduces their plasma exposure following oral dosing [40].

4.5.5 Complexation

Ionic inclusion drug complexes have energy of attraction that is <10 kcal/mol. Ionic inclusion compounds result more from the architecture of the molecules than from their chemical interaction. Herein, one of the constituents of the complex usually is the drug molecule, which is confined in a cage-like molecular structure of the other to yield a thermodynamically stable structure. In that regard, cyclodextrins are widely used complexing agents. These are natural cyclic oligosaccharides that are formed via the enzymatic transformation of starch. The main three cyclodextrins are α-cyclodextrin, β-cyclodextrin, and γ-cyclodextrin (Figure 4.9).

These are composed of six, seven, and eight linked α-D-glucopyranose units, respectively. The cyclodextrin cyclic structure results in a hydrophilic outer shell and a lipophilic core. They have been effectively used in trapping poorly soluble and lipophilic drugs in their molecular envelope and forming an inclusion complex. Interestingly, noncovalent bonds are involved in the complex formation in the aqueous solution where drug molecules bound in the complex are in rapid equilibrium with free solubilized drug molecules. This complex is effective in increasing drug dissolution rate, solubility, as well as the extent and rate of drug absorption [41]. For example, Ridhurkar et al. compared the impact of cyclodextrin complex of aprepitant to Emend®, an oral formulation incorporating drug nanoparticles, on rate and extent of oral absorption in human. Aprepitant is a poorly soluble drug that is approved for the treatment of chemotherapy-induced nausea and vomiting and postoperative nausea and vomiting [42]. *In vitro* studies confirmed the inclusion of aprepitant in the cyclodextrin complex. Also, a significant increase in its dissolution rate and solubility was demonstrated. Following oral dosing in

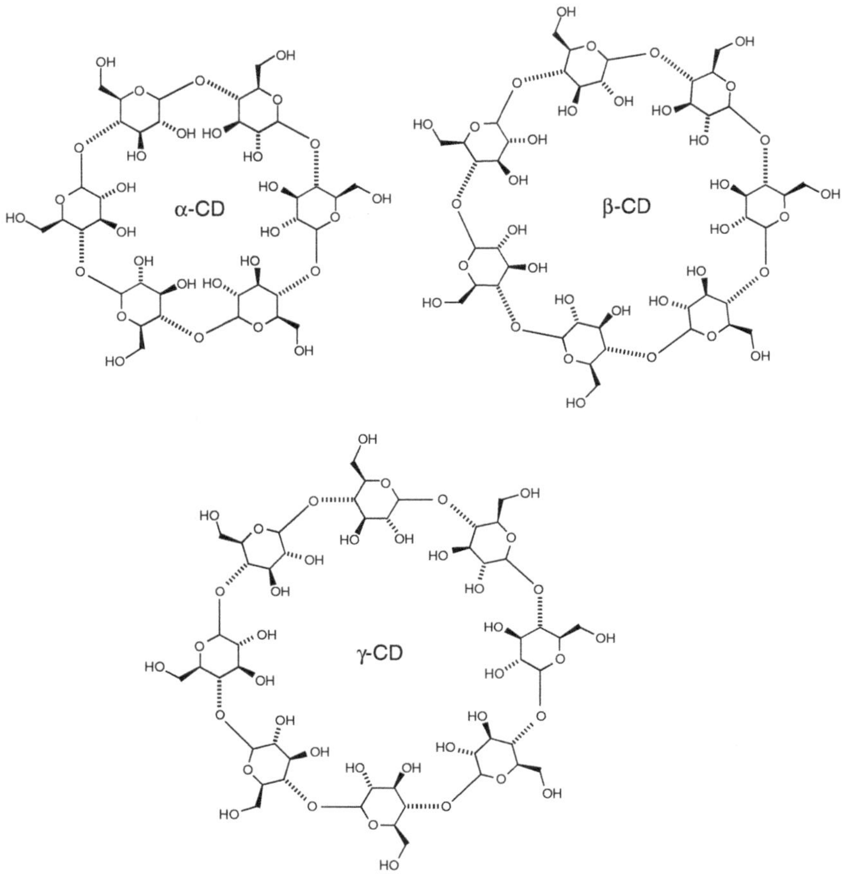

FIGURE 4.9 The chemical structures of α-cyclodextrin, β-cyclodextrin, and γ-cyclodextrin.

human, the rate (C_{max}) and extent of absorption (AUC) of aprepitant from the complex were found to be comparable to that of Emend that has a moderate oral bioavailability in human (67%).

Cyclodextrins can also work as permeation enhancer through one or a combination of the following mechanisms:

1. Cyclodextrins complex components of the intestinal brush border membranes, such as phospholipids, cholesterol, and membrane proteins. This complex leads to increase in the membrane fluidity, which increases drug permeability via transcellular pathway.
2. Cyclodextrins also open cellular tight junctions without significantly compromising cell membrane integrity. The alteration in the tight junction can be pivotal in increasing the permeability of compounds absorbed via paracellular pathway [43].
3. Cyclodextrins can improve drug delivery across intestinal membrane when the permeation resistance of unstirred water layer (UWL) on the luminal side is equal or higher than the intestinal membrane barrier resistance. Indeed, this is mainly driven by one of

the following case scenarios: (i) Cyclodextrins complexation increases the total amount of dissolved drug molecules in the luminal side leading to higher concentration gradient across the intestinal membrane. This increases the amount permeating since drug release from the cyclodextrins complex is typically faster than the flux of drug molecules through the UWL. (ii) Low molecular weight carbohydrates behave as chaotropes at relatively low concentrations; cyclodextrins disrupt the hydrogen-bonded water network in the UWL. This disruption could facilitate penetration of drug/cyclodextrins complexes through the UWL and increase drug availability at the membrane surface. However, there are no indications that cyclodextrins enhance permeation of free drug molecules through the UWL. (iii) Cyclodextrins complexation of drug molecules could possibly prevent interaction with molecules in the UWL, such as mucin, and increase their overall delivery rate to the membrane surface [44].

Taken together, cyclodextrins improve oral absorption and bioavailability of drug molecules via the increase of drug solubility, dissolution rate, and/or intestinal permeability [45–47]. Table 4.1 provides a summary of all the marketed cyclodextrin-containing drug products.

4.6 APPROACHES TO INCREASE DRUG DISSOLUTION

Dissolution rate is defined as the amount of drug substance that goes in solution per unit time under standardized conditions. Dissolution rate is a critical parameter that has a significant impact on drug oral absorption. For example, a drug is considered poorly soluble when its dissolution rate is slow and takes longer than its intestinal transit time (1–4 h). Dissolution rate can be calculated using the Noyes–Whitney equation:

$$\mathrm{DR} = \frac{D_{\mathrm{eff}}}{h_{\mathrm{eff}}} A(C_{\mathrm{s}} - C_{\mathrm{b}}) \tag{4.6}$$

where DR is the drug dissolution rate, A the surface area available for dissolution, D_{eff} the diffusion coefficient of the drug, h_{eff} the thickness of the boundary diffusion layer adjacent to the dissolving drug surface, C_{s} the saturation solubility of the drug in the diffusion layer, and C_{b} the concentration of the drug in the bulk solution at time, t. According to Equation (4.6), various factors affect the dissolution rate of drug molecule. For example, dissolution rate is proportional to the surface area of the drug particle. As a result, decreasing particle size by micronization or nanomilling is associated with increase in surface area, which in turn increases drug dissolution rate as shown in the following sections. In addition, increasing the concentration gradient by increasing saturation solubility relative to the concentration in the bulk solution would also increase drug dissolution rate.

4.7 PHYSICOCHEMICAL AND BIOPHARMACEUTICAL PROPERTIES THAT INFLUENCE DISSOLUTION

4.7.1 Amorphous Form

There are significant differences between crystalline polymorphs and the amorphous form of a drug. Due to the lack of crystalline structure, the amorphous form tends to have significantly higher dissolution rate and solubility compared with their crystalline form counterparts. This may significantly increase their rate and extent of oral absorption.

TABLE 4.1 Marketed Cyclodextrin-Containing Drug Products

Cyclodextrin/Drug	Trade Name	Formulation	Country
α-Cyclodextrin			
Alprostadil (PGE1)	Provastatin, Rigidur	I.V. solution	Japan, Europe, USA
OP-1206	Opalmon	Tablet	Japan
Cefotiam hexetil HCl	Pansporin T	Tablet	Japan
β-Cyclodextrin			
Benexate HCl	Ulgut, Lonmiel	Capsule	Japan
Cephalosporin (ME1207)	Meiact	Tablet	Japan
Chlordiazepoxide	Transillium	Tablet	Japan
Dexamethasone	Glymesason	Ointment	Argentina
Diphenhydramine HCl, chlorophylin	Stada-Travel	Chewing tablet	Europe
Iodine	Mena-Gargle	Solution	Japan
Nicotine	Nicorette, Nicogum	Sublingual tablet, chewing gum	Europe
Nimesulide	Nimedex	Tablet	Europe
Nitroglycerin	Nitropen	Sublingual tablet	Japan
Omeprazole	Omebeta	Tablet	Europe
PGE2	Prostamon E	Sublingual tablet	Japan
Piroxicam	Brexin Flogene, Cicladon	Tablet, suppository, liquid	Europe, Brazil
Diaprofenic acid	Surgamyl	Tablet	Europe
2-Hydroxypropyl-β-cyclodextrin			
Cisapride	Propulsid	Suppository	Europe
Itraconazole	Sporanox	Oral and I.V. solutions	Europe, USA
Mitomycin	Mitozytrex	I.V. Infusion	Europe, USA
Methylated β-cyclodextrin			
Chloramphenicol	Clorocil	Eye drop solution	Europe
17β-Estradiol	Aerodiol	Nasal spray	Europe, USA
Ziprasidone mesylate	Geodon, Zeldox	IM solution	Europe, USA
2-Hydroxypropyl-γ-cyclodextrin			
Diclofenac sodium	Voltaren	Eye drop solution	Europe
Tc-99 Tecoboroxime	Cardiotec	I.V. solution	USA

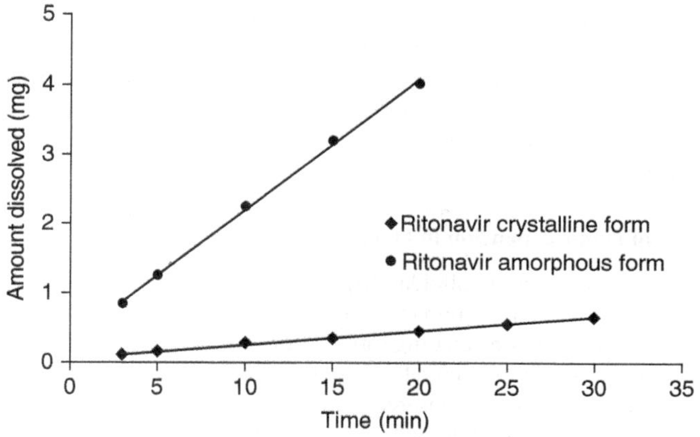

FIGURE 4.10 The dissolution rate of ritonavir in 0.1 N HCl at 37 °C using crystalline (0.03 mg/cm²/min) and amorphous forms (0.3 mg/cm²/min). Law et al. 2004 [48] with permission.

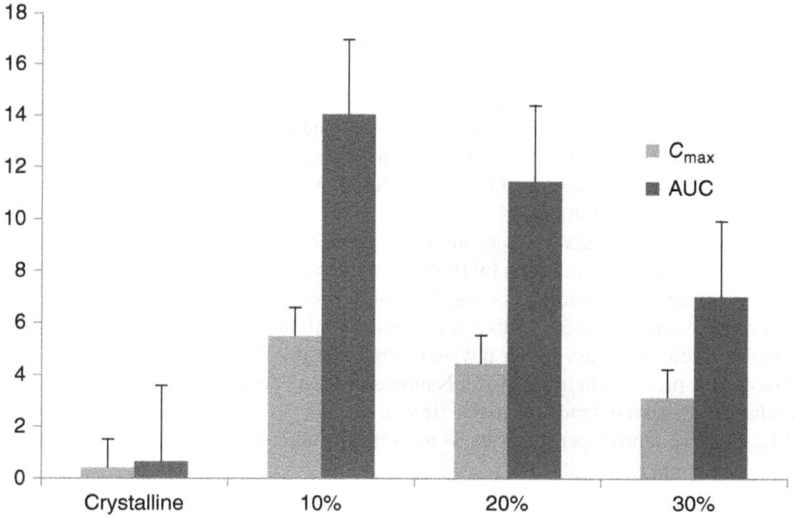

FIGURE 4.11 The C_{max} (μg/mL) and AUC (μg*h/mL) of various ritonavir dispersions in beagle dogs. Law et al. 2004 [48–51]. Reproduced with permission of Elsevier.

Ritonavir is a P-glycoprotein lipophilic substrate that has slow dissolution rate in aqueous media. Low et al. evaluated the effect of enhanced dissolution rate of poly(ethylene glycol) (PEG) amorphous ritonavir solid dispersions with different drug loadings on its oral absorption using beagle dogs as a preclinical model. As illustrated in Figure 4.10, *in vitro* dissolution rate measurements suggested a 10-fold increase in dissolution rate for amorphous ritonavir relative to its crystalline form. In an *in vivo* beagle dog study, results suggested that solid dispersions with different drug loadings (10–30%) had a significantly higher extent (AUC) and rate (C_{max}) of absorption relative to crystalline drug (Figure 4.11). However, higher solid dispersion drug loadings (20% and 30%) were associated with lower AUC and C_{max} relative to 10% loading. The differences in the exposure were consistent with lower dissolution rates at these levels [48].

It should be stressed that the amorphous form is chemically unstable. This instability is mainly attributed to the lack of a three-dimensional crystalline lattice, higher free volume, and greater molecular mobility. The chemical stability of amorphous systems has been discussed in detail elsewhere [52, 53].

4.7.2 The Use of Nanosuspension in Overcoming Poor Dissolution

Nanosuspensions are submicron colloidal dispersions of drug particles that are stabilized in liquid media by surfactants and/or polymers. They have a particle size range between 10 and 1000 nm. The use of nanosuspension drug formulation is an effective approach to improve the solubility and increase the dissolution rate of poorly soluble drug molecules. Nanosuspension formulation enhances the plasma exposure of drug molecules following oral dosing. They are usually associated with reduced inter- and intraindividual variability and lower total oral dose administered [49–51].

Special attention should be made to the physicochemical properties of drug molecules considered for nanosuspension, since they have a significant impact on the rate and extent of absorption (represented by C_{max} and $AUC_{(0\rightarrow t)}$, respectively) of poorly soluble drug in nanosuspension formulations. Recently, Li et al. investigated the influence of these properties on dissolution rate and rat oral plasma exposure of five poorly soluble drug nanosuspensions: fenofibrate, probucol, itraconazole, albendazole, and ravaprazan hydrochloride. The nanosuspensions were formulated using high-pressure homogenization in the same stabilizer, poloxamer 188, and had a similar particle size. The group reported that melting point, log P, and PSA are key factors that affect the dissolution rate in FaSIF media and rat oral plasma profiles of poorly soluble drug nanosuspensions. Indeed, drugs with smaller melting point, log P around 5, and PSA value in the range of 50–60 Å^2 could achieve higher $AUC_{(0\rightarrow t)}$ using nanosuspension formulation [54].

Nanosuspension formulation was shown to be effective in addressing issues that would be encountered during formulation by traditional methods. For example, nanosuspension formulation increases drug loading and can be used to reduce reported side effect and toxicity such as gastric irritation [55]. Unlike some other methodologies (e.g., amorphous formulation), nanosuspension offers higher physicochemical stability, which is a key factor for a long pharmaceutical product shelf time [56]. Nanosuspension is used to increase product intestinal mucoadhesion, which extends residence time at the site of absorption and increases effective half-life. Finally, nanosuspension can be used in manufacturing production and is amenable for large-scale operations [49–51].

4.7.3 Cocrystal Drug

Cocrystal is a multicomponent molecular crystal structure that is constituted of two or more different chemical molecules with a definite stoichiometric ratio. They interact through noncovalent interactions that include hydrogen bonding, ionic interactions, π-stacking, and hydrophobic Van der Waals forces. One component is the active pharmaceutical ingredient and the other is the cocrystal former. The latter should be inert, nontoxic, and part of FDA "Everything Added to Food in the United States" (EAFUS) list, which is composed of safe food additives or approved as Generally Regarded as Safe ingredients (GRAS) [57].

Cocrystals result in major differences in the physical properties of their active pharmaceutical ingredients. These include and are not limited to differences in density, solubility, compressibility, friability, melting point, hygroscopicity, and dissolution rate. These alterations in physical properties may expand solid dosage form options considered in the formulation of new chemical entities. Stanton et al. investigated the impact of 12 carboxylic acid on the dissolution rate and pharmacokinetic profile of AMG517, a poorly soluble weak base, following oral dosing of 100 mg/kg to rats. The group demonstrated that all carboxylic acid cocrystals tested had higher intrinsic and powder dissolution rate of AMG517 in fasted simulated intestinal fluid (FaSIF). They also reported higher rat oral plasma exposure (AUC) compared with the free base. Interestingly, there was a moderate linear relation between dissolution rate and AUC among the 12 cocrystals. This result suggested that *in vitro* dissolution is a pivotal tool that can be effectively used to rank order cocrystal and predict if an improvement in oral exposure would be achieved for poorly soluble active ingredients [58]. In addition, cocrystals provide a thermodynamically stable crystal structure for amorphous active pharmaceutical products, which is pivotal in extending the product shelf life [59].

4.8 THE USE OF PRODRUGS TO INCREASE SOLUBILITY AND PERMEABILITY

Prodrugs are covalently linked complexes. These are biologically active substrates (drugs) that are administered in an inactive or significantly less active chemical form. Once absorbed, the prodrug is metabolized *in vivo* into the active form/parent drug. Prodrugs are prepared by chemical modification of the parent drug molecule via the addition of labile moiety. Figure 4.12 illustrates known functional groups on parent drug that can be used in prodrug design that include and not limited to hydroxyl, amine, or carboxylic acid groups [60, 61].

The labile groups are typically broken by enzymatic/chemical transformations, where the parent drug is free to produce its pharmacological activity.

Prodrug can be used as a viable approach to increase drug solubility and plasma exposure of drug molecules following either intravenous or oral administration. This is achieved by adding ionizable or polar groups such as phosphate or sulfamate. For example, phenytoin is an antiepileptic agent that has poor water solubility, which is difficult to formulate to dose intravenously. To overcome this liability and enable the treatment of acute treatment convulsive status epilepticus, fosphenytoin was developed. This prodrug significantly increased the solubility of phenytoin from 20–25 μg/mL to 140 mg/mL. Fosphenytoin rapidly breaks down to phenytoin by alkaline phosphatases ($t_{1/2}$ = 7–15 min) [62, 63]. Examples on the use of prodrug to improve solubility of drug molecules were reported with sulindac, miproxifene phosphate, fosamprenavir, estramustine phosphate, prednisolone phosphate, fludarabine phosphate [64–70].

Prodrugs increase passive mediated intestinal permeability through various mechanisms. Prodrugs either reduce the parent drug hydrogen-bonding potential or increase its lipophilicity (extent will be determined by the ester moiety added). For acids, prodrugs eliminate ionizable center rendering molecules more permeable [71–73].

Penciclovir is an antiviral agent with poor oral absorption and bioavailability in human (4%). To improve its oral exposure, famciclovir (dimethyl ester of penciclovir) was designed. *In vitro*, famciclovir rapidly breaks down to penciclovir by hydrolase and aldehyde oxidase mediated oxidation of purine to guanide (Figure 4.13). In human, famciclovir increased penciclovir oral absorption and bioavailability to 75% by increasing penciclovir lipophilicity and passive intestinal permeability [74–76].

Prodrugs also increase intestinal permeability by improving affinity to uptake intestinal transporters located at the brush border membrane of the intestine (e.g., Pept1 transporter) [5, 77, 78]. For example, acyclovir is an antiviral agent that is absorbed via paracellular pathway [79]. It has poor oral absorption and low bioavailability in human (15%). To increase its oral plasma exposure, a prodrug of acyclovir (valacyclovir) was developed. Once absorbed, valacyclovir breaks down to the parent drug and provides a superior acyclovir oral plasma exposure. The improvement in overall exposure is ascribed to the better transcellular permeability of valacyclovir that is mediated by Pept1 transporter located at the brush border membrane of the small intestine (Figure 4.14) [80–82]. Similar methodology was applied to enable the controlled release formulation of gabapentin. This molecule is absorbed via system L transporter

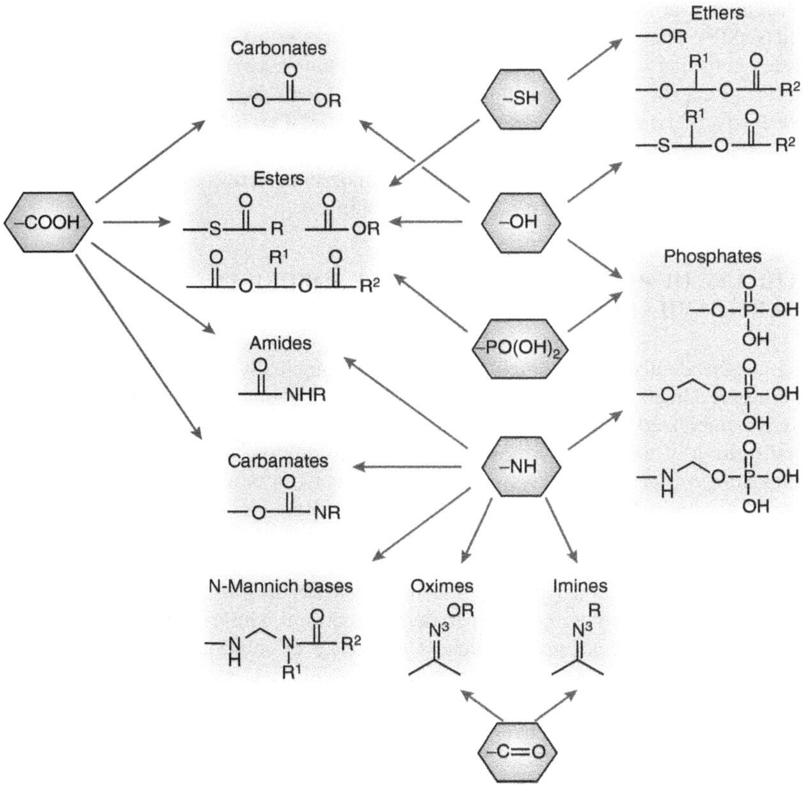

FIGURE 4.12 Known functional groups on parent drugs that are pliable to prodrug design (shown in gray). Rautio et al. 2008 [60]. Reproduced with permission of Nature Publishing Group. *(See color plate section for the color representation of this figure.)*

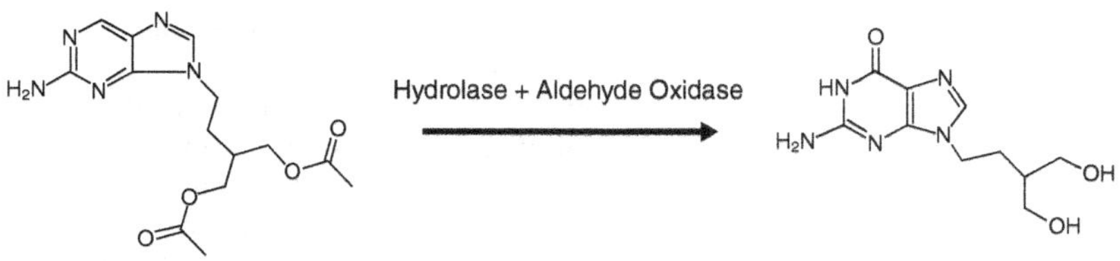

Famciclovir Penciclovir

FIGURE 4.13 Bioconversion pathways of famciclovir to penciclovir. Rautio et al. 2008 [60]. Reproduced with permission of Nature Publishing Group.

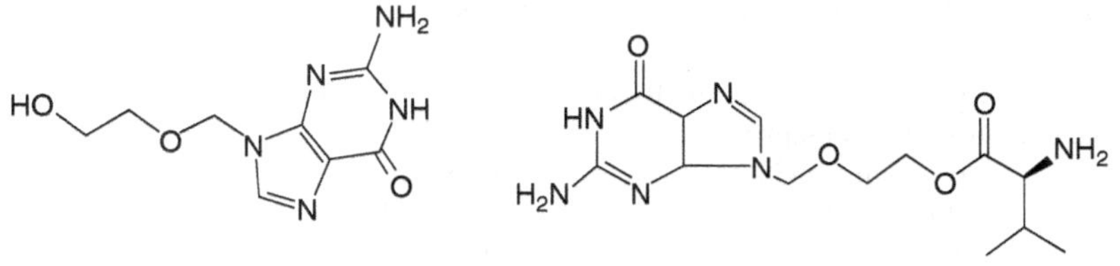

Acyclovir
F% = 15%

Valacyclovir
F% = 54% (Parent drug: Acyclovir)

FIGURE 4.14 The chemical structures of acyclovir and its commercial prodrug valacyclovir, respectively, with acyclovir bioavailability following oral dosing in humans for the two agents [80–82].

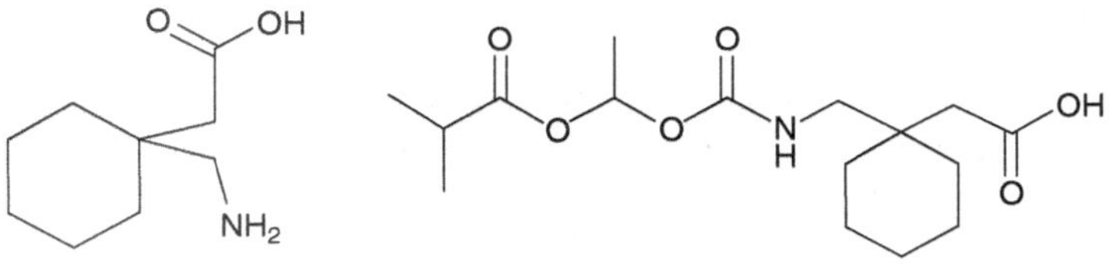

Gabapentin

XP13512

FIGURE 4.15 The chemical structures of gabapentin and its commercial prodrug XP13512 (Gabapentin enacarbil).

and paracellular pathway. In the colon, it has poor permeability since colon tight junction is smaller and system L transporter expression is limited. Therefore, gabapentin is not amenable for controlled release formulation using traditional methodologies targeting the colon. Xenoport scientists developed a gabapentin prodrug (XP13512 also known as gabapentin enacarbil) that is a good substrate for MCT1 transporter, which is a high-capacity low-affinity transporter and expressed in the colon (Figure 4.15). Studies in rat and monkey indicated that colonic delivery of XP13512 has a superior exposure for gabapentin versus giving gabapentin alone [83, 84]. As a result, they developed a viable controlled release formulation of XP13512, which provided a longer half-life and reduced the dosing frequency of gabapentin [85]. Other examples on prodrugs that increased absorption by increasing affinity to transporters are Valganciclovir and Midodrine [86, 87].

In summary, prodrug methodology is effective in the development of antibiotics, statins, sartans, and angiotensin converting enzymes inhibitors (ACEi) [88–92]. Plasma exposure is significantly increased following oral/intravenous dosing by improving their permeability and/or solubility [61, 73].

Interestingly, drug complexes do not always improve drug solubility/permeability. A commonly reported example is the complexation of tetracycline with divalent cations such as calcium or magnesium to form a poorly soluble complex that cannot be absorbed. Therefore, coadministering tetracycline-like analogs with antacids that contain aluminum or magnesium hydroxide and/or calcium carbonate or dairy products such as milk or cheese should be avoided [93–97].

4.9 BIOPHARMACEUTICS CLASSIFICATION SYSTEM (BCS) AND DRUG

In 1995, Amidon et al. introduced the Biopharmaceutical Classification System (BCS) to predict the oral absorption in human based on drug solubility, permeability, and dissolution rate [98]. In order to obtain a BCS-based biowaiver of conducting a bioequivalence study, drug product must fulfill predetermined criteria. For example, the drug product should have high aqueous solubility, where it is soluble in 250 mL or less aqueous media over a pH range of 1–7.5. The drug product should also have high permeability, where its extent of absorption in humans is 90% or more of an administered dose. Finally, the drug product should be rapidly dissolving, where 85% or more of the drug is released within 30 min over the pH range outlined earlier. The assessment should be conducted in either USP dissolution apparatus I (baskets) at 100 rpm or USP dissolution apparatus II (paddles) at 50 rpm. As a result, drug product can be classified based on its permeability, solubility, and dissolution rate into four classes.

4.10 CONCLUSION

Oral bioavailability is a determining factor for the success of the discovery and development of new chemical entities since low oral bioavailability is associated with higher interindividual variability and may compromise medications therapeutic usefulness [99]. As reviewed in this chapter, the key parameters that affect oral bioavailability involve a complex interplay of drug physicochemical and biopharmaceutical properties that influence drug intestinal permeability and solubility [100].

Early application of *in silico* computational models and *in vitro* ADME assays (such as lipophilicity, PSA, permeability, and solubility determinations) can serve as useful guide to determine the factors that may influence oral absorption and bioavailability as well as to facilitate the optimization of desirable ADME properties to drive drug design prospectively.

Recent advances in the use of *in silico* computational models to predict ADME properties of compounds will likely continue to grow rapidly as the benefits they provide in throughput and early application in drug design are realized. In addition, there is an increasing range of models available that are focused on projecting various drug pharmacokinetic parameters including bioavailability [101]. This keen interest is now the driving demand for developments in the component elements of model building, namely, higher quality data sets, better molecular descriptors, and more computational power, and the quality of models is improving rapidly [102–104].

4.11 CHAPTER 4 TUTORIALS

1. Absorption

   Discovery project is seeking a lead compound for the treatment of psychosis. Drug X is the reference compound. It has low solubility ($C_s < 1$ μg/mL "free base") and moderate Caco-2 permeability ($5–10 \times 10^{-6}$ cm/s). It has high metabolic stability in rat, dog, and human hepatocytes with clearance values that are significantly lower than hepatic blood flow (*what does that mean*?).

   The team identified a compound (Drug A) that is structurally similar to Drug X and has good pharmacological activity following oral dosing. However, the compound seems to have low oral bioavailability in rats and has the following rat pharmacokinetics profile:

The PK parameters of Drug A following 5 mg/kg IV dosing to rats

Form	CL (mL/min/kg)	V_{dss} (L/kg)	$t^{1/2}$ (h)
Free base	5 (low clearance)	5 (large distribution)	12 (long half-life)

Note: You should compare the CL value to rat hepatic blood flow ($Q = 70$ mL/min/kg) and assess hepatic extraction ratio ($E_h = CL/Q$). If the E_h is low, that means the compound is metabolically stable.

The PK profile of Drug A (*free base*) following 5 mg/kg PO dosing to rats

Form	C_{max} (ng/mL)	t_{max} (h)	AUC (ng*h/mL)	F%
Free base	15 (low exposure)	5 (slow absorption)	50 (low exposure)	5 (low bioavailability)

Also, the compound has the following *in vitro* profile:

1. p$K_a = 5.5$ (weak base)
2. $C_s = 2$ ng/mL (poor solubility)
3. Caco-2 Permeability (AB) = 10×10^{-6} cm/s (good permeability)
4. Excellent stability in rat, dog, and human hepatocytes and microsomes (good hepatic stability).

Based on the profiles above, the team is asking you the following questions:

1. What is the major factor that is contributing to the poor rat oral bioavailability of Drug A?
2. Do you think similar trends would be observed in upper species?
3. What suggestions would you share with the team to overcome the poor oral bioavailability of the lead compound?
4. If you want to change the lead compound chemical structure, what are the changes you would recommend to increase the deficient property?

2. Imagine that you are part of a discovery team that is planning to identify a backup molecule to the following compound:

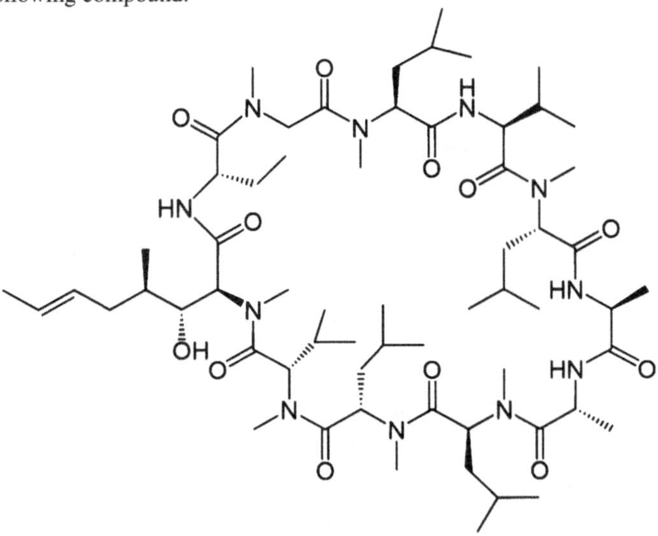

However, it had so many flaws, for example, this compound has high molecular weight ($M_{wt} > 500$ Da), poor solubility ($C_s < 1$ μg/mL), and Caco-2 permeability (AB $< 1 \times 10^{-6}$ cm/s). In addition, this compound is a PGP substrate and effective CYP3A4/OATP1B1 inhibitors with poor oral bioavailability.

The team has the following questions for you:

1. What are the factors that are contributing to the poor oral bioavailability of ketoconazole?
2. What are the structural changes that you recommend the team doing to overcome the poor oral bioavailability? Please draw structures that address these issues.

3. The following compound has low permeability yet good solubility. When assessed in human in FIH, it had the following properties:

Parameters	Values
F	0.05
f_a	0.05
f_g	1
f_h	0.99
Renal (%)	35–50
Biliary (%)	60

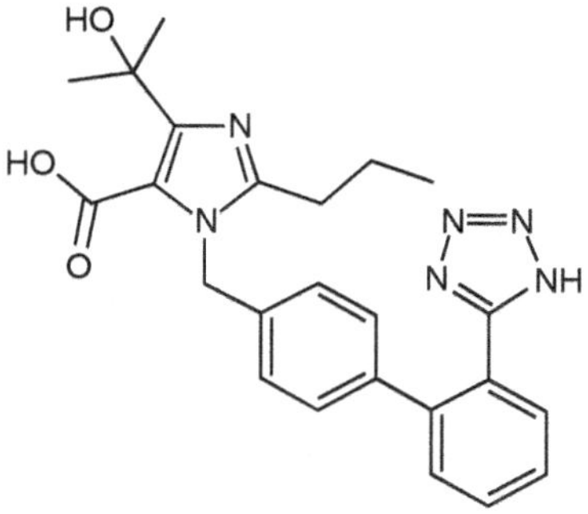

The team has the following questions for you:

1. Do you expect observing significant interindividual variability, if so, how come?
2. How can you mitigate the poor absorption observed with this molecule? What are your recommendations?

4.12 CHAPTER 4 TUTORIALS ANSWERS KEY

1. Absorption

Discovery project is seeking a lead compound for the treatment of psychosis. Drug X is the reference compound. It has low solubility ($C_s < 1$ μg/mL "free base") and moderate Caco-2 permeability ($5–10 \times 10^{-6}$ cm/s). It has high metabolic stability in rat, dog, and human hepatocytes with clearance values that are significantly lower than hepatic blood flow (*what does that mean?*).

Answer: This indicates that the compound has high metabolic stability and hepatic first pass would not limit the bioavailability of the molecule.

The team identified a compound (Drug A) that is structurally similar to Drug X and has good pharmacological activity following oral dosing. However, the compound seems to have low oral bioavailability in rats and has the following rat pharmacokinetics profile:

The PK profile of Drug A following 5 mg/kg IV dosing to rats

Form	CL (mL/min/kg)	V_{dss} (L/kg)	$t^{1/2}$ (h)
Free base	5 (low clearance)	5 (large distribution)	12 (long half-life)

Note: You should compare the CL value to rat hepatic blood flow ($Q = 70$ mL/min/kg) and assess hepatic extraction ratio ($E_h = CL/Q$). If the E_h is low, that means the compound is metabolically stable.

The PK profile of Drug A (*free base*) following 5 mg/kg PO dosing to rats

Form	C_{max} (ng/mL)	t_{max} (h)	AUC (ng*h/mL)	F%
Free base	15 (low exposure)	5 (slow absorption)	50 (low exposure)	5 (low bioavailability)

Also, the compound has the following *in vitro* profile:

1. p$K_a = 5.5$ (weak base)
2. $C_s = 2$ ng/mL (poor solubility)
3. Caco-2 Permeability (AB) $= 10 \times 10^{-6}$ cm/s (good permeability)
4. Excellent stability in rat, dog, and human hepatocytes and microsomes (good hepatic stability).

Based on the profiles above, the team is asking you the following questions:

1. What is the major factor that is contributing to the poor rat oral bioavailability of Drug A?

   Answer: Poor solubility combined with potential poor dissolution is the cause of the poor bioavailability of the molecule in rats.
2. Do you think similar trends would be observed in upper species?

   Answer: Yes, similar trends would be expected in upper species.
3. What suggestions would you share with the team to overcome the poor oral bioavailability of the lead compound?

   Answer: I would suggest considering having a salt of this molecule to improve overall solubility/dissolution and increase oral absorption of the molecule.
4. If you want to change the lead compound chemical structure, what are the changes you would recommend to increase the deficient property?

   Answer: I would work to alter the structure by considering the following options:

   (a) Increase the compound pK_a rendering it more basic

   (b) Adding polar group to increase hydrophilicity and overall polarity.

2. Imagine that you are part of a discovery team that is planning to identify a backup molecule to the following compound:

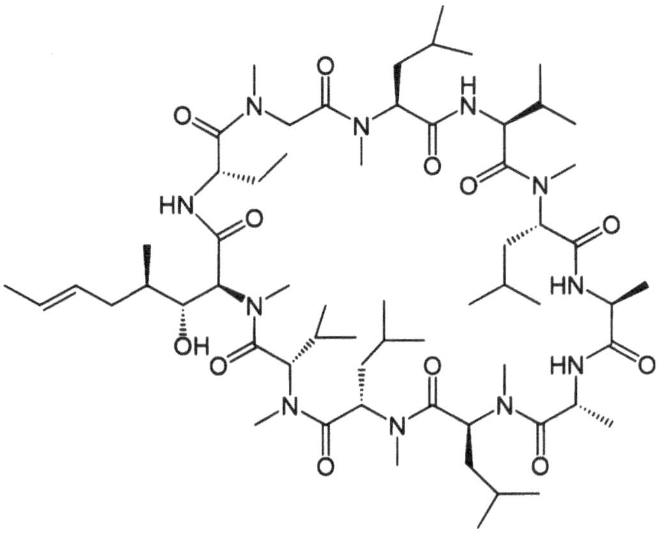

However, it had so many flaws, for example, this compound has high molecular weight ($M_{wt} > 500$ Da), poor solubility ($C_s < 1$ μg/mL), and Caco-2 permeability (AB $< 1 \times 10^{-6}$ cm/s). In addition, this compound is a PGP substrate and effective CYP3A4/OATP1B1 inhibitors with poor oral bioavailability.

The team has the following questions for you:

1. What are the factors that are contributing to the poor oral bioavailability of ketoconazole?

   Answer: The poor bioavailability of this compound is mainly attributed to its overall poor permeability/solubility. Overall, it leads to poor oral absorption.

2. What are the structural changes that you recommend the team doing to overcome the poor oral bioavailability? Please draw structures that address these issues.

   Answer: You may consider lowering the molecular weight and reduce the polarity of the molecule. This should increase the overall solubility of the molecule and provide more favorable absorption profile.

3. The following compound has low permeability yet good solubility. When assessed in human in FIH, it had the following properties:

Parameters	Values
F	0.05
f_a	0.05
f_g	1
f_h	0.99
Renal (%)	35–50
Biliary (%)	60

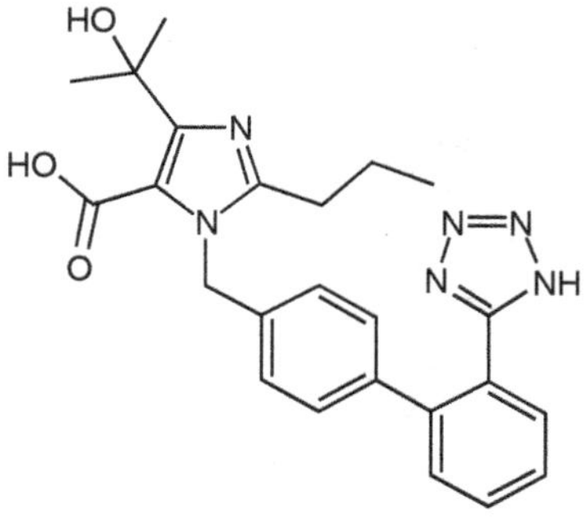

The team has the following questions for you:

1. Do you expect observing significant interindividual variability, if so, how come?

   Answer: Considering the poor absorption of this molecule, it is highly likely it would be associated with higher interindividual as well as intraindividual variability.

2. How can you mitigate the poor absorption observed with this molecule? What are your recommendations?

   Answer: Scientists developed a prodrug of this molecule by shielding the carboxylic acid group moiety, and reduce charge and overall ionization using a medoxomil group. This ended up increasing the overall absorption of the molecule. The structure of olmesartan medoxomil is shown as follows:

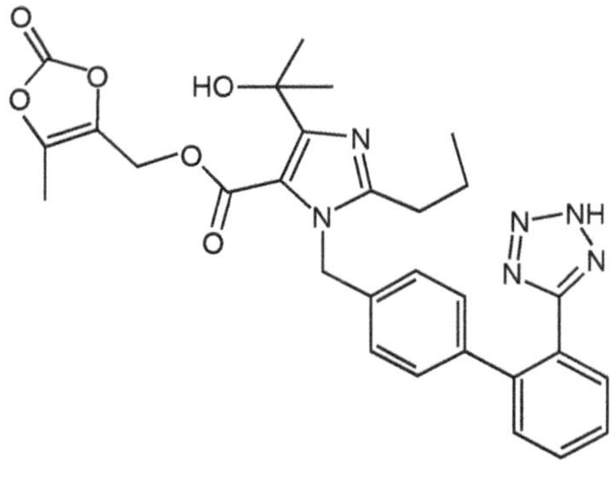

REFERENCES

1. Hurst, S., Loi, C.M., Brodfuehrer, J. and El-Kattan, A., *Impact of physiological, physicochemical and biopharmaceutical factors in absorption and metabolism mechanisms on the drug oral bioavailability of rats and humans.* Expert Opin Drug Metab Toxicol, 2007. 3(4): p. 469–489.
2. Thomas, V.H., Bhattachar, S., Hitchingham, L., Zocharski, P., Naath, M., Surendran, N., Stoner, C.L. and El-Kattan, A., *The road map to oral bioavailability: an industrial perspective.* Expert Opin Drug Metab Toxicol, 2006. 2(4): p. 591–608.
3. Varma, M.V., Ambler, C.M., Ullah, M., Rotter, C.J., Sun, H., Litchfield, J., Fenner, K.S. and El-Kattan, A.F., *Targeting intestinal transporters for optimizing oral drug absorption.* Curr Drug Metab, 2010. 11(9): p. 730–742.

4. Varma, M.V., Obach, R.S., Rotter, C., Miller, H.R., Chang, G., Steyn, S.J., El-Kattan, A. and Troutman, M.D., *Physicochemical space for optimum oral bioavailability: contribution of human intestinal absorption and first-pass elimination*. J Med Chem, 2010. 53(3): p. 1098–1108.
5. Yang, B. and Smith, D.E., *Significance of peptide transporter 1 in the intestinal permeability of valacyclovir in wild-type and PepT1 knockout mice*. Drug Metab Dispos, 2013. 41(3): p. 608–614.
6. Lennernas, H., *Does fluid flow across the intestinal mucosa affect quantitative oral drug absorption? Is it time for a reevaluation?* Pharm Res, 1995. 12(11): p. 1573–1582.
7. Karlsson, J., Ungell, A., Grasjo, J. and Artursson, P., *Paracellular drug transport across intestinal epithelia: influence of charge and induced water flux*. Eur J Pharm Sci, 1999. 9(1): p. 47–56.
8. Gordi, T., Hou, E., Kasichayanula, S. and Berner, B., *Pharmacokinetics of gabapentin after a single day and at steady state following the administration of gastric-retentive-extended-release and immediate-release tablets: a randomized, open-label, multiple-dose, three-way crossover, exploratory study in healthy subjects*. Clin Ther, 2008. 30(5): p. 909–916.
9. Schwartz, S.L., Gordi, T., Hou, E., Cramer, M., Heritier, M. and Cowles, V.E., *Clinical development of metformin extended-release tablets for type 2 diabetes: an overview*. Expert Opin Drug Metab Toxicol, 2008. 4(9): p. 1235–1243.
10. Berner, B. and Cowles, V.E., *Case studies in swelling polymeric gastric retentive tablets*. Expert Opin Drug Deliv, 2006. 3(4): p. 541–548.
11. Avdeef, A., *Physicochemical profiling (solubility, permeability and charge state)*. Curr Top Med Chem, 2001. 1(4): p. 277–351.
12. Lipinski, C.A., Computational Alerts for Potential Absorption Problems: Profiles of Clinically Tested Drugs. in *Tools for Oral Absorption. Part Two. Predicting Human Absorption, in AAPS Meeting*. 1995, AAPS: Miami.
13. Lipinski, C.A., *Drug-like properties and the causes of poor solubility and poor permeability*. J Pharmacol Toxicol Methods, 2000. 44(1): p. 235–249.
14. Pantzar, N., Lundin, S., Wester, L., Westrom, B.R., *Bidirectional small-intestinal permeability in the rat to some common marker molecules* in vitro. Scand J Gastroenterol, 1994. 29(8): p. 703–709.
15. Knipp, G.T., Ho, N.F., Barsuhn, C.L. and Borchardt, R.T., *Paracellular diffusion in Caco-2 cell monolayers: effect of perturbation on the transport of hydrophilic compounds that vary in charge and size*. J Pharm Sci, 1997. 86(10): p. 1105–1110.
16. Hou, T.J., Zhang, W., Xia, K., Qiao, X.B. and Xu, X.J., *ADME evaluation in drug discovery. 5. Correlation of Caco-2 permeation with simple molecular properties*. J Chem Inf Comput Sci, 2004. 44(5): p. 1585–1600.
17. Hou, T.J., Zhang, W., Xia, K., Qiao, X.B. and Xu, X.J., *ADME evaluation in drug discovery. 5. Correlation of Caco-2 permeation with simple molecular properties*. J Chem Inf Comput Sci, 2004. 44(5): p. 1585–1600.
18. Van de Waterbeemd, H., Camenisch, G., Folkers, G., Chretien, J.R. and Raevsky, O.A., *Estimation of blood–brain barrier crossing of drugs using molecular size and shape, and H-bonding descriptors*. J Drug Targeting, 1998. 6(2): p. 151–165.
19. Palm, K., Luthman, K., Ungell, A.L., Strandlund, G. and Artursson, P., *Correlation of drug absorption with molecular surface properties*. J Pharm Sci, 1996. 85(1): p. 32–39.
20. Palm, K., Stenberg, P., Luthman, K. and Artursson, P., *Polar molecular surface properties predict the intestinal absorption of drugs in humans*. Pharm Res, 1997. 14(5): p. 568–571.
21. Kelder, J., Grootenhuis, P.D., Bayada, D.M., Delbressine, L.P. and Ploemen, J.P., *Polar molecular surface as a dominating determinant for oral absorption and brain penetration of drugs*. Pharm Res, 1999. 16(10): p. 1514–1519.
22. Caron, G., Ermondi, G., Scherrer, R.A., John, B.T. and David, J.T., Lipophilicity, Polarity, and Hydrophobicity, in *Comprehensive Medicinal Chemistry II*. 2007, Elsevier: Oxford. p. 425–452.
23. Chandler, D., *Hydrophobicity: two faces of water*. Nature, 2002. 417(6888): p. 491.
24. Tetko, I.V., Livingstone, D.J., John, B.T. and David, J.T., Rule-Based Systems to Predict Lipophilicity, in *Comprehensive Medicinal Chemistry II*. 2007, Elsevier: Oxford. p. 649–668.

25. van Niel, M.B., et al., *Fluorination of 3-(3-(piperidin-1-yl)propyl)indoles and 3-(3-(piperazin-1-yl)propyl)indoles gives selective human 5-HT1D receptor ligands with improved pharmacokinetic profiles.* J Med Chem, 1999. 42(12): p. 2087–2104.

26. Lipinski, C.A., Solubility in Water and DMSO: Issues and Potential Solutions, in *Pharmaceutical Profiling in Drug Discovery for Lead Selection*, Borchardt, R.T., Kerns, E.H., Lipinski, C.A., Thakker, D.R. and Wang, B.W., Editors. 2004, AAPS Press: Washington, DC.

27. Lipinski, C.A., Lombardo, F., Dominy, B.W. and Feeney, P.J., *Experimental and computational approaches to estimate solubility and permeability in drug discovery and development settings.* Adv Drug Deliv Rev, 2001. 46(1–3): p. 3–26.

28. Grant, D.J.W. and Higuchi, T., Solubility Behavior of Organic Compounds, in *Techniques of Chemistry*, Saunders, W.H., Editor. Vol. 21. 1990, John Wiley and Sons: New York.

29. Dehring, K.A., Workman, H.L., Miller, K.D., Mandagere, A. and Poole, S.K., *Automated robotic liquid handling/laser-based nephelometry system for high throughput measurement of kinetic aqueous solubility.* J Pharm Biomed Anal, 2004. 36(3): p. 447–456.

30. Pudipeddi, M. and Serajuddin, A.T., *Trends in solubility of polymorphs.* J Pharm Sci, 2005. 94(5): p. 929–939.

31. Doelker, E., *Crystalline modifications and polymorphism changes during drug manufacture.* Ann Pharm Fr, 2002. 60(3): p. 161–176.

32. Matsuda, Y., Akazawa, R., Teraoka, R. and Otsuka, M., *Pharmaceutical evaluation of carbamazepine modifications: comparative study of photostability for carbamazepine polymorphs by using Fourier-transformed reflection-absorption infrared spectroscopy and calorimetric measurement.* J Pharm Pharmacol, 1993. 46: p. 162–167.

33. Chen, X., Morris, K.R., Griesser, U.J., Byrn, S.R. and Stowell, J.G., *Reactivity differences of indomethacin solid forms with ammonia gas.* J Am Chem Soc, 2002. 124(50): p. 15012–15019.

34. Aoyagi, N., Ogata, H., Kaniwa, N., Ejima, A., Nakata, H., Tsutsumi, J., Fujita, T. and Amada, I., *Bioavailability of indomethacin capsules in humans (III): correlation with bioavailability in beagle dogs.* Int J Clin Pharmacol Ther Toxicol, 1985. 23(11): p. 578–584.

35. Strachan, C.J., Taday, P.F., Newnham, D.A., Gordon, K.C., Zeitler, J.A., Pepper, M. and Rades, T., *Using terahertz pulsed spectroscopy to quantify pharmaceutical polymorphism and crystallinity.* J Pharm Sci, 2005. 94(4): p. 837–846.

36. Yalkowsky, S.H., Solubility and Miscibility, in *Solubility and Solubilization in Aqueous Media*. 1999, Oxford University Press: New York. p. 49–80.

37. Goosen, C., Laing, T.J., du, P.J., Goosen, T.C. and Flynn, G.L., *Physicochemical characterization and solubility analysis of thalidomide and its N-alkyl analogs.* Pharm Res, 2002. 19(1): p. 13–19.

38. Nelson, E., *Comparative dissolution rates of weak acids and their sodium salts.* J Am Pharm Assoc, 1958. 47(4): p. 297–299.

39. Nelson, E., *Solution rate of theophylline salts and effects from oral administration.* J Am Pharm Assoc, 1957. 46(10): p. 607–614.

40. Serajuddin, A.T., *Salt formation to improve drug solubility.* Adv Drug Deliv Rev, 2007. 59(7): p. 603–616.

41. Szejtli, J., *Medicinal applications of cyclodextrins.* Med Res Rev, 1994. 14(3): p. 353–386.

42. Ridhurkar, D.N., Ansari, K.A., Kumar, D., Kaul, N.S., Krishnamurthy, T., Dhawan, S. and Pillai, R., *Inclusion complex of aprepitant with cyclodextrin: evaluation of physico-chemical and pharmacokinetic properties.* Drug Dev Ind Pharm, 2013. 39(11): p. 1783–1792.

43. Shah, P., Jogani, V., Mishra, P., Mishra, A.K., Bagchi, T. and Misra, A., *Modulation of ganciclovir intestinal absorption in presence of absorption enhancers.* J Pharm Sci, 2007. 96(10): p. 2710–2722.

44. Loftsson, T., Vogensen, S.B., Brewster, M.E., Knoraosdottir, F., *Effects of cyclodextrins on drug delivery through biological membranes.* J Pharm Sci, 2007. 96(10): p. 2532–2546.

45. Shimpi, S., Chauhan, B. and Shimpi, P., *Cyclodextrins: application in different routes of drug administration.* Acta Pharm, 2005. 55(2): p. 139–156.

46. Yan, F., Si, L.Q., Huang, J.G. and Li, G., *Advances in the study of excipient inhibitors of intestinal P-glycoprotein*. Yao Xue Xue Bao, 2008. 43(11): p. 1071–1076.
47. Strickley, R.G., *Solubilizing excipients in oral and injectable formulations*. Pharm Res, 2004. 21(2): p. 201–230.
48. Law, D., Schmitt, E.A., Marsh, K.C., Everitt, E.A., Wang, W., Fort, J.J., Krill, S.L. and Qiu, Y., *Ritonavir-PEG 8000 amorphous solid dispersions:* in vitro *and* in vivo *evaluations*. J Pharm Sci, 2004. 93(3): p. 563–570.
49. Zhang, X., Li, L.C. and Mao, S., *Nanosuspensions of poorly water soluble drugs prepared by top-down technologies*. Curr Pharm Des, 2014. 20(3): p. 388–407.
50. Wang, Y., Miao, X., Sun, L., Song, J., Bi, C., Yang, X. and Zheng, Y., *Effects of nanosuspension formulations on transport, pharmacokinetics,* in vivo *targeting and efficacy for poorly water-soluble drugs*. Curr Pharm Des, 2014. 20(3): p. 454–473.
51. Sudhakar, B., NagaJyothi, K. and Murthy, K.V., *Nanosuspensions as a versatile carrier based drug delivery system – an overview*. Curr Drug Deliv, 2014. 11(3): p. 299–305.
52. Gurunath, S., Nanjwade, B.K. and Patila, P.A., *Enhanced solubility and intestinal absorption of candesartan cilexetil solid dispersions using everted rat intestinal sacs*. Saudi Pharm J, 2014. 22(3): p. 246–257.
53. Thombre, A.G., Shah, J.C., Sagawa, K. and Caldwell, W.B., In vitro *and* in vivo *characterization of amorphous, nanocrystalline, and crystalline ziprasidone formulations*. Int J Pharm, 2012. 428(1–2): p. 8–17.
54. Li, W., Quan, P., Zhang, Y., Cheng, J., Liu, J., Cun, D., Xiang, R. and Fang, L., *Influence of drug physicochemical properties on absorption of water insoluble drug nanosuspensions*. Int J Pharm, 2014. 460(1–2): p. 13–23.
55. Peng, W., Jiang, X.Y., Zhu, Y., Omari-Siaw, E., Deng, W.W., Yu, J.N., Xu, X.M. and Zhang, W.M., *Oral delivery of capsaicin using MPEG-PCL nanoparticles*. Acta Pharmacol Sin, 2015. 36(1): p. 139–148.
56. Sharma, M., Sharma, V., Panda, A.K. and Majumdar, D.K., *Development of enteric submicron particle formulation of papain for oral delivery*. Int J Nanomedicine, 2011. 6: p. 2097–2111.
57. Steed, J.W., *The role of co-crystals in pharmaceutical design*. Trends Pharmacol Sci, 2013. 34(3): p. 185–193.
58. Stanton, M.K., Kelly, R.C., Colletti, A., Langley, M., Munson, E.J., Peterson, M.L., Roberts, J. and Wells, M., *Improved pharmacokinetics of AMG 517 through co-crystallization. Part 2: Analysis of 12 carboxylic acid co-crystals*. J Pharm Sci, 2011. 100(7): p. 2734–2743.
59. Miroshnyk, I., Mirza, S. and Sandler, N., *Pharmaceutical co-crystals–an opportunity for drug product enhancement*. Expert Opin Drug Deliv, 2009. 6(4): p. 333–341.
60. Rautio, J., Kumpulainen, H., Heimbach, T., Oliyai, R., Oh, D., Jarvinen, T. and Savolainen, J., *Prodrugs: design and clinical applications*. Nat Rev Drug Discov, 2008. 7(3): p. 255–270.
61. Zhang, Y., Sun, J., Sun, Y., Wang, Y. and He, Z., *Prodrug design targeting intestinal PepT1 for improved oral absorption: design and performance*. Curr Drug Metab, 2013. 14(6): p. 675–687.
62. Inoue, Y., Usui, N., Hiroki, T., Shimizu, K., Kobayashi, S. and Shimasaki, S., *Bioavailability of intravenous fosphenytoin sodium in healthy Japanese volunteers*. Eur J Drug Metab Pharmacokinet, 2013. 38(2): p. 139–148.
63. Browne, T.R., et al., *Bioavailability studies of drugs with nonlinear pharmacokinetics: II. Absolute bioavailability of intravenous phenytoin prodrug at therapeutic phenytoin serum concentrations determined by double-stable isotope technique*. J Clin Pharmacol, 1993. 33(1): p. 89–94.
64. Venuti, M.C., Young, J.M., Maloney, P.J., Johnson, D. and McGreevy, K., *Synthesis and biological evaluation of omega-(N,N,N-trialkylammonium)alkyl esters and thioesters of carboxylic acid nonsteroidal antiinflammatory agents*. Pharm Res, 1989. 6(10): p. 867–873.
65. Stella, V.J., Charman, W.N. and Naringrekar, V.H., *Prodrugs. Do they have advantages in clinical practice?* Drugs, 1985. 29(5): p. 455–473.

66. Morimoto, K., Abe, O. and Kinoshita, H., *Steady state and disappearance of the metabolites of miproxifene phosphate in the treatment of breast cancer*. Gan To Kagaku Ryoho, 1998. 25(10): p. 1565–1573.

67. Falcoz, C., Jenkins, J.M., Bye, C., Hardman, T.C., Kenney, K.B., Studenberg, S., Fuder, H. and Prince, W.T., *Pharmacokinetics of GW433908, a prodrug of amprenavir, in healthy male volunteers*. J Clin Pharmacol, 2002. 42(8): p. 887–898.

68. Heimbach, T., Oh, D.M., Li, L.Y., Rodriguez-Hornedo, N., Garcia, G. and Fleisher, D., *Enzyme-mediated precipitation of parent drugs from their phosphate prodrugs*. Int J Pharm, 2003. 261(1–2): p. 81–92.

69. Stewart, B.H., Amidon, G.L. and Brabec, R.K., *Uptake of prodrugs by rat intestinal mucosal cells: mechanism and pharmaceutical implications*. J Pharm Sci, 1986. 75(10): p. 940–945.

70. Hersh, M.R., Kuhn, J.G., Phillips, J.L., Clark, G., Ludden, T.M. and Von Hoff, D.D., *Pharmacokinetic study of fludarabine phosphate (NSC 312887)*. Cancer Chemother Pharmacol, 1986. 17(3): p. 277–280.

71. Stuurman, F.E., Nuijen, B., Beijnen, J.H. and Schellens, J.H., *Oral anticancer drugs: mechanisms of low bioavailability and strategies for improvement*. Clin Pharmacokinet, 2013. 52(6): p. 399–414.

72. Dahan, A., Khamis, M., Agbaria, R. and Karaman, R., *Targeted prodrugs in oral drug delivery: the modern molecular biopharmaceutical approach*. Expert Opin Drug Deliv, 2012. 9(8): p. 1001–1013.

73. Imai, T. and Ohura, K., *The role of intestinal carboxylesterase in the oral absorption of prodrugs*. Curr Drug Metab, 2010. 11(9): p. 793–805.

74. Vere Hodge, R.A., Sutton, D., Boyd, M.R., Harnden, M.R. and Jarvest, R.L., *Selection of an oral prodrug (BRL 42810; famciclovir) for the antiherpesvirus agent BRL 39123 [9-(4-hydroxy-3-hydroxymethylbut-l-yl)guanine; penciclovir]*. Antimicrob Agents Chemother, 1989. 33(10): p. 1765–1773.

75. Blumer, J., Rodriguez, A., Sanchez, P.J., Sallas, W., Kaiser, G. and Hamed, K., *Single-dose pharmacokinetics of famciclovir in infants and population pharmacokinetic analysis in infants and children*. Antimicrob Agents Chemother, 2010. 54(5): p. 2032–2041.

76. Pue, M.A., Pratt, S.K., Fairless, A.J., Fowles, S., Laroche, J., Georgiou, P. and Prince, W., *Linear pharmacokinetics of penciclovir following administration of single oral doses of famciclovir 125, 250, 500 and 750 mg to healthy volunteers*. J Antimicrob Chemother, 1994. 33(1): p. 119–127.

77. Eriksson, A.H., Varma, M.V., Perkins, E.J. and Zimmerman, C.L., *The intestinal absorption of a prodrug of the mGlu2/3 receptor agonist LY354740 is mediated by PEPT1: in situ rat intestinal perfusion studies*. J Pharm Sci, 2010. 99(3): p. 1574–1581.

78. Terada, T. and Inui, K., *Peptide transporters: structure, function, regulation and application for drug delivery*. Curr Drug Metab, 2004. 5(1): p. 85–94.

79. Merzlikine, A., Rotter, C., Rago, B., Poe, J., Christoffersen, C., Thomas, V.H., Troutman, M. and El-Kattan, A., *Effect of chitosan glutamate, carbomer 974P, and EDTA on the* in vitro *Caco-2 permeability and oral pharmacokinetic profile of acyclovir in rats*. Drug Dev Ind Pharm, 2009. 35(9): p. 1082–1091.

80. Kimberlin, D.F., Weller, S., Whitley, R.J., Andrews, W.W., Hauth, J.C., Lakeman, F. and Miller, G., *Pharmacokinetics of oral valacyclovir and acyclovir in late pregnancy*. Am J Obstet Gynecol, 1998. 179(4): p. 846–851.

81. Acosta, E.P. and Fletcher, C.V., *Valacyclovir*. Ann Pharmacother, 1997. 31(2): p. 185–191.

82. Beutner, K.R., *Valacyclovir: a review of its antiviral activity, pharmacokinetic properties, and clinical efficacy*. Antiviral Res, 1995. 28(4): p. 281–290.

83. Cundy, K.C., et al., *XP13512 [(+/−)-1-([(alpha-isobutanoyloxyethoxy)carbonyl] aminomethyl)-1-cyclohexane acetic acid], a novel gabapentin prodrug: I. Design, synthesis, enzymatic conversion to gabapentin, and transport by intestinal solute transporters*. J Pharmacol Exp Ther, 2004. 311(1): p. 315–323.

84. Cundy, K.C., et al., *XP13512 [(+/−)-1-([(alpha-isobutanoyloxyethoxy)carbonyl] aminomethyl)-1-cyclohexane acetic acid], a novel gabapentin prodrug: II. Improved oral bioavailability, dose proportionality, and colonic absorption compared with gabapentin in rats and monkeys*. J Pharmacol Exp Ther, 2004. 311(1): p. 324–333.

85. Cundy, K.C., Sastry, S., Luo, W., Zou, J., Moors, T.L. and Canafax, D.M., *Clinical pharmacokinetics of XP13512, a novel transported prodrug of gabapentin*. J Clin Pharmacol, 2008. 48(12): p. 1378–1388.

86. Razonable, R.R. and Paya, C.V., *Valganciclovir for the prevention and treatment of cytomegalovirus disease in immunocompromised hosts*. Expert Rev Anti Infect Ther, 2004. 2(1): p. 27–41.

87. Tsuda, M., Terada, T., Irie, M., Katsura, T., Niida, A., Tomita, K., Fujii, N. and Inui, K., *Transport characteristics of a novel peptide transporter 1 substrate, antihypotensive drug midodrine, and its amino acid derivatives*. J Pharmacol Exp Ther, 2006. 318(1): p. 455–460.

88. Tsamandouras, N., Dickinson, G., Guo, Y., Hall, S., Rostami-Hodjegan, A., Galetin, A. and Aarons, L., *Development and application of a mechanistic pharmacokinetic model for simvastatin and its active metabolite simvastatin acid using an integrated population PBPK approach*. Pharm Res, 2015. 32(6): p. 1864–1883.

89. Davidson, M.H., Lukacsko, P., Sun, J.X., Phillips, G., Walters, E., Sterman, A., Niecestro, R. and Friedhoff, L., *A multiple-dose pharmacodynamic, safety, and pharmacokinetic comparison of extended- and immediate-release formulations of lovastatin*. Clin Ther, 2002. 24(1): p. 112–125.

90. Piepho, R.W., *Overview of the angiotensin-converting-enzyme inhibitors*. Am J Health Syst Pharm, 2000. 57(Suppl 1): p. S3–S7.

91. Kamal, M.A., Keep, R.F. and Smith, D.E., *Role and relevance of PEPT2 in drug disposition, dynamics, and toxicity*. Drug Metab Pharmacokinet, 2008. 23(4): p. 236–242.

92. Manoria, P., Manoria, P. and Manoria, P., *Olmesartan medoxomil: a clinical review*. Indian Heart J, 2006. 58(3): p. 282–286.

93. Gugler, R. and Allgayer, H., *Effects of antacids on the clinical pharmacokinetics of drugs. An update*. Clin Pharmacokinet, 1990. 18(3): p. 210–219.

94. Kakemi, K., Sezaki, H., Ogata, H. and Nadai, T., *Absorption and excretion of drugs. XXXVI. Effect of Ca^{2+} on the absorption of tetracycline from the small intestine. (1)*. Chem Pharm Bull, 1968. 16(11): p. 2200–2205.

95. Kakemi, K., Sezaki, H., Hayashi, M. and Nadai, T., *Absorption and excretion of drugs. XXXVII. Effect of Ca^{2+} on the absorption of tetracycline from the small intestine. (2)*. Chem Pharm Bull, 1968. 16(11): p. 2206–2212.

96. Jung, H., Peregrina, A.A., Rodriguez, J.M. and Moreno-Esparza, R., *The influence of coffee with milk and tea with milk on the bioavailability of tetracycline*. Biopharm Drug Dispos, 1997. 18(5): p. 459–463.

97. Johnsgard, M. and Pedersen, A.O., *The consequence of the preparation factors for the effect of the drug*. Tidsskr Nor Laegeforen, 1973. 93(10): p. 698–703.

98. Amidon, G.L., Lennernas, H., Shah, V.P. and Crison, J.R., *A theoretical basis for a biopharmaceutic drug classification: the correlation of* in vitro *drug product dissolution and* in vivo *bioavailability*. Pharm Res, 1995. 12(3): p. 413–420.

99. Hellriegel, E.T., Bjornsson, T.D. and Hauck, W.W., *Interpatient variability in bioavailability is related to the extent of absorption: implications for bioavailability and bioequivalence studies*. Clin Pharmacol Ther, 1996. 60(6): p. 601–607.

100. Martinez, M.N. and Amidon, G.L., *A mechanistic approach to understanding the factors affecting drug absorption: a review of fundamentals*. J Clin Pharm, 2002. 42(6): p. 620–643.

101. Stoner, C.L., Cleton, A., Johnson, K., Oh, D.M., Hallak, H., Brodfuehrer, J., Surendran, N. and Han, H.K., *Integrated oral bioavailability projection using* in vitro *screening data as a selection tool in drug discovery*. Int J Pharm, 2004. 269(1): p. 241–249.

102. Boobis, A., Gundert-Remy, U., Kremers, P., Macheras, P. and Pelkonen, O., In silico *prediction of ADME and pharmacokinetics – report of an expert meeting organised by COST B15.* Eur J Pharm Sci, 2002. 17(4–5): p. 183–193.

103. Gombar, V.K., Silver, I.S. and Zhao, Z., *Role of ADME characteristics in drug discovery and their* in silico *evaluation:* in silico *screening of chemicals for their metabolic stability.* Curr Top Med Chem, 2003. 3(11): p. 1205–1225.

104. van de Waterbeemd, H., *High-throughput and in silico techniques in drug metabolism and pharmacokinetics.* Curr Opin Drug Discov Devel, 2002. 5(1): p. 33–43.

5

PHYSICOCHEMICAL AND BIOPHARMACEUTICAL FACTORS AFFECTING HEPATIC/INTESTINAL FIRST-PASS EFFECT

5.1 INTRODUCTION

Compound attrition continues to be a major burden that halts the progress and development of new molecular entities (NMEs). Compound attrition contributes to the exponential increase in cost of developing new drugs to treat debilitating diseases. Many factors lead to compound attrition such as clinical safety, efficacy, formulation, commercial, toxicology, cost of goods, and pharmacokinetics/bioavailability. In 1991, poor pharmacokinetics and bioavailability contributed to around 40% of reported attrition [1]. In 2001, attrition due to poor pharmacokinetics and bioavailability was reduced to 10% and declined to 1% by 2010 [2]. This achievement is a natural outcome of the major investments undertaken by DMPK departments at large that include and not limited to the following [3, 4]:

1. Better understanding of physicochemical properties that affect drug clearance and structure–activity relationship (SAR) of substrates of main metabolizing enzymes and transporters [5].
2. The development of high-throughput screens that assess the hepatic stability of NMEs using human liver microsomes (HLM) and hepatocytes. These tools were implemented in early drug discovery and eliminated NMEs with poor hepatic stability [6].
3. The development and implementation of reliable scaling techniques from HLM and allometric scaling of clearance and volume of distribution from preclinical species. These techniques allowed a reasonable prediction for human clearance of NMEs that is mediated by cytochrome P450 enzymes or renally eliminated.

Oral Bioavailability Assessment: Basics and Strategies for Drug Discovery and Development, First Edition.
Ayman F. El-Kattan.

Overall, these factors played a key role in the progress of NMEs with good hepatic stability and high oral bioavailability. This reduced attrition attributed to poor pharmacokinetics and first-pass effect and allowed orally bioavailable NMEs to progress in the clinic. This chapter covers two relevant topics: first one reviews known chemical space associated with metabolizing enzymes and the second discusses known approaches that can be used to increase metabolic stability for NMEs.

5.1.1 Known SAR of CYP and UGTs Enzymes

Williams et al. reported that metabolism is the major driver of elimination for the 200 most prescribed drugs in the US market in 2002 (73%). Of which, around 73% and 15% were predominantly metabolized by CYP enzymes and UDP-glucuronosyltransferase (UGTs) enzymes, respectively. Therefore, understanding the SAR of CYP and UGT isoforms is a key for predicting human systemic exposure and pharmacokinetic profiles for NMEs. This is an area of research that continues to be of interest for scientists in both academia and the pharmaceutical industry [7–12].

5.1.1.1 CYP Enzymes As shown in Figure 5.1a, the main CYP isoforms in human liver are CYP1A2, CYP2B6, CYP2C, CYP2D6, CYP2E1, and CYP3A4, whereas the main CYP isoforms in the human intestine are CYP3A4, CYP2C19, CYP2C9, CYP2D6, and CYP2J2 (Figure 5.1b) [13, 14]. David Lewis proposed a decision tree that differentiates among the various CYP isoform substrates based on their physicochemical properties (e.g., planarity, molecular weight, lipophilicity, and volume of distribution) (Figure 5.2). For example, CYP3A4 substrates tend to have a wide molecular weight range and large volume of distribution. Typically, they are neutral or basic in nature. The site of oxidation is typically 5–8 Å from the hydrogen bond acceptor (HBA). Interestingly, the chemical structure of the substrate rather than the topography of CYP3A4 enzyme is what governs the site of oxidation [15].

CYP1A2 enzyme metabolizes planar aromatic heterocyclic amines/amides with two to three hydrogen bond donors such as theophylline or caffeine [16–18]. The topography of the active sites of both CYP2C9 and 2D6 metabolizing enzymes governs the binding of their substrates by ion pairing. In general, CYP2D6 substrates tend to be basic ($pK_a > 7$) with a protonated nitrogen that is 4–7 Å from the site of oxidation. CYP2C9 substrates are lipophilic and are either acidic ($pK_a < 6$) or neutral in nature. In addition, the site of oxidation is 5–7 Å from the hydrogen bond donor/acceptor [19, 20].

5.1.1.2 Glucuronidation Glucuronidation is another key metabolism reaction, where parent drugs or their phase 1 metabolites are metabolized and excreted in the urine and/or bile as glucuronide conjugates. The process is mediated by the UDP-glucuronyl transferases. In humans, UGT1A1, 1A3, 1A4, 1A6, 1A9, 2B4, 2B7, 2B15, and 2B17 have been identified in the liver based on mRNA and protein levels [21, 22]. UGT1A1, UGT1A3, UGT1A6, UGT1A8, and UGT2B7 have been identified in the human small intestine based on protein levels determined using proteomic approaches [22]. The glucuronidation mechanism involves nucleophilic SN2 substitution of the acceptor at the C1 position of the glucuronic acid (Figure 5.3). This is associated with a configurational inversion of the obtained product. It should be emphasized that the nucleophilic site on UGT substrates can be an oxygen atom (e.g., phenols, carboxylic acid, or alcohol), a nitrogen atom (e.g., amine or azaheterocyclic molecules), or an activated carbon atom (e.g., phenylbutazone) [23].

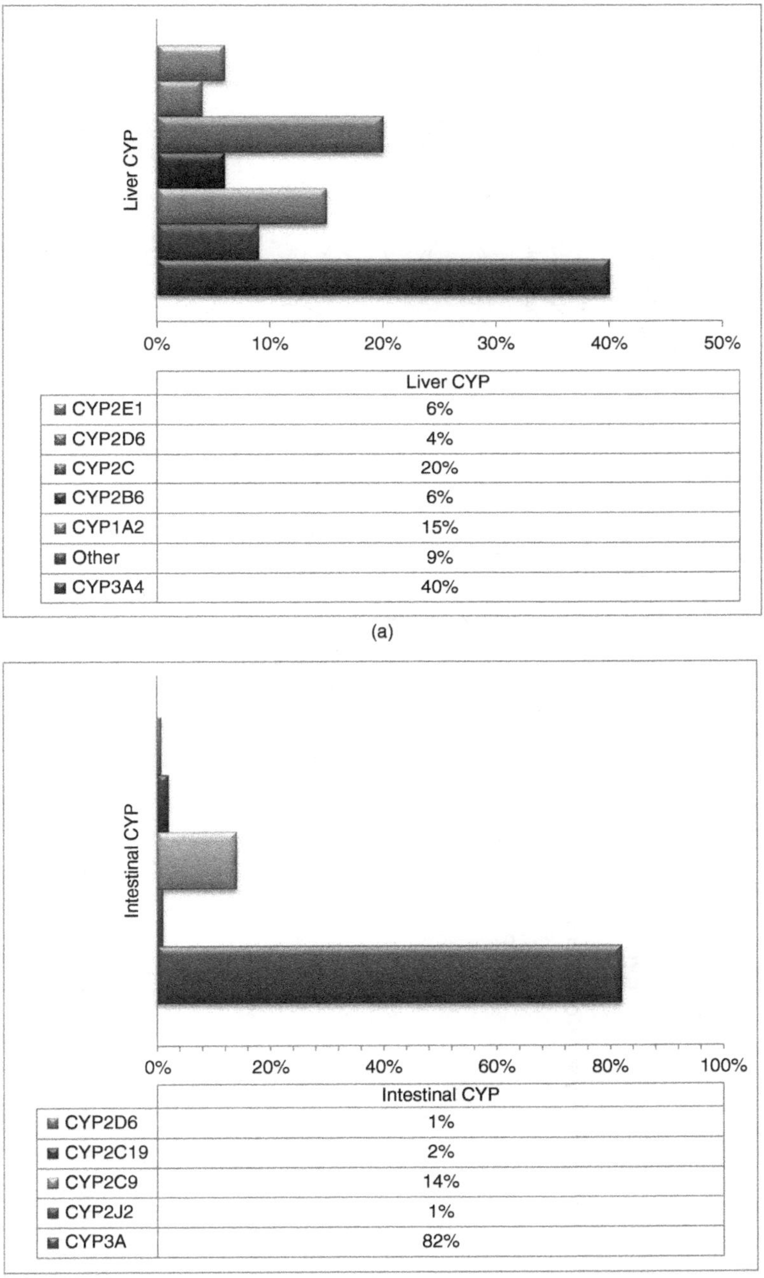

	Liver CYP
CYP2E1	6%
CYP2D6	4%
CYP2C	20%
CYP2B6	6%
CYP1A2	15%
Other	9%
CYP3A4	40%

(a)

	Intestinal CYP
CYP2D6	1%
CYP2C19	2%
CYP2C9	14%
CYP2J2	1%
CYP3A	82%

(b)

FIGURE 5.1 CYP isoforms relative content in (a) human liver and (b) intestine, respectively [13, 14].

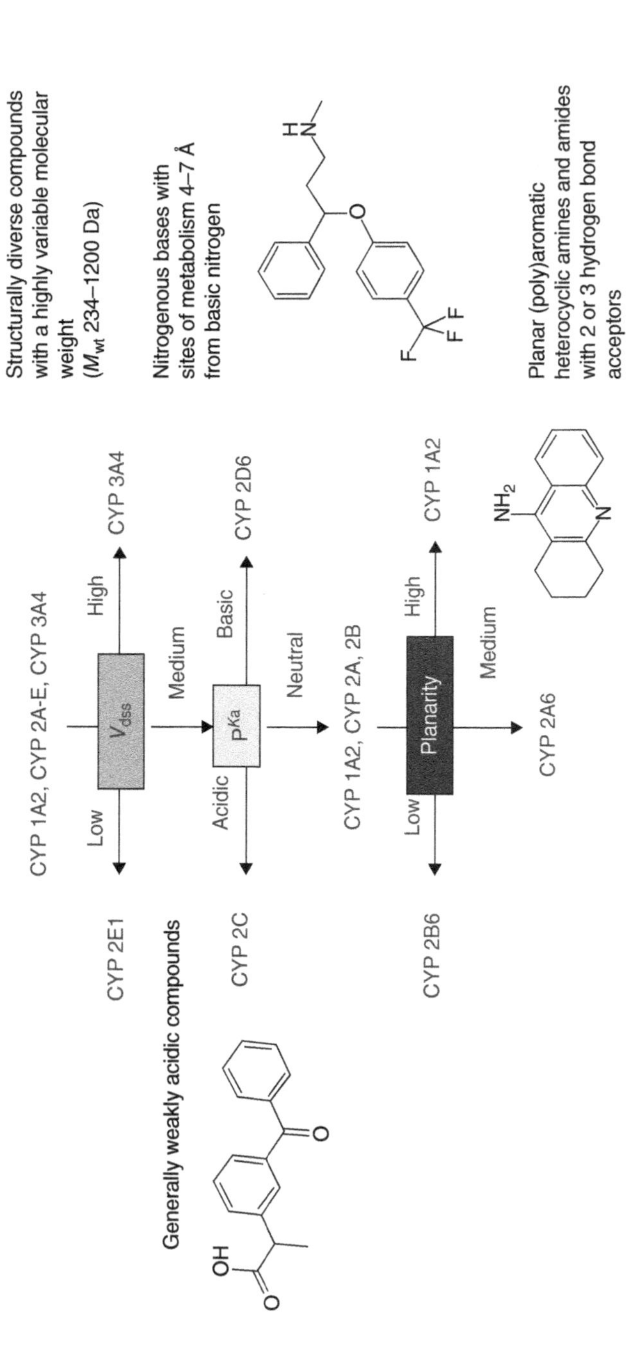

FIGURE 5.2 A decision tree for human CYP450 substrates based on compounds V_{dss}, ionization, and planarity. Lewis 2000 [15]. Reproduced with permission of Elsevier.

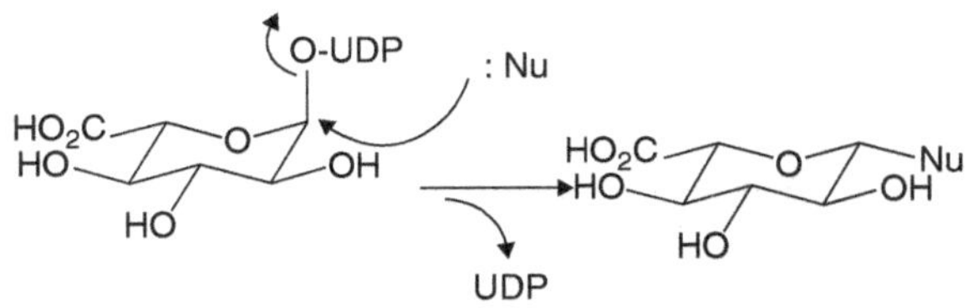

FIGURE 5.3 The mechanism of glucuronide conjugate formation using UDP-glucuronyl transferase. Nu is a nucleophilic site on a drug molecule.

5.2 PHYSICOCHEMICAL APPROACHES TO INCREASE METABOLIC STABILITY

In 2015, Varma et al. introduced the extended clearance classification system (ECCS). Compounds were classified based on their permeability, ionization, and molecular weight. ECCS Class 1 and 2 compounds have a high extent of metabolism (>70%) and high permeability. ECCS Class 3 and 4 compounds have a low extent of metabolism (<70%) and low permeability. These compounds are predominantly eliminated as unchanged drugs in either urine and/or bile [24]. Compounds with *in vitro* permeability higher than 5×10^{-6} cm/s and measured in low efflux MDCK cell line tend to have high extent of absorption ($f_a > 0.80$) and are predominantly BDDCS Class 1 or 2 (Figure 5.4). However, for compounds with permeability lower than 5×10^{-6} cm/s, they are mainly ECCS Class 3 and 4. Hence, drug physicochemical properties that drive low permeability are similar to those that switch the elimination pathway of drug molecules from metabolism to predominant renal and/or biliary-driven elimination [3, 25].

TIP:

Based on ECCS principles, understanding the physicochemical properties that affect the permeability of drug molecules is key for designing molecules with better hepatic stability, in part, by emphasizing reduced lipophilicity and increased polarity.

5.2.1 Decreasing Lipophilicity

One of the most widely used approaches to improve the metabolic stability of NMEs is to reduce drug lipophilicity [26]. Decreasing drug lipophilicity is associated with an increase in the contribution and clearance mediated by renal elimination, whereas increasing compound lipophilicity is associated with an increase in the contribution and clearance mediated by hepatic metabolism. This relationship is also clearly demonstrated in Figure 5.5, where compounds with lipophilicity >3 is associated with a lower fraction that escapes hepatic (f_h) and intestinal first-pass effect (f_g) [27]. In general, these trends are primarily ascribed to the fact the increase in drug lipophilicity is associated with an increase in the affinity of various drugs and xenobiotics to metabolizing enzyme binding sites that typically lead to poor metabolic stability, higher clearance, and hepatic/intestinal first pass. Figure 5.6 demonstrates the relationship between $\log D$ and enzyme affinity (K_m) for a set of CYP3A4 substrates. Increasing

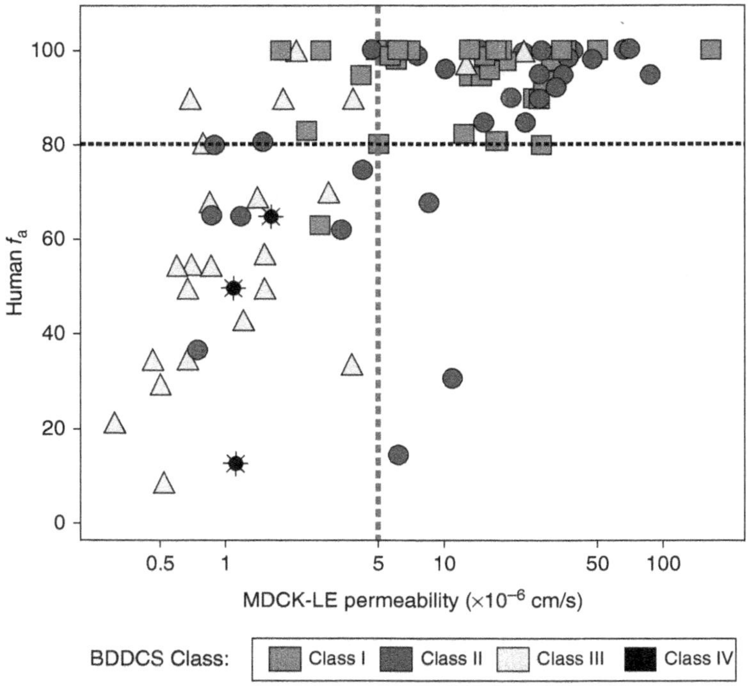

FIGURE 5.4 The relationship between MDCK-LE permeability and human f_a. Human $f_a > 80\%$ (horizontal dotted line) where a permeability value of 5×10^{-6} cm/s cutoff value would differentiate between low and high values. Varma et al. 2012 [3]. Reproduced with permission of American Chemical Society

lipophilicity is associated with higher affinity and therefore higher intrinsic CYP3A4-mediated clearance [26].

A widely used strategy to reduce NMEs' lipophilicity and improve hepatic metabolic stability is by replacing a bulky lipophilic group such as phenyl, benzyl, or cyclic hexyl with groups with lower lipophilicity [48]. For these structural changes to be effective, they should have minimal biological activity. For example, Dragovich et al. replaced a benzyl group in the rhinovirus 3C protease inhibitor lead compound [49] with ethyl [1] and propyl groups [50]. These changes were associated with a significant improvement in the monkey oral exposure while maintaining similar *in vitro* antiviral activity (Figure 5.7) [50].

5.2.2 Introduction of Polarity

The introduction of polarity is a key strategy that is usually considered to reduce hepatic metabolic clearance. As illustrated in Figure 5.8, increasing the polarity of a compound with poor metabolic stability by increasing its polar surface area (PSA) reduces its affinity to metabolizing enzymes, lowers its hepatic clearance, and overcomes high first-pass effect [27]. Indeed, the introduction of polar functional groups, such as pyridine or isosteric atoms such as nitrogen or oxygen, was critical in reducing compound lipophilicity and increasing the metabolic

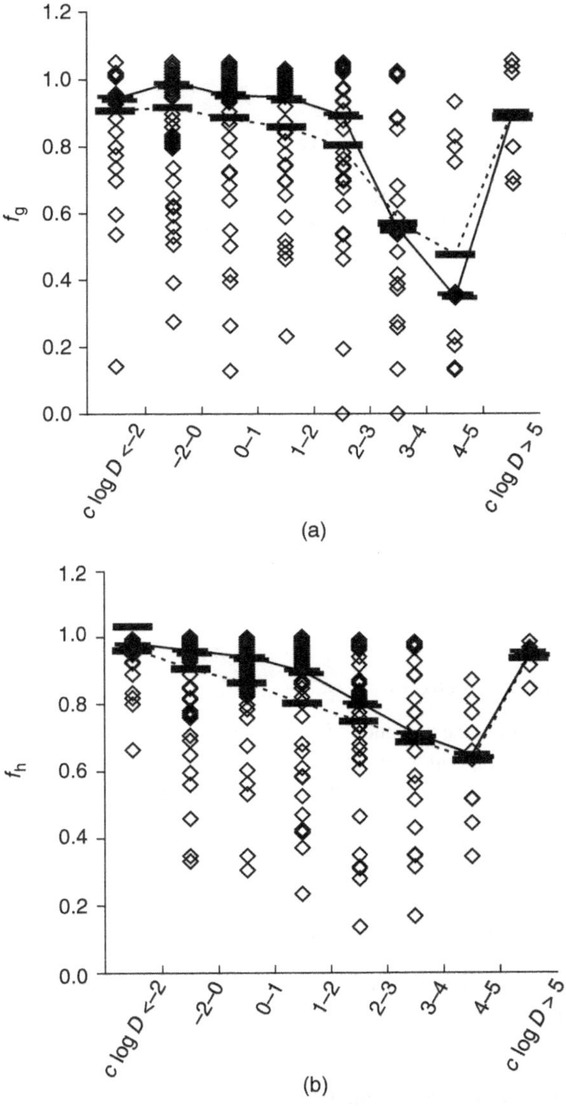

FIGURE 5.5 The relationship between calculated lipophilicity ($c \log D$) and drug fraction that escapes (a) intestinal and (b) hepatic first-pass effects, respectively. Varma et al. 2010 [27]. Reproduced with permission of American Chemical Society.

stability of compounds tested. Tagat et al. replaced the benzamide group in the CCR5 antagonist lead compound with nicotinamide, which was associated with a significant improvement in its metabolic stability and systemic exposure following oral dosing in rats, dogs, and monkeys [51].

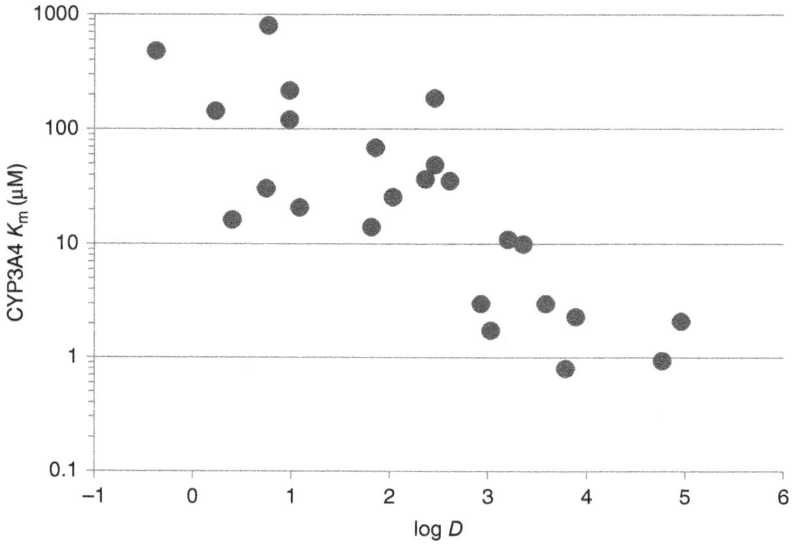

FIGURE 5.6 The relationship between lipophilicity (log *D*) and affinity to CYP3A4 substrates corrected for microsomal protein binding for a set of CYP3A4 substrates [28–47].

5.2.3 Addition of Ionized Group

In addition to lipophilicity, the presence of ionizable groups on the molecule is also an important characteristic of compounds undergoing hepatic and intestinal metabolism. Ionizable groups are important both in terms of contributing to the apparent lipophilicity of the molecule (log *D*) and determining the binding to the active site of metabolizing enzymes, the latter being important in dictating the regioselectivity of metabolism. Tagat et al. implemented various approaches to optimize the pharmacokinetic profile of the CCR5 receptor antagonist lead in their HIV-1 inhibitor discovery program [26]. With their metabolite identification efforts, the group discovered a rapid formation of a single metabolite of nicotinamide in the plasma samples, which was mediated by N-oxidation. The group synthesized the positively charged pyridine N-oxide analog and found it to have excellent potency and oral bioavailability in preclinical species tested [51].

5.2.4 Elimination of Labile Functional Group and Soft Spots

In early drug discovery, the identification of metabolic soft spots is one of the key deliverables of DMPK scientists. This approach allows medicinal chemists to modify NMEs' chemical structure and systematically identify molecules with appropriate pharmacokinetics attributes, high metabolic stability and good oral bioavailability. In drug development, the elucidation of the major biotransformation pathways of NMEs is pivotal to gain a better understanding of biological activities and toxicological liabilities of corresponding metabolites [52]. Mass spectrometry (MS) and nuclear magnetic resonance (NMR) are the main analytical tools that are utilized to characterize and quantify drug metabolites. Indeed, liquid chromatography (LC) combined with atmospheric pressure ionization (API) MS is now the major tool that is utilized for detection, structural elucidation, and quantification of drug metabolites that are found in

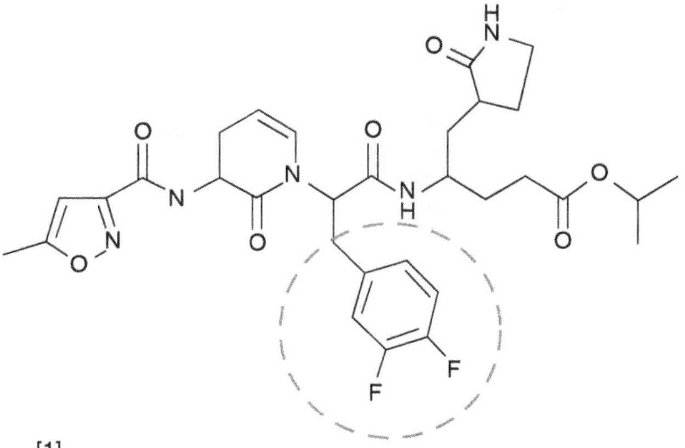

[1]

EC_{50} (μM) = 0.078
C_{7h} (monkey) (μM) = 0.014

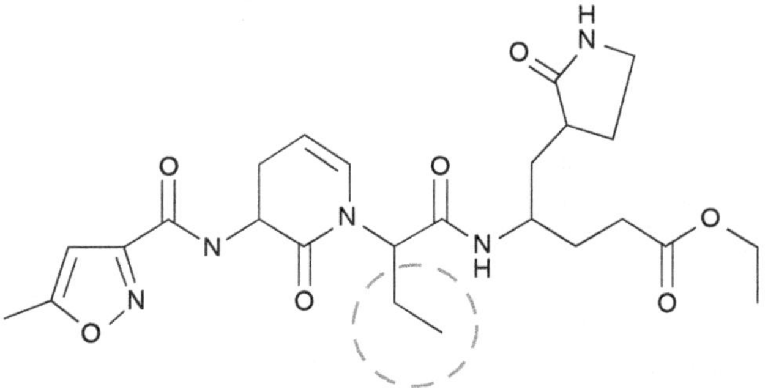

[2]

EC_{50} (μM) = 0.047
C_{7h} (monkey) (μM) = 0.896

FIGURE 5.7 The substitution of bifluorobenzyl group with ethyl group increased plasma exposure observed in monkeys at 7 h following 7 h oral dosing and maintained antiviral activity. Dragovich et al. 2003 [50]. Reproduced with permission of American Chemical Society.

various body matrices. It should be emphasized that MS alone falls short of determining the exact site of metabolism or structure of various metabolites. This is in part attributed to the presence of endogenous materials in biological matrices that usually suppress the ionization of the drug and its metabolites. Overall, it limits the ability of MS to accurately identify drug metabolites. Various techniques are combined with MS to effectively overcome its shortcomings. These techniques span analytical and wet chemistry procedures, such as LC-NMR, enzymatic hydrolysis, chemical derivatization, and hydrogen/deuterium-exchange (H/D-exchange) combined [53].

An approach that is gaining interest from medicinal chemists is the use of *in silico* models of biotransformation such as MetaSite. The MetaSite algorithm is a unique *in silico* tool that predicts the phase I metabolism for human CYP-mediated reaction [7, 54]. It uses both the

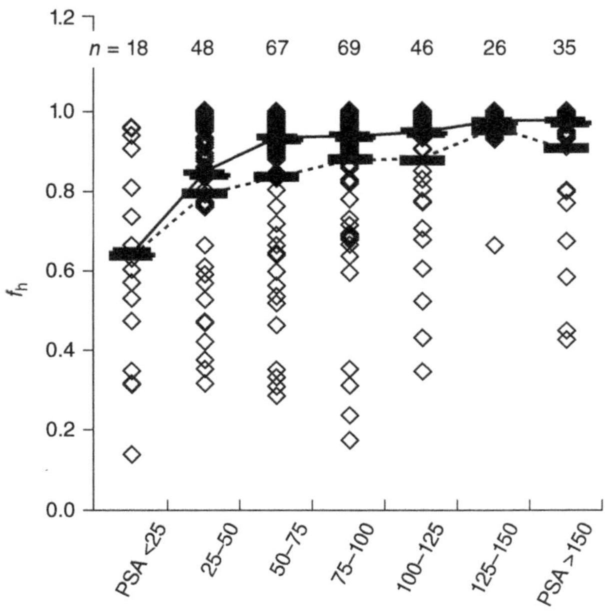

FIGURE 5.8 The relationship between the compound polar surface area (PSA) and the fraction that escapes hepatic first-pass effect (f_h). Varma et al. 2010 [27]. Reproduced with permission of American Chemical Society.

GRID-based representations of the CYP enzymes and computed 3D structure of a compound to identify its potential site of metabolism and soft spots. This probability is measured from the distance between the atoms in the compound investigated to the reaction oxygen in the CYP metabolizing enzyme binding site. Collectively, while it falls short of predicting the rate of hepatic clearance for CYP substrates, MetaSite is an effective approach to identify soft spots, predict structures of likely metabolites, and suggest areas of the drug molecule that may be involved in metabolism, and thus, warn medicinal chemists to avoid poor metabolic stability and quickly assess the suitability of a new structural modification.

5.2.5 Addition of Fluorine or Chlorine

Bioisosteric replacement of hydrogen using halogens such as fluorine or chlorine has been considered since they can influence the physicochemical properties, biological activity, metabolism, pharmacokinetics, and pharmacodynamics of a molecule. Indeed, fluorine and chlorine were effectively used in reducing CYP-mediated aromatic hydroxylation at other sites on a phenyl ring when substituting the hydrogen by fluorine or chlorine on aromatic rings. For example, the first-generation cholesterol absorption inhibitors [2, 55] were reported to have extensive first-pass effect and high hepatic metabolism. The introduction of fluorine blocks the sites of metabolism, improves overall plasma exposure, and increases cholesterol absorption inhibitory (CAI) effect. Indeed, the substitution of fluorine in the phenoxy ring of CAI [55] resulted in a significant increase in biological activity (CAI [2]). Furthermore, blocking metabolic sites in [3] by the introduction of fluorine and optimizing other relevant properties led to ezetimibe [4], which is 55-fold more potent in inhibiting cholesterol in

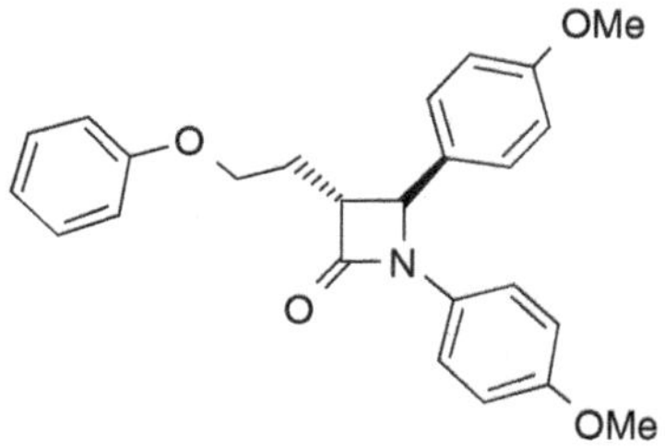

[3]

39% Reduction at 50 mg/kg

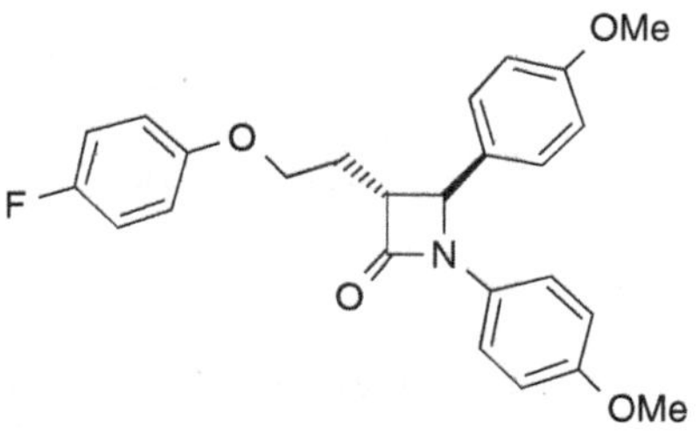

[4]

93% Reduction at 50 mg/kg

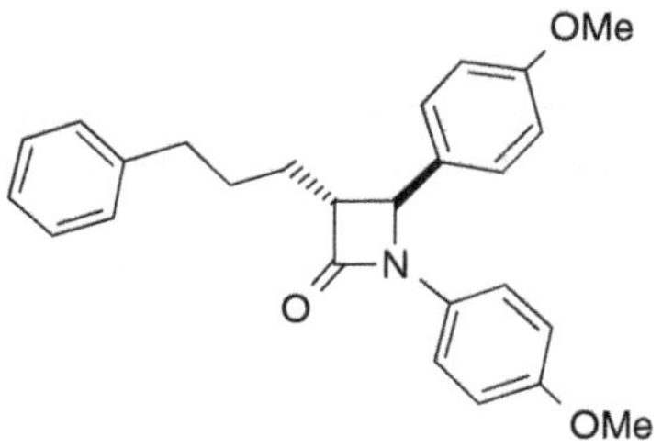

[5]

ED_{30} = 2.2 mg/kg

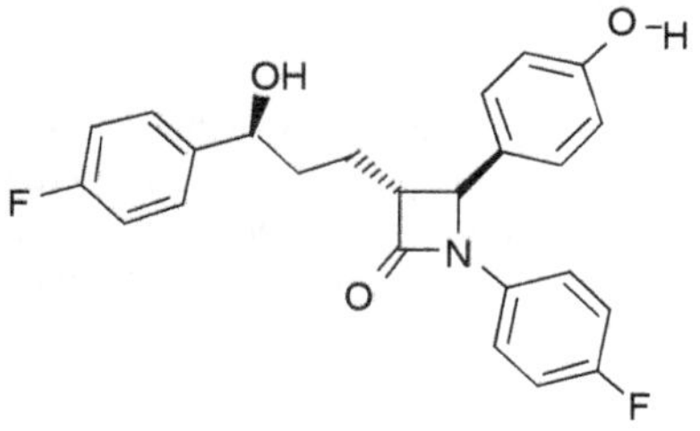

[6] Ezetimibe

ED_{30} = 0.04 mg/kg

FIGURE 5.9 Chemical structure of cholesterol absorption inhibitors including ezetimibe [56–58].

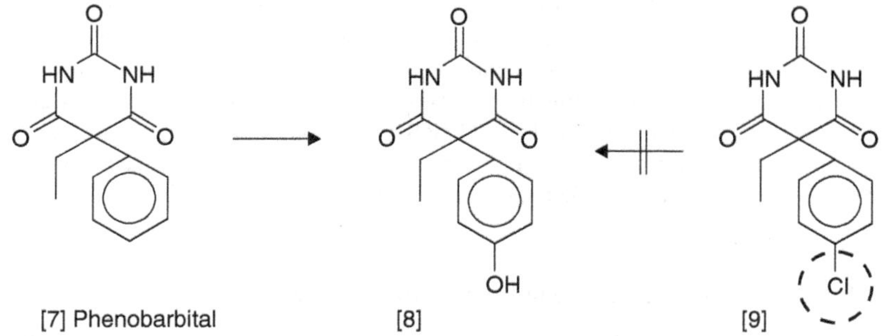

[7] Phenobarbital [8] [9]

FIGURE 5.10 The rate of *para* position hydroxylation of phenobarbital was reduced by chlorination [7, 59].

hamsters than CAI 3 (Figure 5.9) [56–58]. Phenobarbital [5] is an effective treatment for epilepsy and insomnia. It is predominantly metabolized by aromatic hydroxylation at the *para* position that is followed by glucuronidation [6]. A chlorine replacement of hydrogen [7] prevents this metabolism step, further reduces compound hepatic clearance, and increases its half-life and duration of action (Figure 5.10) [59].

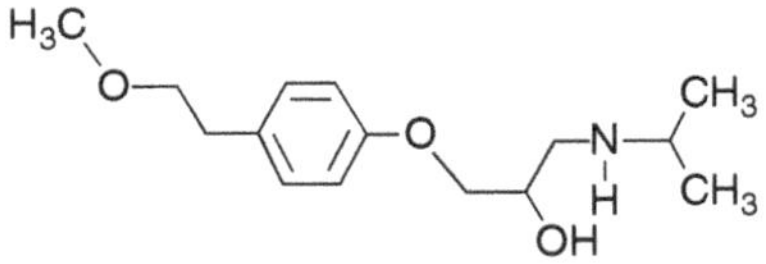

[10] Metoprolol

High hepatic CL = 15 mL/min/kg
Low oral bioavailability = 38%
Half-life = 3 h

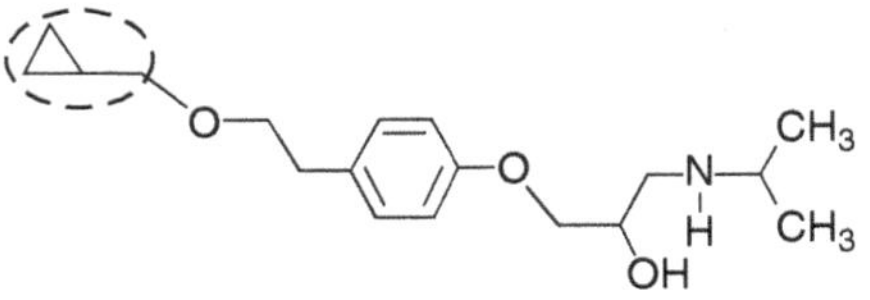

[11] Betaxolol

Low hepatic CL = 5 mL/min/kg
High oral bioavailability = 89%
Half-life = 18 h

FIGURE 5.11 The metabolic rate of O-dealkylation of metoprolol [8] was reduced by replacing the methoxy group with the bulky aliphatic cyclopropyl methoxy group to form betaxolol [9, 62].

It is interesting to note that fluorine is often preferred to chlorine substitution. This is attributed to its ability to mimic the properties of both hydrogen and hydroxyl group. In addition, fluorine has less impact on compound lipophilicity compared to chlorine. Collectively, as a drug design strategy, fluorine substitution is usually effective in overcoming poor metabolic stability while maintaining the SAR of a promising chemical series [60].

5.2.6 Introduction of Steric Hindrance

Understanding the impact of steric effects on the rate of drug metabolism of NME is a practice that is usually considered to overcome compounds' poor metabolic stability and decrease hepatic clearance. Addition of groups such as trifluoromethyl or bulky aliphatic groups (e.g., tertiary butyl, isopropyl, or cyclopropyl groups) next to a soft metabolic spot is usually effective in providing steric hindrance that would reduce the rate of metabolism by limiting the access to the enzyme active binding site [61]. For example, metoprolol is used for the treatment of hypertension. It is primarily metabolized by CYP2D6 and has high hepatic clearance (15 mL/min/kg), low bioavailability, and short half-life in humans. To improve patient compliance by reducing the dosing frequency, metoprolol structure [8] was modified by replacing the methoxy group by a cyclopropyl methoxy group. This reduced CYP2D6-mediated O-dealkylation making betaxolol [9] more metabolically stable with a longer half-life of 18 h. Taken all together, this would improve patient compliance by reducing dosing frequency (Figure 5.11) [62].

5.2.7 Intrinsic Clearance Scaling to Predict Human Clearance

In drug discovery, medicinal chemists determine hepatic intrinsic clearance for their NMEs to eliminate the drug candidates with poor hepatic stability and high first-pass liability. Interestingly, some discovery scientists assess the compound metabolic liability by only categorizing compounds based on their intrinsic clearance values determined in either hepatic microsomes or hepatocytes using Table 5.1. While sound, this practice may misguide discovery scientists and potentially lead to the elimination of NMEs that have good hepatic stability and low-to-moderate hepatic clearance; moreover, it suffers from high intrinsic hepatic clearance. This outcome is typically obtained when you assess compounds with major differences

TABLE 5.1 Classification of Hepatic Metabolic Stability Based on Human and Rat Liver Microsomes and Human Hepatocytes

Systems	Low Clearance	Moderate Clearance	High Clearance
HLM[a]	>94 min <9.2 μL/min/mg	18–94min 9.2–48 μL/min/mg	<18 min >48 μL/min/mg
RLM[b]	>51 min <17 μL/min/mg	10–51 min 17–87 μL/min/mg	<10 min >87 μL/min/mg
HHEP[b]	<20 μL/min/10^6 cells	20–45 μL/min/10^6 cells	>45 μL/min/10^6 cells

HLM, human liver microsome; RLM, rat liver microsome; HHEP, human hepatocyte.
[a]Clearance calculation based on the assumption of *B*/*P* ratio = 1, $f_{u,p} = f_{u,mic} = 1$ and back calculation from 0.3 and 0.7 *in vivo* hepatic extraction ratio.
[b]Clearance calculated from reported low, moderate, high clearance, and hepatic extraction drugs.

TABLE 5.2 *In Vitro* $t_{1/2}$, Intrinsic Clearance, Microsomal and Plasma Fraction Unbound, Blood to Plasma Ratio, and Blood Clearance of a Set of Drugs

Drug	*In Vitro* $t_{1/2}$ (min)	$CL_{int,app}$ (mL/min/kg)	$f_{u,mic}$	$f_{u,p}$	*B*/*P* Ratio	CL_{blood} (mL/min/kg)
Imipramine	66	19	0.18	0.1	1.1	12
Triazolam	33	19	0.78	0.1	0.62	4.7
Chlorpromazine	25	25	0.11	0.05	0.78	11
Ketamine	23	27	0.49	0.88	0.82	20
Midazolam	3.9	160	0.88	0.05	0.53	8.7
Propafenone	8	166	0.26	0.04	0.7	19

Source: Adapted from Obach 1999 [63].

between the microsomal binding and plasma protein binding. As depicted in Equation 5.1:

$$CL_{hep,bl}(mL/min/kg) = \frac{Q_h \cdot f_{u,p}/B/P \text{ Ratio} \cdot CL_{int,app,scaled}/f_{u,mic}}{Q_h + f_{u,p}/B/P \text{ Ratio} \cdot CL_{int,app,scaled}/f_{u,mic}} \qquad (5.1)$$

Blood hepatic clearance ($CL_{hep,bl}$) is determined by five key parameters: hepatic blood flow (Q), unbound plasma fraction ($f_{u,p}$), unbound microsomal fraction ($f_{u,mic}$), blood to plasma ratio (*B*/*P* ratio), and intrinsic *in vivo* clearance ($CL_{int,in\ vivo}$). The 3D graph in Figure 5.12 describes a sensitivity analysis of the impact of changes in $f_{u,mic}$, $f_{u,p}$ on $CL_{hep,bl}$ of an NME. Herein, an increase in the $f_{u,mic}$ combined with low plasma free fraction is associated with a decrease in hepatic blood clearance of NMEs. Table 5.2 summarizes the *in vitro* and blood clearance values of a set of drug molecules. While imipramine has a comparable intrinsic clearance to that of triazolam, the former had a higher blood clearance, which is ascribed to the smaller differences shown between $f_{u,mic}$ and $f_{u,p}$. Similar trends were observed when comparing the blood clearance between chlorpromazine versus ketamine and midazolam versus propafenone [63], where chlorpromazine and midazolam had smaller CL_{blood} than that of ketamine and propafenone, respectively.

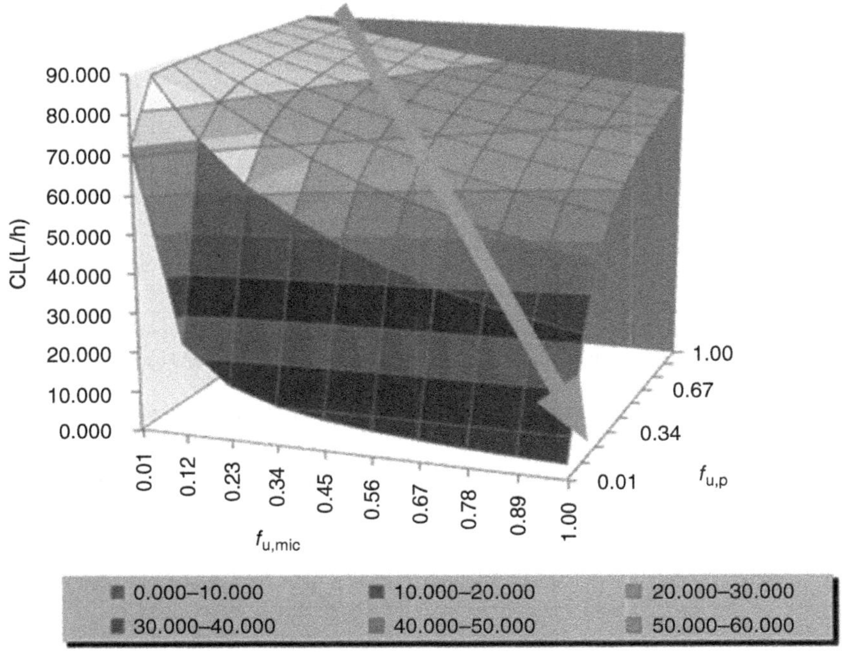

FIGURE 5.12 Sensitivity analysis of the impacts of $f_{u,p}$ and $f_{u,mic}$ on human CL using SimCYP 13.1. *(See color plate section for the color representation of this figure.)*

5.2.8 Species Differences in Hepatic Clearance

TIP:

Stop! Do not drop your lead compound that has good *in vitro* biology profile, if it has poor metabolic stability and high first-pass effect in rats. Similar to remoxipride, reboxetine, indinavir, and many other drugs in the market, a lead compound with poor metabolic stability may have better hepatic stability in upper species. Do yourself a favor and assess its metabolic stability in upper preclinical species and HLMs and hepatocytes [49]. You may find a species difference(s) that may be helpful. If you need to assess its *in vivo* biology, then consider the administration of the compound following subcutaneous or intraperitoneal rather than oral dosing. Subcutaneous dosing enables you to overcome both intestinal and hepatic first-pass effects. For intraperitoneal dosing, this approach enables intestinal first-pass effects to be overcome [55].

In Chapter 2, the impact of species differences on the physiological factors such as hepatic blood flow, enzyme type, and expressions that affect drug metabolism and hepatic clearance was discussed [64, 65]. For example, Nelson et al. reported that so far 14 CYP gene families have been identified in mammals with significant variations in the primary sequence of amino acids across species [64, 65]. However, these members of the superfamily had highly conserved regions of amino acid sequence. Similar findings were also reported with uridine

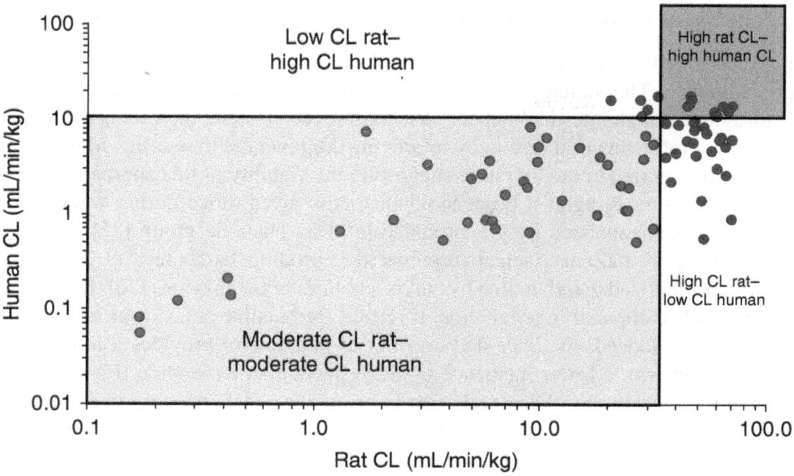

FIGURE 5.13 The relationship between *in vivo* rat clearance and human clearance. Lombardo et al. 2013 [68]. Reproduced with permission of John Wiley & Sons.

diphosphoglucose transferases and carboxylesterases enzymes [66, 67]. Collectively, these differences in the amino acid sequences can lead to significant differences in the substrate affinity and specificity, which translates into differences in the metabolism rate and profile. As depicted in Figure 5.13, compounds with high rat clearance tend to have lower clearance in humans [68]. Similar trends were observed when comparing rat clearance to upper preclinical species such as dogs and monkeys [68]. There are many cited examples that are consistent with this trend. For example, the pharmacokinetics of remoxipride was studied in rodents (mice, rats, hamsters), dogs, and humans [69]. Remoxipride was rapidly and completely absorbed through the intestinal wall in all species evaluated. However, the bioavailability was low in the rodents (<10% in mice and hamsters and <1% in rats) due to extensive first-pass elimination in the hepato-portal system [69]. Blood clearance estimated after the same intravenous doses was high in rodents and similar to or exceeding normal liver blood flow. On the other hand, clearance values in dogs and humans were low and the bioavailability high (>90%) [69]. Other examples of marketed drugs that have poor oral bioavailability in rodents yet good bioavailability in higher species include reboxetine, diazepam, and indinavir to name a few [70–74]. These successful medications would not be on the market if the discovery team solely depended on rat oral bioavailability to evaluate their metabolism in humans.

TIP:

Stop! Do not jump to conclusion when you have an NME with a high intrinsic clearance. It is pivotal to scale your hepatic clearance using Equation 5.1. Failing to do that may lead you to drop potentially viable candidates. A good practice is to have a periodic determination of *B*/*P* ratio, $f_{u,mic}$, $f_{u,p}$, along with $CL_{int,\textit{in vivo}}$ to scale the compound *in vivo* clearance. This will ensure that the medicinal chemists will not drop viable compounds of interest.

5.2.9 The Use of Prodrugs to Overcome Intestinal and Hepatic First-pass Effect

Prodrug strategy is typically used when the lead chemical series is challenged with poor permeability or solubility. This strategy is also considered to improve the oral exposure of compounds that is attributed to high intestinal/hepatic first-pass effects. Herein, the goal is to increase systemic exposure following oral dosing by improving drug metabolic stability. Many successful examples were reported in the literature suggesting the viability of this approach. Terbutaline (12) is a short acting β_2-agonist bronchodilator that is given thrice during waking hours. It is predominantly metabolized by glucuronidation of its phenolic group [75]. To extend its duration of action, AstraZeneca scientists prepared its prodrug, bambuterol (13), which is predominantly metabolized to terbutaline by butyrylcholinesterase enzymes [76]. Bambuterol has a superior profile compared to terbutaline. It releases terbutaline at a slower rate leading to a prolonged action. Indeed, the once-daily dosing of bambuterol provides a relief of asthma that is associated with a lower incidence of side effects than terbutaline (Figure 5.14) [77]. Nalbuphine is a potent opioid analgesic that is primarily metabolized by phenolic hydroxyl group conjugation that leads to significant first-pass effect and low oral bioavailability in preclinical species and humans [78]. In dogs, following oral dosing of nalbuphine, oral bioavailability is around 5–7%. However, acetylsalicylate and anthranilate ester prodrugs increased nalbuphine oral bioavailability to 17–24% and 50%, respectively (Table 5.3). These prodrugs were designed to protect the phenolic group from conjugation [79].

5.2.10 *The Impact of Intestinal Permeability and Metabolic Stability on* f_g

In 2007, Yang et al. proposed a "minimal" Q_{gut} model, which calculates the fraction that escapes intestinal first-pass effect (f_g) (Equation 5.2) [80].

$$f_g = 1 - \frac{fu_{gut} \cdot CLu_{int,g}}{Q_{gut} + fu_{gut} \cdot CLu_{int,g}} \tag{5.2}$$

where fu_{gut} is the fraction of unbound drug in the enterocyte, $CLu_{int,g}$ the net intrinsic intestinal metabolic clearance based on unbound drug concentration in the gut, and Q_{gut} the nominal gut blood flow. This is a hybrid parameter of both permeability through the enterocyte membrane CL_{perm} (a product of effective intestinal permeability P_{eff} and intestinal surface area A) and

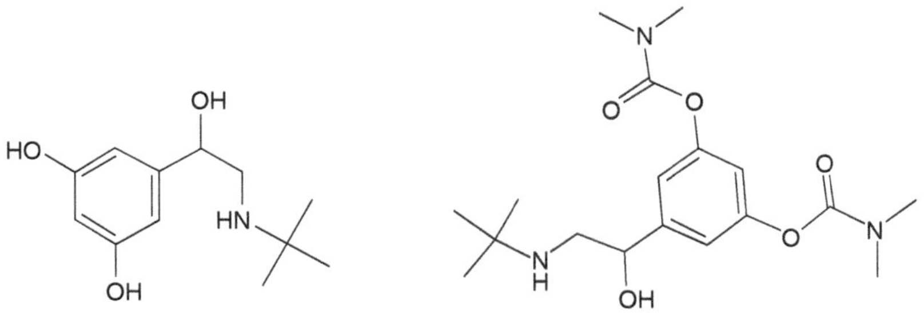

[12] Terbutalin [13] Bambuterol

FIGURE 5.14 The chemical structures of bambuterol and terbutaline.

TABLE 5.3 The Oral Bioavailability of Nalbuphine Following Oral Dosing of the Parent Drug and Its Acetylsalicylate and Anthranilate Esters in Dogs

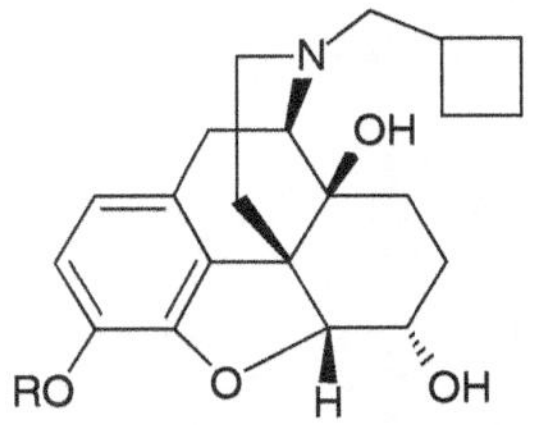

R Group	Dog Oral Bioavailability of Nalbuphine (%)
H [Nalbuphine]	5–7
Acetylsalicylate ester	17–24
Anthranilate	50

Source: Aungst et al. 1987 [79]. Reproduced with permission of Elsevier.

villous blood flow ($Q_{\text{vili}} = 19$ L/h). Q_{gut} is calculated using Equation 5.3.

$$Q_{\text{gut}} = \frac{Q_{\text{vili}} \cdot \text{CL}_{\text{perm}}}{Q_{\text{vili}} + \text{CL}_{\text{perm}}} \tag{5.3}$$

Combining Equations 5.2 and 5.3 yields Equation 5.4, from which f_g can be calculated as follows:

$$f_g = \frac{Q_{\text{vili}}}{Q_{\text{vili}} + \text{CLu}_{int,g} \cdot (1 + Q_{\text{vili}}/\text{CL}_{\text{perm}})} \tag{5.4}$$

As depicted in Figure 5.15, both $\text{CLu}_{\text{int,g}}$ and P_{eff} affect the f_g of a drug molecule, where a decrease in $\text{CLu}_{\text{int,g}}$ combined with an increase in the compound P_{eff} is associated with an increase in the drug f_g and lower intestinal first-pass effect [81]. Interestingly, the impact of $\text{CLu}_{\text{int,g}}$ is more substantial in comparison with P_{eff} of a drug molecule. Overall, NMEs with high permeability and intestinal metabolic stability tend to have better intestinal stability, which contributes to higher overall oral bioavailability.

5.2.11 *The Need to Determine the Major Metabolizing Enzyme when Assessing Drug* f_g

As depicted in Figure 5.1, there are major differences between the relative content of various CYP enzymes in the liver compared to intestine in humans [13, 14]. Indeed, the contribution of CYP3A4 enzyme to the overall content of CYP enzymes is approximately 82% in the intestine and around 40% in the human liver. As a result, it is not unexpected that these major differences in the relative protein content of CYP enzymes have a substantial impact on the f_g of their substrates. In 2014, Hisaka et al. reported a high likelihood of CYP3A substrates to have high intestinal first-pass effect and poor f_g (<0.6) compared to non-CYP3A substrates that have an average f_g of 0.86. This is consistent with the low contribution of non-CYP 3A enzymes to the

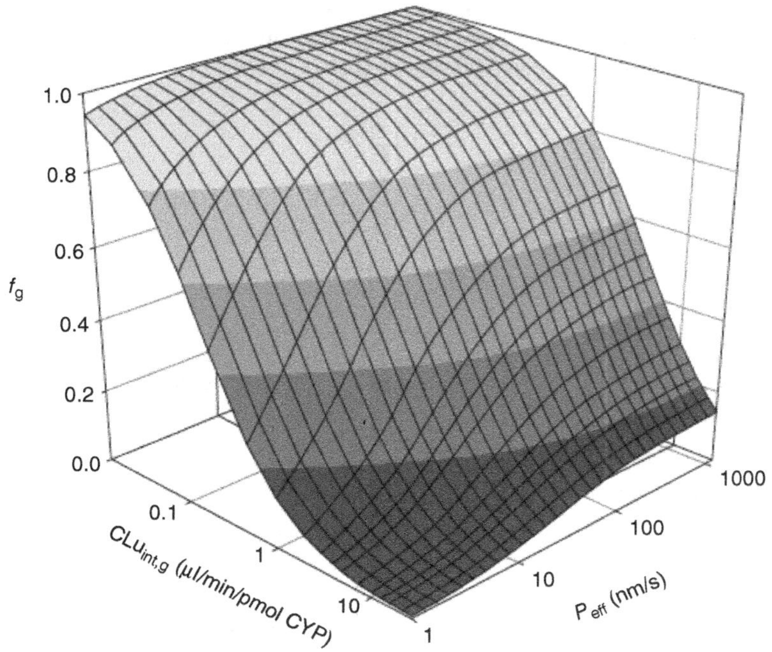

FIGURE 5.15 The relationship between intrinsic *in vitro* intestinal clearance ($CLu_{int,g}$), effective intestinal permeability (P_{eff}), and f_g for the prediction of intestinal extraction from *in vitro* data using Q_{gut} model. Galetin et al. 2008 [81]. Reproduced with permission of Taylor & Francis.

overall CYP enzymes content in human intestine, which is a key driver for better reported f_g values and consistent with Equation 5.4 [82].

TIP:

If your compound is a substrate for CYP3A enzymes, keep a close eye on f_g, your compound has a high likelihood of high intestinal first pass. Further studies to predict the f_g in humans are warranted. As illustrated in Figure 5.15, compounds with low $CLu_{int,g}$ are more likely to have high f_g.

5.2.12 Differences in the Metabolism of Enantiomers

It is well established in the literature that drug-metabolizing enzymes demonstrate stereoselectivity to chiral drug substrates [83–93]. Significant differences in the rate of metabolism of various drugs enantiomers were reported. For example, warfarin is a widely used anticoagulant. The oral clearance of its *R*-enantiomer warfarin is 42% higher than that reported with *S*-enantiomer. This finding is attributed to differences in the CYP isoform responsible for their metabolism, where the *R*-enantiomer is predominantly metabolized by CYP2C9 and the *S*-enantiomer is metabolized by CYP3A4 [94, 95]. Propranolol is a *β*-blocker that is used for the treatment of hypertension. Propranolol demonstrates stereoselective metabolism, where the oral clearance of its *R*-enantiomer is around 40% greater than that reported with its *S*-enantiomer. Verapamil is a widely used calcium channel antagonist for the treatment of hypertension. Its efficacy is predominantly driven by the *S*-enantiomer. Compared to the *R*-enantiomer, the rate of elimination of *S*-enantiomer is route dependent. After intravenous administration, verapamil *S*-enantiomer is mainly metabolized with a systemic clearance that is around 70% higher than that for the *R*-enantiomer [96]. However, following oral dosing, the *S*-enantiomer has an oral clearance that is about fourfold higher than that of the *R*-enantiomer [97]. Therefore, it is not unexpected that verapamil efficacy following oral dosing is poor. If verapamil is administered intravenously, then better efficacy is achieved. This is a good example on the value of understanding the impact of stereoselectivity not only on rate of elimination but also on efficacy. This also underscores the keen interest of medicinal chemists in characterizing the efficacy and pharmacokinetics of enantiomers rather than racemic mixture of NMEs.

5.2.13 Strategies to Overcome Metabolism Mediated by Aldehyde Oxidase

Human aldehyde oxidase (AO) is a soluble complex molybdoflavoprotein that is located in the cytosolic compartment of hepatocytes and other tissue cells. It has wide substrate specificity and primarily oxidizes heteroaromatic rings such as *N*1-methylnicotinamide and *N*-methylphthalazinium. AO also converts aldehyde into carboxylic acid. Known examples on drugs that are substrates for AO include 6-deoxypenciclovir, 6-*O*-benzylguanine, acyclovir, BIBX1382, carbazeran, famciclovir, methotrexate, phenanthridine, phthalazine, zaleplon, ziprasidone, zoniporide, PF-4217903, and PF-945863 [22, 98]. It should be emphasized that there are major species and strain differences in AO, where the highest expression was observed in humans and monkeys followed by rats; interestingly, no AO expression was observed in dogs (see Chapter 2 for more details). Therefore, using single species scaling as a tool to predict the human clearance of AO substrates may misguide DMPK scientists and lead to clinical trial attrition similar to literature reports on Carbazeran, BIBX1382, Zoniporide, RO1, SGX523, and FK3453 (Table 5.4) [98]. The projection of human clearance of NMEs that are substrates for AO should be done by scaling clearance from human cytosol and hepatocyte.

TIP:

It is prudent for a DMPK scientist to follow up when he/she notices a difference in the intrinsic clearance values measured in HLMs versus human hepatocytes. Typically, this may suggest that the metabolism of the NMEs is driven by non-CYP enzymes. Indeed, this is observed when the intrinsic clearance is higher in hepatocytes compared to HLM, since the latter either does express the enzymes (e.g., AO) or has the cofactors that are needed for enzymes to be functional (e.g., UGTs). Further *in vitro* and *in vivo* studies are required to determine the major enzymes responsible for the metabolism of the molecule.

Various drug design strategies have been adapted to minimize or eliminate AO-mediated metabolism. For example, Pryde et al. pursued a strategy that is based on blocking the site of metabolism by replacing the C–H in the pyridine ring with a nitrogen atom to make pyridazine ring. This strategy led to a significantly increased metabolic stability of the molecule in liver cytosol, which suggests the elimination of AO-mediated metabolism. The same group

TABLE 5.4 List of Drug Candidates That Have Dropped in Clinical Trials due to Poor Oral Pharmacokinetic Profiles or Toxicity Ascribed to AO Metabolism

Drug	Structure	Preclinical Data Used for Human PK Predictions	Clinical Outcome
Carbazeran	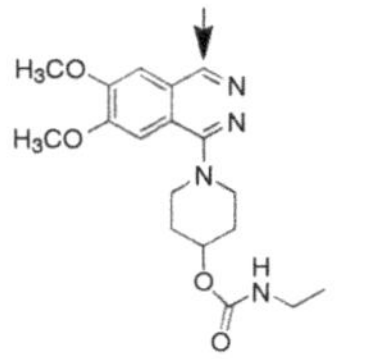	Dog PK	Oral bioavailability <5% – discontinue
BIBX1382	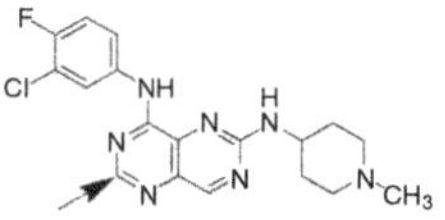	LMs, mouse, rat, and dog PK	Oral bioavailability 5% F – discontinue
Zoniporide	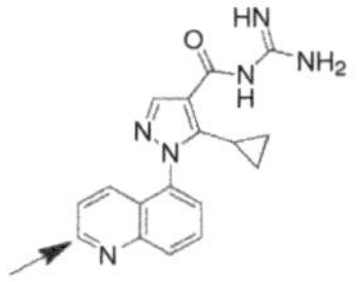	Multiple species PK	High clearance
RO1	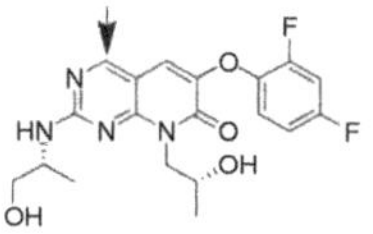	Rat, dog, and monkey PK	<1 h half-life – discontinue
SGX523	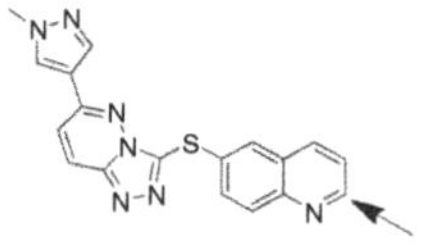	LMs, rat, and dog PK	Obstructive nephropathy – discontinue
FK3453	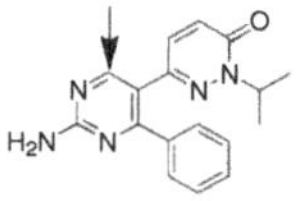	LMs, rat, and dog PK	Unacceptably low exposure – discontinue

Arrows indicate the site of metabolism by AO.
AO, aldehyde oxidase; LM, liver microsome; PK, pharmacokinetics.
Source: Hutzler et al. 2013 [98]. Reproduced with permission of Taylor & Francis.

used bioisosteric group replacement to increase metabolic stability [99–101]. They replaced the pyridine ring with five-membered heterocyclic rings such as isoxazole, thiazole, or oxazole. This significantly reduced AO-mediated metabolism and increased the overall metabolic stability of their molecules. Another strategy that can be considered is the alteration of the electrophilicity of the heteroaromatic ring by either reducing the number of nitrogen atoms or rearranging the nitrogen atoms in the heteroaromatic ring structure. Magee et al. and Linton et al. followed this approach, which played a key role in reducing AO-mediated metabolism of their leading chemical series [102, 103].

5.3 CONCLUSION

Poor metabolism continues to be a major cause of the poor pharmacokinetics and oral bioavailability of NMEs. Appropriate understanding of the factors that affect drug metabolism is a prerequisite for identifying a viable NME that can be used in the clinic to test the mechanism at the site of action. Based on BDDCS principles, physicochemical factors that decrease permeability would lead to reduction in the extent of metabolism and improvement of compound metabolic stability. Medicinal chemists are encouraged to follow a methodical approach to overcome first-pass effect and poor metabolic stability. For example, they can alter the physicochemical properties of NMEs by decreasing drug lipophilicity or increasing polarity. They can also implement various mechanisms to shield soft spots by the use of electrophilic atoms such as fluorine or chlorine or through the addition of steric hindrance to increase overall metabolic stability. It is prudent for DMPK scientists not to jump to conclusions by basing the assessment of the metabolic stability on clearance values observed in preclinical species or without appropriate scaling of *in vitro* intrinsic clearance obtained in human reagents such as HLM or human hepatocytes. In addition, reaction phenotyping for the major CYP enzyme responsible for the metabolism of NMEs is critical especially when we need to project the potential contribution of intestinal first pass to the overall bioavailability of NMEs. Typically, compounds that are CYP3A4 substrates tend to have lower f_g. Finally, CYP enzymes are not the only enzymes that should be considered in light of the shift of chemical space of NMEs toward polar and hydrophilic chemical space. This leads to a shift in the dominant metabolizing enzymes that would affect drug disposition and hence is the interest in understanding the role of AO as an enzyme that would affect drug clearance and disposition in particular those compounds that are polar and hydrophilic.

5.4 TUTORIAL CHAPTER 5

1. True or false

   Compounds that have low permeability tend to have high extent of metabolism.

   Increasing compound lipophilicity or reducing polarity is a key step to increase compound metabolic stability.

   Compounds that have high first-pass effect and clearance in rats will always have the same trend in upper preclinical species including humans.

2. Which of the following changes would increase metabolic stability of the molecule below?

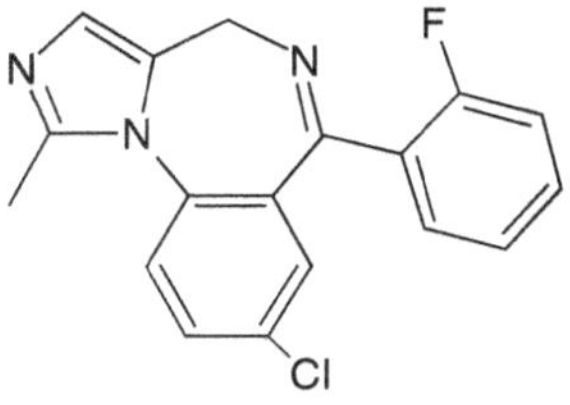

Increasing compound lipophilicity

Reducing compound polarity

Adding a hydroxyl group

Adding steric hindrance

Eliminating labile group.

3. Which compound is expected to have a higher CL_{blood}? Hint: Consider using Equation 5.1 to calculate CL_{blood}. Are you surprised at the outcome? If so, why?

Drug	*In Vitro* $t_{1/2}$(min)	CL_{int} (mL/min/kg)	$f_{u,mic}$	$f_{u,p}$	*B*/*P* Ratio
1	23	27	0.49	0.88	0.82
2	3.9	160	0.88	0.05	0.53

4. Why do compounds that are CYP3A4 substrates tend to have lower f_g (higher intestinal first pass) when compared with other non-CYP3A4 substrates?

5. Below is a compound that undergoes extensive metabolism mediated by AO. What are the structural changes that can be recommended to mitigate the poor metabolic stability?

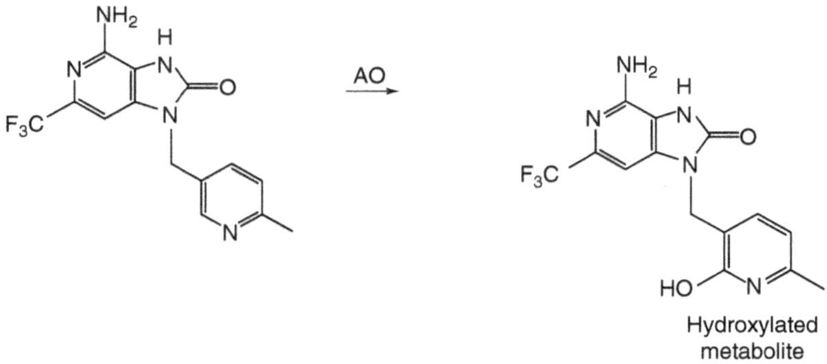

5.5 TUTORIAL CHAPTER 5 ANSWER KEY

1. True or false

   Compounds that have low permeability tend to have high extent of metabolism. **[F]**

   Increasing compound lipophilicity or reducing polarity is a key step to increase compound metabolic stability. **[F]**

   Compounds that have high first-pass effect and clearance in rats will always have the same trend in upper preclinical species including humans. **[F]**

2. Which of the following changes would increase metabolic stability of the molecule below?

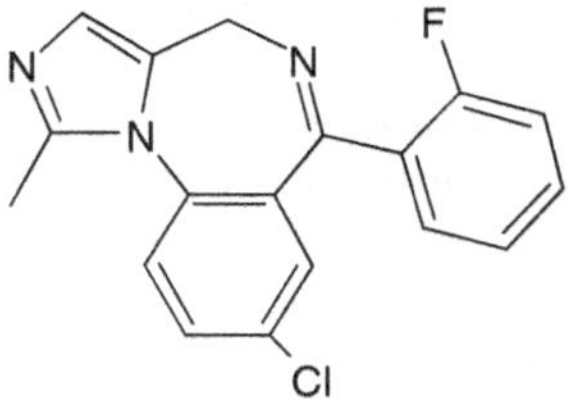

Increasing compound lipophilicity **[No]**
Reducing compound polarity **[No]**
Adding a hydroxyl group **[No]**
Adding steric hindrance **[Yes]**
Eliminating labile group. **[Yes]**.

3. Which compound is expected to have a higher CL_{blood}? Hint: Consider using Equation 5.1 to calculate CL_{blood}. Are you surprised at the outcome? If so, why?

Drug	*In Vitro* $t_{1/2}$ (min)	CL_{int} (mL/min/kg)	$f_{u,mic}$	$f_{u,p}$	*B/P* Ratio
1	23	27	0.49	0.88	0.82
2	3.9	160	0.88	0.05	0.53

Answer: Compound 1 is predicted to have a higher clearance, according to Equation 5.1, that is driven by the difference in the plasma protein binding relative to microsomal protein binding. The outcome is consistent with what is discussed in this chapter.

4. Why do compounds that are CYP3A4 substrates tend to have lower f_g (higher intestinal first pass) when compared with other non-CYP3A4 substrates?

Answer: The highest expressing enzyme in the intestine is CYP3A4. Hence, compounds that are CYP3A4 substrates tend to have lower f_g relative to other molecules that are not substrates.

5. Below is a compound that undergoes extensive metabolism mediated by AO. What are the structural changes that can be recommended to mitigate the poor metabolic stability?

Answer:

Replace the pyridine ring with five-membered heterocyclic rings such as isoxazole, thiazole, or oxazole. This significantly reduced AO-mediated metabolism and increased the overall metabolic stability of their molecules.

Alteration of the electrophilicity of the heteroaromatic ring by either reducing the number of nitrogen atoms or rearranging the nitrogen atoms in the heteroaromatic ring structure.

REFERENCES

1. van De Waterbeemd, H., Smith, D.A., Beaumont, K. and Walker, D.K., *Property-based design: optimization of drug absorption and pharmacokinetics*. J Med Chem, 2001. 44(9): p. 1313–1333.

2. Arrowsmith, J. and Miller, P., *Trial watch: phase II and phase III attrition rates 2011–2012*. Nat Rev Drug Discov, 2013. 12(8): p. 569.
3. Varma, M.V., et al., *pH-Dependent solubility and permeability criteria for provisional biopharmaceutics classification (BCS and BDDCS) in early drug discovery*. Mol Pharm, 2012. 9(5): p. 1199–1212.
4. Jones, H.M., et al., *Mechanistic pharmacokinetic modeling for the prediction of transporter-mediated disposition in humans from sandwich culture human hepatocyte data*. Drug Metab Dispos, 2012. 40(5): p. 1007–1017.
5. Walker, D.K., Alabaster, C.T., Congrave, G.S., Hargreaves, M.B., Hyland, R., Jones, B.C., Reed, L.J. and Smith, D.A., *Significance of metabolism in the disposition and action of the antidysrhythmic drug, dofetilide.* In vitro *studies and correlation with* in vivo *data*. Drug Metab Dispos, 1996. 24(4): p. 447–455.
6. Hosea, N.A., et al., *Prediction of human pharmacokinetics from preclinical information: comparative accuracy of quantitative prediction approaches*. J Clin Pharmacol, 2009. 49(5): p. 513–533.
7. Caron, G., Ermondi, G. and Testa, B., *Predicting the oxidative metabolism of statins: an application of the MetaSite algorithm*. Pharm Res, 2007. 24(3): p. 480–501.
8. Chohan, K.K., Paine, S.W. and Waters, N.J., *Quantitative structure activity relationships in drug metabolism*. Curr Top Med Chem, 2006. 6(15): p. 1569–1578.
9. Walsky, R.L., Gaman, E.A. and Obach, R.S., *Examination of 209 drugs for inhibition of cytochrome P450 2C8*. J Clin Pharmacol, 2005. 45(1): p. 68–78.
10. Riley, R.J. and Grime, K., *Metabolic screening* in vitro: *metabolic stability, CYP inhibition and induction*. Drug Discov Today Technol, 2004. 1(4): p. 365–372.
11. Zuber, R., Anzenbacherova, E. and Anzenbacher, P., *Cytochromes P450 and experimental models of drug metabolism*. J Cell Mol Med, 2002. 6(2): p. 189–198.
12. Lewis, D.F. and Dickins, M., *Substrate SARs in human P450s*. Drug Discov Today, 2002. 7(17): p. 918–925.
13. Wang, H. and Tompkins, L.M., *CYP2B6: new insights into a historically overlooked cytochrome P450 isozyme*. Curr Drug Metab, 2008. 9(7): p. 598–610.
14. Paine, M.F., Hart, H.L., Ludington, S.S., Haining, R.L., Rettie, A.E. and Zeldin, D.C., *The human intestinal cytochrome P450 "pie"*. Drug Metab Dispos, 2006. 34(5): p. 880–886.
15. Lewis, D.F., *On the recognition of mammalian microsomal cytochrome P450 substrates and their characteristics: towards the prediction of human p450 substrate specificity and metabolism*. Biochem Pharmacol, 2000. 60(3): p. 293–306.
16. Ke, A.B., Nallani, S.C., Zhao, P., Rostami-Hodjegan, A., Isoherranen, N. and Unadkat, J.D., *A physiologically based pharmacokinetic model to predict disposition of CYP2D6 and CYP1A2 metabolized drugs in pregnant women*. Drug Metab Dispos, 2013. 41(4): p. 801–813.
17. Mise, M., Hashizume, T. and Komuro, S., *Characterization of substrate specificity of dog CYP1A2 using CYP1A2-deficient and wild-type dog liver microsomes*. Drug Metab Dispos, 2008. 36(9): p. 1903–1908.
18. Brosen, K., *Drug interactions and the cytochrome P450 system. The role of cytochrome P450 1A2*. Clin Pharmacokinet, 1995. 29(Suppl 1): p. 20–25.
19. Bosch, T.M., Meijerman, I., Beijnen, J.H. and Schellens, J.H., *Genetic polymorphisms of drug-metabolising enzymes and drug transporters in the chemotherapeutic treatment of cancer*. Clin Pharmacokinet, 2006. 45(3): p. 253–285.
20. Jones, B.C., Hyland, R., Ackland, M., Tyman, C.A. and Smith, D.A., *Interaction of terfenadine and its primary metabolites with cytochrome P450 2D6*. Drug Metab Dispos, 1998. 26(9): p. 875–882.
21. King, C.D., Rios, G.R., Green, M.D. and Tephly, T.R., *UDP-glucuronosyltransferases*. Curr Drug Metab, 2000. 1(2): p. 143–161.
22. Di, L., *The role of drug metabolizing enzymes in clearance*. Expert Opin Drug Metab Toxicol, 2014. 10(3): p. 379–393.

23. Di, L., et al., *Mechanistic insights from comparing intrinsic clearance values between human liver microsomes and hepatocytes to guide drug design.* Eur J Med Chem, 2012. 57: p. 441–448.

24. Varma, M.V., Steyn, S.J., Allerton, C. and El-Kattan, A.F., *Predicting clearance mechanism in drug discovery: extended clearance classification system (ECCS).* Pharm Res, 2015. 32(12): p. 3785–3802.

25. Di, L., et al., *Development of a new permeability assay using low-efflux MDCKII cells.* J Pharm Sci, 2011. 100(11): p. 4974–4985.

26. Smith, D.A., van de Waterbeemd, H. and Walker, D., *Pharmacokinetics and Metabolism in Drug Design.* 2006, Elsevier.

27. Varma, M.V., Obach, R.S., Rotter, C., Miller, H.R., Chang, G., Steyn, S.J., El-Kattan, A. and Troutman, M.D., *Physicochemical space for optimum oral bioavailability: contribution of human intestinal absorption and first-pass elimination.* J Med Chem, 2010. 53(3): p. 1098–1108.

28. Ladona, M.G., Gonzalez, M.L., Rane, A., Peter, R.M. and de la Torre, R., *Cocaine metabolism in human fetal and adult liver microsomes is related to cytochrome P450 3A expression.* Life Sci, 2000. 68(4): p. 431–443.

29. Yasar, U., Annas, A., Svensson, J.O., Lazorova, L., Artursson, P. and Al-Shurbaji, A., *Ketobemidone is a substrate for cytochrome P4502C9 and 3A4, but not for P-glycoprotein.* Xenobiotica, 2005. 35(8): p. 785–796.

30. Park, J.E., Kim, K.B., Bae, S.K., Moon, B.S., Liu, K.H. and Shin, J.G., *Contribution of cytochrome P450 3A4 and 3A5 to the metabolism of atorvastatin.* Xenobiotica, 2008. 38(9): p. 1240–1251.

31. Roy, P., Yu, L.J., Crespi, C.L. and Waxman, D.J., *Development of a substrate-activity based approach to identify the major human liver P-450 catalysts of cyclophosphamide and ifosfamide activation based on cDNA-expressed activities and liver microsomal P-450 profiles.* Drug Metab Dispos, 1999. 27(6): p. 655–666.

32. Burt, H.J., Pertinez, H., Sall, C., Collins, C., Hyland, R., Houston, J.B. and Galetin, A., *Progress curve mechanistic modeling approach for assessing time-dependent inhibition of CYP3A4.* Drug Metab Dispos, 2012. 40(9): p. 1658–1667.

33. Mirghani, R.A., Yasar, U., Zheng, T., Cook, J.M., Gustafsson, L.L., Tybring, G. and Ericsson, O., *Enzyme kinetics for the formation of 3-hydroxyquinine and three new metabolites of quinine* in vitro*; 3-hydroxylation by CYP3A4 is indeed the major metabolic pathway.* Drug Metab Dispos, 2002. 30(12): p. 1368–1371.

34. Venkatakrishnan, K., von Moltke, L.L., Duan, S.X., Fleishaker, J.C., Shader, R.I. and Greenblatt, D.J., *Kinetic characterization and identification of the enzymes responsible for the hepatic biotransformation of adinazolam and N-desmethyladinazolam in man.* J Pharm Pharmacol, 1998. 50(3): p. 265–274.

35. Klees, T.M., Sheffels, P., Dale, O. and Kharasch, E.D., *Metabolism of alfentanil by cytochrome p4503a (cyp3a) enzymes.* Drug Metab Dispos, 2005. 33(3): p. 303–311.

36. King, C.D., Green, M.D., Rios, G.R., Coffman, B.L., Owens, I.S., Bishop, W.P. and Tephly, T.R., *The glucuronidation of exogenous and endogenous compounds by stably expressed rat and human UDP-glucuronosyltransferase 1.1* Arch Biochem Biophys, 1996. 332(1): p. 92–100.

37. Gentile, D.M., Tomlinson, E.S., Maggs, J.L., Park, B.K. and Back, D.J., *Dexamethasone metabolism by human liver* in vitro*. Metabolite identification and inhibition of 6-hydroxylation.* J Pharmacol Exp Ther, 1996. 277(1): p. 105–112.

38. Chen, M., et al., *Effects of cytochrome P450 2C9 polymorphism on bosentan metabolism.* Drug Metab Dispos, 2014. 42(11): p. 1820–1825.

39. Senda, C., Kishimoto, W., Sakai, K., Nagakura, A. and Igarashi, T., *Identification of human cytochrome P450 isoforms involved in the metabolism of brotizolam.* Xenobiotica, 1997. 27(9): p. 913–922.

40. Guitton, J., Buronfosse, T., Desage, M., Lepape, A., Brazier, J.L. and Beaune, P., *Possible involvement of multiple cytochrome P450S in fentanyl and sufentanil metabolism as opposed to alfentanil.* Biochem Pharmacol, 1997. 53(11): p. 1613–1619.

41. Ward, B.A., Morocho, A., Kandil, A., Galinsky, R.E., Flockhart, D.A. and Desta, Z., *Characterization of human cytochrome P450 enzymes catalyzing domperidone N-dealkylation and hydroxylation* in vitro. Br J Clin Pharmacol, 2004. 58(3): p. 277–287.

42. Li, J., Zhao, M., He, P., Hidalgo, M. and Baker, S.D., *Differential metabolism of gefitinib and erlotinib by human cytochrome P450 enzymes*. Clin Cancer Res, 2007. 13(12): p. 3731–3737.

43. Jonsson, G., Astrom, A. and Andersson, P., *Budesonide is metabolized by cytochrome P450 3A (CYP3A) enzymes in human liver*. Drug Metab Dispos, 1995. 23(1): p. 137–142.

44. Patki, K.C., Von Moltke, L.L. and Greenblatt, D.J., In vitro *metabolism of midazolam, triazolam, nifedipine, and testosterone by human liver microsomes and recombinant cytochromes p450: role of cyp3a4 and cyp3a5*. Drug Metab Dispos, 2003. 31(7): p. 938–944.

45. Nakada, N., et al., *Murine Cyp3a knockout chimeric mice with humanized liver: prediction of the metabolic profile of nefazodone in humans*. Biopharm Drug Dispos, 2016. 37(1): p. 3–14.

46. Walsky, R.L. and Obach, R.S., *Validated assays for human cytochrome P450 activities*. Drug Metab Dispos, 2004. 32(6): p. 647–660.

47. Gertz, M., Houston, J.B. and Galetin, A., *Physiologically based pharmacokinetic modeling of intestinal first-pass metabolism of CYP3A substrates with high intestinal extraction*. Drug Metab Dispos, 2011. 39(9): p. 1633–1642.

48. Nassar, A.E., Kamel, A.M. and Clarimont, C., *Improving the decision-making process in the structural modification of drug candidates: enhancing metabolic stability*. Drug Discov Today, 2004. 9(23): p. 1020–1028.

49. Thomas, V.H., Bhattachar, S., Hitchingham, L., Zocharski, P., Naath, M., Surendran, N., Stoner, C.L. and El-Kattan, A., *The road map to oral bioavailability: an industrial perspective*. Expert Opin Drug Metab Toxicol, 2006. 2(4): p. 591–608.

50. Dragovich, P.S., et al., *Structure-based design, synthesis, and biological evaluation of irreversible human rhinovirus 3C protease inhibitors. 8. Pharmacological optimization of orally bioavailable 2-pyridone-containing peptidomimetics*. J Med Chem, 2003. 46(21): p. 4572–4585.

51. Tagat, J.R., et al., *Piperazine-based CCR5 antagonists as HIV-1 inhibitors. IV. Discovery of 1-[(4,6-dimethyl-5-pyrimidinyl)carbonyl]-4-[4-[2-methoxy-1(R)-4-(trifluoromethyl)phenyl]ethyl-3(S)-methyl-1-piperaz inyl]-4-methylpiperidine (Sch-417690/Sch-D), a potent, highly selective, and orally bioavailable CCR5 antagonist*. J Med Chem, 2004. 47(10): p. 2405–2408.

52. Obach, R.S., *Pharmacologically active drug metabolites: impact on drug discovery and pharmacotherapy*. Pharmacol Rev, 2013. 65(2): p. 578–640.

53. Prakash, C., Shaffer, C.L. and Nedderman, A., *Analytical strategies for identifying drug metabolites*. Mass Spectrom Rev, 2007. 26(3): p. 340–369.

54. Vaz, R.J., Zamora, I., Li, Y., Reiling, S., Shen, J. and Cruciani, G., *The challenges of* in silico *contributions to drug metabolism in lead optimization*. Expert Opin Drug Metab Toxicol, 2010. 6(7): p. 851–861.

55. Nickerson, D.F. and Toler, S.M., *Intraperitoneal and intraportal administration of droloxifene to the Sprague-Dawley rat: assessing the first-pass effect*. Xenobiotica, 1997. 27(6): p. 627–632.

56. Rosenblum, S.B., Huynh, T., Afonso, A., Davis, H.R., Jr., Yumibe, N., Clader, J.W. and Burnett, D.A., *Discovery of 1-(4-fluorophenyl)-(3R)-[3-(4-fluorophenyl)-(3S)-hydroxypropyl]-(4S)-(4-hydroxyphenyl)-2-azetidinone (SCH 58235): a designed, potent, orally active inhibitor of cholesterol absorption*. J Med Chem, 1998. 41(6): p. 973–980.

57. Astles, P.C., et al., *Acyl-CoA:cholesterol O-acyltransferase (ACAT) inhibitors. 2. 2-(1,3-Dioxan-2-yl)-4,5-diphenyl-1H-imidazoles as potent inhibitors of ACAT*. J Med Chem, 1996. 39(7): p. 1423–1432.

58. Clader, J.W., *The discovery of ezetimibe: a view from outside the receptor*. J Med Chem, 2004. 47(1): p. 1–9.

59. Patani, G.A. and LaVoie, E.J., *Bioisosterism: a rational approach in drug design*. Chem Rev, 1996. 96(8): p. 3147–3176.

60. Smith, D.A., *Discovery and ADMET: where are we now*. Curr Top Med Chem, 2011. 11(4): p. 467–481.

61. Park, B.K., Kitteringham, N.R. and O'Neill, P.M., *Metabolism of fluorine-containing drugs*. Annu Rev Pharmacol Toxicol, 2001. 41: p. 443–470.

62. Manoury, P.M., Binet, J.L., Rousseau, J., Lefevre-Borg, F. and Cavero, I.G., *Synthesis of a series of compounds related to betaxolol, a new beta 1-adrenoceptor antagonist with a pharmacological and pharmacokinetic profile optimized for the treatment of chronic cardiovascular diseases*. J Med Chem, 1987. 30(6): p. 1003–1011.

63. Obach, R.S., *Prediction of human clearance of twenty-nine drugs from hepatic microsomal intrinsic clearance data: An examination of* in vitro *half-life approach and nonspecific binding to microsomes*. Drug Metab Dispos, 1999. 27(11): p. 1350–1359.

64. Hurst, S., Loi, C.M., Brodfuehrer, J. and El-Kattan, A., *Impact of physiological, physicochemical and biopharmaceutical factors in absorption and metabolism mechanisms on the drug oral bioavailability of rats and humans*. Expert Opin Drug Metab Toxicol, 2007. 3(4): p. 469–489.

65. Nelson, D.R., et al., *The P450 superfamily: update on new sequences, gene mapping, accession numbers, early trivial names of enzymes, and nomenclature*. DNA Cell Biol, 1993. 12(1): p. 1–51.

66. Burchell, B., Ebner, T., Baird, S., Bin Senafi, S., Clarke, D., Brierley, C. and Sutherland, L., *Use of cloned and expressed human liver UDP-glucuronosyltransferases for analysis of drug glucuronide formation and assessment of drug toxicity*. Environ Health Perspect, 1994. 102(Suppl 9): p. 19–23.

67. Hosokawa, M., Maki, T. and Satoh, T., *Characterization of molecular species of liver microsomal carboxylesterases of several animal species and humans*. Arch Biochem Biophys, 1990. 277(2): p. 219–227.

68. Lombardo, F., et al., *Comprehensive assessment of human pharmacokinetic prediction based on* in vivo *animal pharmacokinetic data, Part 1: volume of distribution at steady state*. J Clin Pharmacol, 2013. 53(2): p. 167–177.

69. Widman, M., Nilsson, L.B., Bryske, B. and Lundstrom, J., *Disposition of remoxipride in different species. Species differences in metabolism*. Arzneimittelforschung, 1993. 43(3): p. 287–297.

70. Cocchiara, G., Battaglia, R., Pevarello, P. and Strolin Benedetti, M., *Comparison of the disposition and of the metabolic pattern of reboxetine, a new antidepressant, in the rat, dog, monkey and man*. Eur J Drug Metab Pharmacokinet, 1991. 16(3): p. 231–239.

71. Worboys, P.D., Bradbury, A. and Houston, J.B., *Kinetics of drug metabolism in rat liver slices. III. Relationship between metabolic clearance and slice uptake rate*. Drug Metab Dispos, 1997. 25(4): p. 460–467.

72. Gao, W., et al., *Effect of indinavir on the intestinal exsorption of amprenavir, saquinavir and nelfinavir after intravenous administration in rats*. Biol Pharm Bull, 2003. 26(2): p. 199–204.

73. Frassetto, L., Thai, T., Aggarwal, A.M., Bucher, P., Jacobsen, W., Christians, U., Benet, L.Z. and Floren, L.C., *Pharmacokinetic interactions between cyclosporine and protease inhibitors in HIV+ subjects*. Drug Metab Pharmacokinet, 2003. 18(2): p. 114–120.

74. Lin, J.H., *Role of pharmacokinetics in the discovery and development of indinavir*. Adv Drug Deliv Rev, 1999. 39(1–3): p. 33–49.

75. Nilsson, H.T., Persson, K. and Tegner, K., *The metabolism of terbutaline in man*. Xenobiotica, 1972. 2(4): p. 363–373.

76. Rosenburg, J., *Clinical-pharmacokinetic aspects of prolonged effect duration as illustrated by beta2-agonists*. Eur J Clin Pharmacol, 2002. 58(4): p. S1–S21.

77. Persson, G., Baas, A., Knight, A., Larsen, B. and Olsson, H., *One month treatment with the once daily oral beta 2-agonist bambuterol in asthmatic patients*. Eur Respir J, 1995. 8(1): p. 34–39.

78. Bullingham, R.E., McQuay, H.J. and Moore, R.A., *Clinical pharmacokinetics of narcotic agonist–antagonist drugs*. Clin Pharmacokinet, 1983. 8(4): p. 332–343.

79. Aungst, B.J., Myers, M.J., Shefter, E. and Shami, E.G., *Prodrugs for improved oral nalbuphine bioavailability: inter-species differences in the disposition of nalbuphine and its acetylsalicylate and anthranilate esters*. Int J Pharm, 1987. 38(1–3): p. 199–209.
80. Yang, J., Jamei, M., Yeo, K.R., Tucker, G.T. and Rostami-Hodjegan, A., *Prediction of intestinal first-pass drug metabolism*. Curr Drug Metab, 2007. 8(7): p. 676–684.
81. Galetin, A., Gertz, M. and Houston, J.B., *Potential role of intestinal first-pass metabolism in the prediction of drug–drug interactions*. Expert Opin Drug Metab Toxicol, 2008. 4(7): p. 909–922.
82. Hisaka, A., Nakamura, M., Tsukihashi, A., Koh, S. and Suzuki, H., *Assessment of intestinal availability (FG) of substrate drugs of cytochrome p450s by analyzing changes in pharmacokinetic properties caused by drug–drug interactions*. Drug Metab Dispos, 2014. 42(10): p. 1640–1645.
83. Niwa, T., Murayama, N. and Yamazaki, H., *Stereoselectivity of human cytochrome p450 in metabolic and inhibitory activities*. Curr Drug Metab, 2011. 12(6): p. 549–569.
84. Howland, R.H., *Understanding chirality and stereochemistry: three-dimensional psychopharmacology*. J Psychosoc Nurs Ment Health Serv, 2009. 47(7): p. 15–18.
85. Chen, C., *Physicochemical, pharmacological and pharmacokinetic properties of the zwitterionic antihistamines cetirizine and levocetirizine*. Curr Med Chem, 2008. 15(21): p. 2173–2191.
86. Hutt, A.J., *Chirality and pharmacokinetics: an area of neglected dimensionality?* Drug Metabol Drug Interact, 2007. 22(2–3): p. 79–112.
87. Eap, C.B., Buclin, T. and Baumann, P., *Interindividual variability of the clinical pharmacokinetics of methadone: implications for the treatment of opioid dependence*. Clin Pharmacokinet, 2002. 41(14): p. 1153–1193.
88. McColl, K.E. and Kennerley, P., *Proton pump inhibitors – differences emerge in hepatic metabolism*. Dig Liver Dis, 2002. 34(7): p. 461–467.
89. Mehvar, R., Brocks, D.R. and Vakily, M., *Impact of stereoselectivity on the pharmacokinetics and pharmacodynamics of antiarrhythmic drugs*. Clin Pharmacokinet, 2002. 41(8): p. 533–558.
90. Takahashi, H. and Echizen, H., *Pharmacogenetics of warfarin elimination and its clinical implications*. Clin Pharmacokinet, 2001. 40(8): p. 587–603.
91. Fleishaker, J.C., *Clinical pharmacokinetics of reboxetine, a selective norepinephrine reuptake inhibitor for the treatment of patients with depression*. Clin Pharmacokinet, 2000. 39(6): p. 413–427.
92. Timmer, C.J., Sitsen, J.M. and Delbressine, L.P., *Clinical pharmacokinetics of mirtazapine*. Clin Pharmacokinet, 2000. 38(6): p. 461–474.
93. Kimko, H.C., Cross, J.T. and Abernethy, D.R., *Pharmacokinetics and clinical effectiveness of methylphenidate*. Clin Pharmacokinet, 1999. 37(6): p. 457–470.
94. Toon, S., Hopkins, K.J., Garstang, F.M., Aarons, L., Sedman, A. and Rowland, M., *Enoxacin–warfarin interaction: pharmacokinetic and stereochemical aspects*. Clin Pharmacol Ther, 1987. 42(1): p. 33–41.
95. Rettie, A.E., Korzekwa, K.R., Kunze, K.L., Lawrence, R.F., Eddy, A.C., Aoyama, T., Gelboin, H.V., Gonzalez, F.J. and Trager, W.F., *Hydroxylation of warfarin by human cDNA-expressed cytochrome P-450: a role for P-4502C9 in the etiology of (S)-warfarin–drug interactions*. Chem Res Toxicol, 1992. 5(1): p. 54–59.
96. Eichelbaum, M., Mikus, G. and Vogelgesang, B., *Pharmacokinetics of (+)-, (−)- and (+/−)-verapamil after intravenous administration*. Br J Clin Pharmacol, 1984. 17(4): p. 453–458.
97. Vogelgesang, B., Echizen, H., Schmidt, E. and Eichelbaum, M., *Stereoselective first-pass metabolism of highly cleared drugs: studies of the bioavailability of* L- *and* D-*verapamil examined with a stable isotope technique*. Br J Clin Pharmacol, 1984. 18(5): p. 733–740.
98. Hutzler, J.M., Obach, R.S., Dalvie, D. and Zientek, M.A., *Strategies for a comprehensive understanding of metabolism by aldehyde oxidase*. Expert Opin Drug Metab Toxicol, 2013. 9(2): p. 153–168.
99. Pryde, D.C., et al., *Medicinal chemistry approaches to avoid aldehyde oxidase metabolism*. Bioorg Med Chem Lett, 2012. 22(8): p. 2856–2860.

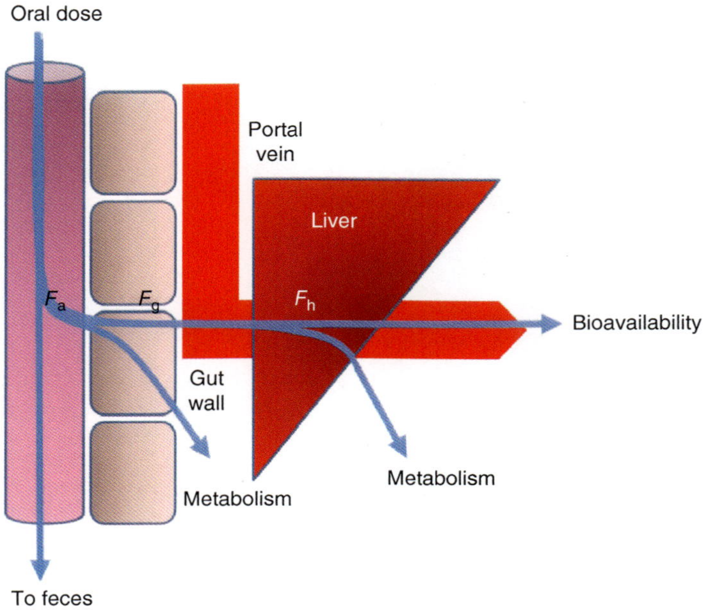

FIGURE 1.11 Oral bioavailability is a multiplicity parameter and a product of f_a, f_g, and f_h.

Oral Bioavailability Assessment: Basics and Strategies for Drug Discovery and Development, First Edition.
Ayman F. El-Kattan.

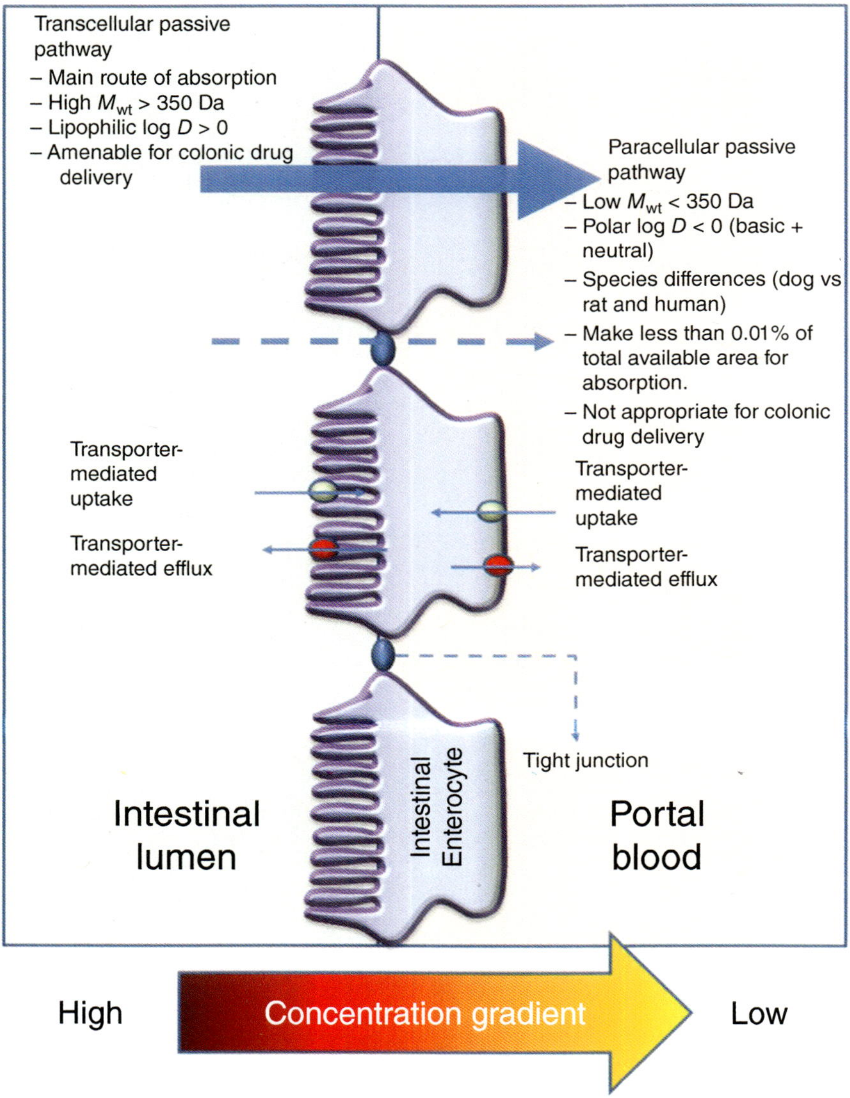

FIGURE 2.4 Possible routes of drug absorption across intestinal enterocytes.

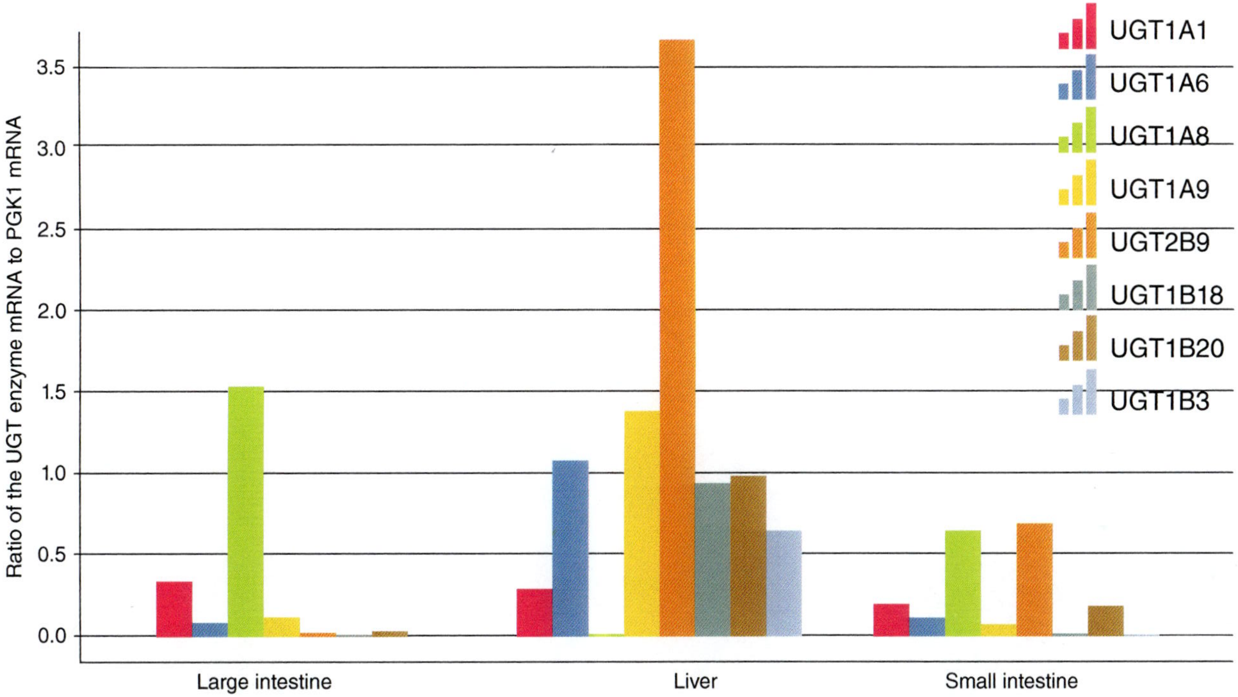

FIGURE 2.9 Expression of UDP-glucuronosyltransferase mRNAs in cynomolgus monkey large intestine, liver, and small intestine. Nishimura et al. 2009 [179]. Reproduced with permission of Elsevier.

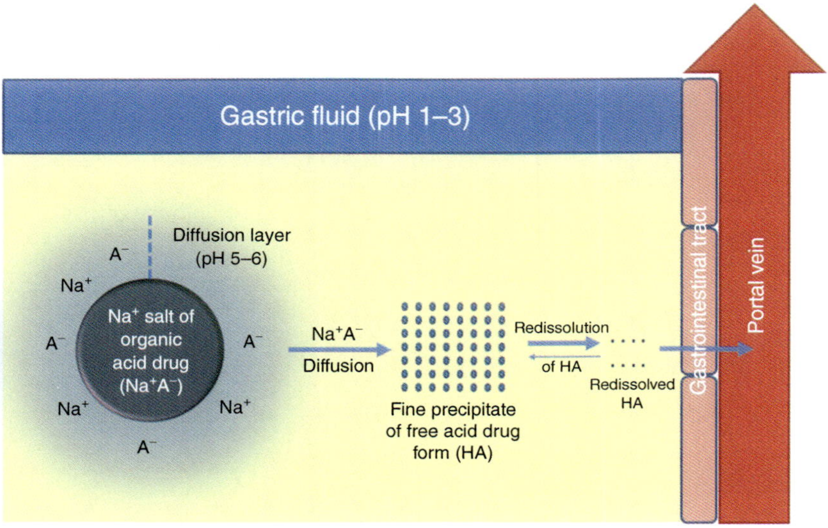

FIGURE 4.8 Dissolution of a salt form of organic acid in gastric fluid.

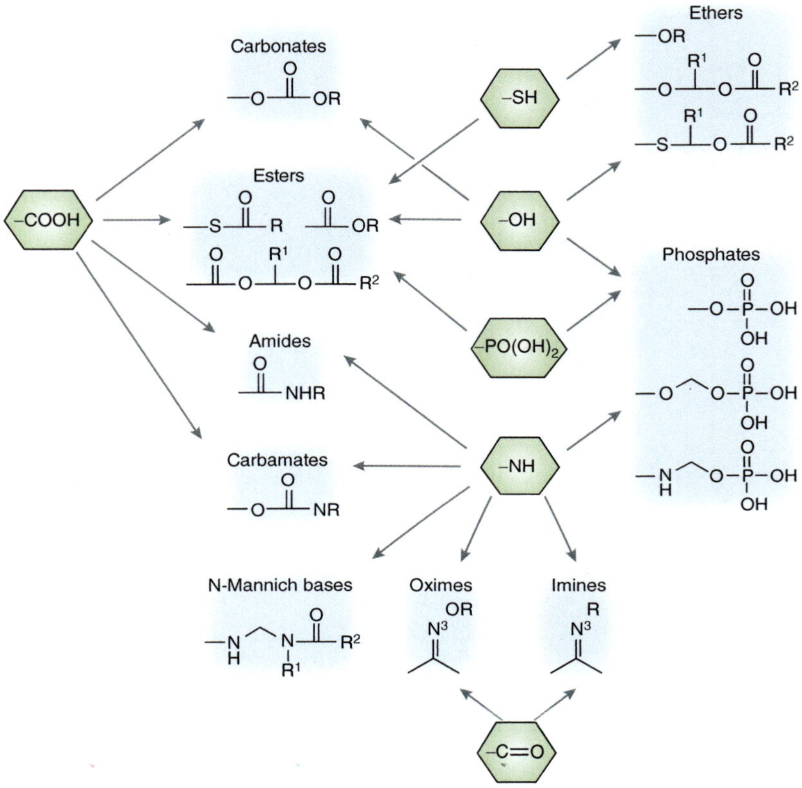

FIGURE 4.12 Known functional groups on parent drugs that are pliable to prodrug design (shown in gray). Rautio et al. 2008 [60]. Reproduced with permission of Nature Publishing Group.

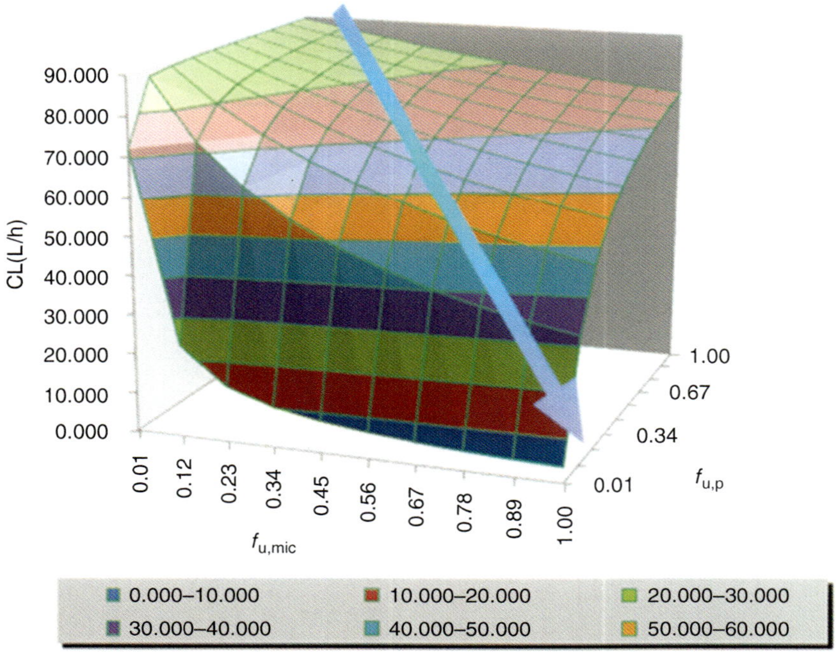

FIGURE 5.12 Sensitivity analysis of the impacts of $f_{u,p}$ and $f_{u,mic}$ on human CL using Sim-CYP 13.1.

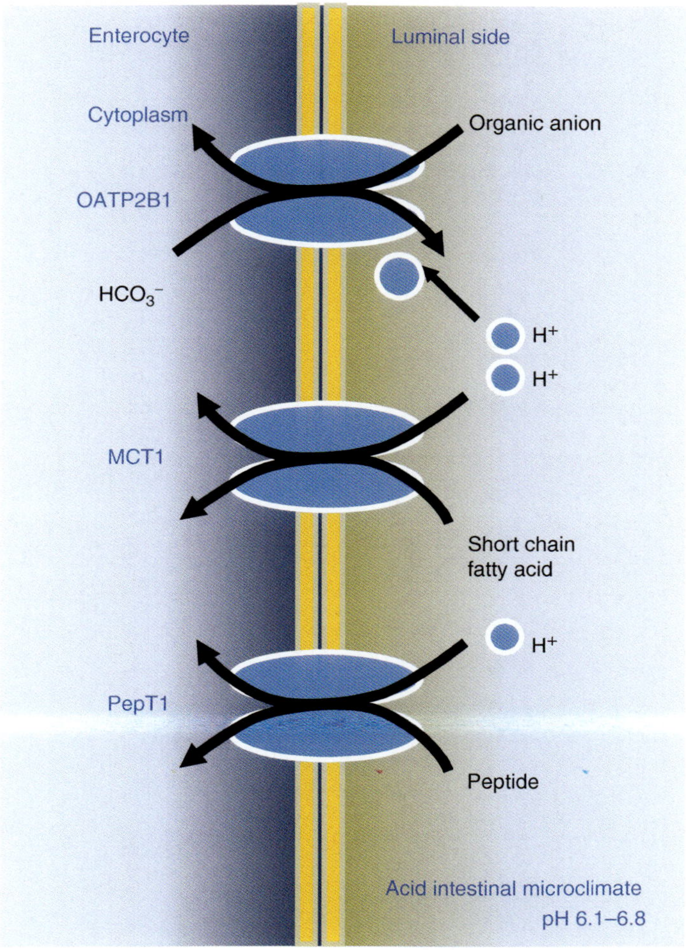

FIGURE 7.1 Known proton-coupled/pH-dependent solute carriers expressed at the brush border membrane of small intestine. Adapted from Anderson and Thwaites 2010 [24].

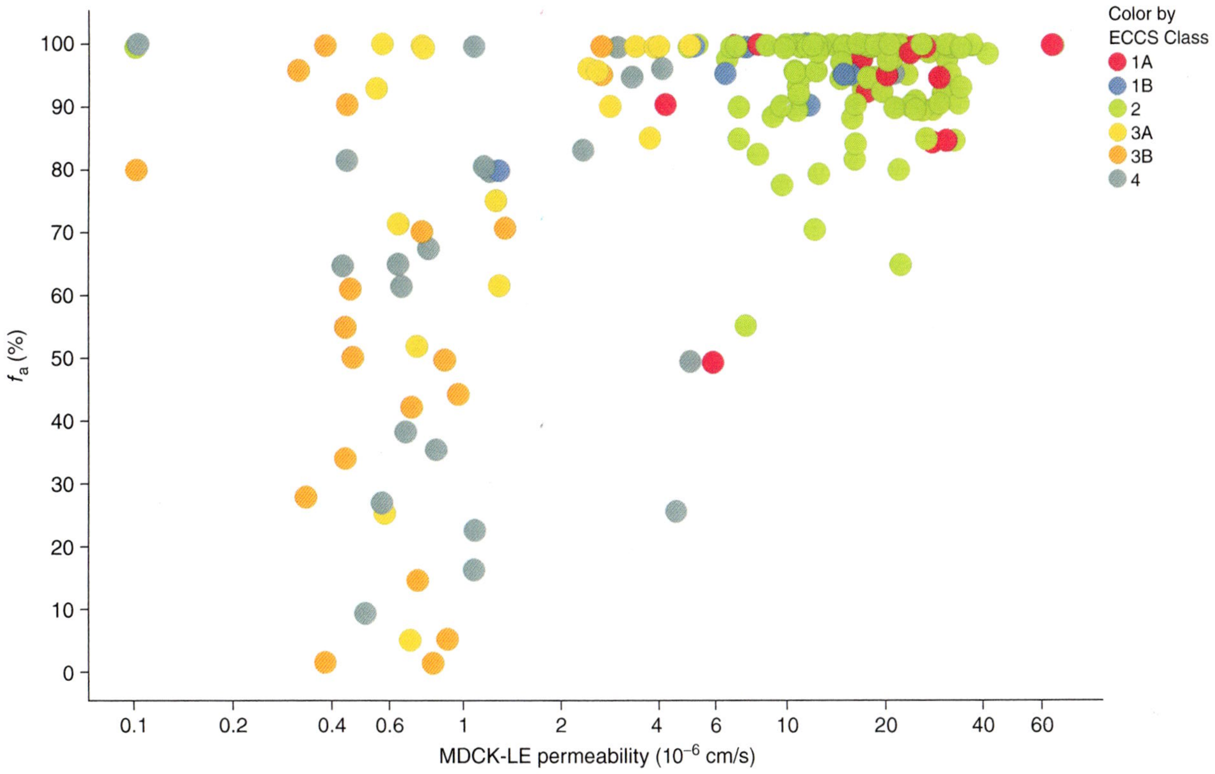

FIGURE 13.1 The relationship between LE-MDCK permeability versus human f_a for set of compounds classified based on ECCS.

100. Jones, P., et al., *Discovery of a highly potent series of TLR7 agonists*. Bioorg Med Chem Lett, 2011. 21(19): p. 5939–5943.

101. Pryde, D.C., Dalvie, D., Hu, Q., Jones, P., Obach, R.S. and Tran, T.D., *Aldehyde oxidase: an enzyme of emerging importance in drug discovery*. J Med Chem, 2010. 53(24): p. 8441–8460.

102. Magee, T.V., et al., *Discovery of azetidinyl ketolides for the treatment of susceptible and multidrug resistant community-acquired respiratory tract infections*. J Med Chem, 2009. 52(23): p. 7446–7457.

103. Linton, A., Kang, P., Ornelas, M., Kephart, S., Hu, Q., Pairish, M., Jiang, Y. and Guo, C., *Systematic structure modifications of imidazo[1,2-a]pyrimidine to reduce metabolism mediated by aldehyde oxidase (AO)*. J Med Chem, 2011. 54(21): p. 7705–7712.

6

IMPACT OF INTESTINAL EFFLUX TRANSPORTERS ON ORAL ABSORPTION

6.1 INTRODUCTION

Benet et al. classified drug molecules based on their extent of metabolism and solubility into four classes. BDDCS Class 1 molecules have high extent of metabolism and solubility; BDDCS Class 2 molecules have high extent of metabolism (extent of metabolism >70%) and low solubility; BDDCS Class 3 molecules have low extent of metabolism and high solubility; BDDCS Class 4 molecules have low extent of metabolism and solubility. Interestingly, compounds with low extent of metabolism also tend to have low intestinal permeability and vice versa. In this classification, compounds that are of Classes 2–4 have oral absorption that would be influenced by efflux transporters such as P-glycoprotein (P-gp) and breast cancer resistance protein (BCRP) transporters that line the apical membrane of the intestinal enterocytes [1].

The goals of this chapter are threefold: first, to review the structure and distribution, mechanism of drug efflux, known substrates and inhibitors, and influence of genetic polymorphism on *in vitro* and *in vivo* activity of P-gp and BCRP substrates. Second, to discuss a sensitivity analysis that was performed to evaluate the impact of changes in K_m, V_{max}, effective intestinal permeability, and solubility on changes in the absorbed fraction with increase in the oral dose using talinolol as a case study. Finally, to review FDA guidance on the assessment if a drug is a P-gp or BCRP substrate/inhibitor and the strategy that would be followed to put these liabilities in perspective.

Oral Bioavailability Assessment: Basics and Strategies for Drug Discovery and Development, First Edition.
Ayman F. El-Kattan.

6.2 INTESTINAL EFFLUX TRANSPORTERS

6.2.1 P-glycoprotein (ABCB1)

When saturable biological processes such as active uptake-mediated absorption, efflux mechanisms, or enterocytic-mediated metabolism are involved, the prediction of human intestinal absorption is challenging and establishing *in vitro* and *in vivo* extrapolations (IVIVE) becomes complex. It is well established in the literature that the interplay of P-glycoprotein (P-gp) and intestinal metabolizing enzymes is a prudent factor that limits the absorption of variety of drug molecules [2]. The following section discusses P-gp structure and distribution, mechanism of drug efflux, known substrates and inhibitors for P-gp, and clinical significance of P-gp efflux along intestine.

6.2.1.1 P-gp Structure and Distribution Human P-gp (*ABCB1*) gene belongs to the human *ABCB (MDR/TAP)* family (ABCB) [3]. It spans 28 exons and its cDNA consists of 3843 base pairs that encodes 1280 amino acid long (molecular weight ~170 kDa). P-gp is composed of a single-chain glycoprotein that is divided into 2 homologous portions of equal length with a total of 12 transmembrane domains. Furthermore, it has two ATP-binding regions that are separated by a flexible linker polypeptide region between the Walker A and B motifs [4, 5]. P-gp plays a key role as a biochemical barrier that limits the toxicity of many xenobiotics. This is mainly attributed to its selective distribution and localization at the ports of entry in various body organs. For example, using monoclonal antibody, immunohistochemical analysis demonstrated the expression of P-gp in capillary endothelial cells of brain and testis, canalicular surface of hepatocytes and apical surface of proximal tubule in kidney [6].

P-gp mRNA is also expressed in the small intestine and colon and immunolocalized at the apical surface of human enterocytes. Oswald et al. quantified the expression of P-gp in four body donors and in Caco-2 [7]. The group reported a substantial expression of P-gp in ileum relative to jejunum (Figure 6.1). This site-dependent expression of P-gp should be

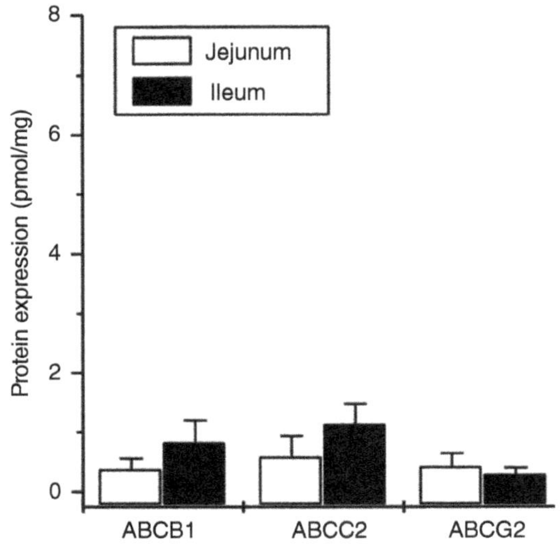

FIGURE 6.1 Intestinal protein expression of intestinal efflux transporters (*ABCB1*, *ABCC2*, and *ABCG2*) in human jejunum and ileum ($N = 4$). Oswald et al. 2013 [7]. Reproduced with permission of Springer.

considered when projecting or assessing the potential impact of P-gp on the permeability of its substrate. Dahan and Amidon used rat intestinal perfusion to investigate the role of site-dependent expression of P-gp on the *in vivo* intestinal absorption of cimetidine and famotidine in the presence or absence of the P-gp inhibitor verapamil. These molecules are BCS Class 3 P-gp substrates where they have low permeability and high solubility. The two molecules demonstrate site-dependent permeability. It is interesting to note that the effective permeability decreased in distal ileum in comparison with proximal regions of the intestine. Furthermore, coadministration of verapamil significantly increased the effective permeability of cimetidine and famotidine in the ileum. No significant change was reported in the effective permeability measured in the jejunum. This study clearly illustrates that site-dependent expression of P-gp should be considered when assessing the potential impact of P-gp on its substrates in particular if they are BCS Class 3.

Similar to PEPT1 transporter, P-gp expression in the Caco-2 was comparable to that shown in the jejunum (Figure 6.2). This suggests that Caco-2 is an effective model to investigate the impact of P-gp on the permeability and absorption of P-gp substrates.

6.2.1.2 Mechanism of Drug Efflux of P-gp Based on X-ray mouse P-gp structure, which has 87% sequence similarity to human P-gp, Aller et al. noted that P-gp protein spans 136 Å perpendicular to and ~70 Å in the plane of the membrane bilayer, in the nucleotide-free inward-facing confirmation [8]. P-gp internal cavity (~6000 Å) within the lipid membrane bilayer is composed of mostly aromatic and hydrophobic residues. This enables the access for entry of hydrophobic compounds from the inner lipid membrane. It is generally accepted that substrates binding to the transmembranes (TMs) on the inner leaflet of the membrane bilayer initiates the transport cycle, which is followed by ATP binding and hydrolysis to efflux the substrate and bringing back the pump conformation to original state. Shapiro and Ling initially proposed different binding sites named H-site and R-site that binds to Hoechst 33342 and rhodamine 123, respectively [9–14]. Later studies showed that prazosin and progesterone stimulate the transport of both Hoechst 33342 and rhodamine 123 suggesting the existence of a third drug-binding site on P-gp with a positive allosteric effect on drug transport by the H- and R-sites [9–14]. Acharya et al. measured both the rate constants for P-gp-mediated transport of substrates and transport inhibition using pairwise combinations and suggested that binding of substrates could be cooperative, either positive or negative [15, 16].

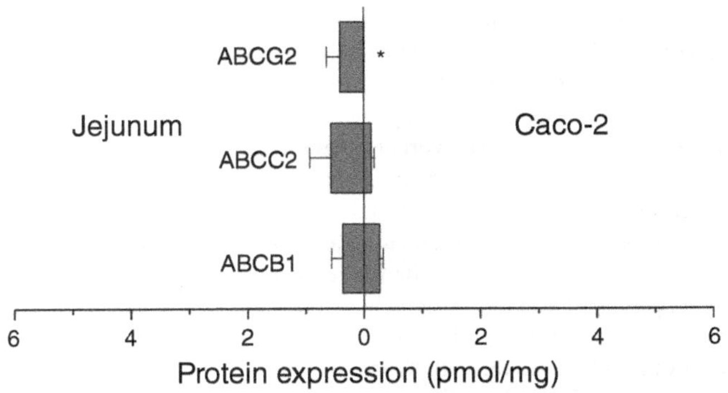

FIGURE 6.2 Protein expression of efflux transporters in human jejunum ($N = 4$) and Caco-2 ($N = 4$). (*Below limit of quantification.). Oswald et al. 2013 [7]. Reproduced with permission of Springer.

6.2.1.3 P-gp Substrates and Inhibitors Screening studies for P-gp-drug interaction identified a number of clinically important drugs as P-gp substrates (Table 6.1). These are as diverse as alpha/beta adrenergic antagonists (acebutolol, alfuzosin, and carvedilol), angiotensin II inhibitors (candesartan and losartan), alkaloids (reserpine, vincristine, and vinblastine), anthracyclines (doxorubicin and daunorubicin), antiarrhythmics (digitoxin and digoxin), antibiotics (azithromycin and clarithromycin), anticholinergics (neostigmine and pirenzepine), anticoagulants and antiplatelets (dabigatran etexilate and dipyridamole), antimalarials (chloroquine and quinine), antipsychotics (amisulpride and fluphenazine), antivirals (famciclovir and sofosbuvir), calcium channel blockers (amlodipine and diltiazem), corticosteroids (betamethasone and budesonide), dipeptidyl peptidase-4 inhibitor (saxagliptin and sitagliptin), H-1 receptor antagonists (astemizole and cetrizine), HMG CoA reductase (atorvastatin, cerivastatin, fluvastatin, and rosuvastatin), immunosuppressants (cyclosporine and tacrolimus), kinase inhibitors (dasatinib and gefitinib), NSAIDs (celecoxib and indomethacin), nucleoside reverse transcriptase inhibitors (abacavir and tenofovir), opioids (methadone, nalbuphine, and oxycodone), antineoplastics (dactinomycin and lenalidomides), protease inhibitors (amprenavir, atazanavir, and darunavir), proton pump inhibitors (lansoprazole and omeprazole), topoisomerase inhibitors (etoposide and irinotecan), and vasodilators (minoxidil) [17–93]. Even dye molecules (rhodamine 123) and pharmaceutical excipients exhibited P-gp substrate activity.

As illustrated in Figure 6.3, P-gp substrates are polar with a median polar surface area (PSA) of 88 $Å^2$ and hydrogen bond donor (HBD) and hydrogen bond acceptors (HBA) of 2 and 5, respectively. P-gp substrates have typically large molecular weight (median $M_{wt} = 406$ Da) and are lipophilic to facilitate their interactions with P-gp hydrophobic core (median measured $\log P$ and $\log D_{pH\,7.4}$ of 2.4 and 1.7, respectively). Indeed, several studies have related the binding affinity (K_m) of P-g for substrates and modulators to their lipid–water partition coefficient. A drug with high lipid–water partition will accumulate to a high concentration within the cytoplasmic membrane and favor binding to P-gp with low apparent binding affinity, while a drug with low partitioning will have a lower membrane concentration and a high K_m value. This trend is captured in Figure 6.4, where increasing compound lipophilicity ($\log P$) is associated with a decrease in its K_m suggesting increase in the compound potency toward P-gp.

Describing the structural requirements for interaction with the transporting binding site with a single pharmacophore model has proven difficult, partly due to overlapping binding sites in P-gp. Ligand-based approaches have demonstrated that P-gp substrates are generally of a lipophilic nature and contain hydrogen bond acceptors or basic functionalities. Shapiro and Ling initially proposed different binding sites named H-site and R-site that binds to Hoechst 33342 and rhodamine 123, respectively [12]. Later studies showed that prazosin and progesterone stimulate the transport of both Hoechst 33342 and rhodamine 123 suggesting the existence of a third drug-binding site on P-gp with a positive allosteric effect on drug transport by the H and R sites [94]. Seelig and coworkers suggested that partitioning into the lipid membrane is the rate-limiting step for the interaction of a substrate with P-gp and that dissociation of the P-glycoprotein–substrate complex is determined by the number and strength of the hydrogen bonds formed between the substrate and the transporter [95]. Three-dimensional structures of a large number of drugs revealed that the minimal common binding element consisting of two or three hydrogen bond acceptor groups in a specific spatial distance. Because the transmembrane sequences of P-gp are rich in HBD groups, they may recognize the hydrogen bond acceptor groups of the substrates through hydrogen bond formation in the lipid membrane.

Improved oral bioavailability and clinical efficacy of various drugs observed by P-gp inhibition lead to the design and development of modulators, which specifically block P-gp efflux and having improved toxicity profiles. The calcium blocker verapamil was the first drug shown to be a modulator of P-gp, and since many different compounds have been shown to modulate

TABLE 6.1 Partial List of Known P-gp Substrates, Their *In Silico* and Measured Physicochemical Properties, Renal Recovery (%), *In Vitro* Transporter Substrate Parameters [17–93]

Generic Name	BDDCS Class	% Excreted Unchanged in Urine	M_{wt}	Measured log P	Measured log $D_{pH\ 7.4}$	c Log P	HBA	HBD	PSA	Efflux Ratio	K_m (μM)	V_{max}	V_{max}/K_m
Abacavir sulfate	1	1.2	286.34	1.20	1.20	0.81	6	3	96	3.1			
Acebutolol HCL	1	10	336.43	1.71	0.19	1.71	5	3	97	7.8	254		
Alfuzosin	1	11	389.46		−1.18	2.55	8	2	109	29			
Alvimopan	3	2	424.54			2.16	5	3	98	2.9			
Ambrisentan	1	5	376.46			3.33	5	1	70	3.6			
Amisulpride	4	50	369.49	1.10		1.80	6	2	108	4.8			
Amlodipine	1	10	408.89	3.00	1.68	3.43	5	2	105	11.8			
Amprenavir	2	1	505.64			3.29	6	3	139	3.4			
Anastrozole	1	10	293.37			1.29	4	0	61	3.4			
Aprepitant	2	0	534.44			4.60	5	2	78	2.3			
Astemizole	2	0.5	458.58	5.70	3.88	6.09	4	1	35	2.22			
Atazanavir sulfate	2	7	704.87			5.92	7	5	186	11.7			
Atorvastatin calcium	2	1	558.66			4.46	5	4	119		110		
Atropine (DL)	3	57	289.38	1.83		1.30	3	1	51	3.4			
Azithromycin	3	6	749.00	4.02		2.64	13	5	186	4.2			
Benznidazole	1		260.25			0.90	4	1	88				
Betamethasone	1	4.8	392.47	1.94	1.94	1.79	5	3	104	6.9			
Bromocriptine	1	2	654.61		4.59	6.58	6	3	118	7.2			
Budesonide	1	0	430.55	3.20	2.70	2.91	6	2	99	4.04			
Bupropion	1	0.5	239.75		3.27	3.21	2	1	33		0.5	6 pmol/ min/ mg	12 μL/ min/ mg

(*continued*)

TABLE 6.1 (*Continued*)

Generic Name	BDDCS Class	% Excreted Unchanged in Urine	M_{wt}	Measured log *P*	Measured log $D_{pH\,7.4}$	*c* Log *P*	HBA	HBD	PSA	Efflux Ratio	K_m (μM)	V_{max}	V_{max}/K_m
Candesartan	4	52	440.47			5.43	7	2	114	3.8			
Carvedilol	2	1	406.49	4.19		4.04	5	3	78	10.6			
Celecoxib	2	2	381.38			4.37	3	1	79	7.5			
Celiprolol	3	20	379.50	1.92	1.14	1.86	5	3	98	6.3			
Cerivastatin	1	24	459.56			3.56	6	3	105	2.4			
Cetirizine	3	50	388.90	1.70	−0.31	2.08	5	1	52	2.6			
Chloroquine	3	61	319.88	4.63	1.54	5.06	3	1	26	2.2			
Clarithromycin	3	35	747.97	3.16		2.37	13	4	190	5.1			
Clemastine	1	5	343.90	5.49	3.04	5.45	2	0	10	3.9			
Colchicine	1		399.45	1.30	1.03	1.20	6	1	88	2.2			
Cyclosporine	2	0.1	1202.64	2.95	2.92	14.36	12	5	290		8.4	2.4 nmol/h/mg	0.286 mL/h/mg
Dactinomycin	3	10	1255.45			8.01	18	5	369	15.9			
Darunavir	2	1.2	547.68			2.89	7	3	148	3.8			
Daunorubicin	2	0.5	527.53	1.83	1.83	0.84	11	5	200	2.5			
Desloratadine	2	5	310.83			3.83	2	1	25	6.5			
Dexamethasone	1	2.6	392.47	1.94	1.83	1.79	5	3	104	2.1			
Dicloxacillin	3	60	470.33	2.91		2.98	5	2	116	35.6			
Digitoxin	3	30	764.96	2.83	2.83	2.85	12	5	193	3.8			
Digoxin	3	60	780.96	1.26	1.26	1.42	13	6	215	2.5			
Diltiazem	1	2	414.53	2.70	2.22	3.65	4	0	56	2			
Diphenhydramine	1	2	255.36	3.27	1.61	3.45	2	0	10	2			
Dipyridamole	2	0.1	504.64		3.71	1.49	12	4	135	98			
Docetaxel	2	5	807.90			4.08	10	5	240	22			

Doxorubicin	1	3.5	543.53	1.27	1.27	0.32	12	6	223		8.97	0.88 pmol/min/mg	0.0981 μL/min/mg
Erythromycin (base)	3	5	733.95	3.06		1.61	13	5	203	3.6			
Etoposide	3	38	588.57	0.60	0.60	0.03	12	3	167	20.1	461		
Everolimus	1	0	958.25			7.10	13	3	213	22.5			
Famciclovir	1	0	321.34			0.09	6	1	113	3.17			
Famotidine	3	67	337.45	−0.64	−1.50	−1.17	8	6	179	9			
Fexofenadine; terfenadine carboxylate	3	25	501.67		2.68	1.96	5	3	87	2			
Flecainide	3	43	414.35	3.78	1.14	3.66	4	2	66				
Fluphenazine hydrochloride	2	0.1	510.45		3.48	4.12	4	1	25	2.3			
Fluticasone propionate	2	3	500.58			3.80	4	1	86	2.3			
Fluvastatin sodium	1	1.5	411.48	3.80		4.05	4	3	87	27.6			
Gefitinib	2	2	446.91	4.85		5.60	7	1	63	8.1			
Glyburide	2	0	494.01		1.41	4.24	5	3	127	2.5			
Guanfacine hydrochloride	3	50	246.10	1.33	1.12	1.37	3	3	84	3			
Hydrochlorothiazide	3	100	297.74	−0.07	−0.07	−0.37	5	3	132	2.1			
Hydroxyzine	1	0.1	374.91	3.50	2.37	4.00	4	1	34	13.7			
Imatinib mesylate	2	5	493.62			4.53	7	2	80	14.2			
Indomethacin	2	15	357.80	4.27	0.77	4.18	4	1	69	2			
Irinotecan	1	16.7	586.69			2.73	6	1	108	4.4			
Ketoconazole	2	3	531.44	4.35	4.05	3.64	6	0	58	2			

(*continued*)

TABLE 6.1 (*Continued*)

Generic Name	BDDCS Class	% Excreted Unchanged in Urine	M_{wt}	Measured log P	Measured log $D_{pH\,7.4}$	c Log P	HBA	HBD	PSA	Efflux Ratio	K_m (μM)	V_{max}	V_{max}/K_m
Labetalol	1	2.5	328.41	1.33	1.08	2.50	4	4	106	8.8			
Lansoprazole	2	0	369.37		2.36	2.60	4	1	62	3.4			
Lapatinib ditosylate	2	1	581.07			5.97	7	2	104	15			
Lenalidomide	4	66	259.27			0.53	4	2	89	1.9			
Loperamide	3	0.5	477.05		4.22	4.66	3	1	43	3.8			
Lopinavir	2	2.2	628.82			6.10	5	4	131	3.2			
Loratadine	2	5	382.89	5.20	5.20	5.05	2	0	38	2			
Lorazepam	1	0.5	321.17	2.39	2.39	2.37	3	2	65	3.7			
Losartan potassium	2	12	461.01			4.10	5	2	87	2.6	232	92.8 nmol/min/cm^2	0.4 mL/cm^2/min
Lovastatin	2	10	404.55	4.26	4.26	4.08	3	1	77	2.2			
Maraviroc	1	8	513.68			3.26	4	1	58	12			
Metformin	3	99	129.17		−5.41	−1.63	5	4	90		0.1	33 pmol/min/mg	330 μL/min/mg
Methadone	1	24	309.46	3.93	2.07	4.17	2	0	19	2.52			
Methylprednisolone	1	4.9	374.48		2.18	1.74	5	3	104	7.1	60.6		
Minoxidil	1	10	209.25	1.24	1.38	−0.72	5	2	89	2.4			
Nalbuphine hydrochloride	1	4	357.45			1.39	5	3	78	2.17			
Nelfinavir	2		567.80			5.84	5	4	112	2.23			
Neostigmine	3	67	223.30			−2.81	1	0	29	2.23			
Nilotinib	2	4	529.53		−1.10	5.84	6	2	88	4			

Nizatidine	3	61	331.46	−0.20	1.00	−0.16	6	2	85	7.7	1200	4 pmol/ cm^2/s	0.003 μL/ cm^2/s
Norethindrone	1	2	298.43	2.97	2.97	2.78	2	1	41	2.2			
Omeprazole	1	0	345.42	2.23	2.23	2.57	5	1	71	2.1			
Oxycodone	1	19	315.37	0.21	1.65	−0.04	5	1	60	2.06			
Paclitaxel	2	5	853.93		6.83	4.73	10	4	236	2.4			
Paliperidone	4	59	426.50		2.52	1.12	5	1	76	3.3			
Pantoprazole sodium	1	0	383.38			2.11	6	1	80	3.6			
Pirenzepine	3	43	351.41	0.10	−0.39	−0.35	5	1	64	3.63			
Pitavastatin	2		421.47		1.50	3.59	5	3	96	4.4			
Prazosin	1	0.5	383.41		0.79	2.03	7	1	96	2.1			
Prednisolone	1	16	360.45	1.62	1.62	1.42	5	3	104	2.5	218		
Quinidine sulfate dihydrate	1	18	782.94	3.44	1.82	2.79	4	1	43	5.2	2.2		
Quinine bisulfate heptahydrate	1	12	548.61	3.44	1.82	2.79	4	1	43	2.5			
Raltegravir potassium	2	9	444.43		0.45	1.16	7	3	150	3.1			
Ranitidine	3	30	314.41	0.27	0.54	0.67	5	2	84	2.2			
Ranolazine	2	6	427.55			1.01	6	2	76		84.6	3 nmol/ cm^2/h	0.035 mL/ cm^2/h
Repaglinide	2	1.5	452.60			5.30	5	2	84	75			
Reserpine	1	1	608.69	3.72	4.14	3.86	8	1	114	3.71			
Rifabutin	2	10	847.03	3.20		4.73	13	5	216	3.3			
Rifampin	2	7	822.96		1.32	3.71	14	6	234	4.2	55	4.3 nmol/ cm^2/h	0.078 mL/ cm^2/h

(continued)

TABLE 6.1 (*Continued*)

Generic Name	BDDCS Class	% Excreted Unchanged in Urine	M_{wt}	Measured log P	Measured log $D_{pH 7.4}$	cLog P	HBA	HBD	PSA	Efflux Ratio	K_m (μM)	V_{max}	V_{max}/K_m
Risperidone	1	3	410.50	3.04	2.52	2.71	4	0	54	2			
Ritonavir	2	3.5	720.96			4.94	6	4	152	2.1			
Romidepsin	2		540.71			3.44	5	4	158	32			
Rosiglitazone maleate	1	0	357.43			3.02	5	1	71		0.084	1.7 pmol/min/mg	20.2 μL/min/mg
Rosuvastatin calcium	3	5	481.55		−0.89	1.90	8	3	145	6.8			
Roxithromycin	4	12	837.07	2.75		2.29	16	5	224	69.8	45.4	1.1 nmol/cm²/h	0.024 mL/cm²/h
Saquinavir methanesulfonate	2	2	766.95	4.70		4.73	7	5	179	5.5	15.4	1.3 nmol/cm²/h	0.084 mL/cm²/h
Saxagliptin	3	24	315.42			0.11	4	2	89	3.1			
Sildenafil	1	7.5	474.59			1.98	7	1	105	2			
Simvastatin	2	10	418.58	4.68	4.68	4.48	3	1	77	2.2			
Sirolimus	2	2	914.20			7.04	12	3	204	17.7			
Sitagliptin	3	78	407.32			0.69	4	1	71	7.38			
Sorafenib tosylate	2	0	464.83			5.46	3	3	99	56.2			
Sparfloxacin	1	10	392.41			−0.61	7	3	103	2			
Sulfasalazine	2	1.5	398.40		−0.78	3.88	8	3	148	160			
Sulindac	2	1	356.42	3.42	−0.66	3.16	3	1	59	3.6			
Sunitinib malate	1	4	398.48		5.40	3.00	3	3	81	2.4			
Tacrolimus	2	0.5	804.04		3.96	5.78	11	3	186	2.1			
Talinolol	3	52.8	363.50			3.15	4	4	93	2			

Telithromycin	2	13	812.03			3.75	11	1	163	18.8	9.8	0.3 nmol/ cm^2/h	0.031 mL/ cm^2/h
Temocapril	1	1.5	476.62			2.10	5	2	102	5.56			
Temsirolimus	1	4.6	1030.31			0.67	14	4	254	79			
Teniposide	2	9	656.67	1.24	3.09	0.72	12	3	167	9.8			
Tenofovir disoproxil	3	82	519.45		1.25	0.80	10	1	176	9.2			
Terfenadine	2	0	471.69	5.69	4.77	6.07	3	2	46	2.5			
Tipranavir	2	0.1	602.68			7.76	5	2	112	11.3			
Topotecan	3	40	421.46			0.73	6	2	103	3.5	170	7.2 nmol/ cm^2/h	0.042 mL/ cm^2/h
Valsartan	4	13	435.53			4.86	6	2	114	4.1			
Vinblastine	1	0.5	811.00	3.70	3.69	5.23	9	3	153	17.9	89.2	1.77 pmol/ cm^2/s	0.02 μL/ cm^2/s
Vincristine	1	15	824.98		2.57	4.04	9	3	171	6.31			
Vinorelbine tartrate	1	11	778.95			5.94	8	2	130	5.9			
Zidovudine	1	2.8	267.25	0.05	0.08	0.04	6	2	127	2.2			

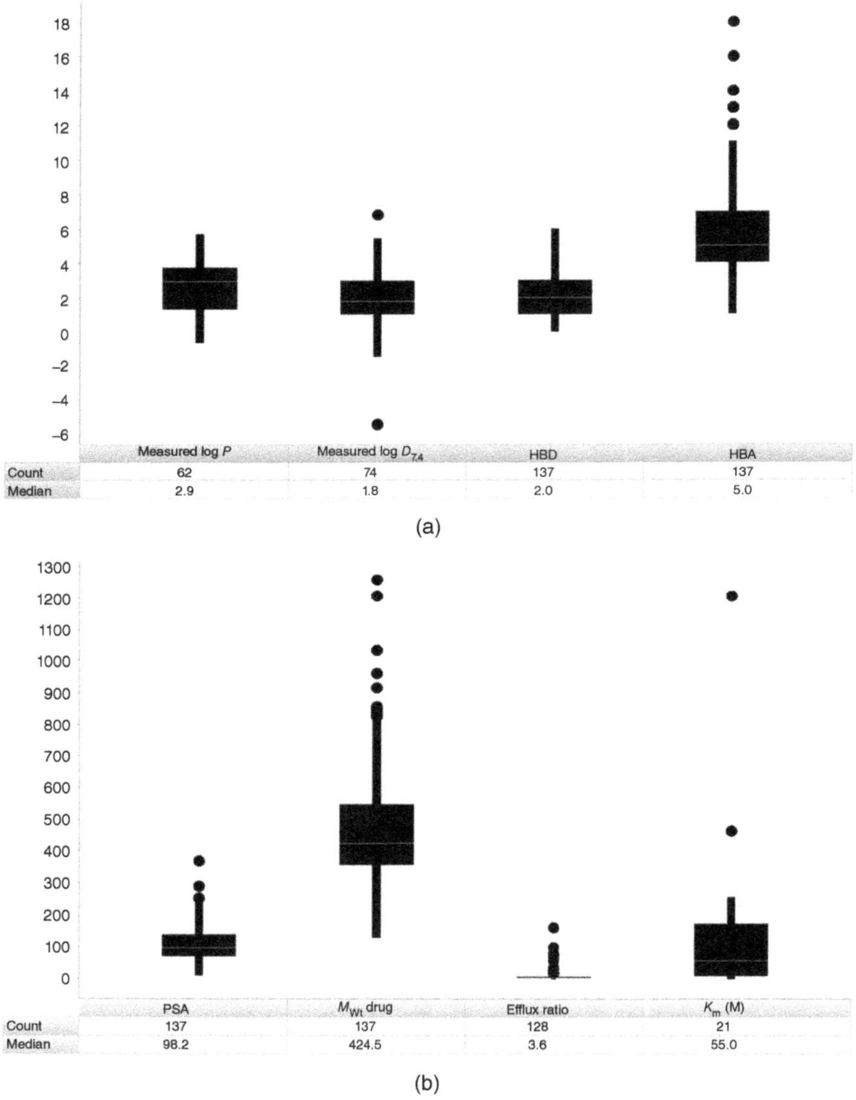

FIGURE 6.3 Physicochemical properties of known P-gp substrates: (a) measured $\log D_{\text{pH 7.4}}$, measured $\log P$, HBD, and HBA of known P-gp substrates; (b) polar surface area (PSA), molecular weight (M_{wt}), efflux ratio, and reported K_{m} of known P-gp substrates [17–93].

P-gp activity. These included particularly immunosuppressors. Cyclosporine A (CSA) was the first immunosuppressor that has been shown to modulate P-gp activity in laboratory models and entered very early into clinical trials for reversal of MDR. The nonimmunosuppressive derivative PSC833 (valspodar) was subsequently developed. This compound showed 10-fold

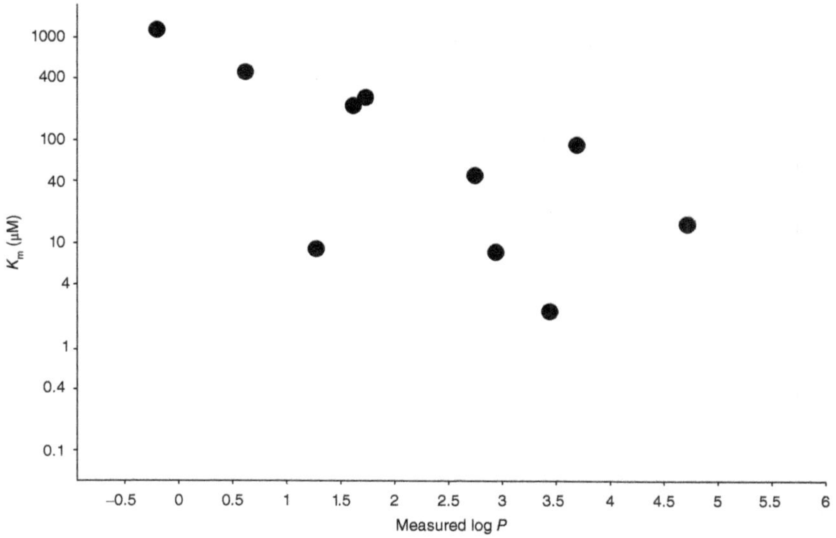

FIGURE 6.4 The relationship between measured log P and K_m values of P-gp substrates. Data obtained from Table 6.1 [17–93].

higher potency in reversal of MDR mediated by P-gp. Other prominent members among the third-generation P-gp modulators are elacridar (GF120918) and tariquidar (XR9576), both containing a dimethoxytetrahydroisoquinoline–ethyl–phenylamine partial structure. A number of new tariquidar analogs have been synthesized and pharmacologically tested, thus supplying good data for a profound structure–activity investigation of this promising class of MDR modulators [96].

Table 6.2 provides a partial list on known P-gp inhibitors with their IC_{50} and K_i values (where available) [24, 37, 50, 54, 55, 58, 59, 97–151]. A physicochemical analysis of P-gp inhibitors and noninhibitors indicated that P-gp inhibitors have twice the number of aromatic bonds (12 vs 6) [117]. Partial least-squares projection to latent structures discriminant analysis (PLS-DA) suggests that log $D_{pH7.4}$, surface area of nonpolar atoms, solvent-accessible surface area of nonpolar atoms and total structure connectivity index are the descriptors with large influence on the model discriminating inhibitors and noninhibitors.

TIP:

It is interesting to note that efflux ratio measured in cell lines such as MDCK/MDR1 or Caco-2 is sometime falsely used to assess the potency of new molecular entities (NMEs) toward P-gp. Indeed, discovery scientists may assume that higher efflux ratios in these cell lines would suggest better potency, that is, lower K_m values and vice versa. As illustrated in Figure 6.5, there is **NO** relationship between compound K_m and its efflux ratio. Therefore, do not jump to conclusions and assume either way. Efflux ratio is a qualitative measure that provides only yes/no answer on whether the compound is a substrate for efflux transporter. To assess if the compound is a good P-gp substrate you need to determine its K_m and other relevant *in vitro* values.

TABLE 6.2 Partial List of Known P-gp Inhibitors and Their *In Vitro* Transporter Substrate Parameters [24, 37, 50, 54, 55, 58, 59, 97–151]

Precipitant	Precipitant Therapeutic Class	Object	K_i (μM)	IC_{50} (μM)	Inhibition (%)
Donepezil	Acetylcholinesterase (AchE) inhibitors	Digoxin		28.6	
Mitomycin	Alkylating agents	Vinblastine		10	
Carvedilol	Alpha/beta adrenergic antagonists	Digoxin		0.15	
Prazosin	Alpha/beta adrenergic antagonists	Calcein AM		240	
Propranolol	Alpha/beta adrenergic antagonists	Digoxin		8.8	
Talinolol	Alpha/beta adrenergic antagonists	Digoxin		157	
Candesartan	Angiotensin receptor blockers	Digoxin		14.7	
Irbesartan	Angiotensin receptor blockers	Digoxin		34	
Losartan	Angiotensin receptor blockers	Digoxin		65	
Telmisartan	Angiotensin receptor blockers	*N*-Methylquinidine		1.6	
Amiodarone	Antiarrhythmics	*N*-Methylquinidine		3.7	
Cymarin	Antiarrhythmics	*N*-Methylquinidine		432	
Digitoxin	Antiarrhythmics	*N*-Methylquinidine		9	
Digoxin	Antiarrhythmics	Edoxaban		51.3	
Dronedarone	Antiarrhythmics	Digoxin		0.7	

Flecainide	Antiarrhythmics	Digoxin	48	
N-Desalkylpropafenone	Antiarrhythmics	Digoxin	21.3	
N-Desethylamiodarone	Antiarrhythmics	Digoxin	1.27	
Propafenone	Antiarrhythmics	Digoxin	6.8	
Quinidine	Antiarrhythmics	Digoxin	0.6	
Montelukast	Antiasthmatics	Digoxin	160	
Azithromycin	Antibiotics	Digoxin	21.8	
Clarithromycin	Antibiotics	Digoxin	4.1	
Erythromycin	Antibiotics	Digoxin	10	
Gemifloxacin	Antibiotics	Erythromycin	123	
Grepafloxacin	Antibiotics	Erythromycin	2266	86.4
Levofloxacin	Antibiotics	Erythromycin	1644	78.9
Rifampin	Antibiotics	Digoxin	169	
Roxithromycin	Antibiotics	Digoxin	15.4	
Sparfloxacin	Antibiotics	Digoxin	529	
Telithromycin	Antibiotics	Digoxin	1.8	
Dipyridamole	Anticoagulants and antiplatelets	Digoxin	40	
Dipyridamole	Anticoagulants and antiplatelets	Digoxin	1.5	
Dipyridamole	Anticoagulants and antiplatelets	Digoxin	4.9	
Dipyridamole	Anticoagulants and antiplatelets	Digoxin	9.7	
Dipyridamole	Anticoagulants and antiplatelets	Digoxin	2.4	
Dipyridamole	Anticoagulants and antiplatelets	Digoxin	10.8	
Dipyridamole	Anticoagulants and antiplatelets	Digoxin	21.5	

(*continued*)

TABLE 6.2 ***(Continued)***

Precipitant	Precipitant Therapeutic Class	Object	K_i (μM)	IC_{50} (μM)	Inhibition (%)
Edoxaban	Anticoagulants and antiplatelets	Digoxin		53.7	
Loperamide	Antidiarrheals	Digoxin		1	
Clotrimazole	Antifungals	Rivaroxaban		13.6	
Itraconazole	Antifungals	Rivaroxaban		0.16	
Ketoconazole	Antifungals	Edoxaban		0.244	
Miconazole	Antifungals	Rivaroxaban		15.1	
Voriconazole	Antifungals	Cyclosporine		2.2	
Artemisinin	Antimalarials	Digoxin		33	
Mefloquine	Antimalarials	*N*-Methylquinidine		72	70
Quinine	Antimalarials	Digoxin		3.3	
N-Desmethyltamoxifen	Antineoplastic hormonal	Rhodamine 123		27	
Tamoxifen	Antineoplastic hormonal	Rhodamine 123		29	
Ivermectin	Antiparasitics	Rhodamine 123		0.1	33
Moxidectin	Antiparasitics	Rhodamine 123		10	38.9
Amisulpride	Antipsychotics	Talinolol		123.5	
Aripiprazole	Antipsychotics	Rhodamine 123		0.78	
Chlorpromazine	Antipsychotics	Digoxin		24	
Chlorprothixene	Antipsychotics	Calcein AM		130	
Clozapine	Antipsychotics	Talinolol		92.78	
Dehydroaripiprazole (OPC-14857)	Antipsychotics	Digoxin		1.3	
Fluphenazine	Antipsychotics	Talinolol		33.88	
Haloperidol	Antipsychotics	Vinblastine		3.6	
Loxapine	Antipsychotics	Digoxin		9.1	
Methotrimeprazine (levomepromazine)	Antipsychotics	Talinolol		23.04	

N-Desmethylclozapine (norclozapine)	Antipsychotics	Talinolol		125.9
N-Desmethylperazine	Antipsychotics	Talinolol		166.1
Paliperidone (9-hydroxyrisperidone)	Antipsychotics	Digoxin		85
Perazine	Antipsychotics	Talinolol		43.26
Perphenazine	Antipsychotics	Talinolol		15.45
Quetiapine	Antipsychotics	Talinolol		0.8
Risperidone	Antipsychotics	Digoxin		27
Sertindole	Antipsychotics	Rhodamine 123		6.5
Thioridazine	Antipsychotics	Calcein AM		45
Oseltamivir	Antivirals	Digoxin		1700
Telaprevir	Antivirals	Vincristine	4.8	7
Midazolam	Benzodiazepines	Digoxin		13
Amlodipine	Calcium channel blockers	Daunorubicin		22
Barnidipine	Calcium channel blockers	Daunorubicin		8.6
Benidipine	Calcium channel blockers	Digoxin		4.96
Bepridil	Calcium channel blockers	Digoxin		3.8
Diltiazem	Calcium channel blockers	Digoxin		4.6
Efonidipine	Calcium channel blockers	Digoxin		13
Felodipine	Calcium channel blockers	Digoxin		4.2
Isradipine	Calcium channel blockers	Digoxin		6
Manidipine	Calcium channel blockers	Daunorubicin		4.6
Mibefradil	Calcium channel blockers	Digoxin		1.2
Nicardipine	Calcium channel blockers	Digoxin		1.1
Nifedipine	Calcium channel blockers	Digoxin		22
Nilvadipine	Calcium channel blockers	Digoxin		18
Nisoldipine	Calcium channel blockers	Digoxin		44.1

(*continued*)

TABLE 6.2 (*Continued*)

Precipitant	Precipitant Therapeutic Class	Object	K_i (μM)	IC_{50} (μM)	Inhibition (%)
Nitrendipine	Calcium channel blockers	Digoxin		6	
Norverapamil	Calcium channel blockers	Digoxin		0.3	
Verapamil	Calcium channel blockers	Edoxaban		0.498	
Ranolazine	Cardiovascular drugs	Digoxin		14.1	
AZD5672	CCR5 receptor antagonists	Digoxin		32	
Linagliptin	Dipeptidyl peptidase-4 inhibitors	Digoxin		9.4	
Spironolactone	Diuretics	Digoxin		6.7	
Macitentan	Endothelin receptor antagonists	Calcein AM		64	
Curcumin	Food products	Rhodamine 123		50.5	
Naringenin	Food products	Talinolol		236	60
Naringin	Food products	Talinolol		2409	
Piperine	Food products	Digoxin		15.5	
Astemizole	H-1 receptor antagonists	Digoxin		0.83	
Azelastine	H-1 receptor antagonists	Daunorubicin		16	
Desmethylazelastine	H-1 receptor antagonists	Daunorubicin		11.8	
Ebastine	H-1 receptor antagonists	Rhodamine 123		1.5	
Dehydroandrographolide	Herbal medications	Digoxin		77.8	71.44
Emodin	Herbal medications	Digoxin		9.42	93.92
Ginsenoside F1	Herbal medications	Digoxin		76.08	55.17
Peruvoside	Herbal medications	*N*-Methylquinidine		214	
Proscillaridin A	Herbal medications	*N*-Methylquinidine		25	
Puerarin	Herbal medications	Digoxin		1.6	
Strophanthidol	Herbal medications	*N*-Methylquinidine		242	

Atorvastatin	HMG CoA reductase inhibitors (Statins)	Digoxin	43.5
Cerivastatin lactone	HMG CoA reductase inhibitors (Statins)	Digoxin	28.2
Lovastatin	HMG CoA reductase inhibitors (Statins)	Digoxin	56.4
Simvastatin	HMG CoA reductase inhibitors (Statins)	Daunorubicin	4.6
Cyclosporine	Immunosuppressants	*N*-Methylquinidine	0.147
Everolimus	Immunosuppressants	Digoxin	3
Tacrolimus	Immunosuppressants	Dabigatran etexilate	0.66
Abacavir dimer C8	Kinase inhibitors	Iodoarylazidoprazosin (IAAP)	0.24
BI 2536	Kinase inhibitors	Calcein AM	5.81
Lapatinib	Kinase inhibitors	Digoxin	1.5
Quizartinib (AC220)	Kinase inhibitors	3,3-Diethyloxacarbocyanine iodide (Dioc2(3))	3
Temsirolimus	Kinase inhibitors	Digoxin	3.5
Memantine	NMDA receptor antagonists	Digoxin	26
HZ08	None	Calcein AM	2.44
Efavirenz	Nonnucleoside reverse transcriptase inhibitors	Calcein AM	123
Etravirine	Nonnucleoside reverse transcriptase inhibitors	Digoxin	16.4
Rilpivirine	Nonnucleoside reverse transcriptase inhibitors	Digoxin	4.48
Celecoxib	NSAIDS	Vinblastine	30
Etoricoxib	NSAIDS	Digoxin	57
Alfentanil	Opioids	Digoxin	112

(continued)

TABLE 6.2 (*Continued*)

Precipitant	Precipitant Therapeutic Class	Object	K_i (μM)	IC_{50} (μM)	Inhibition (%)
Fentanyl	Opioids	Digoxin		6.5	
Fentanyl	Opioids	Digoxin		18.7	
Methadone	Opioids	Rhodamine 123	2.77	2.17	
Sufentanil	Opioids	Digoxin		4.5	
Domperidone	Other antiemetics	Talinolol		45.64	
Reserpine	Other antihypertensives	*N*-Methylquinidine		0.62	
Ubenimex (bestatin)	Other antineoplastics	Digoxin		227	
Indinavir	Protease inhibitors	Digoxin		44	
Nelfinavir	Protease inhibitors	Calcein AM		0.25	
Ritonavir	Protease inhibitors	*N*-Methylquinidine		0.33	
Saquinavir	Protease inhibitors	Calcein AM		0.2	
Lansoprazole	Proton pump inhibitors	Digoxin		62.8	
Omeprazole	Proton pump inhibitors	Digoxin		17.7	
Pantoprazole	Proton pump inhibitors	Digoxin		17.9	
Fluvoxamine	Selective serotonin reuptake inhibitors	Talinolol		146.7	
Troglitazone	Thiazolidinediones	*N*-Methylquinidine		10.44	
Etoposide	Topoisomerase inhibitors	Irinotecan		1185	
Elacridar (GF120918)	Transporter modulators	Digoxin		0.0156	
HM30181	Transporter modulators	Paclitaxel		0.0354	
Laniquidar (R101933)	Transporter modulators	Digoxin		0.443	
Tariquidar (XR9576)	Transporter modulators	Digoxin		0.0129	
Valspodar (PSC 833)	Transporter modulators	Calcein AM		0.013	
Zosuquidar (LY335979)	Transporter modulators	Digoxin		0.019	
Amitriptyline	Tricyclics and tetracyclics	Digoxin		59.1	
Conivaptan	Vasopressin antagonist	Digoxin		16.6	
Tolvaptan	Vasopressin antagonist	Digoxin		15.9	
Vinblastine	Vinca alkaloids	*N*-Methylquinidine		1.29	

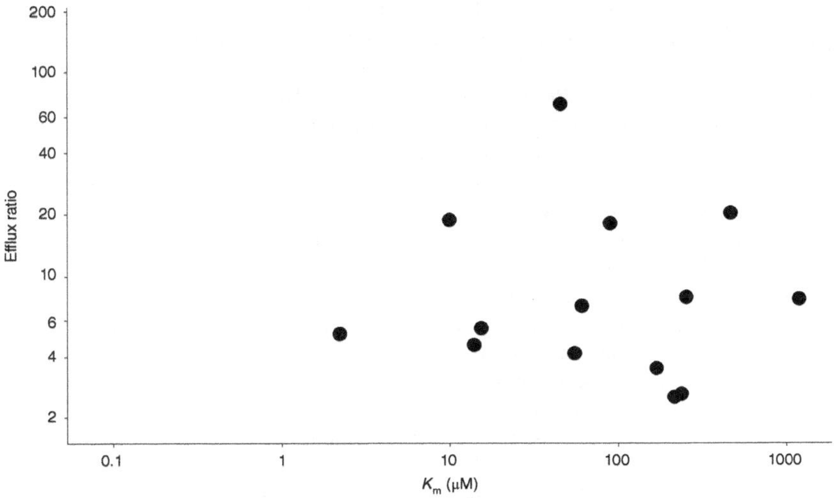

FIGURE 6.5 The relationship between efflux ratio and K_m of P-gp substrates. Data obtained from Table 6.1.

6.2.1.4 Influence of Genetic Polymorphism on In Vitro ***and*** In Vivo ***Activity of P-gp*** P-gp is encoded by *ABCB1* gene that is located on chromosome 7 and spans 28 exons. Genetic polymorphisms in *ABCB1* gene and their association with transport function/expression have been well characterized in the literature. The major SNPs in the *ABCB1* gene discovered are summarized in Table 6.4. Hoffmeyer and coworkers discovered 15 different genetic variations in a Caucasian population. The group indicated that the polymorphism that located in exon 26 (3435C>T) impacts digoxin pharmacokinetics [153]. Indeed, the three most frequent SNPs in the *ABCB1* gene based on three databases are 1236C>T in exon 12, 2677G>T/A in exon 21, and 3435C>T in exon 26. It is interesting to note that similar to other transporters, there are major ethnic differences in the frequency of functionality of 3435C>T genetic variants (Table 6.4). P-gp genetic variations have a significant impact on the pharmacokinetics of P-gp substrates. Hoffmeyer et al. was the first group that suggested the link between *ABCB1* polymorphism and pharmacokinetics of P-gp substrates [153]. While 3435C>T transition is not associated with change in the amino acid encoded, this variant is linked with profound reduction in the P-gp expression. This is ascribed to the potential alteration of P-gp mRNA stability [154], or the impact of timing of cotranslational folding and insertion of P-gp into the membrane [155]. Consistent with reported changes in P-gp expression, the maximum plasma concentration (C_{max}) of digoxin in 3435C>T carriers was 38% higher than that in 3435CC (wild-type) carriers. This indicates the potential increase in the intestinal absorption of P-gp substrates in individuals carrying 3435C>T allele [79]. Subjects with 2677TT and 3435TT genotypes were found to be associated with 70% increase of the area under the blood concentration–time curve (AUC) and 32% decrease of renal clearance of digoxin [156]. Similar findings were reported with other drug molecules that are P-gp substrates [152]. *ABCB1* haplotypes are associated with pharmacokinetics (PK) and efficacy differences of HMG-CoA reductase inhibitors atorvastatin and simvastatin [24, 25]. However, a large number of conflicting results on the clinical relevance has precluded a clear understanding of the *ABCB1* polymorphism in clinical situations. For example, digoxin exposure in Japanese population was significantly lower (35%) in the 3435TT homozygote subjects than in group carrying 3435CC homozygote [157]. Gerloff et al. found no functional influence of 3435C>T or 2677G>T/A on the oral PK of digoxin [27]. The *ABCB1* 1236CC-2677GG-3435CC and 1236TT-2677TT-3435TT haplotypes had no significant effect on the PK of fluvastatin, pravastatin, lovastatin, and rosuvastatin, although these statins are transported by P-gp *in vitro* [28]. Meta-analyses reports on the impact of *ABCB1* polymorphism suggested no effect of 3435C>T on PK parameters of digoxin [29] and cyclosporine [30]. Collectively, clinical studies on PK and efficacy have yielded conflicting information. The lack of distinct observations with the clinical endpoints may be due to several reasons that include the following: genetic mutations translate to minor alterations in function or expression levels, possible induction of *ABCB1* protein expression caused by the P-gp substrates, complex interplay between CYP3A4-dependent metabolism and MDR1-mediated transport in disposition, minimal P-gp contribution to intestinal absorption of substrate drugs at therapeutic doses due to transporter saturation, influence of other uptake and efflux transporters in the overall disposition, the possible haplotype constellations among different ethnicities, and small subject numbers among the studies cited [152]. In addition, the most widely studied drug, digoxin, is a poor marker of P-gp function, as overall systemic exposure increased less than twofold by potent inhibitors.

6.2.1.5 Clinical Significance of P-gp Efflux in Intestinal Absorption: P-gp versus BDDCS

In 2005, Benet and Wu introduced the concept of biopharmaceutics drug disposition classification system (BDDCS) [1]. It classifies molecules into four categories based on their extent of metabolism and solubility. BDDCS Class 1 molecules have high extent of metabolism (>70%) and solubility at physiologically relevant values of pH for the highest strength oral dose. BDDCS Class 2 molecules have high extent of metabolism, yet low solubility. BDDCS Class 3 molecules have low extent of metabolism and high solubility. Class 3 molecules are eliminated as unchanged drugs in urine and/or bile. BDDCS Class 4 molecules have both low extent of metabolism as well as low solubility. It should be emphasized that compounds with high extent of metabolism typically have high extent of absorption and vice versa [158]. As illustrated in Figure 6.7, the impact of P-gp (ABCB1) pharmacogenomics is BDDCS Class dependent [156, 159–193]. The median changes in the AUC between pharmacogenetic variant and wild-type carrier volunteers for BDDCS Class 1, 2, and 3 molecules that are P-gp substrates were 29%, 39%, and 54%, respectively. However, Class 4 molecules had the highest change that is equal to 66%. The results are consistent with the profiles of various BDDCS classes [194, 195]. Completely absorbed drugs or those showing the features of BDDCS Class 1, where drug absorption is not limited by their solubility and/or permeability, P-gp activity is less likely to influence their absorbed fraction (f_a). Therefore, even compounds such as quinidine and verapamil, whose permeability is significantly attenuated by P-gp-mediated efflux activity, are not influenced by P-gp pharmacogenomics, induction, or inhibition *in vivo* [196]. However, these compounds have strong affinity to P-gp and inhibit P-gp-mediated efflux and may show clinically significant drug–drug interactions (DDI) while increasing the intestinal absorption of various P-gp substrates on coadministration [197]. P-gp substrates belonging to Class 2 (e.g., cyclosporine) are fairly permeable and may be well absorbed into systemic circulation. However, low dissolution rate may limit the concentration of drug at the site of absorption thereby leading to less passive diffusion and further preventing saturation of efflux transporter [198]. Thus, saturation of P-gp by providing high drug concentration at the site of absorption or inhibiting P-gp transport may significantly improve absorption of drugs such as cyclosporine.

TIP:

P-gp should be considered when developing sustained release formulations for P-gp substrates. Typically, sustained release formulations target lower intestine such as ileum and upper colon where P-gp transporter expression is higher [199]. Slower drug release from these formulations combined with higher transporter expression would reduce the likelihood of transporter saturation rendering it more effective in lowering intestinal flux and reducing overall absorption and bioavailability. For example, Tubic et al. compared the pharmacokinetics of talinolol, a known P-gp substrates following oral administration of immediate release and controlled release formulation to healthy human volunteers. It is interesting to note that the oral plasma exposure (AUC) of the immediate release tablet was twofold that reported following controlled release tablet at similar strength (Figure 6.6). Furthermore, talinolol plasma concentrations were overall higher following the administration of two controlled release tablets relative to one. Interestingly, the increase in AUC was less than dose proportional between the two controlled release dose strengths. Similar trends were observed when comparing the cumulative amounts of talinolol excreted unchanged among various formulations. In part, this is attributed to lower release rate of the controlled release tablet compared to immediate release tablet. This significantly lowers the likelihood of P-gp saturation leading to lower oral exposure [200]. Take home message, when you need to consider developing sustained releaes formulation to increase half life and improve patient compliance be mindful of P-gp if your compound is a substrate. It may end up lowering the overall exposure of your molecule. Furthermore, doubling the dose may not be associated with proportional increase in exposure, which may further complicate the study design.

Compounds can also be classified in reference to their P-gp activity into high, low, or nonresponders (Table 6.3) [201–204]. For high responders, intestinal absorption is significantly impacted by interindividual differences in P-gp expression levels or by coadministration of P-gp substrates/inhibitors. On the other hand, the impact of changes in P-gp function on the oral absorption of low responders is limited [153]. High responders typically belong to BDDCS Class 3 (fexofenadine and digoxin) or BDDCS Class 4 (paclitaxel). Consistent with their classification, these molecules have inherent low intestinal permeability, which is in part attributed to P-gp-mediated efflux. Coadministration of fexofenadine and ketoconazole was associated with an increase in fexofenadine C_{max} and area under plasma concentration–time curve (AUC) with no change in half-life. This suggests the inhibition of P-gp-mediated efflux along the GIT, since fexofenadine is not subjected to any metabolism [205]. Similar trend of increase in the oral bioavailability of digoxin and paclitaxel was reported with coadministration with P-gp inhibitors [206, 207]. *In vivo* absorption of low responders indinavir, sulindac, and ranitidine have minimal efflux activity due to substrate recognition, while incomplete absorption of nonresponders is attributable to their intrinsic solubility and permeability limitations and P-gp has no significance. Overall, P-gp activity may play a significant role in limiting intestinal permeability of BDDCS Class 2–4 P-gp substrates based on their efflux activity, but not for BDDCS Class 1 compounds irrespective of P-gp activity. It should be emphasized that compounds that have high intestinal permeability are predominantly absorbed in the jejunum that has short transit time (~40 min) and lower P-gp expression. Therefore, P-gp-mediated efflux has limited impact on the absorption of these molecules (BDDCS Class 1). However, the intestinal absorption of compounds with limited solubility/permeability (BDDCS Classes 2–4) is shifted to the ileum that has higher P-gp expression and an intestinal transit time of ~140 min [208]. Overall, these molecules make the moderately absorbed P-gp substrates more susceptible to P-gp-mediated efflux (BDDCS Classes 3–4) (Figure 6.7).

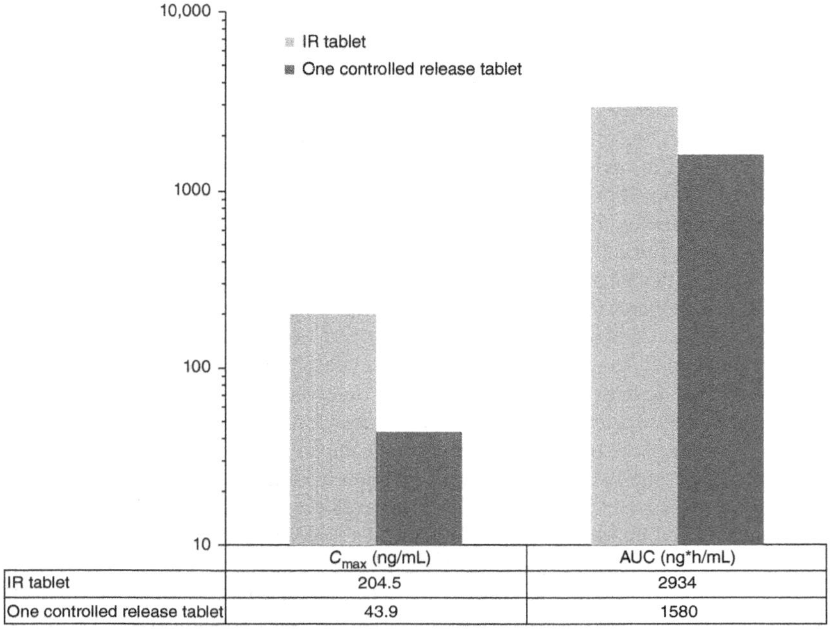

FIGURE 6.6 The C_{max} and AUC of talinolol following single doses of 100-mg talinolol immediate-release, 100-mg talinolol controlled-release. Tubic et al. 2006 [200]. Reproduced with permission of Elsevier.

TABLE 6.3 Classification Scheme of the Drugs Based on the Relevance of P-gp-Mediated Efflux Transport on the Intestinal Drug Absorption *In Vivo* [201–203]

Class	Moderate Permeability			High Permeability
Subclass	High Responders[a]	Low Responders[b]	Nonresponders[c]	
Examples from the data set	Digoxin Paclitaxel Fexofenadine Talinolol	Indinavir Sulindac Ranitidine Lovastatin	Furosemide Hydrochlorthiazide	Cyclosporine Verapamil Quinidine Imipramine Propranolol

[a]Highly influenced by the level of expression and induction/inhibition of P-gp. Criteria ⇒ AQ > 0.5 and HIA enhancement ratio > 1.5.

[b]Reasonably influenced by the level of expression and induction/inhibition of P-gp. Criteria ⇒ 0.1 < AQ <0.5 and 1 < HIA enhancement ratio < 1.5.

[c]Not influenced by P-gp. Criteria ⇒ AQ < 0.1 and HIA enhancement ratio < 1.1.

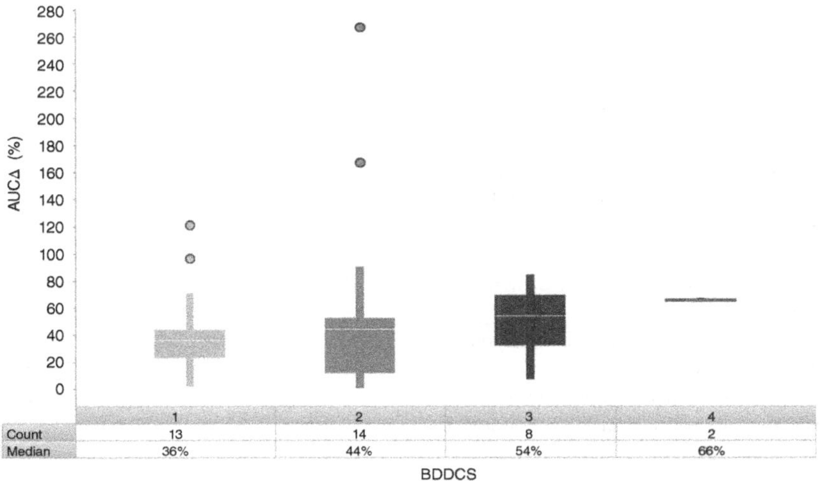

FIGURE 6.7 The median changes in the AUC between pharmacogenetic variant and wild-type carrier (ABCB1) volunteers based on BDDCS Classes [156, 159–193].

6.2.1.6 Prediction of Nonlinear PK of CYP3A4 and P-gp Substrates: The Use of Linearity Index (LIN) CYP3A4 and P-gp are expressed in the enterocyte and their interplay is a major factor that limits the intestinal availability of their substrates following oral dosing [209, 210]. This interplay increases the likelihood of a substrate to be metabolized by CYP3A4 by extending its residence time at the enterocyte through the cycle of absorption and P-gp-mediated efflux. To predict whether intestinal P-gp/CYP3A4 is saturated at a therapeutic dose range, Sugiyama et al. proposed an empirical method that establishes the relationship between dose-normalized AUC and Dose/K_m value. This is termed linearity index (LIN) and is calculated using Equation 6.1 [211]:

$$\text{LIN} = \frac{\text{Dose}}{K_m} \tag{6.1}$$

where K_m is the P-gp/CYP3A4 substrate Michaelis–Menten affinity constant. If the K_m value was not available, drug IC_{50} or K_i was used instead. The normalized AUC/dose ratio was calculated as the ratio of AUC/dose to that at the minimum oral dose reported. If AUC is not available, the C_{max}/dose ratio was calculated using the same methodology. A set of 38 CYP3A4 and/or P-gp substrates was used to build this relationship, from which 16 substrates demonstrated nonlinear pharmacokinetics and the remaining substrates had linear pharmacokinetics. Substrates with large LIN demonstrated nonlinear pharmacokinetics, and vice versa. Interestingly, LIN values ≥2.8 L or 0.77 L were associated with nonlinear pharmacokinetics for CYP3A4 and P-gp substrates, respectively. Using LIN and drug $f_a * f_g$, the group proposed a decision tree for predicting nonlinear pharmacokinetics for CYP3A4 and P-gp substrates. As illustrated in Figure 6.10, compounds with $\text{LIN}_{3A4} < 2.8$ L or $\text{LIN}_{\text{P-gp}} < 0.77$ L, are projected to have linear pharmacokinetics. However, if the drug molecule has a $\text{LIN}_{3A4} > 2.8$ L or $\text{LIN}_{\text{P-gp}} > 0.77$ L and $f_a * f_g > 0.8$, it is projected to have nonlinear pharmacokinetics that is mediated by CYP3A4 or P-gp saturation, respectively.

TIP:

When assessing the potential of non-linear pharmacokinetics mediated by P-gp saturation, DMPK scientists should also consider the potential impact of P-gp saturation on the disposition of co-med that are P-gp substrates. For example, talinolol is a good P-gp substrate ($K_m = 9\ \mu M$, Table 6.1). Increasing its oral dose is associated with non-linear pharmacokinetics that is attributed to P-gp saturation. Westphal et al. investigated the impact of talinolol administered at an oral dose of 100 mg on the oral disposition of digoxin (0.5 mg) in healthy volunteers. Talinolol coadministration significantly increased the oral exposure and C_{max} of digoxin by 23% and 45%, respectively ($p < 0.05$) [212]. This is a profound increase espacially when we consider that digoxin has narrow therapeutic index that would compromise the wellbeing of the patient.

Take home message, don't limit your thinking to only the molecule you are working on. Proactively consider the potential impact on known co-medications disposition and pharmacokinetics. In essence, understanding the disposition of your co-medications is a prerequiste for successful development of your compound in the clinic.

6.2.2 Breast Cancer Resistance Protein (BCRP)

6.2.2.1 BCRP Structure and Distribution BCRP (ABCP/MXR; *ABCG2*), a member of the ABC family of transporters, is considered a half-transporter with six TM domains and one ATP-binding domain at the amino terminus and is believed to homodimerize in order to function [213]. It is composed of 655 amino acids with a M_{wt} of 72 kDa [214]. An atomic model of BCRP was predicted by homology modeling based on the crystal structure of the bacterial multidrug exporter Sav1866, which suggested that BCRP had multiple drug-binding sites [215, 216]. Recently, BCRP expressed in *Pichia pastoris* was purified to >90% homogeneity, and subjected to two-dimensional crystallization with and without mitoxantrone [217]. Analysis of projection maps without mitoxantrone revealed an asymmetric structure with ring-shaped density features appearing more open and less symmetric than the map with mitroxantrone. Cai et al. suggested that the basic residues, His(457) and Arg(465), near the TM helix II of BCRP might be directly involved in substrate binding [218]. BCRP expression can be traced to placenta, kidney, liver, testis, brain, mammary tissue, and intestine [7, 219]. The mRNA level of BCRP is notably higher than other efflux transporters such as P-gp and MRP2 in the human intestine [220]. Oswald et al. quantified BCRP protein using advanced proteomics technology and indicated that BCRP expression along the length of the small intestine does not vary significantly, which is different from trends reported with P-gp transporters (Figure 6.1) [221]. The group also quantified BCRP protein expression in Caco-2 cell lines and interestingly, the expression was below limit of quantification (Figure 6.2). This would undermine the utility of Caco-2 in investigating the impact of BCRP on the intestinal permeability of BCRP substrates [7].

Since BCRP is highly expressed on the apical membrane of enterocytes and effluxing substrates back into the lumen, it has been noted to play an important role as a detoxification efflux transporter and limiting drug absorption in the gastrointestinal tract [222].

6.2.2.2 BCRP Substrates and Inhibitors BCRP exhibits broad substrate specificity and accepts diverse chemical space, as do other ABC transporters. Substrates to BCRP include (Table 6.5) alpha/beta adrenergic antagonists (prazosin), angiotensin converting enzyme inhibitors (temocapril), anthracyclines (daunorubicin), antibiotics (erythromycin, norfloxacin, ciprofloxacin, and nitrofurantoin), anticoagulants (dipyridamole), diuretics (furosemide and chlorothiazide), erectile dysfunction treatments (sildenafil and vardenafil), HMG CoA reductase inhibitors (cerivastatin, pitavastatin, rosuvastatin, atorvastatin, and fluvastatin), kinase inhibitors (gefitinib, lapatinib, and imatinib), NSAIDs (diclofenac and sulfasalazine), and topoisomerase inhibitor (topotecan and irinotecan) [18, 25, 35, 60, 87, 223–249].

As illustrated in Figure 6.8, BCRP substrates are polar with a median PSA of 105.1 Å^2 and HBD and HBA of 3 and 6, respectively. BCRP substrates have typically large molecular weight (median M_{wt} = 435.9 Da) and interestingly are less lipophilic compared to P-gp substrates (median measured log P and log $D_{pH\ 7.4}$ of 0.1 and 0.3, respectively) (Figure 6.3). Substrates of

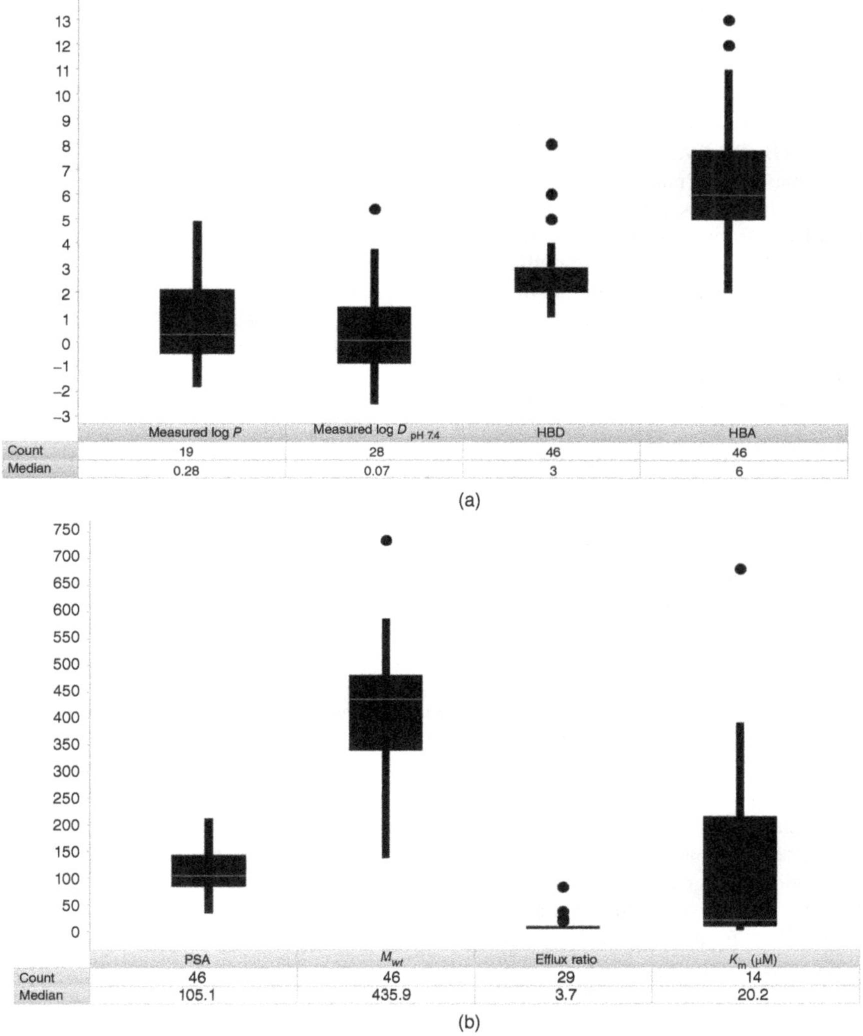

FIGURE 6.8 Physicochemical properties of known BCRP substrates: (a) measured log $D_{pH\ 7.4}$, measured log P, HBD, and HBA of known BCRP substrates; (b) polar surface area (PSA), molecular weight (M_{wt}), efflux ratio, and reported K_m of known BCRP substrates [18, 25, 35, 60, 87, 223–249].

BCRP can be either negatively or positively charged and either unconjugated or conjugated. Several attempts were made to establish SAR for BCRP interaction; however, many analysis methods were based on the data sets of inhibitors [117, 226, 250, 251]. Yoshikawa et al. studied BCRP substrate specificity of 14 camptothecin (CPT) analogs, and noted that CPT analogs that showed ATP-dependent transport in BCRP-overexpressing membrane vesicles possess one – hydroxy or – amino group [252]. Also CPT analogs showed a good correlation between polarity and BCRP-association, where highly polar compounds showed substrate specificity. It is likely that the presence of hydroxyl and amino functional groups facilitates hydrogen bonding with the amino acid residues at the binding site of BCRP. Furthermore, the presence of a negative electrostatic potential area at position 10 for SN-38 and SN-398, but not in SN-22, suggests that CPT analogs with this feature are potential substrates for BCRP [253]. BCRP substrate specificity of a set of pyrrolobenzodiazepine (PBD) derivatives showed a good correlation with the electrostatic potential and aromaticity [254]. PBDs with a greater number of HBA (O and N atoms) and the electronegativity and aromaticity of the C2 substitution show affinity to BCRP. Evidently, BCRP-mediated efflux could be circumvented by limiting C2 aryl substituents and the number of aromatic rings. In general, BCRP substrates share a same set of molecular properties as that of substrates to P-gp and other efflux pumps [255–257]. Some of the BCRP substrates are also effectively effluxed by P-gp. For example, etoposide, irinotecan, and tamoxifen are substrates for both BCRP and P-gp.

Table 6.6 provides a partial list on known BCRP inhibitors with their IC_{50} and K_i values (where available) parameters [20, 37, 43, 60, 117, 130, 139, 140, 143, 148–150, 224, 225, 229, 230, 237, 251, 258–318]. These inhibitors include alpha/beta antagonists (prazosin), androgens (dehydroepiandrosterone sulfate), anthracyclines (doxorubicin and daunorubicin), antibiotics (novobiocin and rifampin), antimalarials (amodiaquine and artemisinin), antiparasitics (oxfendazole and ivermectin), antipsychotics (chlorprothixene, thioridazine, and aripiprazole), antivirals (telaprevir and GS-9191), calcium channel blockers (amlodipine, nifedipine, and nicardipine), corticosteroids (beclomethasone and dexamethasone), diuretics (chlorothiazide), food products (apigenin, flavone, and tangeretin), HIV-integrase strand transfer inhibitors (dolutegravir and raltegravir), HMG coA reductase inhibitors (atorvastatin, cerivastatin, and fluvastatin), immunosuppressants (tacrolimus and cyclosporine), kinase inhibitors (imatinib, erlotinib, and sunitinib), nonnucleoside reverse transcriptase inhibitors (delavirdine and efavirenz), NSAIDs (diclofenac and celecoxib), protease pump inhibitors (omeprazole, lansoprazole, and pantoprazole), and vasodilators (nicergoline and papaverine).

6.2.2.3 Influence of Genetic Polymorphism on* In Vitro *and* In Vivo *Activity of BCRP BCRP is encoded by *ABCG2* and a member of the ABC family [152]. Due to its location at the apical surface of enterocyte, BCRP could limit the intestinal absorption by decreasing uptake of its substrates from the gastrointestinal tract leading to lower oral bioavailability. It could also mediate substrate extrusion in tumor cells and regional organs, which would reduce its pharmacodynamic activity and lead to the development of drug resistance that limits the activity of anticancer agents (Table 6.5). In the literature, more than 80 naturally occurring genetic variations have been identified for the *ABCG2* gene. High interethnic variability was reported with the allele frequency of 421C>A that has low frequency in African Americans (1–4%) and slightly higher in European Americans (8–11%). The allele frequency of 421C>A is predominant in Asian populations with a frequency of about 41% (Table 6.4). Patients with 421CA genotype compared with 421CC carriers had a marked increase of plasma exposure of sulfasalazine (AUC $\Delta = 2.4$-fold). It is interesting to note that BCRP protein and mRNA expression in the intestine did not significantly differ between the populations [319]. In addition, investigation into the impact of genetic polymorphisms of *ABCG2* yielded significant differences in AUC of sulfasalazine among groups with 421CC, 421CA, and 421AA genotypes ($AUC_{0\rightarrow 48h}$ 171, 330, and 592 μg h/mL, respectively) [319, 320]. Collectively, sulfasalazine was suggested to be a useful probe substrate for evaluating the impact of BCRP on intestinal

TABLE 6.4 Nonsynonymous Nucleotide Polymorphisms in the *ABCB1* Gene and Known Ethnic Differences in the Frequency of Genetic Variants

Transporters	SNP	Amino Acid Exchange	rs Number	Chromosome: Base	% Allelic Frequency (*N*)			Database[a,b,c]
					European American	African American	Asian	
MDR1 (ABCB1)	61A > G	Asn21Asp	rs9282564	7:87229440	8.0(200)	2.5(200)	1.7 (60)	UCSF
					11.5(226)	2.6(114)	NA	HapMap
					9.8(7020)	2.7(3738)	NA	EVS
	1236C > T	Gly412Gly	rs1128503	7:87179601	45.9(194)	20.9(196)	68.5 (54)	UCSF
					45.1(226)	20.2(114)	NA	HapMap
					42.9(7020)	21.3(3738)	NA	EVS
	2677G > T/A	Ala893Ser/Thr	rs2032582	7:87160618	46.4(196)	10.0(200)	45.0 (60)	UCSF
					46.9(226)	10.5(114)	NA	HapMap
					42.8(7020)	11.0(3738)	NA	EVS
	3421T > A	Ser1141Thr	rs2229107	7:87138659	0 (198)	11.1(198)	0 (60)	UCSF
					0 (128)	5.3(114)	NA	HapMap
					0 (7020)	5.9(3738)	NA	EVS
	3435C > T	Ile1145Ile	rs1045642	7:87138645	56.1(198)	20.2(198)	40.0 (60)	UCSF
					57.1(226)	20.5(112)	NA	HapMap
					52.2(7020)	22.8(3738)	NA	EVS
BCRP (ABCG2)	34G > A	Val12Met	rs2231137	4:89061114	6.6(152)	7.7(156)	27.2 (114)	UCSF
					1.5(130)	NA	NA	HapMap
					3.7(7020)	5.9(3738)	NA	EVS
	421C > A	Gln141Lys	rs2231142	4:89052323	8.1(160)	1.3(160)	40.8 (120)	UCSF
					11.1(226)	4.4(114)	NA	HapMap
					10.9(7020)	3.1(3738)	NA	EVS

NA, not available.

[a]HapMap, The International HapMap Project.

[b]UCSF, The UCSF Pharmacogenetics of Membrane Transporters (PMT) Project [3].

[c]EVS: Exome Variant Server, NHLBI Exome Sequencing Project (ESP), Seattle, WA [3] [accessed 12-2011].

Source: Lai et al. 2012. [152]. Reproduced with permission of Taylor & Francis.

absorption. However, a recent study in Chinese subjects ($N = 36$) found that high intersubject variability in systemic exposure of sulfasalazine was observed both across and within genotypic cohorts, with AUC ranging 81-, 66-, and 21-fold for the CC, CA, and AA cohorts, respectively. Although mean changes for the CA and AA cohorts were decreased by 48% and increased by 95%, respectively, the changes were not statistically significant (and inconsistent in direction). This study was further complicated by the high frequency of 34G>A mutations (14 of 36 subjects). The authors conclude that sulfasalazine is not a sufficiently sensitive probe substrate for assessing *ABCG2* genetic variation [321]. Further investigation is needed with respect to sulfasalazine selectivity as a probe for BCRP function. BCRP polymorphism significantly affects the pharmacokinetics of several HMG-CoA reductase inhibitors (statins), including atrovastatin, rosuvastatin, fluvastatin, and simvastatin lactone but has no significant effect on pravastatin or simvastatin acid (Table 6.5) [240, 322–325]. For example, rosuvastatin AUC was 100% and 144% greater in the c.421AA genotype population than in those with c.421CA and the c.421CC genotypes, respectively. Although, few clinical studies have been reported on the role of BCRP in the intestinal absorption, several studies using BCRP knockout mice suggest significant impact [222, 232, 326, 327]. Furthermore, the *ABCG2* polymorphism has a significant impact on the pharmacokinetics, pharmacodynamics, and toxicity for several anticancer drug substrates [152]. For example, the nonsynonymous 421C>A SNP in exon 5 of *ABCG2* gene is associated with decreased transporter expression and/or drug transport activity [152]. Patients with 421C>A polymorphism are associated with increased risk of gefitinib-induced diarrhea [152].

6.2.3 Sensitivity Analysis

In 2010, Sugiyama et al. performed a sensitivity analysis to assess the impact of K_m, V_{max}, effective intestinal permeability (P_{eff}), and solubility on dose-dependent alterations in human $f_a * f_g$ using Talinolol inputs as a frame of reference (Figure 6.9a–d) [328]. As the potency toward P-gp increases (decrease in the K_m), the degree of its nonlinearity increases. However, increasing the transporter capacity (illustrated as an increase in the V_{max}) is associated with a shift in the dose nonlinearity to the right. Interestingly, increasing P_{eff} is associated with a decrease in the nonlinearity, which is shown with compounds that are BDDCS Class 3 and 4. Nonlinearity is apparent with compounds with lower solubility, which is consistent with BDDCS Class 2 and 4. These principles can also be extrapolated to the impact of other efflux transporters such as BCRP and MRP2 on the absorption NMEs (Figure 6.10).

TIP:

In our scientific community we have scientists who believe that we should not worry about the potential impact of P-gp on the intestinal dispositon of compounds especially if the oral dose is higher than 100 mg. They believe at this high dose, the intestinal drug concentration is high and would lead to saturation of P-gp, which would reduce its potential impact on drug intestinal flux and absorption. They tend to generalize this theme and use it as a guiding principle to ignore the potential impact of P-gp on the absorption of NMEs. This is only valid if the NME has high P_{eff} and solubility (BDDCS Class 1) as shown in Figure 6.7, and in addition, if the compound is not a potent substrate for P-gp. Therefore, before you jump to conclusions and decide to ignore P-gp, do consider the overall package and use advanced PBPK tools (GastroPlus or SimCYP) to appropriately predict the potential impact of P-gp on your compound absorption and disposition. It should also be emphasized that when P-gp is saturated the impact on absorption will also span co-medications that are substrates for this transporter similar to the case of talinolol and digoxin discussed earlier [212]. Therefore, understanding co-medications disposition and transporters that infleunce their profiles is key for successful development of NMEs.

TABLE 6.5 Partial List of Known BCRP Substrates, Their *In Silico* and Measured Physicochemical Properties, Renal Recovery (%), *In Vitro* Transporter Substrate Parameters [18, 25, 35, 60, 87, 223–249]

Generic Name	BDDCS Class	% Excreted Unchanged in Urine	M_{wt}	Measured log P	Measured log $D_{pH\,74}$	c log P	HBA	HBD	PSA	Efflux Ratio	K_m (μM)	V_{max}	V_{max}/K_m
Albendazole sulfoxide	2	1	281.3			3.46	3	2	65.6	1.8			
Allopurinol	2	12	136.1	−0.55	−0.55	0.63	3	2	70.5				
Atorvastatin calcium	2	1	558.7			4.46	5	4	119.1	1.7			
Bupropion	1	0.5	239.7		3.27	3.21	2	1	32.8		3	30 pmol/min/mg	10 μL/min/mg
Cerivastatin	1	24	459.6			3.56	6	3	105.0				
Chlorothiazide	4	92	295.7	−0.24	−1.1	−1.0	6	2	126.9	36	334.6		
Cimetidine	3	62	252.3	0.4	0.33	0.19	5	3	81.7	3.4			
Ciprofloxacin	4	65	331.3	0.28	−1.21	−0.73	6	2	75.1	2.9			
Daunorubicin	2	0.5	527.5	1.83	1.83	0.84	11	5	200.3	2.9			
Daunorubicin	2	0.5	527.5	1.83	1.83	0.84	11	5	200.3	4.9			
Diclofenac	1	0.5	296.2	4.51	1.13	4.73	3	2	54.8	1.25			
Dipyridamole	2	0.1	504.6		3.71	1.49	12	4	134.7				
Erlotinib hydrochloride	2	0.3	393.4			4.34	7	1	70.3				
Erythromycin (base)	3	5	733.9	3.06		1.61	13	5	203.3	8.1			
Fluvastatin sodium	1	1.5	411.5	3.8		4.05	4	3	86.5	4.9			
Furosemide	4	66	330.7	2.03	−1.54	1.9	5	3	130.0	17.5			
Gefitinib	2	2	446.9	4.85		5.6	7	1	62.6				

(continued)

TABLE 6.5 (*Continued*)

Generic Name	BDDCS Class	% Excreted Unchanged in Urine	M_{wt}	Measured log P	Measured log $D_{pH\,74}$	$c\log P$	HBA	HBD	PSA	Efflux Ratio	K_m (μM)	V_{max}	V_{max}/K_m
Glyburide	2	0	494.0		1.41	4.24	5	3	126.9		13.07	101 pmol/min/mg	7.7 μL/min/mg
Glyburide	2	0	494.0		1.41	4.24	5	3	126.9	9.3			
Imatinib mesylate	2	5	493.6			4.53	7	2	79.6	2.4			
Irinotecan	1	16.7	586.7			2.73	6	1	108.4	3.7			
Lamivudine	3	67	229.3	−0.93		−1.46	5	2	88.1	3	216.5	340 pmol/min/10^6 cells	1.6 μL/min/10^6 cells
Lapatinib ditosylate	2	1	581.1			5.97	7	2	104.3	2.6			
Methotrexate	3	81	454.4	−1.85	−2.52	−0.53	12	5	211.7		681	2384 pmol/min/mg	3.5 μL/min/mg
Methotrexate	3	81	454.4	−1.85	−2.52	−0.53	12	5	211.7	6.78			
Mitoxantrone	3	7	444.5		0.7	2.3	10	8	185.1		18		
Mitoxantrone	3	7	444.5		0.7	2.3	10	8	185.1	6.1			
Nitrofurantoin	4	47	238.2	−0.47	−0.19	−0.47	5	1	118.7		69.4	14 nmol/cm^2/h	0.2 μL/cm^2/h
Nitrofurantoin	4	47	238.2	−0.47	−0.19	−0.47	5	1	118.7	24.1			
Norfloxacin	4	29	319.3	−1.03	−2	−0.78	6	2	75.1	3			
Ofloxacin	3	64	361.4	−0.39	−0.40	−0.51	7	1	70.8	1.4			
Pemetrexed disodium	3	80	427.4			−1.17	9	6	198.6		390	4725 pmol/min/mg	12.1 μL/min/mg

Pitavastatin	2		421.5		1.50	3.59	5	3	96.2		5.73	1106 pmol/min/mg	193 μL/min/mg
Pravastatin	3	20	424.5	2.18	−0.23	2.05	6	4	135.6				
Prazosin	1	0.5	383.4		0.79	2.03	7	1	96.1	11.1			
Raltegravir potassium	2	9	444.4		0.45	1.16	7	3	150.5				
Rosuvastatin calcium	3	5	481.5		−0.89	1.9	8	3	144.8		2.02	304 pmol/min/mg	150.5 μL/min/mg
Rosuvastatin calcium	3	5	481.5		−0.89	1.9	8	3	144.8	83.2			
Sildenafil	1	7.5	474.6			1.98	7	1	105.2	3.59	22.4	37 pmol/cm^2/min	1.7 μL/min/cm^2
Sorafenib tosylate	2	0	464.8			5.46	3	3	99.3	2.7			
Sulfasalazine	2	1.5	398.4		−0.78	3.88	8	3	148.2		0.7	52 nmol/min/mg	74.3 μL/min/mg
Sunitinib malate	1	4	398.5		5.4	3	3	3	80.5	1.9			
Temocapril	1	1.5	476.6			2.1	5	2	101.9	5.56			
Topotecan	3	40	421.5			0.73	6	2	102.8		213.28	127 pmol/cm^2/min	0.6 μL/cm^2/min
Topotecan	3	40	421.5			0.73	6	2	102.8	6.9			
Vardenafil	1	4	488.6			2.23	7	1	104.9	4.12	14.3	20 pmol/cm^2/min	1.54 μL/min/cm^2

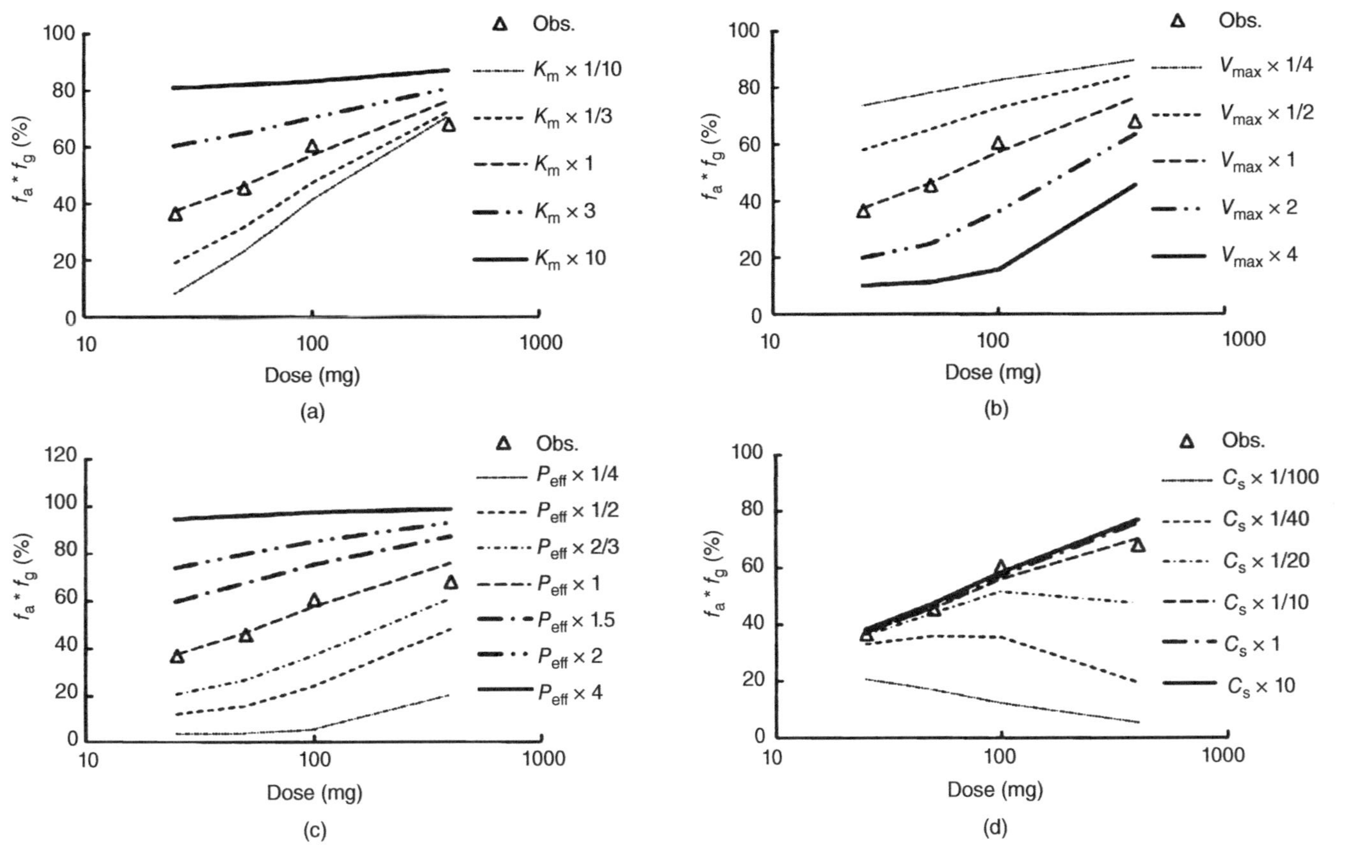

FIGURE 6.9 The relationship between $F_a * F_g$ and the talinolol dose in humans. The triangles represent actual observed $F_a * F_g$ at various oral doses. The lines represent fitted or simulated results based on changes in input parameters. Sensitivity analysis was run for (a) K_m, (b) V_{max}, (c) P_{eff}, and (d) solubility. Tachibana et al. 2010 [328]. Reproduced with permission of Springer.

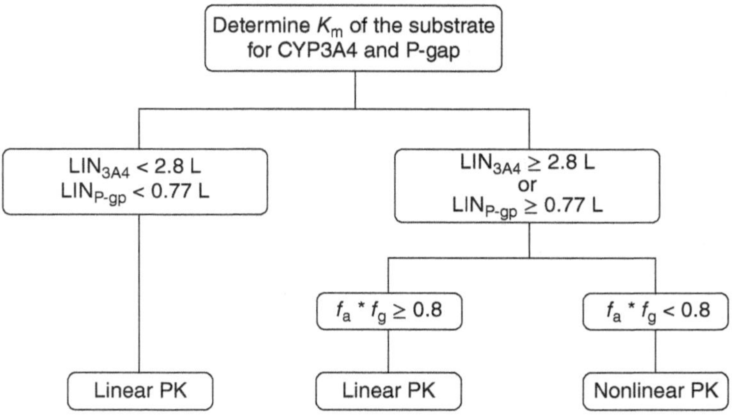

FIGURE 6.10 A proposed decision tree for predicting nonlinear pharmacokinetics of P-gp/CYP3A4 substrates. Tachibana et al. 2012 [211]. Reproduced with permission of Springer.

6.2.4 FDA Guidance on Assessing if a Drug is a P-gp or BCRP Substrate

BCRP and P-gp are efflux transporters that line the apical membrane of intestinal enterocyte and play a key role in limiting the oral absorption of xenobiotics. FDA guidance mandates the need to test if NMEs are substrates for these transporters using *in vitro* tools such as Caco-2 or overexpressed cell lines such MDCK/MDR1 or MDCK/BCRP. If the NME is deemed to be a substrate for these transporters (efflux ratio ≥ 2), it should be evaluated in human *in vivo* by assessing the impact of known inhibitors of these transporters on their absorption and oral exposure. Consistent with the sensitivity analysis discussed earlier (Figure 6.9), NME that is both highly permeable and soluble (BDDCS Class 1), it may be appropriate to waive the need to conduct an *in vivo* evaluation in humans using P-gp/BCRP inhibitors.

6.2.5 FDA Guidance on Assessing if a Drug is a P-gp or BCRP Inhibitor

Determining if the NME as an inhibitor for efflux transporters such as P-gp and BCRP is crucial, in particular, if the molecule will be coadministered with drugs that are substrates for P-gp or BCRP and have narrow therapeutic index. Examples include digoxin and rosuvastatin that are substrates for P-gp and BCRP, respectively (Tables 6.1 and 6.6). First, the assessment is performed in *in vitro* bidirectional transport assay (e.g., Caco-2 or MDCK/MDR1) using a probe P-gp substrate such as digoxin. If the net flux ratio of the probe substrate decreases with an increase in the NME concentration, this may suggest that the NME is a P-gp substrate. This would require a determination of the inhibitor K_i or IC_{50}. The NME will require an *in vivo* drug interaction study with a P-gp substrate (e.g., digoxin) if one of the following two conditions is met:

$[I]_1/IC_{50}$ (or K_i) ≥ 0.1, where $[I]_1$ is the maximum steady-state total concentration (C_{max}) observed following oral dosing of the highest clinical dose.

$[I]_2/IC_{50}$ (or K_i) ≥ 10, where $[I]_2$ is the intestinal concentration of the drug calculated by dividing the oral dose by 250 mL. Note that the concentration should be expressed in millimolar.

TABLE 6.6 Partial List of Known BCRP Inhibitors and Their *In Vitro* Transporter Substrate Parameters [20, 37, 43, 60, 117, 130, 139, 140, 143, 148–150, 224, 225, 229, 230, 237, 251, 258–318]

Precipitant	Precipitant Therapeutic Class	Object	K_i (µM)	IC_{50} (µM)
2,4-Dinitrophenyl-*S*-glucuronide	None	Estrone-3-sulfate		
3-Hydroxyflavone	Food products	Mitoxantrone		8.1
4-Methylumbelliferone glucuronide	None	Estrone-3-sulfate		
4-Methylumbelliferone sulfate	None	Estrone-3-sulfate		6
6-(1,1-Dimethylallyl)chrysin	None	Mitoxantrone		0.78
6,7-Dihydroxybergamottin (DHB)	Food products	Dasatinib		5.2
6,8-Digeranylchrysin	None	Mitoxantrone		2.1
6,8-Diprenylnaringenin	Food products	Methotrexate		0.41
6,8-Diprenylnaringenin	Food products	Mitoxantrone		
6-Geranylchrysin	None	Mitoxantrone		1
6-Prenylchrysin	None	Mitoxantrone		0.29
6-Prenylnaringenin	Food products	Methotrexate		0.538
6-Prenylnaringenin	Food products	Mitoxantrone		
7-Ethyl-10-hydroxycamptothecin (SN-38)	Topoisomerase inhibitors	Estrone-3-sulfate		1.6
7-Ethyl-10-hydroxycamptothecin (SN-38)	Topoisomerase inhibitors	Estrone-3-sulfate		
7-Ethyl-10-hydroxycamptothecin (SN-38)	Topoisomerase inhibitors	Methotrexate		
7-Hydroxyflavone	Food products	Mitoxantrone		7.1
8-(1,1-Dimethylallyl)chrysin	None	Mitoxantrone		1.4
8-Prenylchrysin	None	Mitoxantrone		0.89
8-Prenylnaringenin	Food products	Methotrexate		0.043
8-Prenylnaringenin	Food products	Mitoxantrone		
Abacavir	Nucleoside reverse transcriptase inhibitors (NRTIs)	Pheophorbide A		385

Acemetacin	NSAIDS	Methotrexate		
Afatinib	Kinase inhibitors	Estrone-3-sulfate		0.75
Amlodipine	Calcium channel blockers	Mitoxantrone		15.68
Amodiaquine	Antimalarials	Mitoxantrone		
Amprenavir	Protease inhibitors	Pheophorbide A		181
Apigenin	Food products	Mitoxantrone		16
Aripiprazole	Antipsychotics	Methotrexate		3.5
Artemisinin	Antimalarials	Estrone sulfate		
Atazanavir	Protease inhibitors	Pheophorbide A		69.1
Atorvastatin	HMG CoA reductase inhibitors (statins)	Estrone-3-sulfate	14.3	
Atovaquone	Antimalarials	Estrone sulfate		0.23
Axitinib	Kinase inhibitors	Topotecan		4.4
AZD9056	Treatments of pain and inflammation	Estrone-3-sulfate		32
Barasertib (AZD1152)	Kinase inhibitors	Methotrexate		
Beclomethasone	Corticosteroids	2-Amino-1-methyl-6-phenylimidazo(4,5-*B*) pyridine (Phip)		
Benzbromarone	Antigout and uricosuric agents	Mitoxantrone		
Benzoflavone	None	Purpurin-18		
Bergamottin (GF-I-2)	Food products	Dasatinib		3.19
BI 2536	Kinase inhibitors	Pheophorbide A		44.92
Biochanin A	Food products	Mitoxantrone		
Bitter melon extract	Food products	Doxorubicin		
Boceprevir	Antivirals	Methotrexate		81
Bohemine	Kinase inhibitors	Glyburide (glibenclamide)		
Cabazitaxel	Taxanes	Methotrexate		41.8

(continued)

TABLE 6.6 (*Continued*)

Precipitant	Precipitant Therapeutic Class	Object	K_i (μM)	IC_{50} (μM)
Cannabidiol	Cannabinoids	Mitoxantrone		
Celecoxib	NSAIDS	Calcein AM		24.1
Cerivastatin	HMG CoA reductase inhibitors (statins)	Estrone-3-sulfate	18.1	
Chloroquine	Antimalarials	Estrone sulfate		
Chlorothiazide	Diuretics	Estrone-3-sulfate		212.3
Chlorprothixene	Antipsychotics	Mitoxantrone		42
Cholic acid	Other	Estradiol-17-beta-glucuronide		
Chrysin	Herbal medications	Mitoxantrone		4.6
Cobicistat (GS-9350)	None	Hoechst 33342		59
CP-778875	PPAR agonists	Pitavastatin		
Crenolanib	Kinase inhibitors	Pheophorbide A		
Curcumin	Food products	Methotrexate		1.04
Cyclosporine	Immunosuppressants	Estrone-3-sulfate	6.7	
Dantrolene	Muscle relaxants	Estrone-3-sulfate		
Darunavir	Protease inhibitors	Estradiol 3-sulfate		53
Daunorubicin	Anthracyclines	Methotrexate		28
Degarelix	Cancer treatments	Estrone-3-sulfate		48
Dehydroaripiprazole (OPC-14857)	Antipsychotics	Methotrexate		0.52
Dehydroepiandrosterone Sulfate	Androgens	Estrone-3-sulfate		55
Delavirdine	Nonnucleoside reverse transcriptase inhibitors (NNRTIs)	Pheophorbide A		18.7
Dexamethasone	Corticosteroids	2-Amino-1-methyl-6-phenylimidazo (4,5-*B*)pyridine (Phip)		

Diclofenac	NSAIDS	Methotrexate		71
Dihydroartemisinin	Antimalarials	Estrone sulfate		
Dipyridamole	Anticoagulants and antiplatelets	Epimedin A		
Dolutegravir	HIV-integrase strand transfer inhibitors	Cimetidine		
Domperidone	Other antiemetics	Mitoxantrone		11.15
Dovitinib	Kinase inhibitors	Pheophorbide A		10.3
Doxorubicin	Anthracyclines	Estrone-3-sulfate		
Dronabinol	Cannabinoids	Imatinib		
Efavirenz	Nonnucleoside reverse transcriptase inhibitors (NNRTIs)	Pheophorbide A		20.6
Elacridar (GF120918)	Transporter modulators	Mitoxantrone		0.31
Eliglustat	Glucosylceramide synthase inhibitors	Prazosin		126
Eltrombopag	Miscellaneous agents	Rosuvastatin		
Elvitegravir (GS-9137)	HIV-integrase strand transfer inhibitors	Pheophorbide A		88.9
Empagliflozin	Sodium-dependent glucose cotransporter 2 (SGLT2) inhibitors	Probe substrate (not provided)		114
Erlotinib	Kinase inhibitors	Estrone-3-sulfate	0.15	0.13
Estradiol 3-sulfate	Estrogens	Estrone-3-sulfate		14
Estrone	None	Estrone-3-sulfate		
Estrone-3-sulfate	Estrogens	Daidzein		
Flavone	Food products	Mitoxantrone		2.8
Flavopiridol (Alvocidib)	Kinase inhibitors	Mitoxantrone		

(continued)

TABLE 6.6 (*Continued*)

Precipitant	Precipitant Therapeutic Class	Object	K_i (μM)	IC_{50} (μM)
Fluvastatin	HMG CoA reductase inhibitors (statins)	Estrone-3-sulfate	5.43	
Fosinopril	Angiotensin converting enzyme (ACE) inhibitors	Mitoxantrone		11.56
Fumitremorgin C (FTC)	None	Methotrexate	0.3	
Galangin	Food products	Mitoxantrone		19
Gefitinib	Kinase inhibitors	Estrone-3-sulfate		1.01
Genistein	Food products	Mitoxantrone		24
Gimatecan	Topoisomerase inhibitors	Methotrexate		
Glycocholic acid	None	Estradiol-17-beta-glucuronide		
GS-9191	Antivirals	Hoechst 33342		
GSK1292263	Other antilipemics	Cimetidine		
GW 583340	Kinase inhibitors	Methotrexate		
GW2974	Kinase inhibitors	Methotrexate		
Hydroxysuvorexant	Hypnotics–sedatives	Methotrexate		
Imatinib	Kinase inhibitors	Topotecan		
Isoscutellarin	Food products	Methotrexate		20
Isoxanthohumol	Food products	Methotrexate		0.077
Ivermectin	Antiparasitics	Estrone-3-sulfate	1.4	
Kaempferol	Food products	Dasatinib		26.6
Ko143	Transporter modulators	Pheophorbide A		0.01
Lamellarin O	None	Mitoxantrone		4.7
Lansoprazole	Proton pump inhibitors	Methotrexate		14.4
Lapatinib	Kinase inhibitors	Cimetidine		0.025
Ledipasvir	Antivirals	Hoechst 33342		
Linagliptin	Dipeptidyl peptidase-4 inhibitors	Estrone-3-sulfate		

Precipitant	Precipitant Therapeutic Class	Object	K_i (μM)	IC_{50} (μM)
Loperamide	Antidiarrheals	Mitoxantrone		100
Lopinavir	Protease inhibitors	Pheophorbide A		7.66
LSN355984	Transporter modulators	Dipyridamole		
Lumefantrine	Antimalarials	Estrone sulfate		
LY294002	Kinase inhibitors	Mitoxantrone		
Macitentan	Endothelin receptor antagonists	Pheophorbide A		13.2
Macitentan	Endothelin receptor antagonists	Pheophorbide A		75
Maraviroc	CCR5 receptor antagonists	2-Ethylidene-1,5-dimethyl-3,3-diphenylpyrrolidine (EDDP)		
MBLI-87	None	Mitoxantrone	0.141	
Mefloquine	Antimalarials	Estrone sulfate		18
Melatonin	Other	Mitoxantrone		
Methylprednisolone	Corticosteroids	2-Amino-1-methyl-6-phenylimidazo(4,5-*B*) pyridine (Phip)		
Milk thistle (silymarin; silybum marianum)	Herbal medications	Rosuvastatin	97.9	46
Mitoxantrone	Other antineoplastics	Estrone-3-sulfate		61
Mitoxantrone	Other antineoplastics	Methotrexate		
MK571	None	Estrone-3-sulfate		
MK572	None	Estrone-3-sulfate		
MK573	None	Genistein		
MK574	None	Sulfasalazine		
Monohydroxy-Vorapaxar (M20)	Anticoagulants and antiplatelets	Methotrexate		1.6
Naringenin	Food products	Mitoxantrone		37

(*continued*)

TABLE 6.6 (*Continued*)

Precipitant	Precipitant Therapeutic Class	Object	K_i (μM)	IC_{50} (μM)
Nelfinavir	Protease inhibitors	Pheophorbide A		13.5
Netupitant	Antiemetics	Estrone-3-sulfate		6
Netupitant N-oxide metabolite M2	Antiemetics	Estrone-3-sulfate		22.6
Nevirapine	Nonnucleoside reverse transcriptase inhibitors (NNRTIs)	Pheophorbide A		
Nicardipine	Calcium channel blockers	Methotrexate		12
Nicergoline	Vasodilators	Mitoxantrone		37.06
Nifedipine	Calcium channel blockers	Methotrexate		60
Nilotinib	Kinase inhibitors	Methotrexate	0.69	
Nisoldipine	Calcium channel blockers	Mitoxantrone		25.93
Nobiletin	Food products	Dasatinib		1.04
Novobiocin	Antibiotics	Estrone-3-sulfate		0.063
NSC73306	Cancer treatments	Purpurin-18		
Olodaterol	Alpha/beta adrenergic agonists	Estrone-3-sulfate		
Olomoucine II	Kinase inhibitors	Glyburide (glibenclamide)		
Omeprazole	Proton pump inhibitors	Methotrexate		13
Oxfendazole	Antiparasitics	2-Amino-1-methyl-6-phenylimidazo (4,5-*B*)pyridine (Phip)		
Paclitaxel	Taxanes	Methotrexate		
Pantoprazole	Proton pump inhibitors	Methotrexate		5.5
Papaverine	Vasodilators	Mitoxantrone		55.08
Pheophorbide A	None	Methotrexate		
Pioglitazone	Thiazolidinediones	Pheophorbide A	28.6	

Pitavastatin	HMG CoA reductase inhibitors (statins)	Estrone-3-sulfate	2.92	
p-Nitrophenyl sulfate	None	Estrone-3-sulfate		53
Ponatinib	Kinase inhibitors	Cladribine		0.013
Pravastatin	HMG CoA reductase inhibitors (statins)	Estrone-3-sulfate		
Prazosin	Alpha/beta adrenergic antagonists	Methotrexate		10
Prochlorperazine	Antipsychotics	Mitoxantrone		12.44
Proguanil	Antimalarials	Estrone sulfate		118
Promethazine	Antipsychotics	Mitoxantrone		89.87
Purvalanol A	Kinase inhibitors	Glyburide (glibenclamide)		
Quercetin	Food products	Methotrexate		0.6
Quinine	Antimalarials	Estrone sulfate		
Quizartinib (AC220)	Kinase inhibitors	Pheophorbide A		0.5
Rabeprazole	Proton pump inhibitors	Methotrexate		8.5
Raltegravir	HIV-integrase strand transfer inhibitors	Methotrexate		
Regorafenib	Kinase inhibitors	Topotecan		0.0447
Reserpine	Other antihypertensives	Mitoxantrone		
Rifampin	Antibiotics	Rosuvastatin		14
Rilpivirine	Nonnucleoside reverse transcriptase inhibitors (NNRTIs)	Pheophorbide A		1.5
Riociguat	Soluble guanylate cyclase (sGC) stimulators	Pheophorbide A		46.2
Ritonavir	Protease inhibitors	Prazosin		
Rosiglitazone	Thiazolidinediones	Pheophorbide A	24	
Rosuvastatin	HMG CoA reductase inhibitors (statins)	Estrone-3-sulfate	15.4	

(continued)

TABLE 6.6 (*Continued*)

Precipitant	Precipitant Therapeutic Class	Object	K_i (μM)	IC_{50} (μM)
Ruxolitinib	Kinase inhibitors	2-Amino-1-methyl-6-phenylimidazo (4,5-*B*)pyridine (Phip)		48
Saquinavir	Protease inhibitors	Pheophorbide A		27.4
Scutellarin	Food products	Methotrexate		10
Seliciclib (R-roscovitine)	Kinase inhibitors	Glyburide (glibenclamide)		
Simvastatin acid	HMG CoA reductase inhibitors (statins)	Estrone-3-sulfate	18	
Sofosbuvir	Antivirals	Estrone-3-sulfate		
Sorafenib	Kinase inhibitors	Mitoxantrone		
Sulfasalazine	NSAIDS	Estrone-3-sulfate		0.46
Sunitinib	Kinase inhibitors	Estrone-3-sulfate	0.32	
Suvorexant	Hypnotics–sedatives	Methotrexate		
Tacrolimus	Immunosuppressants	Methotrexate		6
Tangeretin	Food products	Dasatinib		1.19
Tariquidar (XR9576)	Transporter modulators	Purpurin-18		
Taurocholic acid	None	Estrone-3-sulfate		
Taurolithocholate	Other	Estrone-3-sulfate		
Taurolithocholate sulfate	Other	Estrone-3-sulfate		37
Tectochrysin	None	Mitoxantrone		3
Tedizolid	Antibiotics	Genistein		51.1
Telaprevir	Antivirals	Methotrexate	94.8	31
Telatinib	Kinase inhibitors	Estradiol-17-beta-glucuronide		
Telmisartan	Angiotensin II inhibitors (angiotensin receptor blockers or ARBs)	Pheophorbide A	14.3	
Teriflunomide	Other immunomodulators	Methotrexate		0.146

Thioridazine	Antipsychotics	Mitoxantrone	35.68
Thioridazine	Antipsychotics	Mitoxantrone	41
Tipranavir	Protease inhibitors	Estradiol 3-sulfate	16
Trametinib	Kinase inhibitors	Cimetidine	1.1
Triclabendazole sulfone	Antiparasitics	Danofloxacin	
Triclabendazole sulfoxide	Antiparasitics	Danofloxacin	
Trifluoperazine	Antipsychotics	Mitoxantrone	7.56
Urolithin A	Herbal medications	Mitoxantrone	
Valspodar (PSC 833)	Transporter modulators	Glyburide (glibenclamide)	
Vandetanib	Kinase inhibitors	2-Amino-1-methyl-6-phenylimidazo (4,5-*B*)pyridine (Phip)	21.5
Verapamil	Calcium channel blockers	Calcein AM	3.65
Vismodegib	Cancer treatments	Prazosin	2.4
Vorapaxar	Anticoagulants and antiplatelets	Methotrexate	2.5
Xanthohumol	Food products	Methotrexate	0.359
Zidovudine	Nucleoside reverse transcriptase nhibitors (NRTIs)	Pheophorbide A	

6.3 CONCLUSION

As discussed, P-gp and BCRP are important efflux transporters that line the apical membrane of enterocytes and play a crucial role in limiting the oral absorption of xenobiotics. Their impact is clearly illustrated for compounds that are BDDCS Class 2, 3, and 4, where they would have poor solubility and/or poor intestinal permeability. In drug discovery, if the prototype molecule is provisionally classified as BDDCS Class 2–4, the discovery team may consider optimizing the properties (K_m, V_{max}, P_{eff}, and solubility) to reduce the potential impact of P-gp on its absorption. As a result, maximizing the value of available *in silico/in vitro* models to optimize this interaction is helpful in mitigating this liability. If the molecule has been selected as a candidate for drug development, it is prudent to assess the likelihood of saturation of P-gp that is typically associated with nonlinearity as well as potential for drug–drug interaction. The same goes for understanding if the NME is a P-gp/BCRP inhibitor. Therefore, discovery scientists are expected to have good comprehension for the disposition of comedications to determine potential interactions. This is prudent in particular if some of these molecules have narrow therapeutic index and are associated with high risk drug interactions such as rosuvastatin and digoxin.

6.4 CHAPTER 6 TUTORIALS

1. True or false
   - P-gp and BCRP are effective uptake transporters that limit many xenobiotics exposure to various body organs. ATP hydrolysis is the main source of energy.
   - All P-gp and BCRP substrates are associated with poor oral absorption and bioavailability.
   - The affinity (K_m), solubility, and permeability of P-gp substrates play key roles in determining the potential impact of P-gp on the absorption and oral plasma exposure of its substrates.
2. What are the variables that should be considered when considering the potential impact of P-gp on the oral plasma exposure of its substrates dosed as sustained release formulation?
3. Discuss the potential impact of P-gp on the oral plasma exposure of its substrate from BDDCS perspective.
4. Elaborate on the impact of P-gp substrates K_m, V_{max}, and solubility on their $f_a * f_g$ in relation to oral dose.

6.5 CHAPTER 6 TUTORIAL ANSWER KEY

1. True or false
   - P-gp and BCRP are effective uptake transporters that limit many xenobiotics exposure to various body organs. ATP hydrolysis is the main source of energy. **[F]**
   - All P-gp and BCRP substrates are associated with poor oral absorption and bioavailability. **[F]**
   - The affinity (K_m), solubility, and permeability of P-gp substrates play key roles in determining the potential impact of P-gp on the absorption and oral plasma exposure of its substrates. **[T]**

2. What are the variables that should be considered when considering the potential impact of P-gp on the oral plasma exposure of its substrates dosed as sustained release formulation?

Answer: As discussed with the case study of talinolol, it is pivotal to understand and define if tested article has the following properties:

a. Assess if the tested article is P-gp substrate that is determined by assessing the tested article K_m and V_{max}, and efflux ratio using MDR1 transfected cell line and compare with probe substrates such as talinolol or digoxin.

b. It is important to determine the compound permeability. Compounds with high permeability tend to be less influenced by P-gp. According to Varma et al., compounds with permeability $<5 \times 10^{-6}$ cm/s measured in low efflux MDCK cell line tend to have poor absorption (f_a) < 85%. It is important that you calibrate your permeability *in vitro* model with compounds with known human f_a and follow a procedure similar to the Varma et al. paper [158].

c. Need also to determine the compound solubility and its dissolution rate.

3. Discuss the potential impact of P-gp on the oral plasma exposure of its substrate from BDDCS perspective.

Answer: BDDCS classifies compounds into four classes based on their permeability and solubility. Low permeability compounds such as BDDCS Class 3 and 4, their $f_a * f_g$ is more prone to be influenced by intestinal permeability. On the other hand, compounds that are high permeability, that is, BDDCS Class 1–2, their $f_a * f_g$ is less likely to be influenced by P-gp. The extent of impact on $f_a * f_g$ is also influenced by the compound's solubility since compounds with low solubility are prone not to saturate P-gp. Overall, they have higher likelihood to have absorption that would be limited by P-gp.

4. Elaborate on the impact of P-gp substrates K_m, V_{max}, and solubility on their $f_a * f_g$ in relation to oral dose

Answer: As illustrated in Figure 6.9, for compounds that are P-gp substrates, increasing oral dose is typically associated with increasing their $f_a * f_g$, the extent of this increase is driven by compounds P-gp K_m, V_{max}, and solubility. For example, compounds that are potent substrates (low K_m) increasing their oral dose is going to be associated with a significant impact, that is, significant increase in $f_a * f_g$ compared with compounds that have poor potency. Similar trend is observed with compounds with poor permeability/solubility, where the increase in the $f_a * f_g$ is more significant to that with high permeability/solubility compound.

REFERENCES

1. Wu, C.Y. and Benet, L.Z., *Predicting drug disposition via application of BCS: transport/absorption/elimination interplay and development of a biopharmaceutics drug disposition classification system*. Pharm Res, 2005. 22(1): p. 11–23.
2. Benet, L.Z., *The drug transporter-metabolism alliance: uncovering and defining the interplay*. Mol Pharm, 2009. 6(6): p. 1631–1643.
3. Juliano, R.L. and Ling, V., *A surface glycoprotein modulating drug permeability in Chinese hamster ovary cell mutants*. Biochim Biophys Acta, 1976. 455(1): p. 152–162.
4. Ambudkar, S.V., Dey, S., Hrycyna, C.A., Ramachandra, M., Pastan, I. and Gottesman, M.M., *Biochemical, cellular, and pharmacological aspects of the multidrug transporter*. Annu Rev Pharmacol Toxicol, 1999. 39: p. 361–398.

5. Schinkel, A.H., Kemp, S., Dolle, M., Rudenco, G. and Wagenaar, E., *N-glycosylation and deletion mutants of the human MDR1 P-glycoprotein*. J Biol Chem, 1993. 268: p. 7474–7481.
6. Thiebut, F., Tsuruo, T., Hamada, H., Gottesman, M.M. and Pastan, I., *Celluar localisation of the multidrug resistance gene product P-glycoprotien in normal human tissues*. Proc Natl Acad Sci USA, 1987. 84: p. 7735–7738.
7. Oswald, S., Groer, C., Drozdzik, M. and Siegmund, W., *Mass spectrometry-based targeted proteomics as a tool to elucidate the expression and function of intestinal drug transporters*. AAPS J, 2013. 15(4): p. 1128–1140.
8. Whitlock, G.A., Brennan, P.E., Roberts, L.R. and Stobie, A., *Potent and selective alpha1A adrenoceptor partial agonists – novel imidazole frameworks*. Bioorg Med Chem Lett, 2009. 19(11): p. 3118–3121.
9. Shapiro, A.B. and Ling, V., *The mechanism of ATP-dependent multidrug transport by P-glycoprotein*. Acta Physiol Scand Suppl, 1998. 643: p. 227–234.
10. Shapiro, A.B. and Ling, V., *Stoichiometry of coupling of rhodamine 123 transport to ATP hydrolysis by P-glycoprotein*. Eur J Biochem, 1998. 254(1): p. 189–193.
11. Shapiro, A.B. and Ling, V., *Transport of LDS-751 from the cytoplasmic leaflet of the plasma membrane by the rhodamine-123-selective site of P-glycoprotein*. Eur J Biochem, 1998. 254(1): p. 181–188.
12. Shapiro, A.B. and Ling, V., *Positively cooperative sites for drug transport by P-glycoprotein with distinct drug specificities*. Eur J Biochem, 1997. 250(1): p. 130–137.
13. Shapiro, A.B. and Ling, V., *Extraction of Hoechst 33342 from the cytoplasmic leaflet of the plasma membrane by P-glycoprotein*. Eur J Biochem, 1997. 250(1): p. 122–129.
14. Shapiro, A.B. and Ling, V., *Effect of quercetin on Hoechst 33342 transport by purified and reconstituted P-glycoprotein*. Biochem Pharmacol, 1997. 53(4): p. 587–596.
15. Acharya, P., O'Connor, M.P., Polli, J.W., Ayrton, A., Ellens, H. and Bentz, J., *Kinetic identification of membrane transporters that assist P-glycoprotein-mediated transport of digoxin and loperamide through a confluent monolayer of MDCKII-hMDR1 cells*. Drug Metab Dispos, 2008. 36(2): p. 452–460.
16. Acharya, P., Tran, T.T., Polli, J.W., Ayrton, A., Ellens, H. and Bentz, J., *P-Glycoprotein (P-gp) expressed in a confluent monolayer of hMDR1-MDCKII cells has more than one efflux pathway with cooperative binding sites*. Biochemistry, 2006. 45(51): p. 15505–15519.
17. Shaik, N., Giri, N., Pan, G. and Elmquist, W.F., *P-glycoprotein-mediated active efflux of the anti-HIV1 nucleoside abacavir limits cellular accumulation and brain distribution*. Drug Metab Dispos, 2007. 35(11): p. 2076–2085.
18. Benet, L.Z., Broccatelli, F. and Oprea, T.I., *BDDCS applied to over 900 drugs*. AAPS J, 2011. 13(4): p. 519–547.
19. Troutman, M.D. and Thakker, D.R., *Efflux ratio cannot assess P-glycoprotein-mediated attenuation of absorptive transport: asymmetric effect of P-glycoprotein on absorptive and secretory transport across Caco-2 cell monolayers*. Pharm Res, 2003. 20(8): p. 1200–1209.
20. Zhao, R., Raub, T.J., Sawada, G.A., Kasper, S.C., Bacon, J.A., Bridges, A.S. and Pollack, G.M., *Breast cancer resistance protein interacts with various compounds* in vitro, *but plays a minor role in substrate efflux at the blood–brain barrier*. Drug Metab Dispos, 2009. 37(6): p. 1251–1258.
21. Morland, J. and Olsen, H., *Metabolism of sulfadimidine, sulfanilamide, p-aminobenzoic acid, and isoniazid in suspensions of parenchymal and nonparenchymal rat liver cells*. Drug Metab Dispos, 1977. 5(6): p. 511–517.
22. Sobocinski, P.Z., Canterbury, W.J., Jr. and Powanda, M.C., *Differential effect of parenteral zinc on the course of various bacterial infections*. Proc Soc Exp Biol Med, 1977. 156(2): p. 334–339.
23. Schmitt, U., et al., *Cyclosporine A (CsA) affects the pharmacodynamics and pharmacokinetics of the atypical antipsychotic amisulpride probably via inhibition of P-glycoprotein (P-gp)*. J Neural Transm, 2006. 113(7): p. 787–801.

24. Poirier, A., Cascais, A.C., Bader, U., Portmann, R., Brun, M.E., Walter, I., Hillebrecht, A., Ullah, M. and Funk, C., *Calibration of* in vitro *multidrug resistance protein 1 substrate and inhibition assays as a basis to support the prediction of clinically relevant interactions* in vivo. Drug Metab Dispos, 2014. 42(9): p. 1411–1422.

25. Lin, X., Skolnik, S., Chen, X. and Wang, J., *Attenuation of intestinal absorption by major efflux transporters: quantitative tools and strategies using a Caco-2 model.* Drug Metab Dispos, 2011. 39(2): p. 265–274.

26. Miyajima, M., Kusuhara, H., Takahashi, K., Takashima, T., Hosoya, T., Watanabe, Y. and Sugiyama, Y., *Investigation of the effect of active efflux at the blood–brain barrier on the distribution of nonsteroidal aromatase inhibitors in the central nervous system.* J Pharm Sci, 2013. 102(9): p. 3309–3319.

27. Maguire, M.E., Ross, E.M. and Gilman, A.G., *Beta-adrenergic receptor: ligand binding properties and the interaction with adenylyl cyclase.* Adv Cyclic Nucleotide Res, 1977. 8: p. 1–83.

28. Mahar Doan, K.M., Humphreys, J.E., Webster, L.O., Wring, S.A., Shampine, L.J., Serabjit-Singh, C.J., Adkison, K.K. and Polli, J.W., *Passive permeability and P-glycoprotein-mediated efflux differentiate central nervous system (CNS) and non-CNS marketed drugs.* J Pharmacol Exp Ther, 2002. 303(3): p. 1029–1037.

29. Kis, O., Zastre, J.A., Hoque, M.T., Walmsley, S.L. and Bendayan, R., *Role of drug efflux and uptake transporters in atazanavir intestinal permeability and drug–drug interactions.* Pharm Res, 2013. 30(4): p. 1050–1064.

30. Boyd, R.A., Stern, R.H., Stewart, B.H., Wu, X., Reyner, E.L., Zegarac, E.A., Randinitis, E.J. and Whitfield, L., *Atorvastatin coadministration may increase digoxin concentrations by inhibition of intestinal P-glycoprotein-mediated secretion.* J Clin Pharmacol, 2000. 40(1): p. 91–98.

31. Togami, K., Chono, S. and Morimoto, K., *Transport characteristics of clarithromycin, azithromycin and telithromycin, antibiotics applied for treatment of respiratory infections, in Calu-3 cell monolayers as model lung epithelial cells.* Pharmazie, 2012. 67(5): p. 389–393.

32. Rigalli, J.P., et al., *Regulation of biotransformation systems and ABC transporters by benznidazole in HepG2 cells: involvement of pregnane X-receptor.* PLoS Negl Trop Dis, 2012. 6(12): p. e1951.

33. Yates, C.R., Chang, C., Kearbey, J.D., Yasuda, K., Schuetz, E.G., Miller, D.D., Dalton, J.T. and Swaan, P.W., *Structural determinants of P-glycoprotein-mediated transport of glucocorticoids.* Pharm Res, 2003. 20(11): p. 1794–1803.

34. Dilger, K., Schwab, M. and Fromm, M.F., *Identification of budesonide and prednisone as substrates of the intestinal drug efflux pump P-glycoprotein.* Inflamm Bowel Dis, 2004. 10(5): p. 578–583.

35. Hemauer, S.J., Patrikeeva, S.L., Wang, X., Abdelrahman, D.R., Hankins, G.D., Ahmed, M.S. and Nanovskaya, T.N., *Role of transporter-mediated efflux in the placental biodisposition of bupropion and its metabolite, OH-bupropion.* Biochem Pharmacol, 2010. 80(7): p. 1080–1086.

36. Zhou, L., Chen, X., Gu, Y. and Liang, J., *Transport characteristics of candesartan in human intestinal Caco-2 cell line.* Biopharm Drug Dispos, 2009. 30(5): p. 259–264.

37. Pagliarulo, V., Ancona, P., Niso, M., Colabufo, N.A., Contino, M., Cormio, L., Azzariti, A. and Pagliarulo, A., *The interaction of celecoxib with MDR transporters enhances the activity of mitomycin C in a bladder cancer cell line.* Mol Cancer, 2013. 12: p. 47.

38. Huang, J., Si, L., Jiang, L., Fan, Z., Qiu, J. and Li, G., *Effect of pluronic F68 block copolymer on P-glycoprotein transport and CYP3A4 metabolism.* Int J Pharm, 2008. 356(1–2): p. 351–353.

39. Kivisto, K.T., et al., *Characterisation of cerivastatin as a P-glycoprotein substrate: studies in P-glycoprotein-expressing cell monolayers and mdr1a/b knock-out mice.* Naunyn Schmiedebergs Arch Pharmacol, 2004. 370(2): p. 124–130.

40. Obradovic, T., Dobson, G.G., Shingaki, T., Kungu, T. and Hidalgo, I.J., *Assessment of the first and second generation antihistamines brain penetration and role of P-glycoprotein.* Pharm Res, 2007. 24(2): p. 318–327.

41. Schwab, D., Fischer, H., Tabatabaei, A., Poli, S. and Huwyler, J., *Comparison of* in vitro *P-glycoprotein screening assays: recommendations for their use in drug discovery.* J Med Chem, 2003. 46(9): p. 1716–1725.

42. Coyle, J.T., *Biochemical aspects of neurotransmission in the developing brain.* Int Rev Neurobiol, 1977. 20: p. 65–103.
43. Lepist, E.I., Phan, T.K., Roy, A., Tong, L., Maclennan, K., Murray, B. and Ray, A.S., *Cobicistat boosts the intestinal absorption of transport substrates, including HIV protease inhibitors and GS-7340,* in vitro. Antimicrob Agents Chemother, 2012. 56(10): p. 5409–5413.
44. Chen, C., Hanson, E., Watson, J.W. and Lee, J.S., *P-glycoprotein limits the brain penetration of nonsedating but not sedating H1-antagonists.* Drug Metab Dispos, 2003. 31(3): p. 312–318.
45. Crowe, A. and Tan, A.M., *Oral and inhaled corticosteroids: differences in P-glycoprotein (ABCB1) mediated efflux.* Toxicol Appl Pharmacol, 2012. 260(3): p. 294–302.
46. Hughes, J. and Crowe, A., *Inhibition of P-glycoprotein-mediated efflux of digoxin and its metabolites by macrolide antibiotics.* J Pharmacol Sci, 2010. 113(4): p. 315–324.
47. Chu, X.Y., et al., *Transport of the dipeptidyl peptidase-4 inhibitor sitagliptin by human organic anion transporter 3, organic anion transporting polypeptide 4C1, and multidrug resistance P-glycoprotein.* J Pharmacol Exp Ther, 2007. 321(2): p. 673–683.
48. Wils, P., Phung-Ba, V., Warnery, A., Lechardeur, D., Raeissi, S., Hidalgo, I.J. and Scherman, D., *Polarized transport of docetaxel and vinblastine mediated by P-glycoprotein in human intestinal epithelial cell monolayers.* Biochem Pharmacol, 1994. 48(7): p. 1528–1530.
49. Sheu, M.J., Teng, Y.N., Chen, Y.Y. and Hung, C.C., *The functional influences of common ABCB1 genetic variants on the inhibition of P-glycoprotein by* Antrodia cinnamomea *extracts.* PLoS One, 2014. 9(2): p. e89622.
50. Vadlapatla, R.K., Vadlapudi, A.D., Kwatra, D., Pal, D. and Mitra, A.K., *Differential effect of P-gp and MRP2 on cellular translocation of gemifloxacin.* Int J Pharm, 2011. 420(1): p. 26–33.
51. Crowe, A. and Lemaire, M., In vitro *and* in situ *absorption of SDZ-RAD using a human intestinal cell line (Caco-2) and a single pass perfusion model in rats: comparison with rapamycin.* Pharm Res, 1998. 15(11): p. 1666–1672.
52. Dahan, A. and Amidon, G.L., *Segmental dependent transport of low permeability compounds along the small intestine due to P-glycoprotein: the role of efflux transport in the oral absorption of BCS class III drugs.* Mol Pharm, 2009. 6(1): p. 19–28.
53. Horie, A., Ishida, K., Shibata, K., Taguchi, M., Ozawa, A., Hirono, K., Ichida, F. and Hashimoto, Y., *Pharmacokinetic variability of flecainide in younger Japanese patients and mechanisms for renal excretion and intestinal absorption.* Biopharm Drug Dispos, 2014. 35(3): p. 145–153.
54. El Ela, A.A., Hartter, S., Schmitt, U., Hiemke, C., Spahn-Langguth, H. and Langguth, P., *Identification of P-glycoprotein substrates and inhibitors among psychoactive compounds – implications for pharmacokinetics of selected substrates.* J Pharm Pharmacol, 2004. 56(8): p. 967–975.
55. Gnoth, M.J., Buetehorn, U., Muenster, U., Schwarz, T. and Sandmann, S., In vitro *and* in vivo *P-glycoprotein transport characteristics of rivaroxaban.* J Pharmacol Exp Ther, 2011. 338(1): p. 372–380.
56. Berginc, K., Zakelj, S. and Kristl, A., In vitro *interactions between aged garlic extract and drugs used for the treatment of cardiovascular and diabetic patients.* Eur J Nutr, 2010. 49(6): p. 373–384.
57. Huang, L., Berry, L., Ganga, S., Janosky, B., Chen, A., Roberts, J., Colletti, A.E. and Lin, M.H., *Relationship between passive permeability, efflux, and predictability of clearance from* in vitro *metabolic intrinsic clearance.* Drug Metab Dispos, 2010. 38(2): p. 223–231.
58. Elsby, R., Surry, D.D., Smith, V.N. and Gray, A.J., *Validation and application of Caco-2 assays for the* in vitro *evaluation of development candidate drugs as substrates or inhibitors of P-glycoprotein to support regulatory submissions.* Xenobiotica, 2008. 38(7–8): p. 1140–1164.
59. Pauli-Magnus, C., Rekersbrink, S., Klotz, U. and Fromm, M.F., *Interaction of omeprazole, lansoprazole and pantoprazole with P-glycoprotein.* Naunyn Schmiedebergs Arch Pharmacol, 2001. 364(6): p. 551–557.
60. Polli, J.W., Humphreys, J.E., Harmon, K.A., Castellino, S., O'Mara, M.J., Olson, K.L., John-Williams, L.S., Koch, K.M. and Serabjit-Singh, C.J., *The role of efflux and uptake transporters in [N-{3-chloro-4-[(3-fluorobenzyl)oxy]phenyl}-6-[5-({[2-(methylsulfonyl)ethyl]amino*

]methyl)-2-furyl]-4-quinazolinamine (GW572016, lapatinib) disposition and drug interactions. Drug Metab Dispos, 2008. 36(4): p. 695–701.

61. Hofmeister, C.C., et al., *Phase I trial of lenalidomide and CCI-779 in patients with relapsed multiple myeloma: evidence for lenalidomide-CCI-779 interaction via P-glycoprotein.* J Clin Oncol, 2011. 29(25): p. 3427–3434.
62. Feng, B., Mills, J.B., Davidson, R.E., Mireles, R.J., Janiszewski, J.S., Troutman, M.D. and de Morais, S.M., In vitro *P-glycoprotein assays to predict the* in vivo *interactions of P-glycoprotein with drugs in the central nervous system.* Drug Metab Dispos, 2008. 36(2): p. 268–275.
63. Moaddel, R., Hamid, R., Patel, S., Bullock, P.L. and Wainer, I.W., *Identification of P-glycoprotein substrates using open tubular chromatography on an immobilized P-glycoprotein column: comparison of chromatographic results with Caco-2 permeability.* Anal Chim Acta, 2006. 578(1): p. 25–30.
64. Soldner, A., Benet, L.Z., Mutschler, E. and Christians, U., *Active transport of the angiotensin-II antagonist losartan and its main metabolite EXP 3174 across MDCK-MDR1 and caco-2 cell monolayers.* Br J Pharmacol, 2000. 129(6): p. 1235–1243.
65. Walker, D.K., Abel, S., Comby, P., Muirhead, G.J., Nedderman, A.N. and Smith, D.A., *Species differences in the disposition of the CCR5 antagonist, UK-427,857, a new potential treatment for HIV.* Drug Metab Dispos, 2005. 33(4): p. 587–595.
66. Hemauer, S.J., Patrikeeva, S.L., Nanovskaya, T.N., Hankins, G.D. and Ahmed, M.S., *Role of human placental apical membrane transporters in the efflux of glyburide, rosiglitazone, and metformin.* Am J Obstet Gynecol, 2010. 202(4): p. 383.e1–383.e7.
67. Hassan, H.E., Myers, A.L., Coop, A. and Eddington, N.D., *Differential involvement of P-glycoprotein (ABCB1) in permeability, tissue distribution, and antinociceptive activity of methadone, buprenorphine, and diprenorphine:* in vitro *and* in vivo *evaluation.* J Pharm Sci, 2009. 98(12): p. 4928–4940.
68. Korzekwa, K.R., Nagar, S., Tucker, J., Weiskircher, E.A., Bhoopathy, S. and Hidalgo, I.J., *Models to predict unbound intracellular drug concentrations in the presence of transporters.* Drug Metab Dispos, 2012. 40(5): p. 865–876.
69. Judkiewicz, L., *[Disseminated intravascular coagulation in experimental metabolic acidosis].* Pol Tyg Lek, 1977. 32(42): p. 1643–1644.
70. Dahan, A., Sabit, H. and Amidon, G.L., *The H2 receptor antagonist nizatidine is a P-glycoprotein substrate: characterization of its intestinal epithelial cell efflux transport.* AAPS J, 2009. 11(2): p. 205–213.
71. Kim, W.Y. and Benet, L.Z., *P-glycoprotein (P-gp/MDR1)-mediated efflux of sex-steroid hormones and modulation of P-gp expression* in vitro. Pharm Res, 2004. 21(7): p. 1284–1293.
72. Hellinger, E., et al., *Comparison of brain capillary endothelial cell-based and epithelial (MDCK-MDR1, Caco-2, and VB-Caco-2) cell-based surrogate blood–brain barrier penetration models.* Eur J Pharm Biopharm, 2012. 82(2): p. 340–351.
73. Hassan, H.E., Myers, A.L., Lee, I.J., Coop, A. and Eddington, N.D., *Oxycodone induces overexpression of P-glycoprotein (ABCB1) and affects paclitaxel's tissue distribution in Sprague Dawley rats.* J Pharm Sci, 2007. 96(9): p. 2494–2506.
74. Troutman, M.D. and Thakker, D.R., *Novel experimental parameters to quantify the modulation of absorptive and secretory transport of compounds by P-glycoprotein in cell culture models of intestinal epithelium.* Pharm Res, 2003. 20(8): p. 1210–1224.
75. Shirasaka, Y., Suzuki, K., Shichiri, M., Nakanishi, T. and Tamai, I., *Intestinal absorption of HMG-CoA reductase inhibitor pitavastatin mediated by organic anion transporting polypeptide and P-glycoprotein/multidrug resistance 1.* Drug Metab Pharmacokinet, 2011. 26(2): p. 171–179.
76. Kunimatsu, S., Mizuno, T., Fukudo, M. and Katsura, T., *Effect of P-glycoprotein and breast cancer resistance protein inhibition on the pharmacokinetics of sunitinib in rats.* Drug Metab Dispos, 2013. 41(8): p. 1592–1597.
77. Hoque, M.T., Kis, O., De Rosa, M.F. and Bendayan, R., *Raltegravir permeability across blood–tissue barriers and the potential role of drug efflux transporters.* Antimicrob Agents Chemother, 2015. 59(5): p. 2572–2582.

78. Patelski, J., *Participation of phospholipids in arterial metabolism of cholesterol esters*. Adv Exp Med Biol, 1977. 82: p. 882–887.

79. Polli, J.W., Wring, S.A., Humphreys, J.E., Huang, L., Morgan, J.B., Webster, L.O. and Serabjit-Singh, C.S., *Rational use of* in vitro *P-glycoprotein assays in drug discovery*. J Pharmacol Exp Ther, 2001. 299(2): p. 620–628.

80. Polli, J.W., Jarrett, J.L., Studenberg, S.D., Humphreys, J.E., Dennis, S.W., Brouwer, K.R. and Woolley, J.L., *Role of P-glycoprotein on the CNS disposition of amprenavir (141W94), an HIV protease inhibitor*. Pharm Res, 1999. 16(8): p. 1206–1212.

81. Bacon, J.P. and Altman, J.S., *A silver intensification method for cobalt-filled neurones in wholemount preparations*. Brain Res, 1977. 138(2): p. 359–363.

82. Li, J., Wang, Y. and Hidalgo, I.J., *Kinetic analysis of human and canine P-glycoprotein-mediated drug transport in MDR1-MDCK cell model: approaches to reduce false-negative substrate classification*. J Pharm Sci, 2013. 102(9): p. 3436–3446.

83. Pachot, J.I., Botham, R.P., Haegele, K.D. and Hwang, K., *Experimental estimation of the role of P-Glycoprotein in the pharmacokinetic behaviour of telithromycin, a novel ketolide, in comparison with roxithromycin and other macrolides using the Caco-2 cell model*. J Pharm Pharm Sci, 2003. 6(1): p. 1–12.

84. Shepherd, J., Gotto, A.M., Jr., Taunton, O.D., Caslake, M.J. and Farish, E., *The* in vitro *interaction of human apolipoprotein A-I and high density lipoproteins*. Biochim Biophys Acta, 1977. 489(3): p. 486–501.

85. Askari, A.D., *Pericardial tamponade with hemorrhagic fluid in systemic lupus erythematosus*. J Am Med Womens Assoc, 1978. 33(3): p. 111–114.

86. de Lange, E.C., Marchand, S., van den Berg, D., van der Sandt, I.C., de Boer, A.G., Delon, A., Bouquet, S. and Couet, W., In vitro *and* in vivo *investigations on fluoroquinolones; effects of the P-glycoprotein efflux transporter on brain distribution of sparfloxacin*. Eur J Pharm Sci, 2000. 12(2): p. 85–93.

87. Tang, S.C., Lagas, J.S., Lankheet, N.A., Poller, B., Hillebrand, M.J., Rosing, H., Beijnen, J.H. and Schinkel, A.H., *Brain accumulation of sunitinib is restricted by P-glycoprotein (ABCB1) and breast cancer resistance protein (ABCG2) and can be enhanced by oral elacridar and sunitinib coadministration*. Int J Cancer, 2012. 130(1): p. 223–233.

88. Shirasaka, Y., Kuraoka, E., Spahn-Langguth, H., Nakanishi, T., Langguth, P. and Tamai, I., *Species difference in the effect of grapefruit juice on intestinal absorption of talinolol between human and rat*. J Pharmacol Exp Ther, 2010. 332(1): p. 181–189.

89. Ohura, K., Nozawa, T., Murakami, K. and Imai, T., *Evaluation of transport mechanism of prodrugs and parent drugs formed by intracellular metabolism in Caco-2 cells with modified carboxylesterase activity: temocapril as a model case*. J Pharm Sci, 2011. 100(9): p. 3985–3994.

90. Vaccari, A., Valcana, T. and Timiras, P.S., *Effects of hypothyroidism on the enzymes for biogenic amines in the developing rat brain*. Pharmacol Res Commun, 1977. 9(8): p. 763–780.

91. Crowe, A. and Wright, C., *The impact of P-glycoprotein mediated efflux on absorption of 11 sedating and less-sedating antihistamines using Caco-2 monolayers*. Xenobiotica, 2012. 42(6): p. 538–549.

92. Rott, R. and Avi-Dor, Y., *Effect of ionophoric compounds on aqueous suspensions of purple membrane*. FEBS Lett, 1977. 81(2): p. 267–270.

93. Horie, K., Tang, F. and Borchardt, R.T., *Isolation and characterization of Caco-2 subclones expressing high levels of multidrug resistance protein efflux transporter*. Pharm Res, 2003. 20(2): p. 161–168.

94. Shapiro, A.B., Fox, K., Lam, P. and Ling, V., *Stimulation of P-glycoprotein-mediated drug transport by prazosin and progesterone. Evidence for a third drug-binding site*. Eur J Biochem, 1999. 259(3): p. 841–850.

95. Seelig, A. and Landwojtowicz, E., *Structure–activity relationship of P-glycoprotein substrates and modifiers*. Eur J Pharm Sci, 2000. 12(1): p. 31–40.

96. Ashokraj, Y., Singh, I., Kaur, K.J., Kohli, G., Bhade, S.R., Varma, M.V., Kaul, C.L. and Panchagnula, R., *Establishment of a reference formulation for bioequivalence assessment of rifampicin-containing FDCs: an essential step towards improving tuberculosis treatment*. Int J Tuberc Lung Dis, 2005. 9(7): p. 791–796.
97. Pauli-Magnus, C., von Richter, O., Burk, O., Ziegler, A., Mettang, T., Eichelbaum, M. and Fromm, M.F., *Characterization of the major metabolites of verapamil as substrates and inhibitors of P-glycoprotein*. J Pharmacol Exp Ther, 2000. 293(2): p. 376–382.
98. Katoh, M., Nakajima, M., Yamazaki, H. and Yokoi, T., *Inhibitory potencies of 1,4-dihydropyridine calcium antagonists to P-glycoprotein-mediated transport: comparison with the effects on CYP3A4*. Pharm Res, 2000. 17(10): p. 1189–1197.
99. Katoh, M., Nakajima, M., Yamazaki, H. and Yokoi, T., *Inhibitory effects of CYP3A4 substrates and their metabolites on P-glycoprotein-mediated transport*. Eur J Pharm Sci, 2001. 12(4): p. 505–513.
100. Ekins, S., et al., *Three-dimensional quantitative structure–activity relationships of inhibitors of P-glycoprotein*. Mol Pharmacol, 2002. 61(5): p. 964–973.
101. Wandel, C., Kim, R., Wood, M. and Wood, A., *Interaction of morphine, fentanyl, sufentanil, alfentanil, and loperamide with the efflux drug transporter P-glycoprotein*. Anesthesiology, 2002. 96(4): p. 913–920.
102. Luo, F.R., Paranjpe, P.V., Guo, A., Rubin, E. and Sinko, P., *Intestinal transport of irinotecan in Caco-2 cells and MDCK II cells overexpressing efflux transporters Pgp, cMOAT, and MRP1*. Drug Metab Dispos, 2002. 30(7): p. 763–770.
103. Takara, K., et al., *Effects of 12 Ca2+ antagonists on multidrug resistance, MDR1-mediated transport and MDR1 mRNA expression*. Eur J Pharm Sci, 2002. 16(3): p. 159–165.
104. Bhardwaj, R.K., Glaeser, H., Becquemont, L., Klotz, U., Gupta, S.K. and Fromm, M.F., *Piperine, a major constituent of black pepper, inhibits human P-glycoprotein and CYP3A4*. J Pharmacol Exp Ther, 2002. 302(2): p. 645–650.
105. Kakumoto, M., Takara, K., Sakaeda, T., Tanigawara, Y., Kita, T. and Okumura, K., *MDR1-mediated interaction of digoxin with antiarrhythmic or antianginal drugs*. Biol Pharm Bull, 2002. 25(12): p. 1604–1607.
106. Verstuyft, C., et al., *Dipyridamole enhances digoxin bioavailability via P-glycoprotein inhibition*. Clin Pharmacol Ther, 2003. 73(1): p. 51–60.
107. Bekaii-Saab, T.S., Perloff, M.D., Weemhoff, J.L., Greenblatt, D.J. and von Moltke, L.L., *Interactions of tamoxifen, N-desmethyltamoxifen and 4-hydroxytamoxifen with P-glycoprotein and CYP3A*. Biopharm Drug Dispos, 2004. 25(7): p. 283–289.
108. Sikri, V., Pal, D., Jain, R., Kalyani, D. and Mitra, A.K., *Cotransport of macrolide and fluoroquinolones, a beneficial interaction reversing P-glycoprotein efflux*. Am J Ther, 2004. 11(6): p. 433–442.
109. Burk, O., Arnold, K.A., Nussler, A.K., Schaeffeler, E., Efimova, E., Avery, B.A., Avery, M.A., Fromm, M.F. and Eichelbaum, M., *Antimalarial artemisinin drugs induce cytochrome P450 and MDR1 expression by activation of xenosensors pregnane X receptor and constitutive androstane receptor*. Mol Pharmacol, 2005. 67(6): p. 1954–1965.
110. Fahrmayr, C., Konig, J., Auge, D., Mieth, M. and Fromm, M.F., *Identification of drugs and drug metabolites as substrates of multidrug resistance protein 2 (MRP2) using triple-transfected MDCK-OATP1B1-UGT1A1-MRP2 cells*. Br J Pharmacol, 2012. 165(6): p. 1836–1847.
111. Griffin, J., Fletcher, N., Clemence, R., Blanchflower, S. and Brayden, D.J., *Selamectin is a potent substrate and inhibitor of human and canine P-glycoprotein*. J Vet Pharmacol Ther, 2005. 28(3): p. 257–265.
112. Sakaeda, T., et al., *Effects of acid and lactone forms of eight HMG-CoA reductase inhibitors on CYP-mediated metabolism and MDR1-mediated transport*. Pharm Res, 2006. 23(3): p. 506–512.
113. Keogh, J.P. and Kunta, J.R., *Development, validation and utility of an* in vitro *technique for assessment of potential clinical drug–drug interactions involving P-glycoprotein*. Eur J Pharm Sci, 2006. 27(5): p. 543–554.

114. Iwaki, K., et al., *Haloperidol is an inhibitor but not substrate for MDR1/P-glycoprotein*. J Pharm Pharmacol, 2006. 58(12): p. 1617–1622.

115. de Castro, W.V., Mertens-Talcott, S., Derendorf, H. and Butterweck, V., *Grapefruit juice-drug interactions: grapefruit juice and its components inhibit P-glycoprotein (ABCB1) mediated transport of talinolol in Caco-2 cells*. J Pharm Sci, 2007. 96(10): p. 2808–2817.

116. Balimane, P.V., Marino, A. and Chong, S., *P-gp inhibition potential in cell-based models: which "calculation" method is the most accurate?* AAPS J, 2008. 10(4): p. 577–586.

117. Matsson, P., Pedersen, J.M., Norinder, U., Bergstrom, C.A. and Artursson, P., *Identification of novel specific and general inhibitors of the three major human ATP-binding cassette transporters P-gp, BCRP and MRP2 among registered drugs*. Pharm Res, 2009. 26(8): p. 1816–1831.

118. Hoffmann, G., Funk, C., Fowler, S., Otteneder, M.B., Breidenbach, A., Rayner, C.R., Chu, T. and Prinssen, E.P., *Nonclinical pharmacokinetics of oseltamivir and oseltamivir carboxylate in the central nervous system*. Antimicrob Agents Chemother, 2009. 53(11): p. 4753–4761.

119. Kwak, J.O., Lee, S.H., Lee, G.S., Kim, M.S., Ahn, Y.G., Lee, J.H., Kim, S.W., Kim, K.H. and Lee, M.G., *Selective inhibition of MDR1 (ABCB1) by HM30181 increases oral bioavailability and therapeutic efficacy of paclitaxel*. Eur J Pharmacol, 2010. 627(1–3): p. 92–98.

120. Kamiyama, E., Nakai, D., Mikkaichi, T., Okudaira, N. and Okazaki, O., *Interaction of angiotensin II type 1 receptor blockers with P-gp substrates in Caco-2 cells and hMDR1-expressing membranes*. Life Sci, 2010. 86(1–2): p. 52–58.

121. Cook, J.A., et al., *Refining the* in vitro *and* in vivo *critical parameters for P-glycoprotein, [I]/IC50 and [I2]/IC50, that allow for the exclusion of drug candidates from clinical digoxin interaction studies*. Mol Pharm, 2010. 7(2): p. 398–411.

122. Shoaf, S.E., Ohzone, Y., Ninomiya, S., Furukawa, M., Bricmont, P., Kashiyama, E. and Mallikaarjun, S., In vitro *P-glycoprotein interactions and steady-state pharmacokinetic interactions between tolvaptan and digoxin in healthy subjects*. J Clin Pharmacol, 2011. 51(5): p. 761–769.

123. Elsby, R., Gillen, M., Butters, C., Imisson, G., Sharma, P., Smith, V. and Surry, D.D., *The utility of* in vitro *data in making accurate predictions of human P-glycoprotein-mediated drug–drug interactions: a case study for AZD5672*. Drug Metab Dispos, 2011. 39(2): p. 275–282.

124. Reitman, M.L., et al., *Rifampin's acute inhibitory and chronic inductive drug interactions: experimental and model-based approaches to drug–drug interaction trial design*. Clin Pharmacol Ther, 2011. 89(2): p. 234–242.

125. Broccatelli, F., Carosati, E., Neri, A., Frosini, M., Goracci, L., Oprea, T.I. and Cruciani, G., *A novel approach for predicting P-glycoprotein (ABCB1) inhibition using molecular interaction fields*. J Med Chem, 2011. 54(6): p. 1740–1751.

126. Sugimoto, H., Matsumoto, S., Tachibana, M., Niwa, S., Hirabayashi, H., Amano, N. and Moriwaki, T., *Establishment of* in vitro *P-glycoprotein inhibition assay and its exclusion criteria to assess the risk of drug–drug interaction at the drug discovery stage*. J Pharm Sci, 2011. 100(9): p. 4013–4023.

127. Oswald, S., et al., *Impact of efavirenz on intestinal metabolism and transport: insights from an interaction study with ezetimibe in healthy volunteers*. Clin Pharmacol Ther, 2012. 91(3): p. 506–513.

128. Reed, A., Huie, K., Perloff, E.S., Cassella, J.V. and Takahashi, L.H., *Loxapine P-glycoprotein interactions* in vitro. Drug Metab Lett, 2012. 6(1): p. 26–32.

129. Minocha, M., Khurana, V., Qin, B., Pal, D. and Mitra, A.K., *Co-administration strategy to enhance brain accumulation of vandetanib by modulating P-glycoprotein (P-gp/Abcb1) and breast cancer resistance protein (Bcrp1/Abcg2) mediated efflux with m-TOR inhibitors*. Int J Pharm, 2012. 434(1–2): p. 306–314.

130. Nagasaka, Y., Oda, K., Iwatsubo, T., Kawamura, A. and Usui, T., *Effects of aripiprazole and its active metabolite dehydroaripiprazole on the activities of drug efflux transporters expressed both in the intestine and at the blood–brain barrier*. Biopharm Drug Dispos, 2012. 33(6): p. 304–315.

131. Park, S.J., et al., *Pharmacokinetic effect of voriconazole on cyclosporine in the treatment of aspergillosis after renal transplantation*. Clin Nephrol, 2012. 78(5): p. 412–418.

132. Gozalpour, E., Wittgen, H.G., van den Heuvel, J.J., Greupink, R., Russel, F.G. and Koenderink, J.B., *Interaction of digitalis-like compounds with p-glycoprotein*. Toxicol Sci, 2013. 131(2): p. 502–511.

133. Chen, C.H., Uang, Y.S., Wang, S.T., Yang, J.C. and Lin, C.J., *Interaction between red yeast rice and CYP450 enzymes/P-glycoprotein and its implication for the clinical pharmacokinetics of lovastatin*. Evid Based Complement Alternat Med, 2012. 2012: p. 127043.

134. Hung, C.C., Chiou, M.H., Teng, Y.N., Hsieh, Y.W., Huang, C.L. and Lane, H.Y., *Functional impact of ABCB1 variants on interactions between P-glycoprotein and methadone*. PLoS One, 2013. 8(3): p. e59419.

135. Werner, M., Atil, B., Sieczkowski, E., Chiba, P. and Hohenegger, M., *Simvastatin-induced compartmentalisation of doxorubicin sharpens up nuclear topoisomerase II inhibition in human rhabdomyosarcoma cells*. Naunyn Schmiedebergs Arch Pharmacol, 2013. 386(7): p. 605–617.

136. Dickens, D., Owen, A., Alfirevic, A. and Pirmohamed, M., *ABCB1 single nucleotide polymorphisms (1236C>T, 2677G>T, and 3435C>T) do not affect transport activity of human P-glycoprotein*. Pharmacogenet Genomics, 2013. 23(6): p. 314–323.

137. Ellens, H., et al., *Application of receiver operating characteristic analysis to refine the prediction of potential digoxin drug interactions*. Drug Metab Dispos, 2013. 41(7): p. 1367–1374.

138. Heredi-Szabo, K., et al., *A P-gp vesicular transport inhibition assay – optimization and validation for drug–drug interaction testing*. Eur J Pharm Sci, 2013. 49(4): p. 773–781.

139. Wu, C.P., Hsiao, S.H., Sim, H.M., Luo, S.Y., Tuo, W.C., Cheng, H.W., Li, Y.Q., Huang, Y.H. and Ambudkar, S.V., *Human ABCB1 (P-glycoprotein) and ABCG2 mediate resistance to BI 2536, a potent and selective inhibitor of Polo-like kinase 1*. Biochem Pharmacol, 2013. 86(7): p. 904–913.

140. Bhullar, J., Natarajan, K., Shukla, S., Mathias, T.J., Sadowska, M., Ambudkar, S.V. and Baer, M.R., *The FLT3 inhibitor quizartinib inhibits ABCG2 at pharmacologically relevant concentrations, with implications for both chemosensitization and adverse drug interactions*. PLoS One, 2013. 8(8): p. e71266.

141. Huo, X., Liu, Q., Wang, C., Meng, Q., Sun, H., Peng, J., Ma, X. and Liu, K., *Enhancement effect of P-gp inhibitors on the intestinal absorption and antiproliferative activity of bestatin*. Eur J Pharm Sci, 2013. 50(3–4): p. 420–428.

142. Moss, D.M., Liptrott, N.J., Curley, P., Siccardi, M., Back, D.J. and Owen, A., *Rilpivirine inhibits drug transporters ABCB1, SLC22A1, and SLC22A2* in vitro. Antimicrob Agents Chemother, 2013. 57(11): p. 5612–5618.

143. Fujita, Y., Noguchi, K., Suzuki, T., Katayama, K. and Sugimoto, Y., *Biochemical interaction of anti-HCV telaprevir with the ABC transporters P-glycoprotein and breast cancer resistance protein*. BMC Res Notes, 2013. 6: p. 445.

144. Kishimoto, W., Ishiguro, N., Ludwig-Schwellinger, E., Ebner, T. and Schaefer, O., In vitro *predictability of drug–drug interaction likelihood of P-glycoprotein-mediated efflux of dabigatran etexilate based on [I]2/IC50 threshold*. Drug Metab Dispos, 2014. 42(2): p. 257–263.

145. Liu, Q., Wang, C., Meng, Q., Huo, X., Sun, H., Peng, J., Ma, X., Sun, P. and Liu, K., *MDR1 and OAT1/OAT3 mediate the drug–drug interaction between puerarin and methotrexate*. Pharm Res, 2014. 31(5): p. 1120–1132.

146. Li, X., Hu, J., Wang, B., Sheng, L., Liu, Z., Yang, S. and Li, Y., *Inhibitory effects of herbal constituents on P-glycoprotein* in vitro *and* in vivo*: herb–drug interactions mediated via P-gp*. Toxicol Appl Pharmacol, 2014. 275(2): p. 163–175.

147. Mikkaichi, T., Yoshigae, Y., Masumoto, H., Imaoka, T., Rozehnal, V., Fischer, T., Okudaira, N. and Izumi, T., *Edoxaban transport via P-glycoprotein is a key factor for the drug's disposition*. Drug Metab Dispos, 2014. 42(4): p. 520–528.

148. Lepist, E.I., Gillies, H., Smith, W., Hao, J., Hubert, C., St Claire, R.L., 3rd, Brouwer, K.R. and Ray, A.S., *Evaluation of the endothelin receptor antagonists ambrisentan, bosentan, macitentan, and sitaxsentan as hepatobiliary transporter inhibitors and substrates in sandwich-cultured human hepatocytes*. PLoS One, 2014. 9(1): p. e87548.

149. Rijpma, S.R., van den Heuvel, J.J., van der Velden, M., Sauerwein, R.W., Russel, F.G. and Koenderink, J.B., *Atovaquone and quinine anti-malarials inhibit ATP binding cassette transporter activity*. Malar J, 2014. 13: p. 359.

150. Revalde, J.L., Li, Y., Hawkins, B.C., Rosengren, R.J. and Paxton, J.W., *Heterocyclic cyclohexanone monocarbonyl analogs of curcumin can inhibit the activity of ATP-binding cassette transporters in cancer multidrug resistance*. Biochem Pharmacol, 2015. 93(3): p. 305–317.

151. Hu, Z., Zhou, Z., Hu, Y., Wu, J., Li, Y. and Huang, W., *HZ08 reverse P-glycoprotein mediated multidrug resistance* in vitro *and* in vivo. PLoS One, 2015. 10(2): p. e0116886.

152. Lai, Y., et al., *Impact of drug transporter pharmacogenomics on pharmacokinetic and pharmacodynamic variability – considerations for drug development*. Expert Opin Drug Metab Toxicol, 2012. 8(6): p. 723–743.

153. Hoffmeyer, S., et al., *Functional polymorphisms of the human multidrug-resistance gene: multiple sequence variations and correlation of one allele with P-glycoprotein expression and activity* in vivo. Proc Natl Acad Sci USA, 2000. 97(7): p. 3473–3478.

154. Wang, D. and Sadee, W., *Searching for polymorphisms that affect gene expression and mRNA processing: example ABCB1 (MDR1)*. AAPS J, 2006. 8(3): p. E515–E520.

155. Kimchi-Sarfaty, C., Oh, J.M., Kim, I.W., Sauna, Z.E., Calcagno, A.M., Ambudkar, S.V. and Gottesman, M.M., *A "silent" polymorphism in the MDR1 gene changes substrate specificity*. Science, 2007. 315(5811): p. 525–528.

156. Kurata, Y., et al., *Role of human MDR1 gene polymorphism in bioavailability and interaction of digoxin, a substrate of P-glycoprotein*. Clin Pharmacol Ther, 2002. 72(2): p. 209–219.

157. Sakaeda, T., et al., *MDR1 genotype-related pharmacokinetics of digoxin after single oral administration in healthy Japanese subjects*. Pharm Res, 2001. 18(10): p. 1400–1404.

158. Varma, M.V., et al., *pH-Dependent solubility and permeability criteria for provisional biopharmaceutics classification (BCS and BDDCS) in early drug discovery*. Mol Pharm, 2012. 9(5): p. 1199–1212.

159. Kim, R.B., et al., *Identification of functionally variant MDR1 alleles among European Americans and African Americans*. Clin Pharmacol Ther, 2001. 70(2): p. 189–199.

160. Mathijssen, R.H., Marsh, S., Karlsson, M.O., Xie, R., Baker, S.D., Verweij, J., Sparreboom, A. and McLeod, H.L., *Irinotecan pathway genotype analysis to predict pharmacokinetics*. Clin Cancer Res, 2003. 9(9): p. 3246–3253.

161. Skarke, C., Jarrar, M., Schmidt, H., Kauert, G., Langer, M., Geisslinger, G. and Lotsch, J., *Effects of ABCB1 (multidrug resistance transporter) gene mutations on disposition and central nervous effects of loperamide in healthy volunteers*. Pharmacogenetics, 2003. 13(11): p. 651–660.

162. Katz, D.A., et al., *CYP3A5 genotype has a dose-dependent effect on ABT-773 plasma levels*. Clin Pharmacol Ther, 2004. 75(6): p. 516–528.

163. Plasschaert, S.L., et al., *Influence of functional polymorphisms of the MDR1 gene on vincristine pharmacokinetics in childhood acute lymphoblastic leukemia*. Clin Pharmacol Ther, 2004. 76(3): p. 220–229.

164. Saitoh, A., Singh, K.K., Powell, C.A., Fenton, T., Fletcher, C.V., Brundage, R., Starr, S. and Spector, S.A., *An MDR1-3435 variant is associated with higher plasma nelfinavir levels and more rapid virologic response in HIV-1 infected children*. AIDS, 2005. 19(4): p. 371–380.

165. Putnam, W.S., Woo, J.M., Huang, Y. and Benet, L.Z., *Effect of the MDR1 C3435T variant and P-glycoprotein induction on dicloxacillin pharmacokinetics*. J Clin Pharmacol, 2005. 45(4): p. 411–421.

166. Lepper, E.R., et al., *Effect of common CYP3A4 and CYP3A5 variants on the pharmacokinetics of the cytochrome P450 3A phenotyping probe midazolam in cancer patients*. Clin Cancer Res, 2005. 11(20): p. 7398–7404.

167. Yamaguchi, H., et al., *Genetic variation in ABCB1 influences paclitaxel pharmacokinetics in Japanese patients with ovarian cancer*. Int J Gynecol Cancer, 2006. 16(3): p. 979–985.

168. Taubert, D., von Beckerath, N., Grimberg, G., Lazar, A., Jung, N., Goeser, T., Kastrati, A., Schomig, A. and Schomig, E., *Impact of P-glycoprotein on clopidogrel absorption*. Clin Pharmacol Ther, 2006. 80(5): p. 486–501.

169. Slanar, O., Nobilis, M., Kvetina, J., Matouskova, O., Idle, J.R. and Perlik, F., *Pharmacokinetics of tramadol is affected by MDR1 polymorphism C3435T*. Eur J Clin Pharmacol, 2007. 63(4): p. 419–421.

170. la Porte, C.J., Li, Y., Beique, L., Foster, B.C., Chauhan, B., Garber, G.E., Cameron, D.W. and van Heeswijk, R.P., *The effect of ABCB1 polymorphism on the pharmacokinetics of saquinavir alone and in combination with ritonavir*. Clin Pharmacol Ther, 2007. 82(4): p. 389–395.

171. Kuypers, D.R., de Jonge, H., Naesens, M., Lerut, E., Verbeke, K. and Vanrenterghem, Y., *CYP3A5 and CYP3A4 but not MDR1 single-nucleotide polymorphisms determine long-term tacrolimus disposition and drug-related nephrotoxicity in renal recipients*. Clin Pharmacol Ther, 2007. 82(6): p. 711–725.

172. Weiner, M., Burman, W., Luo, C.C., Peloquin, C.A., Engle, M., Goldberg, S., Agarwal, V. and Vernon, A., *Effects of rifampin and multidrug resistance gene polymorphism on concentrations of moxifloxacin*. Antimicrob Agents Chemother, 2007. 51(8): p. 2861–2866.

173. Ufer, M., et al., *Influence of CYP3A4, CYP3A5, and ABCB1 genotype and expression on budesonide pharmacokinetics: a possible role of intestinal CYP3A4 expression*. Clin Pharmacol Ther, 2008. 84(1): p. 43–46.

174. Lal, S., Wong, Z.W., Sandanaraj, E., Xiang, X., Ang, P.C., Lee, E.J. and Chowbay, B., *Influence of ABCB1 and ABCG2 polymorphisms on doxorubicin disposition in Asian breast cancer patients*. Cancer Sci, 2008. 99(4): p. 816–823.

175. Miura, M., Satoh, S., Inoue, K., Kagaya, H., Saito, M., Inoue, T., Habuchi, T. and Suzuki, T., *Influence of CYP3A5, ABCB1 and NR1I2 polymorphisms on prednisolone pharmacokinetics in renal transplant recipients*. Steroids, 2008. 73(11): p. 1052–1059.

176. Keskitalo, J.E., Kurkinen, K.J., Neuvoneni, P.J. and Niemi, M., *ABCB1 haplotypes differentially affect the pharmacokinetics of the acid and lactone forms of simvastatin and atorvastatin*. Clin Pharmacol Ther, 2008. 84(4): p. 457–461.

177. Yin, O.Q., Tomlinson, B. and Chow, M.S., *Effect of multidrug resistance gene-1 (ABCB1) polymorphisms on the single-dose pharmacokinetics of cloxacillin in healthy adult Chinese men*. Clin Ther, 2009. 31(5): p. 999–1006.

178. Chen, B., Zhang, W., Fang, J., Jin, Z., Li, J., Yu, Z. and Cai, W., *Influence of the MDR1 haplotype and CYP3A5 genotypes on cyclosporine blood level in Chinese renal transplant recipients*. Xenobiotica, 2009. 39(12): p. 931–938.

179. Cho, H.Y., Yoo, H.D. and Lee, Y.B., *Influence of ABCB1 genetic polymorphisms on the pharmacokinetics of levosulpiride in healthy subjects*. Neuroscience, 2010. 169(1): p. 378–387.

180. Tapaninen, T., Neuvonen, P.J. and Niemi, M., *Effect of ABCB1 haplotypes on the pharmacokinetics and renin-inhibiting effect of aliskiren*. Eur J Clin Pharmacol, 2010. 66(9): p. 865–870.

181. Weiner, M., et al., *Effects of tuberculosis, race, and human gene SLCO1B1 polymorphisms on rifampin concentrations*. Antimicrob Agents Chemother, 2010. 54(10): p. 4192–4200.

182. Yoo, H.D., Lee, S.N., Kang, H.A., Cho, H.Y., Lee, I.K. and Lee, Y.B., *Influence of ABCB1 genetic polymorphisms on the pharmacokinetics of risperidone in healthy subjects with CYP2D6*10/*10*. Br J Pharmacol, 2011. 164(2b): p. 433–443.

183. Chew, S.C., Singh, O., Chen, X., Ramasamy, R.D., Kulkarni, T., Lee, E.J., Tan, E.H., Lim, W.T. and Chowbay, B., *The effects of CYP3A4, CYP3A5, ABCB1, ABCC2, ABCG2 and SLCO1B3 single nucleotide polymorphisms on the pharmacokinetics and pharmacodynamics of docetaxel in nasopharyngeal carcinoma patients*. Cancer Chemother Pharmacol, 2011. 67(6): p. 1471–1478.

184. Xiang, Q., Cui, Y.M., Zhao, X., Yan, L. and Zhou, Y., *The influence of MDR1 G2677T/a genetic polymorphisms on the pharmacokinetics of repaglinide in healthy Chinese volunteers*. Pharmacology, 2012. 89(1–2): p. 105–110.

185. Zhou, Q., Ruan, Z.R., Yuan, H. and Zeng, S., *CYP2C9*3(1075A>C), MDR1 G2677T/A and MDR1 C3435T are determinants of inter-subject variability in fluvastatin pharmacokinetics in healthy Chinese volunteers*. Arzneimittelforschung, 2012. 62(11): p. 519–524.

186. Krivoy, N., Zuckerman, T., Elkin, H., Froymovich, L., Rowe, J.M. and Efrati, E., *Pharmacokinetic and pharmacogenetic analysis of oral busulfan in stem cell transplantation: prediction of poor drug metabolism to prevent drug toxicity.* Curr Drug Saf, 2012. 7(3): p. 211–217.

187. Chan, S.W., Hu, M., Ko, S.S., Tam, C.W., Fok, B.S., Yin, O.Q., Chow, M.S. and Tomlinson, B., *CYP2C19 genotype has a major influence on labetalol pharmacokinetics in healthy male Chinese subjects.* Eur J Clin Pharmacol, 2013. 69(4): p. 799–806.

188. Wang, S.Y., et al., *Effect of quercetin on P-glycoprotein transport ability in Chinese healthy subjects.* Eur J Clin Nutr, 2013. 67(4): p. 390–394.

189. Zhou, Q., Ruan, Z.R., Yuan, H., Xu, D.H. and Zeng, S., *ABCB1 gene polymorphisms, ABCB1 haplotypes and ABCG2 c.421c>A are determinants of inter-subject variability in rosuvastatin pharmacokinetics.* Pharmazie, 2013. 68(2): p. 129–134.

190. Zhou, Q., Chen, Q.X., Ruan, Z.R., Yuan, H., Xu, H.M. and Zeng, S., *CYP2C9*3(1075A > C), ABCB1 and SLCO1B1 genetic polymorphisms and gender are determinants of inter-subject variability in pitavastatin pharmacokinetics.* Pharmazie, 2013. 68(3): p. 187–194.

191. Lai, M.L., Tien, Y.E., Huang, Y.S. and Huang, J.D., *Studies on pharmacokinetic mechanism of phenytoin resistance in refractory epilepsy.* J Pharm Sci, 2013. 102(9): p. 3189–3195.

192. Li, C.Y., Zhang, J., Chu, J.H., Xu, M.J., Ju, W.Z., Liu, F. and Jian-Dong, Z., *A correlative study of polymorphisms of CYP2C19 and MDR1 C3435T with the pharmacokinetic profiles of lansoprazole and its main metabolites following single oral administration in healthy adult Chinese subjects.* Eur J Drug Metab Pharmacokinet, 2014. 39(2): p. 121–128.

193. Kim, K.A., Joo, H.J., Lee, H.M. and Park, J.Y., *Influence of ABCB1 and CYP3A5 genetic polymorphisms on the pharmacokinetics of quetiapine in healthy volunteers.* Pharmacogenet Genomics, 2014. 24(1): p. 35–42.

194. Shugarts, S. and Benet, L.Z., *The role of transporters in the pharmacokinetics of orally administered drugs.* Pharm Res, 2009. 26(9): p. 2039–2054.

195. Estudante, M., Morais, J.G., Soveral, G. and Benet, L.Z., *Intestinal drug transporters: an overview.* Adv Drug Deliv Rev, 2013. 65(10): p. 1340–1356.

196. Sandstrom, R., Karlsson, A., Knutson, L. and Lennernas, H., *Jejunal absorption and metabolism of R/S-verapamil in humans.* Pharm Res, 1998. 15(6): p. 856–862.

197. Kharasch, E.D., Hoffer, C., Altuntas, T.G. and Whittington, D., *Quinidine as a probe for the role of p-glycoprotein in the intestinal absorption and clinical effects of fentanyl.* J Clin Pharmacol, 2004b. 44(3): p. 224–233.

198. Panchagnula, R., Bansal, T., Varma, M.V. and Kaul, C.L., *Reversed-phase liquid chromatography with ultraviolet detection for simultaneous quantitation of indinavir and propranolol from* ex-vivo *rat intestinal permeability studies.* J Chromatogr B Analyt Technol Biomed Life Sci, 2004. 806(2): p. 277–282.

199. Groer, C., Bruck, S., Lai, Y., Paulick, A., Busemann, A., Heidecke, C.D., Siegmund, W. and Oswald, S., *LC-MS/MS-based quantification of clinically relevant intestinal uptake and efflux transporter proteins.* J Pharm Biomed Anal, 2013. 85: p. 253–261.

200. Tubic, M., Wagner, D., Spahn-Langguth, H., Weiler, C., Wanitschke, R., Bocher, W.O. and Langguth, P., *Effects of controlled-release on the pharmacokinetics and absorption characteristics of a compound undergoing intestinal efflux in humans.* Eur J Pharm Sci, 2006. 29(3–4): p. 231–239.

201. Varma, M.V., Sarkar, M., Kapoor, N. and Panchagnula, R., *pH-Dependent functional activity of P-glycoprotein in limiting intestinal absorption of protic drugs 1. Simultaneous determination of quinidine and permeability markers in rat in situ perfusion samples.* J Chromatogr B Analyt Technol Biomed Life Sci, 2005. 816(1–2): p. 243–249.

202. Varma, M.V., Sateesh, K. and Panchagnula, R., *Functional role of P-glycoprotein in limiting intestinal absorption of drugs: contribution of passive permeability to P-glycoprotein mediated efflux transport.* Mol Pharm, 2005. 2(1): p. 12–21.

203. Varma, M.V.S. and Panchagnula, R., *Prediction of* in vivo *intestinal absorption enhancement on P-glycoprotein inhibition, from rat* in situ *permeability.* J Pharm Sci, 2005: 94(8): p. 1694–1704.

204. Varma, M.V. and Panchagnula, R., *pH-Dependent functional activity of P-glycoprotein in limiting intestinal absorption of protic drugs: kinetic analysis of quinidine efflux in situ.* J Pharm Sci, 2005. 94(12): p. 2632–2643.

205. Cvetkovic, M., Leake, B., Fromm, M.F., Wilkinson, G.R. and Kim, R.B., *OATP and P-glycoprotein transporters mediate the cellular uptake and excretion of fexofenadine.* Drug Metab Dispos, 1999. 27(8): p. 866–871.

206. Westphal, K., et al., *Oral bioavailability of digoxin is enhanced by talinolol: evidence for involvement of intestinal P-glycoprotein.* Clin Pharmacol Ther, 2000. 68(1): p. 6–12.

207. Meerum Terwogt, J.M., Beijnen, J.H., ten Bokkel Huinink, W.W., Rosing, H. and Schellens, J.H., *Co-administration of cyclosporin enables oral therapy with paclitaxel.* Lancet, 1998. 352(9124): p. 285.

208. Kaus, L.C., Gillespie, W.R., Hussain, A.S. and Amidon, G.L., *The effect of* in vivo *dissolution, gastric emptying rate, and intestinal transit time on the peak concentration and area-under-the-curve of drugs with different gastrointestinal permeabilities.* Pharm Res, 1999. 16(2): p. 272–280.

209. Cummins, C.L., Salphati, L., Reid, M.J. and Benet, L.Z., In vivo *modulation of intestinal CYP3A metabolism by P-glycoprotein: studies using the rat single-pass intestinal perfusion model.* J Pharmacol Exp Ther, 2003. 305(1): p. 306–314.

210. Cummins, C.L., Jacobsen, W. and Benet, L.Z., *Unmasking the dynamic interplay between intestinal P-glycoprotein and CYP3A4.* J Pharmacol Exp Ther, 2002. 300(3): p. 1036–1045.

211. Tachibana, T., Kato, M. and Sugiyama, Y., *Prediction of nonlinear intestinal absorption of CYP3A4 and P-glycoprotein substrates from their* in vitro *Km values.* Pharm Res, 2012. 29(3): p. 651–668.

212. Chiou, W.L., Ma, C., Chung, S.M. and Wu, T.C., *An alternative hypothesis to involvement of intestinal P-glycoprotein as the cause for digoxin oral bioavailability enhancement by talinolol.* Clin Pharmacol Ther, 2001. 69(1): p. 79–81.

213. Staud, F. and Pavek, P., *Breast cancer resistance protein (BCRP/ABCG2).* Int J Biochem Cell Biol, 2005. 37(4): p. 720–725.

214. Graf, G.A., Yu, L., Li, W.P., Gerard, R., Tuma, P.L., Cohen, J.C. and Hobbs, H.H., *ABCG5 and ABCG8 are obligate heterodimers for protein trafficking and biliary cholesterol excretion.* J Biol Chem, 2003. 278(48): p. 48275–48282.

215. Hazai, E. and Bikadi, Z., *Homology modeling of breast cancer resistance protein (ABCG2).* J Struct Biol, 2008. 162(1): p. 63–74.

216. Muenster, U., Grieshop, B., Ickenroth, K. and Gnoth, M.J., *Characterization of substrates and inhibitors for the* in vitro *assessment of Bcrp mediated drug–drug interactions.* Pharm Res, 2008. 25(10): p. 2320–2326.

217. Rosenberg, M.F., Bikadi, Z., Chan, J., Liu, X., Ni, Z., Cai, X., Ford, R.C. and Mao, Q., *The human breast cancer resistance protein (BCRP/ABCG2) shows conformational changes with mitoxantrone.* Structure, 2010. 18(4): p. 482–493.

218. Cai, X., Bikadi, Z., Ni, Z., Lee, E.W., Wang, H., Rosenberg, M.F. and Mao, Q., *Role of basic residues within or near the predicted transmembrane helix 2 of the human breast cancer resistance protein in drug transport.* J Pharmacol Exp Ther, 2010. 333(3): p. 670–681.

219. Doyle, L.A. and Ross, D.D., *Multidrug resistance mediated by the breast cancer resistance protein BCRP (ABCG2).* Oncogene, 2003. 22(47): p. 7340–7358.

220. Taipalensuu, J., et al., *Correlation of gene expression of ten drug efflux proteins of the ATP-binding cassette transporter family in normal human jejunum and in human intestinal epithelial Caco-2 cell monolayers.* J Pharmacol Exp Ther, 2001. 299(1): p. 164–170.

221. Bruyere, A., et al., *Effect of variations in the amounts of P-glycoprotein (ABCB1), BCRP (ABCG2) and CYP3A4 along the human small intestine on PBPK models for predicting intestinal first pass.* Mol Pharm, 2010. 7(5): p. 1596–1607.

222. Zaher, H., Khan, A.A., Palandra, J., Brayman, T.G., Yu, L. and Ware, J.A., *Breast cancer resistance protein (Bcrp/abcg2) is a major determinant of sulfasalazine absorption and elimination in the mouse.* Mol Pharm, 2006. 3(1): p. 55–61.

223. Kim, H.S., et al., *The effect of ABCG2 V12M, Q141K and Q126X, known functional variants* in vitro, *on the disposition of lamivudine*. Br J Clin Pharmacol, 2007. 64(5): p. 645–654.

224. Volk, E.L. and Schneider, E., *Wild-type breast cancer resistance protein (BCRP/ABCG2) is a methotrexate polyglutamate transporter*. Cancer Res, 2003. 63(17): p. 5538–5543.

225. Xia, C.Q., Liu, N., Miwa, G.T. and Gan, L.S., *Interactions of cyclosporin a with breast cancer resistance protein*. Drug Metab Dispos, 2007. 35(4): p. 576–582.

226. Matsson, P., Englund, G., Ahlin, G., Bergstrom, C.A., Norinder, U. and Artursson, P., *A global drug inhibition pattern for the human ATP-binding cassette transporter breast cancer resistance protein (ABCG2)*. J Pharmacol Exp Ther, 2007. 323(1): p. 19–30.

227. Xiao, Y., Davidson, R., Smith, A., Pereira, D., Zhao, S., Soglia, J., Gebhard, D., de Morais, S. and Duignan, D.B., *A 96-well efflux assay to identify ABCG2 substrates using a stably transfected MDCK II cell line*. Mol Pharm, 2006. 3(1): p. 45–54.

228. Wright, J.A., Haslam, I.S., Coleman, T. and Simmons, N.L., *Breast cancer resistance protein BCRP (ABCG2)-mediated transepithelial nitrofurantoin secretion and its regulation in human intestinal epithelial (Caco-2) layers*. Eur J Pharmacol, 2011. 672(1–3): p. 70–76.

229. Pavek, P., Merino, G., Wagenaar, E., Bolscher, E., Novotna, M., Jonker, J.W. and Schinkel, A.H., *Human breast cancer resistance protein: interactions with steroid drugs, hormones, the dietary carcinogen 2-amino-1-methyl-6-phenylimidazo(4,5-b)pyridine, and transport of cimetidine*. J Pharmacol Exp Ther, 2005. 312(1): p. 144–152.

230. Merino, G., Jonker, J.W., Wagenaar, E., Pulido, M.M., Molina, A.J., Alvarez, A.I. and Schinkel, A.H., *Transport of anthelmintic benzimidazole drugs by breast cancer resistance protein (BCRP/ABCG2)*. Drug Metab Dispos, 2005. 33(5): p. 614–618.

231. Zhang, Y., Gupta, A., Wang, H., Zhou, L., Vethanayagam, R.R., Unadkat, J.D. and Mao, Q., *BCRP transports dipyridamole and is inhibited by calcium channel blockers*. Pharm Res, 2005. 22(12): p. 2023–2034.

232. Merino, G., Alvarez, A.I., Pulido, M.M., Molina, A.J., Schinkel, A.H. and Prieto, J.G., *Breast cancer resistance protein (BCRP/ABCG2) transports fluoroquinolone antibiotics and affects their oral availability, pharmacokinetics, and milk secretion*. Drug Metab Dispos, 2006. 34(4): p. 690–695.

233. Li, J., Cusatis, G., Brahmer, J., Sparreboom, A., Robey, R.W., Bates, S.E., Hidalgo, M. and Baker, S.D., *Association of variant ABCG2 and the pharmacokinetics of epidermal growth factor receptor tyrosine kinase inhibitors in cancer patients*. Cancer Biol Ther, 2007. 6(3): p. 432–438.

234. Kitamura, S., Maeda, K., Wang, Y. and Sugiyama, Y., *Involvement of multiple transporters in the hepatobiliary transport of rosuvastatin*. Drug Metab Dispos, 2008. 36(10): p. 2014–2023.

235. Li, H., Jin, H.E., Kim, W., Han, Y.H., Kim, D.D., Chung, S.J. and Shim, C.K., *Involvement of P-glycoprotein, multidrug resistance protein 2 and breast cancer resistance protein in the transport of belotecan and topotecan in Caco-2 and MDCKII cells*. Pharm Res, 2008. 25(11): p. 2601–2612.

236. Marchetti, S., de Vries, N.A., Buckle, T., Bolijn, M.J., van Eijndhoven, M.A., Beijnen, J.H., Mazzanti, R., van Tellingen, O. and Schellens, J.H., *Effect of the ATP-binding cassette drug transporters ABCB1, ABCG2, and ABCC2 on erlotinib hydrochloride (Tarceva) disposition in* in vitro *and* in vivo *pharmacokinetic studies employing Bcrp1-/-/Mdr1a/1b-/- (triple-knockout) and wild-type mice*. Mol Cancer Ther, 2008. 7(8): p. 2280–2287.

237. Lagas, J.S., van der Kruijssen, C.M., van de Wetering, K., Beijnen, J.H. and Schinkel, A.H., *Transport of diclofenac by breast cancer resistance protein (ABCG2) and stimulation of multidrug resistance protein 2 (ABCC2)-mediated drug transport by diclofenac and benzbromarone*. Drug Metab Dispos, 2009. 37(1): p. 129–136.

238. Zhou, L., Schmidt, K., Nelson, F.R., Zelesky, V., Troutman, M.D. and Feng, B., *The effect of breast cancer resistance protein and P-glycoprotein on the brain penetration of flavopiridol, imatinib mesylate (Gleevec), prazosin, and 2-methoxy-3-(4-(2-(5-methyl-2-phenyloxazol-4-yl) ethoxy)phenyl)propanoic acid (PF-407288) in mice*. Drug Metab Dispos, 2009. 37(5): p. 946–955.

239. Jani, M., Szabo, P., Kis, E., Molnar, E., Glavinas, H. and Krajcsi, P., *Kinetic characterization of sulfasalazine transport by human ATP-binding cassette G2*. Biol Pharm Bull, 2009. 32(3): p. 497–499.

240. Keskitalo, J.E., Zolk, O., Fromm, M.F., Kurkinen, K.J., Neuvonen, P.J. and Niemi, M., *ABCG2 polymorphism markedly affects the pharmacokinetics of atorvastatin and rosuvastatin*. Clin Pharmacol Ther, 2009. 86(2): p. 197–203.

241. Lagas, J.S., van Waterschoot, R.A., Sparidans, R.W., Wagenaar, E., Beijnen, J.H. and Schinkel, A.H., *Breast cancer resistance protein and P-glycoprotein limit sorafenib brain accumulation*. Mol Cancer Ther, 2010. 9(2): p. 319–326.

242. Pollex, E.K., Anger, G., Hutson, J., Koren, G. and Piquette-Miller, M., *Breast cancer resistance protein (BCRP)-mediated glyburide transport: effect of the C421A/Q141K BCRP single-nucleotide polymorphism*. Drug Metab Dispos, 2010. 38(5): p. 740–744.

243. Li, L., Sham, Y.Y., Bikadi, Z. and Elmquist, W.F., *pH-Dependent transport of pemetrexed by breast cancer resistance protein*. Drug Metab Dispos, 2011. 39(9): p. 1478–1485.

244. Choi, M.K. and Song, I.S., *Characterization of efflux transport of the PDE5 inhibitors, vardenafil and sildenafil*. J Pharm Pharmacol, 2012. 64(8): p. 1074–1083.

245. Beery, E., et al., *ABCG2 modulates chlorothiazide permeability –* in vitro*-characterization of its interactions*. Drug Metab Pharmacokinet, 2012. 27(3): p. 349–353.

246. Li, J., Wang, Y., Zhang, W., Huang, Y., Hein, K. and Hidalgo, I.J., *The role of a basolateral transporter in rosuvastatin transport and its interplay with apical breast cancer resistance protein in polarized cell monolayer systems*. Drug Metab Dispos, 2012. 40(11): p. 2102–2108.

247. Hashiguchi, Y., Hamada, A., Shinohara, T., Tsuchiya, K., Jono, H. and Saito, H., *Role of P-glycoprotein in the efflux of raltegravir from human intestinal cells and CD4+ T-cells as an interaction target for anti-HIV agents*. Biochem Biophys Res Commun, 2013. 439(2): p. 221–227.

248. Varma, M.V., Scialis, R.J., Lin, J., Bi, Y.A., Rotter, C.J., Goosen, T.C. and Yang, X., *Mechanism-based pharmacokinetic modeling to evaluate transporter–enzyme interplay in drug interactions and pharmacogenetics of glyburide*. AAPS J, 2014. 16(4): p. 736–748.

249. Wen, C.C., et al., *Genome-wide association study identifies ABCG2 (BCRP) as an allopurinol transporter and a determinant of drug response*. Clin Pharmacol Ther, 2015. 97(5): p. 518–525.

250. Nicolle, E., Boumendjel, A., Macalou, S., Genoux, E., Ahmed-Belkacem, A., Carrupt, P.A. and Di Pietro, A., *QSAR analysis and molecular modeling of ABCG2-specific inhibitors*. Adv Drug Deliv Rev, 2009. 61(1): p. 34–46.

251. Saito, H., et al., *A new strategy of high-speed screening and quantitative structure–activity relationship analysis to evaluate human ATP-binding cassette transporter ABCG2–drug interactions*. J Pharmacol Exp Ther, 2006. 317(3): p. 1114–1124.

252. Yoshikawa, M., et al., *Novel camptothecin analogues that circumvent ABCG2-associated drug resistance in human tumor cells*. Int J Cancer, 2004. 110(6): p. 921–927.

253. Nakagawa, H., Saito, H., Ikegami, Y., Aida-Hyugaji, S., Sawada, S. and Ishikawa, T., *Molecular modeling of new camptothecin analogues to circumvent ABCG2-mediated drug resistance in cancer*. Cancer Lett, 2006. 234(1): p. 81–89.

254. Kaliszczak, M., Antonow, D., Patel, K.I., Howard, P., Jodrell, D.I., Thurston, D.E. and Guichard, S.M., *Optimization of the antitumor activity of sequence-specific pyrrolobenzodiazepine derivatives based on their affinity for ABC transporters*. AAPS J, 2010 12(4): p. 617–627.

255. Begley, D.J., *ABC transporters and the blood–brain barrier*. Curr Pharm Des, 2004. 10(12): p. 1295–1312.

256. Kunta, J.R. and Sinko, P.J., *Intestinal drug transporters:* in vivo *function and clinical importance*. Curr Drug Metab, 2004. 5(1): p. 109–124.

257. Takano, M., Yumoto, R. and Murakami, T., *Expression and function of efflux drug transporters in the intestine*. Pharmacol Ther, 2006. 109(1–2): p. 137–161.

258. Suzuki, M., Suzuki, H., Sugimoto, Y. and Sugiyama, Y., *ABCG2 transports sulfated conjugates of steroids and xenobiotics*. J Biol Chem, 2003. 278(25): p. 22644–22649.

259. Chen, Z.S., Robey, R.W., Belinsky, M.G., Shchaveleva, I., Ren, X.Q., Sugimoto, Y., Ross, D.D., Bates, S.E. and Kruh, G.D., *Transport of methotrexate, methotrexate polyglutamates, and*

17beta-estradiol 17-(beta-D-glucuronide) by ABCG2: effects of acquired mutations at R482 on methotrexate transport. Cancer Res, 2003. 63(14): p. 4048–4054.

260. Imai, Y., Asada, S., Tsukahara, S., Ishikawa, E., Tsuruo, T. and Sugimoto, Y., *Breast cancer resistance protein exports sulfated estrogens but not free estrogens.* Mol Pharmacol, 2003. 64(3): p. 610–618.
261. Nakanishi, T., Doyle, L.A., Hassel, B., Wei, Y., Bauer, K.S., Wu, S., Pumplin, D.W., Fang, H.B. and Ross, D.D., *Functional characterization of human breast cancer resistance protein (BCRP, ABCG2) expressed in the oocytes of* Xenopus laevis. Mol Pharmacol, 2003. 64(6): p. 1452–1462.
262. Houghton, P.J., Germain, G.S., Harwood, F.C., Schuetz, J.D., Stewart, C.F., Buchdunger, E. and Traxler, P., *Imatinib mesylate is a potent inhibitor of the ABCG2 (BCRP) transporter and reverses resistance to topotecan and SN-38* in vitro. Cancer Res, 2004. 64(7): p. 2333–2337.
263. Breedveld, P., et al., *Mechanism of the pharmacokinetic interaction between methotrexate and benzimidazoles: potential role for breast cancer resistance protein in clinical drug–drug interactions.* Cancer Res, 2004. 64(16): p. 5804–5811.
264. Yanase, K., Tsukahara, S., Asada, S., Ishikawa, E., Imai, Y. and Sugimoto, Y., *Gefitinib reverses breast cancer resistance protein-mediated drug resistance.* Mol Cancer Ther, 2004. 3(9): p. 1119–1125.
265. Ahmed-Belkacem, A., Pozza, A., Munoz-Martinez, F., Bates, S.E., Castanys, S., Gamarro, F., Di Pietro, A. and Perez-Victoria, J.M., *Flavonoid structure-activity studies identify 6-prenylchrysin and tectochrysin as potent and specific inhibitors of breast cancer resistance protein ABCG2.* Cancer Res, 2005. 65(11): p. 4852–4860.
266. Hirano, M., Maeda, K., Matsushima, S., Nozaki, Y., Kusuhara, H. and Sugiyama, Y., *Involvement of BCRP (ABCG2) in the biliary excretion of pitavastatin.* Mol Pharmacol, 2005. 68(3): p. 800–807.
267. Weiss, J., Rose, J., Storch, C.H., Ketabi-Kiyanvash, N., Sauer, A., Haefeli, W.E. and Efferth, T., *Modulation of human BCRP (ABCG2) activity by anti-HIV drugs.* J Antimicrob Chemother, 2007. 59(2): p. 238–245.
268. Marchetti, S., et al., In vitro *transport of gimatecan (7-t-butoxyiminomethylcamptothecin) by breast cancer resistance protein, P-glycoprotein, and multidrug resistance protein 2.* Mol Cancer Ther, 2007. 6(12 Pt 1): p. 3307–3313.
269. Wang, Q., Strab, R., Kardos, P., Ferguson, C., Li, J., Owen, A. and Hidalgo, I.J., *Application and limitation of inhibitors in drug-transporter interactions studies.* Int J Pharm, 2008. 356(1–2): p. 12–18.
270. Suzuki, K., Doki, K., Homma, M., Tamaki, H., Hori, S., Ohtani, H., Sawada, Y. and Kohda, Y., *Co-administration of proton pump inhibitors delays elimination of plasma methotrexate in high-dose methotrexate therapy.* Br J Clin Pharmacol, 2009. 67(1): p. 44–49.
271. Tiwari, A.K., Sodani, K., Wang, S.R., Kuang, Y.H., Ashby, C.R., Jr., Chen, X. and Chen, Z.S., *Nilotinib (AMN107, Tasigna) reverses multidrug resistance by inhibiting the activity of the ABCB1/Pgp and ABCG2/BCRP/MXR transporters.* Biochem Pharmacol, 2009. 78(2): p. 153–161.
272. Kawahara, H., Noguchi, K., Katayama, K., Mitsuhashi, J. and Sugimoto, Y., *Pharmacological interaction with sunitinib is abolished by a germ-line mutation (1291T>C) of BCRP/ABCG2 gene.* Cancer Sci, 2010. 101(6): p. 1493–1500.
273. Jani, M., Makai, I., Kis, E., Szabo, P., Nagy, T., Krajcsi, P. and Lespine, A., *Ivermectin interacts with human ABCG2.* J Pharm Sci, 2011. 100(1): p. 94–97.
274. Elsby, R., Fox, L., Stresser, D., Layton, M., Butters, C., Sharma, P., Smith, V. and Surry, D., In vitro *risk assessment of AZD9056 perpetrating a transporter-mediated drug–drug interaction with methotrexate.* Eur J Pharm Sci, 2011. 43(1–2): p. 41–49.
275. Elsby, R., Smith, V., Fox, L., Stresser, D., Butters, C., Sharma, P. and Surry, D.D., *Validation of membrane vesicle-based breast cancer resistance protein and multidrug resistance protein 2 assays to assess drug transport and the potential for drug–drug interaction to support regulatory submissions.* Xenobiotica, 2011. 41(9): p. 764–783.
276. Hofman, J., Ahmadimoghaddam, D., Hahnova, L., Pavek, P., Ceckova, M. and Staud, F., *Olomoucine II and purvalanol A inhibit ABCG2 transporter* in vitro *and* in situ *and synergistically potentiate cytostatic effect of mitoxantrone.* Pharmacol Res, 2012. 65(3): p. 312–319.

277. Xiao, G., Black, C., Hetu, G., Sands, E., Wang, J., Caputo, R., Rohde, E. and Gan, L.S., *Cerebrospinal fluid can be used as a surrogate to assess brain exposures of breast cancer resistance protein and P-glycoprotein substrates*. Drug Metab Dispos, 2012. 40(4): p. 779–787.

278. Mease, K., Sane, R., Podila, L. and Taub, M.E., *Differential selectivity of efflux transporter inhibitors in Caco-2 and MDCK-MDR1 monolayers: a strategy to assess the interaction of a new chemical entity with P-gp, BCRP, and MRP2*. J Pharm Sci, 2012. 101(5): p. 1888–1897.

279. Sodani, K., Tiwari, A.K., Singh, S., Patel, A., Xiao, Z.J., Chen, J.J., Sun, Y.L., Talele, T.T. and Chen, Z.S., *GW583340 and GW2974, human EGFR and HER-2 inhibitors, reverse ABCG2- and ABCB1-mediated drug resistance*. Biochem Pharmacol, 2012. 83(12): p. 1613–1622.

280. Barrera, B., Otero, J.A., Egido, E., Prieto, J.G., Seelig, A., Alvarez, A.I. and Merino, G., *The anthelmintic triclabendazole and its metabolites inhibit the membrane transporter ABCG2/BCRP*. Antimicrob Agents Chemother, 2012. 56(7): p. 3535–3543.

281. Yoshida, K., Maeda, K. and Sugiyama, Y., *Transporter-mediated drug–drug interactions involving OATP substrates: predictions based on* in vitro *inhibition studies*. Clin Pharmacol Ther, 2012. 91(6): p. 1053–1064.

282. Wei, Y., Ma, Y., Zhao, Q., Ren, Z., Li, Y., Hou, T. and Peng, H., *New use for an old drug: inhibiting ABCG2 with sorafenib*. Mol Cancer Ther, 2012. 11(8): p. 1693–1702.

283. Beery, E., et al., *ABCG2 modulates chlorothiazide permeability* in vitro *– characterization of the interaction*. Drug Metab Pharmacokinet, 2011 27(3): p. 349–353.

284. Gao, C., Zhang, H., Guo, Z., You, T., Chen, X. and Zhong, D., *Mechanistic studies on the absorption and disposition of scutellarin in humans: selective OATP2B1-mediated hepatic uptake is a likely key determinant for its unique pharmacokinetic characteristics*. Drug Metab Dispos, 2012. 40(10): p. 2009–2020.

285. Sostelly, A., et al., *A template model for studying anticancer drug efflux transporter inhibitors* in vitro. Fundam Clin Pharmacol, 2013. 27(5): p. 544–556.

286. Ishiguro, N., Shimizu, H., Kishimoto, W., Ebner, T. and Schaefer, O., *Evaluation and prediction of potential drug–drug interactions of linagliptin using* in vitro *cell culture methods*. Drug Metab Dispos, 2013. 41(1): p. 149–158.

287. Polli, J.W., Hussey, E., Bush, M., Generaux, G., Smith, G., Collins, D., McMullen, S., Turner, N. and Nunez, D.J., *Evaluation of drug interactions of GSK1292263 (a GPR119 agonist) with statins: from* in vitro *data to clinical study design*. Xenobiotica, 2013. 43(6): p. 498–508.

288. Marchetti, S., Pluim, D., van Eijndhoven, M., van Tellingen, O., Mazzanti, R., Beijnen, J.H. and Schellens, J.H., *Effect of the drug transporters ABCG2, Abcg2, ABCB1 and ABCC2 on the disposition, brain accumulation and myelotoxicity of the aurora kinase B inhibitor barasertib and its more active form barasertib-hydroxy-QPA*. Invest New Drugs, 2013. 31(5): p. 1125–1135.

289. Weiss, J., Theile, D., Ruppell, M.A., Speck, T., Spalwisz, A. and Haefeli, W.E., *Interaction profile of macitentan, a new non-selective endothelin-1 receptor antagonist,* in vitro. Eur J Pharmacol, 2013. 701(1–3): p. 168–175.

290. Bouchet, S., Dulucq, S., Pasquet, J.M., Lagarde, V., Molimard, M. and Mahon, F.X., *From* in vitro *to* in vivo*: intracellular determination of imatinib and nilotinib may be related with clinical outcome*. Leukemia, 2013. 27(8): p. 1757–1759.

291. Pan, Y., Chothe, P.P. and Swaan, P.W., *Identification of novel breast cancer resistance protein (BCRP) inhibitors by virtual screening*. Mol Pharm, 2013. 10(4): p. 1236–1248.

292. Weiss, J. and Haefeli, W.E., *Potential of the novel antiretroviral drug rilpivirine to modulate the expression and function of drug transporters and drug-metabolising enzymes* in vitro. Int J Antimicrob Agents, 2013. 41(5): p. 484–487.

293. Gonzalez-Sarrias, A., Miguel, V., Merino, G., Lucas, R., Morales, J.C., Tomas-Barberan, F., Alvarez, A.I. and Espin, J.C., *The gut microbiota ellagic acid-derived metabolite urolithin A and its sulfate conjugate are substrates for the drug efflux transporter breast cancer resistance protein (ABCG2/BCRP)*. J Agric Food Chem, 2013. 61(18): p. 4352–4359.

294. Sonesson, A. and Rasmussen, B.B., In vitro *and* in vivo *human metabolism of degarelix, a gonadotropin-releasing hormone receptor blocker*. Drug Metab Dispos, 2013. 41(7): p. 1339–1346.

295. Kalgutkar, A.S., et al., *Elucidation of the biochemical basis for a clinical drug–drug interaction between atorvastatin and 5-(N-(4-((4-ethylbenzyl)thio)phenyl)sulfamoyl)-2-methyl benzoic acid (CP-778875), a subtype selective agonist of the peroxisome proliferator-activated receptor alpha.* Xenobiotica, 2013. 43(11): p. 963–972.

296. Martin, V., Sanchez-Sanchez, A.M., Herrera, F., Gomez-Manzano, C., Fueyo, J., Alvarez-Vega, M.A., Antolin, I. and Rodriguez, C., *Melatonin-induced methylation of the ABCG2/BCRP promoter as a novel mechanism to overcome multidrug resistance in brain tumour stem cells.* Br J Cancer, 2013. 108(10): p. 2005–2012.

297. Reyner, E.L., Sevidal, S., West, M.A., Clouser-Roche, A., Freiwald, S., Fenner, K., Ullah, M., Lee, C.A. and Smith, B.J., In vitro *characterization of axitinib interactions with human efflux and hepatic uptake transporters: implications for disposition and drug interactions.* Drug Metab Dispos, 2013. 41(8): p. 1575–1583.

298. Mandikova, J., Volkova, M., Pavek, P., Cesnek, M., Janeba, Z., Kubicek, V. and Trejtnar, F., *Interactions with selected drug renal transporters and transporter-mediated cytotoxicity in antiviral agents from the group of acyclic nucleoside phosphonates.* Toxicology, 2013. 311(3): p. 135–146.

299. Bakhsheshian, J., Hall, M.D., Robey, R.W., Herrmann, M.A., Chen, J.Q., Bates, S.E. and Gottesman, M.M., *Overlapping substrate and inhibitor specificity of human and murine ABCG2.* Drug Metab Dispos, 2013. 41(10): p. 1805–1812.

300. Tournier, N., et al., *Effects of selected OATP and/or ABC transporter inhibitors on the brain and whole-body distribution of glyburide.* AAPS J, 2013. 15(4): p. 1082–1090.

301. Feinshtein, V., Erez, O., Ben-Zvi, Z., Eshkoli, T., Sheizaf, B., Sheiner, E. and Holcberg, G., *Cannabidiol enhances xenobiotic permeability through the human placental barrier by direct inhibition of breast cancer resistance protein: an* ex vivo *study.* Am J Obstet Gynecol, 2013. 209(6): p. 573.e1–573.e15.

302. To, K.K. and Tomlinson, B., *Targeting the ABCG2-overexpressing multidrug resistant (MDR) cancer cells by PPARgamma agonists.* Br J Pharmacol, 2013. 170(5): p. 1137–1151.

303. Kwatra, D., Venugopal, A., Standing, D., Ponnurangam, S., Dhar, A., Mitra, A. and Anant, S., *Bitter melon extracts enhance the activity of chemotherapeutic agents through the modulation of multiple drug resistance.* J Pharm Sci, 2013. 102(12): p. 4444–4454.

304. Kobayashi, S., Shinohara, M., Nagai, T. and Konishi, Y., *Transport mechanisms for soy isoflavones and microbial metabolites dihydrogenistein and dihydrodaidzein across monolayers and membranes.* Biosci Biotechnol Biochem, 2013. 77(11): p. 2210–2217.

305. Rizk, M.L., Houle, R., Chan, G.H., Hafey, M., Rhee, E.G. and Chu, X., *Raltegravir has a low propensity to cause clinical drug interactions through inhibition of major drug transporters: an* in vitro *evaluation.* Antimicrob Agents Chemother, 2014. 58(3): p. 1294–1301.

306. Custodio, J.M., et al., *Pharmacokinetics of cobicistat boosted-elvitegravir administered in combination with rosuvastatin.* J Clin Pharmacol, 2014. 54(6): p. 649–656.

307. Telbisz, A., Hegedus, C., Varadi, A., Sarkadi, B. and Ozvegy-Laczka, C., *Regulation of the function of the human ABCG2 multidrug transporter by cholesterol and bile acids: effects of mutations in potential substrate and steroid binding sites.* Drug Metab Dispos, 2014. 42(4): p. 575–585.

308. Chen, Y., Wang, Y., Zhou, J., Gao, X., Qu, D. and Liu, C., *Study on the mechanism of intestinal absorption of epimedins a, B and C in the Caco-2 cell model.* Molecules, 2014. 19(1): p. 686–698.

309. Takeuchi, K., et al., *Interaction of novel platelet-increasing agent eltrombopag with rosuvastatin via breast cancer resistance protein in humans.* Drug Metab Dispos, 2014. 42(4): p. 726–734.

310. Sodani, K., Patel, A., Anreddy, N., Singh, S., Yang, D.H., Kathawala, R.J., Kumar, P., Talele, T.T. and Chen, Z.S., *Telatinib reverses chemotherapeutic multidrug resistance mediated by ABCG2 efflux transporter* in vitro *and* in vivo. Biochem Pharmacol, 2014. 89(1): p. 52–61.

311. Prueksaritanont, T., et al., *Pitavastatin is a more sensitive and selective organic anion-transporting polypeptide 1B clinical probe than rosuvastatin.* Br J Clin Pharmacol, 2014. 78(3): p. 587–598.

312. Rickert, V., Haefeli, W.E. and Weiss, J., *Pharmacokinetic interaction profile of riociguat, a new soluble guanylate cyclase stimulator,* in vitro. Pulm Pharmacol Ther, 2014. 28(2): p. 130–137.

313. Huang, X.C., Xiao, X., Zhang, Y.K., Talele, T.T., Salim, A.A., Chen, Z.S. and Capon, R.J., *Lamellarin O, a pyrrole alkaloid from an Australian marine sponge, Ianthella sp., reverses BCRP mediated drug resistance in cancer cells*. Mar Drugs, 2014. 12(7): p. 3818–3837.

314. Tan, K.W., Cooney, J., Jensen, D., Li, Y., Paxton, J.W., Birch, N.P. and Scheepens, A., *Hop-derived prenylflavonoids are substrates and inhibitors of the efflux transporter breast cancer resistance protein (BCRP/ABCG2)*. Mol Nutr Food Res, 2014. 58(11): p. 2099–2110.

315. Fleisher, B., Unum, J., Shao, J. and An, G., *Ingredients in fruit juices interact with dasatinib through inhibition of BCRP: a new mechanism of beverage–drug interaction*. J Pharm Sci, 2015. 104(1): p. 266–275.

316. Weiss, J., Theile, D., Dvorak, Z. and Haefeli, W.E., *Interaction potential of the multitargeted receptor tyrosine kinase inhibitor dovitinib with drug transporters and drug metabolising enzymes assessed* in vitro. Pharmaceutics, 2014. 6(4): p. 632–650.

317. Mathias, T.J., Natarajan, K., Shukla, S., Doshi, K.A., Singh, Z.N., Ambudkar, S.V. and Baer, M.R., *The FLT3 and PDGFR inhibitor crenolanib is a substrate of the multidrug resistance protein ABCB1 but does not inhibit transport function at pharmacologically relevant concentrations*. Invest New Drugs, 2015. 33(2): p. 300–309.

318. Campbell, S.D., Gadel, S., Friedel, C., Crafford, A., Regina, K.J. and Kharasch, E.D., *Influence of HIV antiretrovirals on methadone N-demethylation and transport*. Biochem Pharmacol, 2015. 95(2): p. 115–125.

319. Urquhart, B.L., et al., *Breast cancer resistance protein (ABCG2) and drug disposition: intestinal expression, polymorphisms and sulfasalazine as an* in vivo *probe*. Pharmacogenet Genomics, 2008. 18(5): p. 439–448.

320. Yamasaki, Y., et al., *Pharmacogenetic characterization of sulfasalazine disposition based on NAT2 and ABCG2 (BCRP) gene polymorphisms in humans*. Clin Pharmacol Ther, 2008. 84(1): p. 95–103.

321. Adkison, K.K., et al., *Oral sulfasalazine as a clinical BCRP probe substrate: pharmacokinetic effects of genetic variation (C421A) and pantoprazole coadministration*. J Pharm Sci, 2010. 99(2): p. 1046–1062.

322. Keskitalo, J.E., Pasanen, M.K., Neuvonen, P.J. and Niemi, M., *Different effects of the ABCG2 c.421C>A SNP on the pharmacokinetics of fluvastatin, pravastatin and simvastatin*. Pharmacogenomics, 2009. 10(10): p. 1617–1624.

323. Bailey, K.M., et al., *Hepatic metabolism and transporter gene variants enhance response to rosuvastatin in patients with acute myocardial infarction: the GEOSTAT-1 study*. Circ Cardiovasc Genet. 3(3): p. 276–285.

324. Huang, L., Wang, Y. and Grimm, S., *ATP-dependent transport of rosuvastatin in membrane vesicles expressing breast cancer resistance protein*. Drug Metab Dispos, 2006. 34(5): p. 738–742.

325. Ieiri, I., Higuchi, S. and Sugiyama, Y., *Genetic polymorphisms of uptake (OATP1B1, 1B3) and efflux (MRP2, BCRP) transporters: implications for inter-individual differences in the pharmacokinetics and pharmacodynamics of statins and other clinically relevant drugs*. Expert Opin Drug Metab Toxicol, 2009. 5(7): p. 703–729.

326. Seamon, J.A., Rugg, C.A., Emanuel, S., Calcagno, A.M., Ambudkar, S.V., Middleton, S.A., Butler, J., Borowski, V. and Greenberger, L.M., *Role of the ABCG2 drug transporter in the resistance and oral bioavailability of a potent cyclin-dependent kinase/Aurora kinase inhibitor*. Mol Cancer Ther, 2006. 5(10): p. 2459–2467.

327. Yamagata, T., Kusuhara, H., Morishita, M., Takayama, K., Benameur, H. and Sugiyama, Y., *Improvement of the oral drug absorption of topotecan through the inhibition of intestinal xenobiotic efflux transporter, breast cancer resistance protein, by excipients*. Drug Metab Dispos, 2007. 35(7): p. 1142–1148.

328. Tachibana, T., Kato, M., Takano, J. and Sugiyama, Y., *Predicting drug–drug interactions involving the inhibition of intestinal CYP3A4 and P-glycoprotein*. Curr Drug Metab, 2010. 11(9): p. 762–777.

7

IMPACT OF INFLUX TRANSPORTERS ON DRUG ABSORPTION

7.1 INTRODUCTION

Drug discovery and development is on the verge of a major transformation, where increasingly, new molecular entities (NMEs) that are selected as potential drug candidates are not fully compliant with Lipinski's rule of five [1, 2]. Instead, these candidates tend to be hydrophilic (calculated $\log P$, $c\log P < 3$) and polar (polar surface area, PSA > 75 Å^2). Although this trend can be attributed to many factors, there is an increasing awareness of how the physicochemistry of an NME may influence drug safety and pharmacokinetics [1, 3, 4]. Historically, high-throughput screening (HTS) approaches that were adopted throughout the pharmaceutical industry were geared toward enabling rapid identification of biologically relevant NMEs by primarily focusing on maximizing potency against the biological target [5, 6]. The consequence of this approach was the selection of substantial numbers of highly lipophilic (and less polar) drug candidates. Despite the high biological potency against their targets, these compounds were commonly associated with major liabilities, in particular toxicity [6–9]. More recently, a group of Pfizer scientists evaluated the relationship between physicochemical drug properties and *in vivo* toxicity [1, 10]. With a data set of animal *in vivo* toleration (IVT) studies on 245 preclinical Pfizer compounds, it was apparent that compounds with low $c\log P$ ($c\log P < 3$)/high PSA (PSA > 75 Å^2) are ~2.5 less likely to show any *in vivo* toxicity findings at a fixed plasma concentration of 10 μM (total) or 1 μM (free); precisely the reverse holds when these properties are inverted. This trend exists across a wide range of toxicity types and a broad swath of chemical space. Another contributing factor is the keen interest by the pharmaceutical industry to optimize the physicochemical space of NMEs to minimize their CYP-mediated metabolic liability. Evidently, reducing lipophilicity and increasing polarity would lead to reduced CYP-mediated clearance and potential for CYP-mediated drug–drug interactions (DDI). Consequently, these factors are driving the selection of NMEs that are both hydrophilic and polar in nature [11–13].

Oral Bioavailability Assessment: Basics and Strategies for Drug Discovery and Development, First Edition.
Ayman F. El-Kattan.

The outcome and strategy described above result in greater numbers of compounds presenting with low permeability characteristics, a class of compounds identified as Class III and Class IV according to the Biopharmaceutics Classification System (BCS) [14]. Indeed, these compounds with low metabolic clearance are also identified as Class III and Class IV of Biopharmaceutics Drug Disposition Classification System (BDDCS) [15–17]. While this paradigm shift brings with it the reduction of *in vivo* toxicity findings and better potency toward certain biological targets, it also presents the research community with a more difficult chemical space for addressing oral drug delivery. Since these molecules are polar and hydrophilic, they tend to lose some of the favorable properties that enable passive transcellular permeability. There are far fewer options for permeation enhancement available to the pharmaceutical scientist, especially those that can improve intestinal permeability for Class III and IV compounds.

7.2 INTESTINAL DRUG TRANSPORTERS

SLC transporters that are expressed at the intestinal apical surface of epithelial cells and play a role in the intestinal absorption of Class III and IV molecules include peptide transporter (PEPT1; *SLC15A1*), organic anion polypeptide transporters (OATP1A2, *SLCO1A2*; OATP2B1, *SLCO2B1*), monocarboxylate transporters (MCT1; *SLC16A1*), sodium-multivitamin transport (SMVT; *SLC5A6*), and organic cation/zwitterion transporters (OCTN1, *SLC22A4*; OCTN2, *SLC22A5*). L-type amino acid transporters (LAT1 and LAT2) and proton-dependent amino acid transporters (hPAT1, *SLC36A1*) are intestinal absorptive transporters for amino acids and amino acid mimetics and are known to mediate transport of drugs such as melphalan, baclofen, gabapentin, pregabalin, and gaboxadol [18–20]. Several other SLC transporters including organic anion or cation transporters (OATs or OCTs; *SLC22*) have also been identified in the intestine, but seem to be of less importance in oral drug absorption [21–23].

By understanding the transport mechanisms and required structural activity relationship (SAR) involved in the influx of drug molecules, it may allow the medicinal chemist to use rational drug design to discover molecules that take advantage of these processes and improve the oral absorption of NMEs. Within this review, the most relevant intestinal transporters involved in the drug cellular uptake (PepT1, OATPs, and MCT1) are discussed (Figure 7.1 and Table 7.1). For these transporters, information on the structure, distribution, and mechanism of transport is provided along with the clinical relevance of targeting these transport proteins and a detailed discussion on the reported SAR, where available. Recommendations for investment in transporter strategy as related to BDDCS class are also presented [15, 25–27]. Although, other transporters are reported to play a significant role in the intestinal disposition, this review does not discuss the details of all, because of their limited relevance in drug absorption process and lack of SAR knowledge in the literature.

7.2.1 Organic Anion Transporting Polypeptide (OATP2B1) (SLC21A9)

OATP2B1, a high-affinity/low-capacity transporter, was originally isolated from human brain and named OATP-B [28, 29] or SLC21A9 [30]. It is localized to the apical side of intestinal epithelial cells along with other uptake transporters such as PEPT1 and MCT1 (Figure 7.1). OATP2B1 is involved in electrogenic, Na^+-independent and pH-gradient-dependent transport of xenobiotics (Figure 7.1 and Table 7.1).

7.2.1.1 OATP2B1 Structure and Distribution Human OATP2B1 (*SLCO2B1* also known as *SLC21A9*) gene is classified as part of solute carrier family 21A (SLC21A) [28, 29, 31].

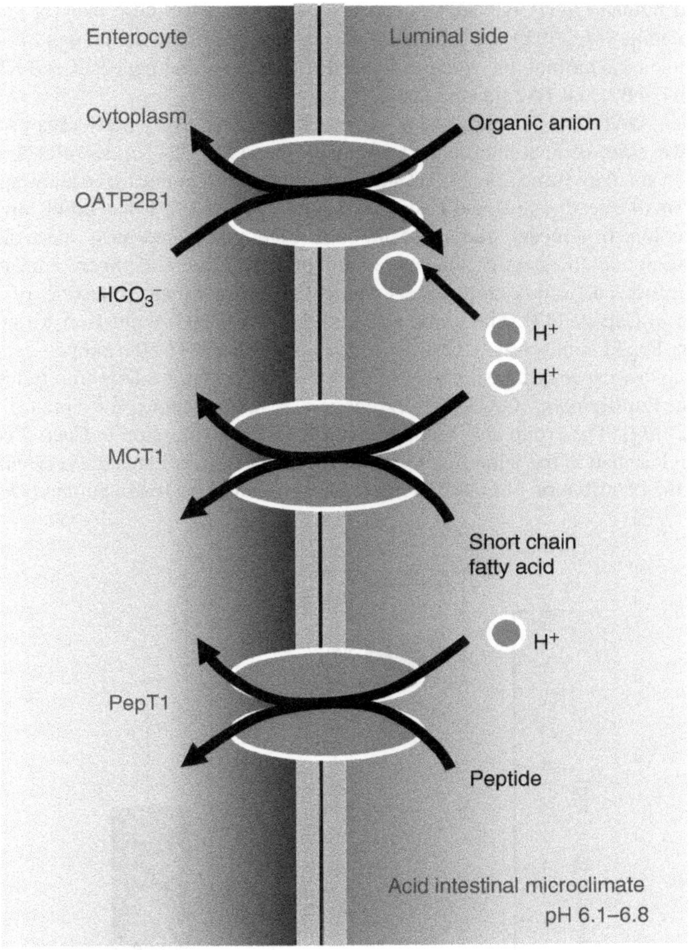

FIGURE 7.1 Known proton-coupled/pH-dependent solute carriers expressed at the brush border membrane of small intestine. Adapted from Anderson and Thwaites 2010 [24]. *(See color plate section for the color representation of this figure.)*

TABLE 7.1 Major Uptake Intestinal Transporter Uptake Mechanism, Expression, GIT Segment Expression, Affinity, and Capacity

Transporter	Uptake Mechanism	Expression	GIT Segment	Affinity	Capacity
OATP2B1	pH dependent	Apical membrane	Small intestine	High affinity	Low capacity
PepT1	pH dependent	Apical membrane	Small intestine	Low affinity	High capacity
MCT1	pH dependent	Apical membrane	Small and large intestines	Low affinity	High capacity

The cloned human OATP2B1 cDNA sequence comprises 4108 base pair (bp) and encodes 709 amino acids (M_{wt} 79 kDa). It consists of 12 transmembrane domains with cytoplasmic carboxy and amino termini. Its structure showed relatively shorter helices X and XII compared to OATP1B1, 1B3, and 1A2 transporters.

In human, OATP2B1 is ubiquitously expressed in almost every body organ such as liver, placenta, eye, skin, skeletal muscle, and mammary gland [28, 32–36]. Its mRNA is expressed in the human small intestine [28, 29, 31] and its protein is immunolocalized at the apical surface of human small intestine [32] and Caco-2 cell monolayers [31]. It has no or low expression in human colon. In a recent study, Oswald et al. used a validated mass spectrometry-based targeted proteomic to quantify the expression of OATP2B1 and other clinically relevant intestinal influx and efflux transporters along the entire human intestine of eight body donors and in Caco-2 [37]. The group reported that OATP2B1 expression was significantly lower than PepT1 transporter. This is consistent with OATP2B1 properties of being a high-affinity/low-capacity transporter, while PepT1 transporter is a low-affinity/high-capacity transporter. Furthermore, OATP1B1 expression is homogenous in jejunum and ileum (Figure 7.2) [37]. The group also reported that OATP2B1 expression in Caco-2 cell line was fourfold higher than in the jejunum. As a result, Caco-2 may not be a good *in vitro* model to determine the contribution of OATP2B1 to the absorption of NMEs in human (Figure 7.3).

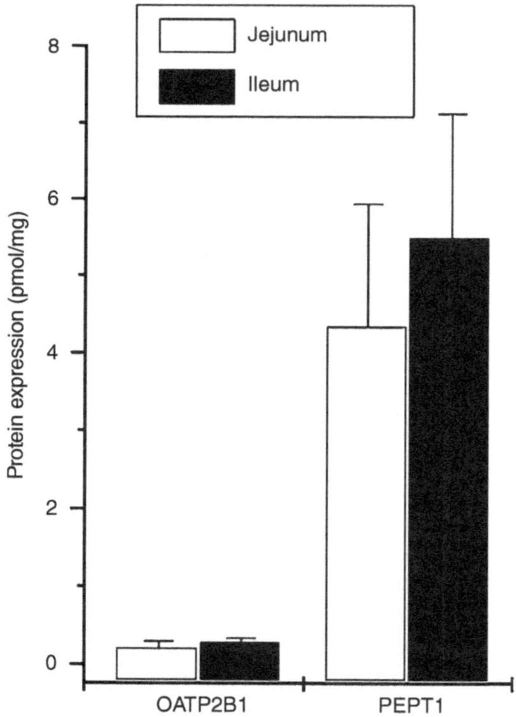

FIGURE 7.2 Intestinal protein expression of clinically relevant uptake (OATP2B1, PepT1) transporters in human jejunum and ileum from four donors quantified using validated mass spectrometry-based targeted proteomic. Oswald et al. 2013 [37]. Reproduced with permission of Springer.

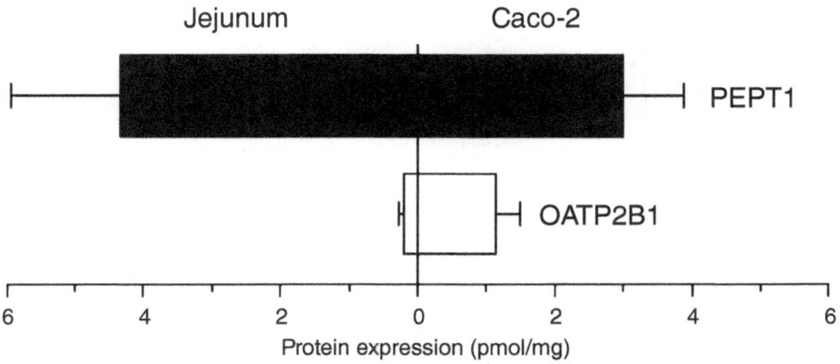

FIGURE 7.3 Protein expression of clinically relevant intestinal influx transporters in human jejunum ($N = 4$) and in Caco-2 cells ($N = 4$). Mean ± SD are given. Oswald et al. 2013 [37]. Reproduced with permission of Springer.

7.2.1.2 Mechanism of Drug Uptake of OATP2B1 The OATP2B1-mediated transport occurs in a bidirectional fashion and is driven by the substrate concentration gradient across the cell membrane. Heterologous expression of OATP2B1 produces Na^{+}-independent and pH-gradient-dependent transport characteristics with a relatively narrow substrate specificity compared to other OATPs [38]. *In vitro* acidification of the extracellular medium promoted OATP2B1-mediated solute uptake. This is consistent with the expression of this transporter and the lower pH within the microclimate at the enterocyte membrane surface. This is a trait that is shared with other clinically relevant intestinal uptake transporters such as MCT1 and PepT1 (Figure 7.1) [39–41]. Kobayashi et al. studied the impact of the pH on the uptake of both estrone-3-sulfate and pravastatin on their uptake in OATP-2B1-transfected HEK 293 cells. The group reported that the uptake of both compounds was pH dependent, where higher uptake at pH 5.5 relative to that at pH 7.4 was observed. It is interesting to note that the increase was only observed with V_{max} when the pH was decreased from 7.4 to 5.0. On the other hand, negligible change in affinity (K_m) was noted [32]. Similar findings were reported by our group, where we studied the impact of OATP2B1 and pH on the uptake of statins (rosuvastatin, fluvastatin, pravastatin, atorvastatin, cerivastatin, pitavastatin, lovastatin, and simvastatin) and estrone-3-sulfate *in vitro* using Caco-2 and OATP2B1-transfected HEK293 cells (Figure 7.4). OATP2B1 demonstrated broader substrate specificity and improved transporter activity at acidic pH for all compounds tested except simvastatin and lovastatin, where the uptake was comparable at all pHs (Figure 7.5). This is not unexpected since simvastatin and lovastatin are lactone (carboxylic acid prodrug) not acid and neutral at all pHs tested. Therefore, no major change in the OATP2B1-mediated uptake is observed. Other statins have carboxylic acid groups with higher unionized fraction at acidic pH. This suggests that the unionized fraction is the main driver for OATP2B1-mediated uptake. It should be emphasized that OATP2B1-driven uptake is substantial for statins that have inherent poor permeability, hydrophilic, and polar in nature such as pravastatin and rosuvastatin (Figure 7.5). However, absorption of statins that are lipophilic with high permeability tend not to be driven by active OATP2B1-mediated uptake, for example, atorvastatin, fluvastatin, cerivastatin, lovastatin, and simvastatin (Figure 7.5). This is not unexpected and in concordance with clinical reports on the impact of grapefruit juice on the plasma exposure of these molecules. As illustrated in Figure 7.6, grapefruit juice increased rather than decreased the plasma exposure of atorvastatin acid, losartan, lovastatin, and simvastatin in healthy volunteers. It is well established in the

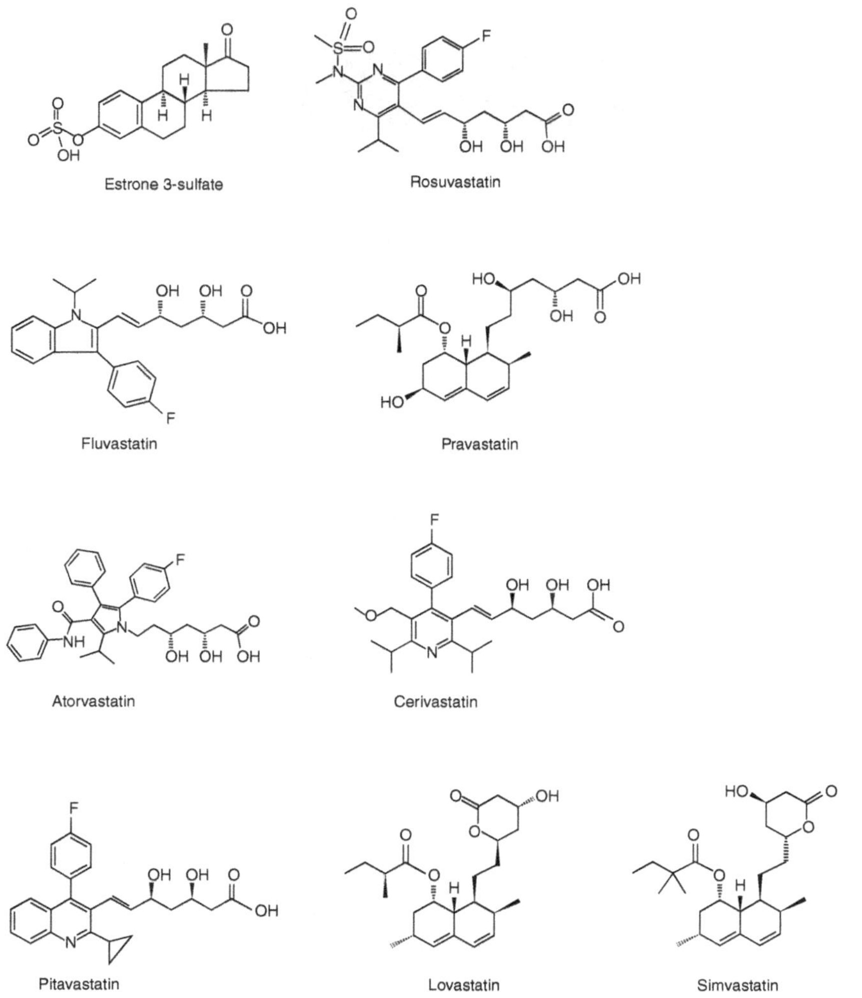

FIGURE 7.4 Chemical structures of estrone-3-sulfate and statins.

literature that grapefruit juice is an inhibitor of CYP3A4, P-glycoprotein, and OATP2B1 [51]. Since it increased the plasma exposure of these molecules, this suggests that OATP2B1 is not a major driver for these molecules absorption.

Structural modeling of OATP1B3 and OATP2B1 suggested that OATPs transport solutes across membranes through a rocker-switch type of mechanism similar to that shown with bacterial antiporter, GlpT [52], where the transport protein is believed to have two major alternating conformations: inward- and outward-facing. At any moment, a single binding site in a pore is accessible from only one side of the membrane. Interconversion between the two conformations is only possible via a substrate-bound form of the transport protein [53].

Tamai et al. used estrone sulfate (ES) to demonstrate that OATP2B1 has biphasic saturation kinetics characterized with high- and low-affinity binding sites and K_m values of 0.1 and 30 μM,

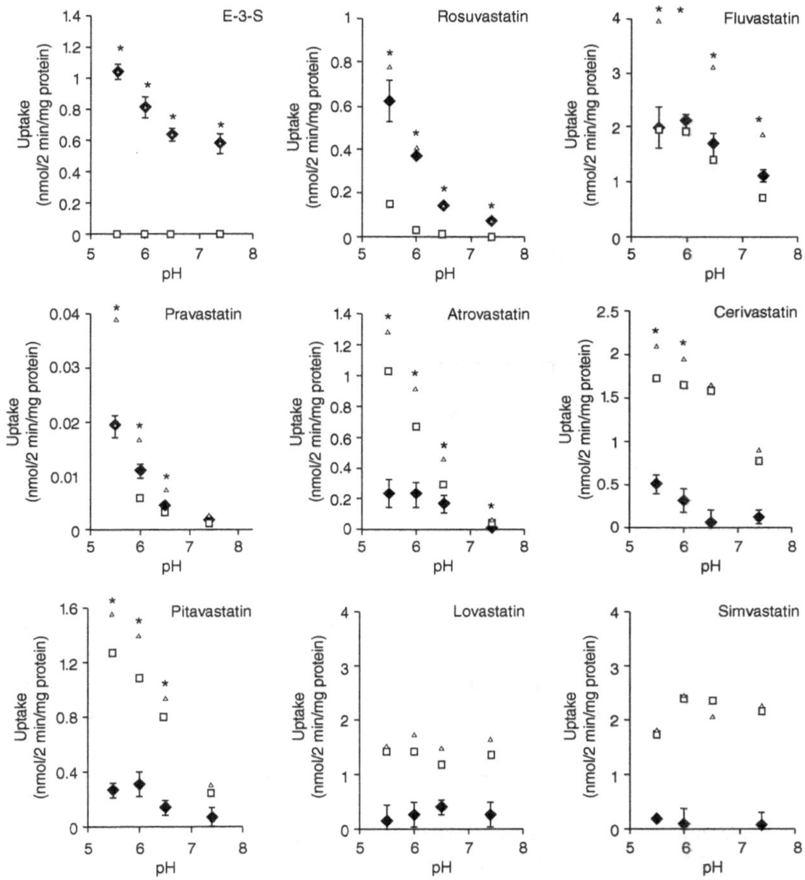

FIGURE 7.5 pH-Dependent uptake of E-3-S and statins by the mock-transfected and OATP2B1-transfected cells. Uptake by mock-transfected (□) and OATP2B1-transfected (Δ) cells was measured over 2 min, at various extracellular pH values in the range of 5.5 and 7.4. OATP2B1-mediated uptake (♦) was obtained by subtracting the uptake in the mock-transfected cells from that by the OATP2B1-transfected cells. Varma et al. 2011 [42]. Reproduced with Permission of American Chemical Society.

respectively. Food and fruit juice have different effects on these binding sites. For example, naringin (a main ingredient of grapefruit juice) had a significant impact on OATP2B1-mediated uptake of ES only when low concentrations of ES were investigated. This suggests that naringin only influenced the high-affinity site of ES binding [54]. In a recent study, the same group investigated whether the multiple binding sites of OATP2B1 can explain the difference in the impact of grapefruit juice on fexofenadine (significant reduction in the oral plasma profile) versus pravastatin (minimal impact on oral plasma profile). Both fexofenadine and pravastatin exhibited biphasic kinetics suggesting the presence of multiple binding sites. Interestingly, grapefruit juice only inhibited the uptake of pravastatin high-affinity site. No significant inhibition of the low affinity site was noted. In contrast, grapefruit juice significantly inhibited the

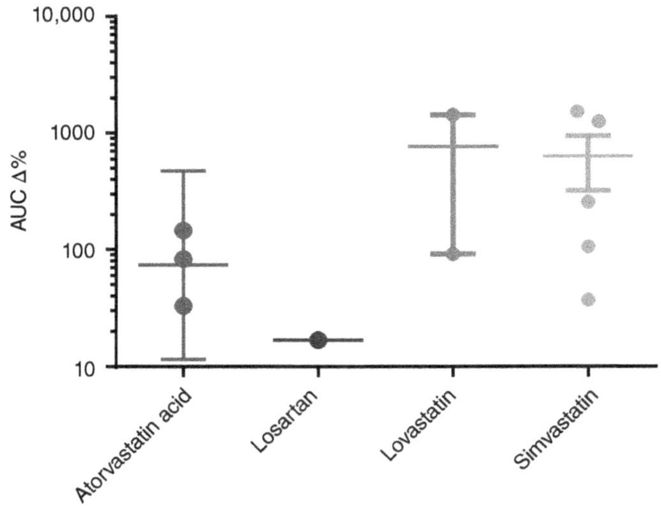

FIGURE 7.6 The impact of grapefruit juice on atorvastatin acid, losartan, lovastatin, and simvastatin plasma exposure $AUC_{\Delta}\%$ in healthy volunteers [43–50].

low-affinity site of fexofenadine with modest inhibition for the high-affinity site. Collectively, at therapeutically relevant concentrations, both compounds are transported by low-affinity site. However, grapefruit juice only inhibited that of fexofenadine. This may explain the significant impact of grapefruit juice on the oral plasma exposure of fexofenadine compared to pravastatin.

TIP:

Before conducting a grapefruit juice study in healthy volunteers to investigate the potential impact of OATP2B1 on the overall oral disposition of an NME, it is prudent to consider the following *in vitro* studies:

1. Assess the intestinal permeability and project the absorption fraction of NMEs. If the compound absorption is high ($f_a \geq 0.85$) then the likelihood of significant impact of grapefruit juice on the absorption of the molecule should be limited.
2. If not, determine if the NME is an OATP2B1 substrate using OATP2B1 transfected cell, for example, Oocytes of HEK293 cell lines.
3. If so, characterize if its OATP2B1 uptake follows biphasic kinetics principles. Define relevant *in vitro* kinetics parameters for the two sites (if applicable), that is, K_m and V_{max}.
4. Compare K_m values of the two binding sites (if applicable) with therapeutically relevant concentrations.
5. Investigate if grapefruit juice affects the kinetics of the high- and/or low-affinity sites.
6. Project if the interactions may occur by determining if the NCE and grapefruit juice share the same OATP2B1 binding site.

7.2.1.3 OATP2B1 Substrates and Inhibitors OATP2B1 transports physiological sulfate-conjugated steroids, estrone-3-sulfate, and dehydroepiandrosterone sulfate (DHEAS) [28, 29] at lower pH [38]. This suggests that OATP2B1 may play a role in the enterohepatic circulation of both bile acids and estrogen [31]. Partial lists of known OAPT2B1 substrates and inhibitors are shown in Tables 7.2 and 7.3. As illustrated in Figure 7.7a and b, OATP2B1 substrates tend to have moderate molecular weight (M_{wt}), be polar with an average topological PSA of 113 Å^2. They have low permeability (calculated MDCK-Low $< 2.45 \times 10^{-6}$ cm/s) and are also hydrophilic with an average $c \log D$ of 0.84 (Figure 7.7) [95, 96]. OATP2B1 substrates are typically acidic in nature, which is consistent with the physicochemical properties reported with OATP1B1 and OATP1B3 substrates [97].

TABLE 7.2 Partial List of Known Drugs that are in Clinical Use and Substrates of OATP2B1 [38, 55–75]

Object	K_m (μM)	V_{max} (pmol/min/mg)	V_{max}/K_m (μL/min/mg)
Aliskiren	72		
Amiodarone	66.8	9.9 nmol/min/mg	0.15
Atorvastatin	2.8	24.3	8.55
Atorvastatin acid	0.2	30.5	152.5
Bosentan	202	2.3 pmol/min/μg	0.011 μL/min/μg
Celiprolol			
Cerivastatin			
CP-671,305	4		
Eltrombopag			
Estradiol-17-β-glucuronide			
Estrone-3-sulfate	14	169	12.1
Etoposide			
Ezetimibe glucuronide			
Fexofenadine			
Fluvastatin	0.8		
Glyburide (glibenclamide)	6.3	112	17.9
Isoscutellarin	43	113	2.6
Lovastatin			
Mesalamine (5-ASA)	77.4	85.1	1.1
Methotrexate			
Montelukast			
Pemetrexed	307.1	325.2	1.1
Pitavastatin	1.2	7.4	6.3
Pravastatin	2250	4.2	0.002
Pregnenolone sulfate			
Rosuvastatin	2.4	1.3	0.5
Scutellarin	1.8	19.7	11.1
Simvastatin			
Sulfasalazine	1.7	9.7	5.6
Talinolol	629	1.2 pmol/min/oocyte	0.002 μL/min/oocyte
Taurocholic acid	71.8	546.5	7.6
Telmisartan acylglucuronide	1.1	22.3	20.5

TABLE 7.3 Partial List of Known Inhibitors of OATP2B1 [31, 58, 61–63, 65, 75–94]

Inhibitor	Object	K_i (μM)	IC_{50} (μM)
Cyclosporine	Rosuvastatin		0.07
Mk571	Bromosulfophthalein (BSP)		0.2
Ezetimibe glucuronide	Flutamide Flu-6 metabolite		0.26
Bromosulfophthalein (BSP)	Estrone-3-sulfate		0.282
Milk thistle	Estrone-3-sulfate		0.3
Atorvastatin acid	Estrone-3-sulfate	0.69	0.39
Erlotinib	Estrone-3-sulfate		0.55
Lopinavir	Estrone-3-sulfate		0.72
Tipranavir	Estrone-3-sulfate		0.77
Macitentan	Estrone-3-sulfate		0.8
Silybin B	Estrone-3-sulfate		0.8
Nelfinavir	Estrone-3-sulfate		0.9
Ritonavir	Estrone-3-sulfate		0.93
Montelukast	Bromosulfophthalein (BSP)		1
Silibinin	Bromosulfophthalein (BSP)		1
Phloretin	Estradiol 3-sulfate		1.31
Cp-778875	Rosuvastatin		1.62
Hesperidin	Estradiol 3-sulfate		1.92
Glyburide (glibenclamide)	Bromosulfophthalein (BSP)	0.5	2
Rifampicin	Estrone-3-sulfate	1.6	2.1
Mifepristone (RU 486)	DHEAS		2.2
Glimepiride	Bromosulfophthalein (BSP)		2.4
Rifamycin	Estrone-3-sulfate		2.6
Sulfasalazine	Estrone-3-sulfate		3
Pregnenolone sulfate	Estrone-3-sulfate		3.5
Silycristin	Estrone-3-sulfate		3.6
Indinavir	Estrone-3-sulfate	3	3.9
Silybin A	Estrone-3-sulfate		4.5
Naringin	Estradiol 3-sulfate		4.63
Atazanavir	Estrone-3-sulfate		5.2
Repaglinide	Bromosulfophthalein (BSP)		5.2
Rosiglitazone	Bromosulfophthalein (BSP)		5.2
Baicalin	Estrone sulfate		5.6
Ezetimibe	Flutamide Flu-6 metabolite		5.9
Gemfibrozil	Rosuvastatin		8
Dehydroepiandrosterone sulfate (DHEAS)	Estrone-3-sulfate		8.6
Saquinavir	Estrone-3-sulfate		9.4
Quercetin	Estradiol 3-sulfate		9.47
Efavirenz	Estrone-3-sulfate		9.6
Apigenin	Atorvastatin		13.9
Quercetin	Atorvastatin		14.1
Cyclosporine	Bromosulfophthalein (BSP)		20
Kaempferol	Atorvastatin		20.7
Testosterone	Estrone-3-sulfate		21.2
Phlorizin (phloridzin)	Estradiol 3-sulfate		23.2

TABLE 7.3 (*Continued*)

Inhibitor	Object	K_i (μM)	IC_{50} (μM)
Paclitaxel	Bromosulfophthalein (BSP)		25
Darunavir	Estrone-3-sulfate		26
Nateglinide	Bromosulfophthalein (BSP)		28.5
Darunavir	Estrone-3-sulfate		29
(−)-Epicatechin gallate	Estrone-3-sulfate		35.9
Cyclosporine	Estrone-3-sulfate		37
Progesterone	Bromosulfophthalein (BSP)		42.1
Naringenin	Estradiol 3-sulfate		49.2
Hoechst 33342	Estrone-3-sulfate		51
Cerivastatin acid	Estrone-3-sulfate		66.2
Hesperetin	Estradiol 3-sulfate		67.6
Ticlopidine	Estrone-3-sulfate		76.6
Simvastatin acid	Estrone-3-sulfate		84.7
Pravastatin	Estrone-3-sulfate		190
Estrone-3-sulfate	Telmisartan acylglucuronide		223
Doxorubicin	Estrone-3-sulfate		240
Digoxin	Estrone-3-sulfate		300
Penicillin G (benzylpenicillin)	Estrone-3-sulfate		16,000
Rifamycin	Bromosulfophthalein (BSP)	3	
Fandosentan (CI-1034)	Bromosulfophthalein (BSP)	3.3	
Silibinin	Estrone-3-sulfate	3.64	
Eltrombopag	Estrone-3-sulfate	8.48	
Quercetin	Bromosulfophthalein (BSP)	8.7	
Raltitrexed	Pemetrexed	70	
Methotrexate	Pemetrexed	600	

7.2.1.4 OATP2B1 Structural Activity Relationship (SAR) The sequence alignment of OATP2B1 with OATP1B1 and OATP1B3 showed the homology of their amino acid sequences is 27% and 28%, respectively, much less than the homology within OATP1 family, which is 79%. However, all three transporters share the same topological structure with 12 putative transmembrane domains. Comparing to OATP1 family, OATP2B1 structure showed relative short helices X and XII, and also slightly different packing of the large extracellular loop between transmembrane domains IX and X. Previous studies [98] found that this large extracellular loop, especially 10 conserved cysteines in the loop, is a determining factor for the transport activity of OATP2B1; however, the conserved cysteine motif has also been found in OATP1 family and other OATPs. Therefore, the region is less likely to impact the structure–activity relationship of OATP2B1 specifically. A more reasonable interpretation of the specificity of OATP2B1 substrates should be correlated to its unique substrate-binding site residues located within the putative positive-charged pore region surrounded by helices VIII, IX, and X. The investigation on the interactions between these residues and substrates could thus help us define the structure–activity relationship of OATP2B1.

Indeed, the VIII and IX transmembrane domains of OATP1B1 were found to affect the transport of estrone-3-sulfate and estradiol-17β-D-glucuronide [99]. In addition, amino acids Y537, S545, and T550 located in the transmembrane domain X of OATP1B3 were identified

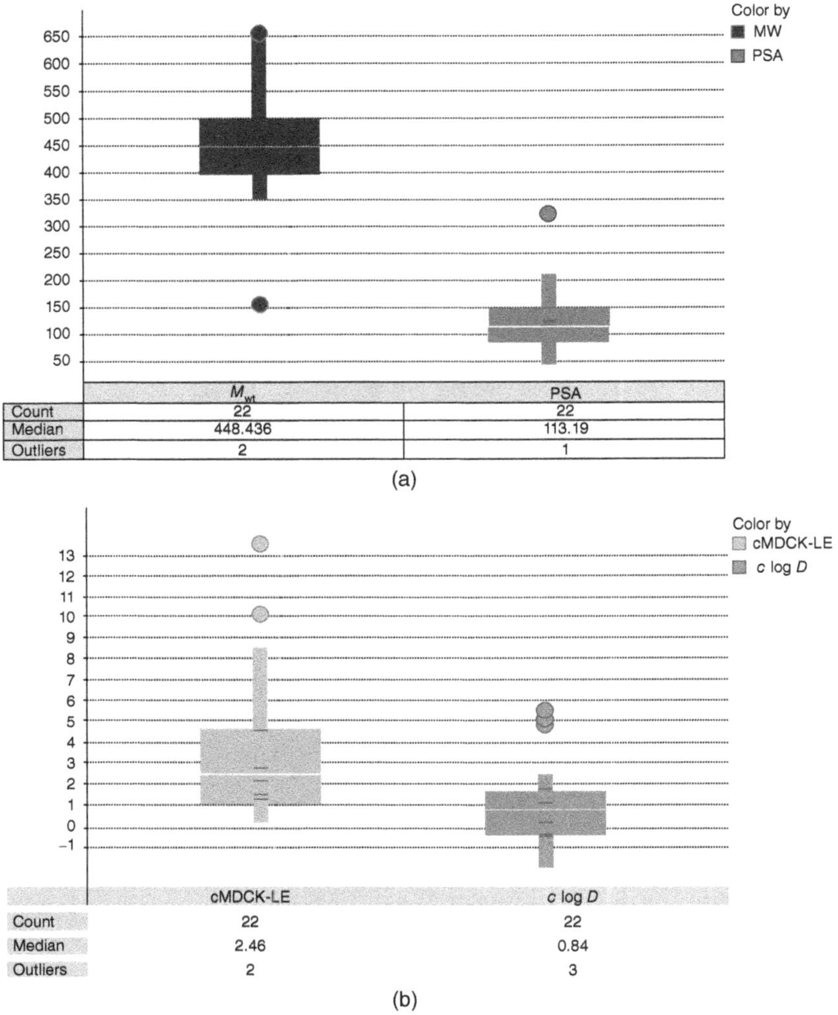

FIGURE 7.7 The physicochemical properties of OATP2B1 substrates (a) M_{wt} and PSA, (b) cMDCK-LE and $c\log D$. Di et al. 2011 [95, 96]. Reproduced with permission of Elsevier.

as the key residues for the transport of cholecystokinin octapeptide (CCK-8) by site-directed mutagenesis [100, 101]. They may potentially form the hydrogen bonds with the substrates though their side chain hydroxyl groups. In OATP1B1, these residues are replaced by a phenylalanine and two leucines, which lack the hydrogen donor and acceptor groups, and thus could explain the dramatically reduced transport of CCK-8. For OATP2B1, the corresponding residue to S545 is the same, but the other two residues are phenylalanine and leucine instead, which may indicate the potential decrease of the transport activity of CCK-8. The site-directed mutagenesis in replacing lysine and other positively charged amino acids with noncharged alanine or glycine residues reduces the activity significantly. The same results might apply to OATP1B1, which has the same residues in these positions. This might not apply to OATP2B1 because the K41 corresponding residue is a glutamine instead. In addition, R181 in OATP1B1 and OATP1B3, and H579 in OATP2B1 are likely to contribute the positive potential of the pore. They are located within the similar distance to both membranes that could access either side of member and also face the pore, which can be the important switching residues during the transport [53]. Therefore, the binding mode and affinity of substrates or inhibitors to the above-mentioned polar and charged residues will help us determine the structure–activity relationship of OATP2B1.

7.2.1.5 Influence of Genetic Polymorphism on In Vitro ***and*** In Vivo ***Activity of OATP2B1*** Table 7.4 and Figure 7.8 summarize the major SNPs in the *SLCO2B1* gene discovered. Tamai et al. was the first group to report the polymorphism of OATP2B1. They identified two genetic variations and investigated their allelic frequencies and functional activities (SLCO2B1*2 and SLCO2B1*3) [29, 106]. SLCO2B1*2 has the mutation of SLCO2B1c.1175C>T, which causes the amino acid change T392I. This variant was not identified in the Japanese population investigated. Further studies are needed to determine its frequency in other ethnicities. SLCO2B1*3 has the mutation of SLCO2B1c.1457C>T, which leads to the amino acid change S486F [54]. It has high allelic frequency of 31% in Japanese population [54]. The active uptake [^{3}H]estrone-3-sulfate (ES) in term of the *in vitro* kinetic parameters of K_m (substrate affinity) and V_{max} (translocation ability) was compared over the concentration range from 4 nM to 10 μM. HEK293 cells transfected with SLCO2B1*1 (wild type), SLCO2B1*2, and SLCO2B1*3 alleles were used. The K_m values of ES were comparable, whereas significant differences were observed in the V_{max} values of ES in SLCO2B1*1-, SLCO2B1*2-, and SLCO2B1*3-transfected HEK293 cells with values of 100, 71.1, and 42.5, respectively. This suggests that OATP2B1*3 is associated with lower expression of OATP2B1 protein, which reduces its intrinsic transport activity to 43% of that of wild type after normalizing for protein expression. Furthermore, this may affect the physiological function and/or pharmacological effect of OATP2B1 substrates *in vivo*. Indeed, Imanaga et al. investigated the impact of OATP2B1 polymorphism on the pharmacokinetics of fexofenadine following oral dosing in Japanese population. As depicted in Figure 7.9, fexofenadine plasma exposure ($AUC_{(0\to\infty)}$ and C_{max}) is SLCO2B1*3 dose dependent. At an oral dose of 60 mg, fexofenadine area under plasma concentration time curve (AUC) in CT, and TT genotypes were 64% and 62% of that observed in CC genotype of 1762 ng h/mL, respectively. Similarly, the peak plasma concentration (C_{max}) of fexofenadine in CT and TT genotypes were 65% and 52% of that observed in the wild-type CC genotype of 343 ng/mL, respectively. Overall, this study suggests the critical role of OATP2B1 in the oral absorption of fexofenadine. Similar trends were also reported when the impact of OATP2B1 polymorphism on the pharmacokinetics of celiprolol and talinolol was investigated [55, 108].

SLCO2B1c.935G>A is another genetic variant that was reported to have an impact on the absorption of OATP2B1 substrates and causes amino acid change of R312Q. In Caucasian, African American, and Korean populations, the allelic frequencies of SLCO2B1c.935G>A is relatively common with values of 8.2%, 13%, and 42.6%, respectively. A number of conflicting findings on the clinical relevance of SLCO2B1c.935G>A impeded our understanding of its role in clinical settings. For example, in two separate research articles, Mougey et al. reported that c.935G>A variant was associated with lower montelukast plasma concentrations and poor biological response in adolescent and young adult patients with asthma when compared with the wild-type OATP2B1 variant [56, 109]. Interestingly, no homozygotes for the SNP (c.935AA) were included in either of these studies. However, Tapaninen et al. reported no impact of *SLCO2B1* c.935G>A variant on the disposition of montelukast or aliskiren in healthy volunteers [110]. This is not consistent with the outcome of Mougey et al. studies. These conflicting findings are attributed to major differences in the clinical study design.

TABLE 7.4 Nonsynonymous Nucleotide Polymorphisms in the *SLCO2B1* Gene [102–105]

Transporter (Old Name)	SNP	mRNA	Amino Acid Exchange	Common Haplotype	rs Number	Exon	% Allelic Frequency (N)			MAF[a]	Comment
							Caucasians	Africans/African Americans	Asians		
OATP2B1 (OATPB)	c.1175C>T	1570	T392I	SLCO2B1*2	rs1621378	10			0	NA	Reduced activity
	c.1457C>T	1852	S486F	SLCO2B1*3	rs2306168	11	2.8		30.9[b] (26.87[c])	18.8	Reduced activity
	c.935G>A	1330	R312Q		rs12422149	8	8.2	13	42.73[c]	18.1	Reduced activity
	c.43C>T	438	P15S		rs56837383	3			100[c]	1.2	
	c.601G>A	996	V201M		rs35199625	6	2.1		7.27[c]	1.8	Reduced activity
	c.1240G>A	1635	G414S		rs111782322	10				NA	
	c.1526G>A	1921	V201M R509H		rs140407559	11				NA	Reduced activity
	c.1624G>A	2019	V542M		rs143480565	12				NA	
	c.1638C>A	2033	N546K		rs145875125	13				NA	
	c.1642G>A	2037	V548M		rs149242910	12				NA	
	c.2071G>A	2466	V691I		rs149765874	15				NA	
	g.282G>A				rs2712807	Promoter				NA	

NA: Not applicable.
[a]Global minor allele frequency (MAF) is obtained from dbSNFP and shown in %.
[b]Japanese population [105].
[c]Korean population [105].

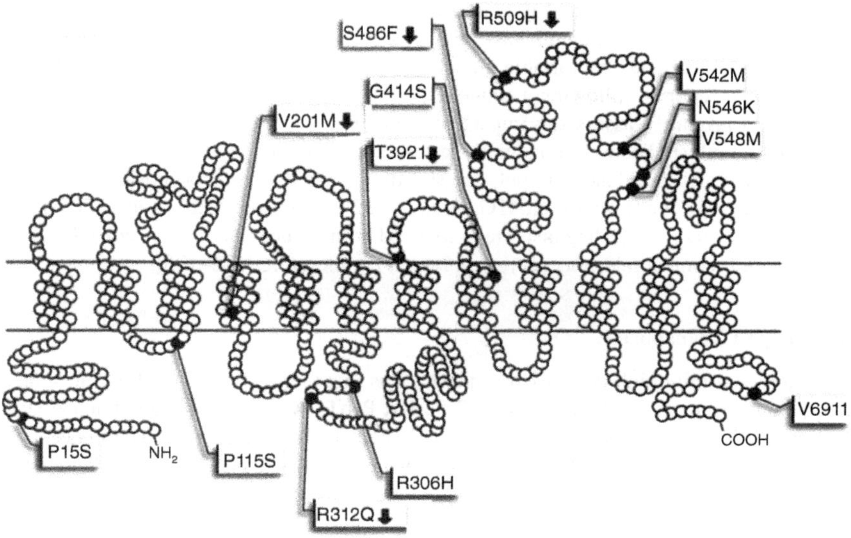

FIGURE 7.8 Schematic representation of the single-nucleotide polymorphisms in structure of OATP2B1. Arrows indicate alteration of *in vitro* functional activity of the variant protein. ↓: Reduced activity, ↑: increased activity. Nakanishi and Tamai 2012 [104]. Reproduced with permission of Elsevier.

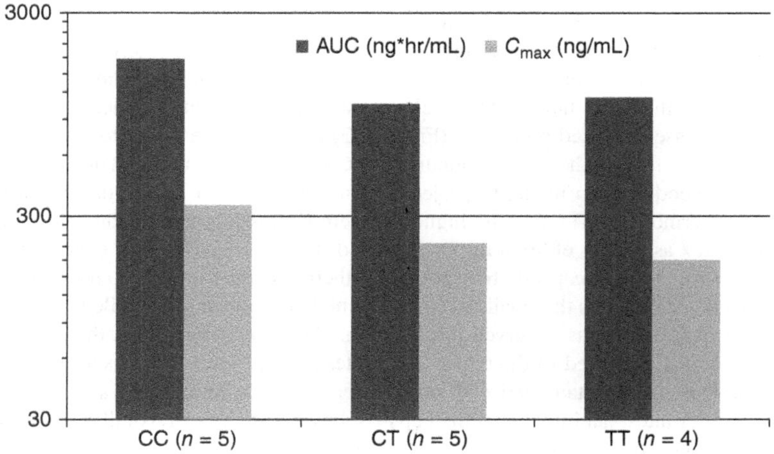

FIGURE 7.9 The $AUC_{(0\to\infty)}$ and C_{max} parameters of fexofenadine following oral administration of 60 mg orally with water in relation to the *SLCO2B1* c.1457C>T polymorphism. Imanaga et al. 2011 [107]. Reproduced with permission of Wolters Kluwer Health, Inc.

For example, a Mougey et al. study conducted in 2009 was in part to compare the efficacy of low-dose theophylline, montelukast, and placebo. Furthermore, the patient population compliance is thought to be variable since the plasma exposure at the morning dose was lower than that reported following a single dose. As for Mougey et al.'s second study, it was conducted to primarily assess the impact of fruit juice on the disposition of montelukast. Furthermore, both studies were conducted retrospectively using patients with different ethnic backgrounds, which may lead to population stratification and cause false positives.

Other genetic variance such as c.601G>A and c.1526G>A cause V201M and R509H amino acid change, respectively. They also are associated with reduced activity *in vitro*. No *in vivo* studies on their clinical relevance were reported in the literature [102–105].

7.2.2 PepT1 Transporter

PepT1, a low-affinity/high-capacity transporter, was first cloned from the rabbit intestine and subsequently from rat and human (Table 7.1). Rat PepT1 was localized to the apical side of intestinal epithelial cells and in early regions (S1 segments) of apical proximal tubules. PepT1 is involved in electrogenic, H+-dependent transport of small peptides as well as various peptidomimetics drugs.

7.2.2.1 PepT1 Structure and Distribution PepT1 (*SLC15A1*) transporter is a proton-dependent dipeptide transporter. The cloned human PepT1 cDNA sequence encodes a 708 amino acid protein (M_{wt} 79 kDa) with an isoelectric point of 8.6. PepT1 has 12 putative α-helical membrane-spanning domains and large extracellular loop between the ninth and tenth transmembrane domains with intracellularly localized N- and C-termini [111, 112]. Herrera-Ruiz et al. evaluated the spatial expression patterns of PepT1 in the brush border membrane of human and rat gastrointestinal tracts (GIT) [113]. The group reported that PepT1 appears to be localized predominantly in the duodenum, with decreasing expression in the jejunum and ileum. Furthermore, it was also shown to be widely expressed in the apical membrane of the enterocytes in rats. In 2013, Oswald et al. quantified the expression of PepT1 along the entire human intestine of eight body donors and in Caco-2 using a mass spectrometry-based targeted proteomic (Figure 7.3) [37]. It is interesting to note that PepT1 protein expression in both human jejunum and Caco-2 is comparable. This would render Caco-2 as a good *in vitro* model to project the impact of PepT1 on the oral absorption of NMEs. Furthermore, PepT1 had the highest protein expression when compared with other intestinal influx as well as efflux transporters tested (Figure 7.2). This is in concordance with being a low-affinity/high-capacity transporter. Furthermore, the same group noted that PepT1 is exclusively expressed in the small intestine with higher abundance in the ileum (Figure 7.2). No mRNA expression was observed in the colon. This may suggest that the absorption of PepT1 substrates is limited to the upper small intestine and low or no absorption occurs in the colon. This was substantiated with clinical reports that showed only a significant drug absorption after intestinal infusion of the PepT1 substrates such as amoxicillin and benazepril to the small intestine but not to colon [114, 115]. Similar findings were reported in mouse that has a PepT1 expression in the GIT that is comparable to that observed in human, where expression is exclusively limited to small intestine and no expression is observed in the colon [116]. Posada and Smith investigated whether differences existed in the regional activity of PepT1 by measuring the effective permeability of cefadroxil in four regions of the intestines in wild-type and PepT1 knockout mice. The effective permeability of cefadroxil was significantly lower in duodenum, jejunum, and ileum of PepT1 knockout mice compared to wild-type mice. In the colon, no substantial differences in cefadroxil effective permeability were observed between the two genotypes, which corroborate the poor expression of PepT1 in colon [117].

Renal PepT1 is localized predominantly in S1 segments of the early convoluted proximal tubule in rat. PepT1 mRNA is expressed at lower level in the rabbit and human liver and human pancreas [118]. PepT1 protein expression was not found in brain [119].

7.2.2.2 ***PepT1 Mechanism of Drug Uptake*** PepT1 was shown to be independent of Na^+ and similar to OATP2B1 and MCT1 uses a proton gradient and membrane potential to provide the necessary driving force for substrate translocation (Figure 7.1). At the brush border membrane of the enterocytes, an inward protein gradient is generated through the activity of an electroneutral proton/cation exchanger, that is, Na^+/H^+ antiporter. Then the uptake of PepT1 transporter substrates is coupled by the influx of protons back into the enterocytes [120].

Similar to other intestinal influx transporters, the uptake of the PepT1 substrates is strongly dependent on the extracellular pH, where a pH of 4.5–6.5 depending on the net charge of the substrate is needed for optimal transport activity. In an interesting study, Irie et al. investigated the transport mechanisms of PepT1 for neutral and charged substrates by experimental studies and computational simulation [121]. The group reported that the uptake studies revealed that the Michaelis–Menten constant (K_m) of glycylsarcosine (Gly-Sar), a neutral substrate, decreased with a fall in pH from 7.4 to 5.5. The K_m value of an anionic substrate, ceftibuten, declined steadily with decreasing pH. The group hypothesized the transport mechanisms in which (i) H^+ binds to both the H^+ and substrate-binding sites, and (ii) H^+ at the substrate-binding site inhibits the interaction of neutral and cationic substrates but is necessary for that of anionic substrates.

7.2.2.3 ***PepT1 Substrates and Inhibitors*** PepT1 has a broad substrate specificity and play a unique role in the absorption and distribution of peptidomimetics that include β-lactam antibiotics such as cephalosporins and penicillins, angiotensin converting enzyme inhibitor such as captopril, and ester prodrugs enalapril and fosinopril, selected rennin inhibitors, antitumor agents such as bestatin, dopamine receptor antagonists such as sulpiride, prodrugs of acyclovir (e.g., valacyclovir), levodopa or 3,4-dihydroxy-l-phenylalanine (L-dopa), azidothymidine or zidovudine (ZT), and various amino acid ester prodrugs [112, 122]. Table 7.5 lists known PepT1 substrates, their *in vitro* kinetics and oral doses. Unlike OATP2B1, PepT1 is low-affinity and high-capacity transporters, which is consistent with its high protein expression in human small intestine. This allowed the dosing of large oral doses up to 4 g a day of β-lactam and cephalosporin antibiotics (Table 7.5).

Bretschneider et al. investigated the impact of PEPT1 substrate affinity on their absorption rate and bioavailability. The group reported a strong relation between the affinity of PepT1 substrates and their oral absorption. Penicillins and cephalosporins compounds with high *in vitro* uptake rates and affinity (K_m) are typically associated with higher absorption rates [133]. Interestingly, a threshold K_i value of 14 mM with respect to Gly-Sar (probe substrate) uptake is required to enable good absorption across the enterocytes.

TIP:

Considering the exclusive expression of PepT1 in small intestine, it should be noted that PepT1 substrates are not amenable for controlled release formulation using traditional techniques targeting colon. Instead, gastro-retentive-controlled release technologies would be a viable option that can be considered to extend the half-life of the molecule and improve patient compliance. However, the typical high oral doses of PepT1 substrates would have a significant challenge for developing such technology and should be investigated prior to major investment [134, 135].

TABLE 7.5 Partial List of Known Drugs that are in Clinical Use and Substrates of PepT1 Transporter [123–132]

Object	K_m (μM)	V_{max} (nmol/min/cm^2)	Oral Dose/Day (mg)
Amoxicillin			1,500
Ampicillin			2,000
Benazepril	900–1,000		40
Bestatin	500–4,000		
Captopril			300
Cefaclor	4,000–11,000		1,500
Cefadroxil	1,000–7,000		1,000
Cefditoren			800
Cefixime	1,900	1.2	400
Ceftibuten	300–1,500		400
Cephalexin	4,000–15,000		2,000
Ceronapril	600–1,000		
Cloxacillin	3,000		4,000
Cyclacillin	100–600		
Enalapril	2,000–20,000		40
Fluvastatin	300		80
Fosinopril	30–200		40
Glibenclamide	25		20
Irbesartan	200		150
JBP485	1,700		
Levovirin			
Losartan	20–50		100
LY544344			
NP-647	12,000		
Oseltamivir	6,540	45.6	75
Quinapril	400–1,000		80
Spirapril	2,200		
Temocapril			
Valacyclovir			1,000
Valsartan	400–600		40
Zofenopril	47		

A number of PepT1 inhibitors were identified by screening for transporter substrates. It is worth noting that new inhibitors are rather designed by a rational approach based on the current understanding of the SAR of the PepT1 transporter. Table 7.6 lists known inhibitors of PepT1 transporter. Clinical inhibitors of PepT1 transporters involve compounds from the orally active sulfonylurea antidiabetic drugs such as nateglinide and glibenclamide, tolbutamide, and chlorpropamide and compounds related to cephalosporins and ACE inhibitor-ester prodrugs, for example, quinalapril and quinalaprilat, fosinoprilat, and enalaprilat (Table 7.6). In the clinic, various research groups reported multiple PepT1-mediated drug–drug interactions. For example, Garrigues et al. investigated the impact of cephalexin on the oral exposure of cefadroxil. Indeed, cephalexin delayed cefadroxil T_{max} and decreased its oral exposure (AUC) (Figure 7.10) [142]. This suggests that the oral absorption of cefadroxil is dose dependent and driven by a saturable mechanism mediated by PepT1 transporter.

TABLE 7.6 Partial List of Known Drugs that are in Clinical Use and Inhibitors of PepT1 Transporter [123–132, 136–141]

Inhibitor	Object	K_i (μM)	IC_{50} (μM)
6-Aminopenicillanic acid	Glycylsarcosine	50,000	
7-Aminocephalosporanic acid	Glycylsarcosine	15,000	
Amoxicillin	Glycylsarcosine	25,000	
Amoxicillin	Glycylsarcosine		
Bepridil	Cefixime		
Cefaclor	Glycylsarcosine	4,520	
Cefadroxil	Glycylsarcosine		
Cephalexin	Glycylsarcosine	1,100	
Cephalexin	Jbp485	260	
Cephradine	Glycylsarcosine		
Cloxacillin	Glycylsarcosine	3,000	
Dicloxacillin	Glycylsarcosine	7,200	
Diltiazem	Cefixime		
Eprosartan	Glycylsarcosine	2,200	
Etodolac	Glycylsarcosine		
Glycylsarcosine	Jbp485	16,900	
Ibuprofen	Glycylsarcosine	400	
Irbesartan	Glycylsarcosine	230	
Jbp485	Glycylsarcosine	36,000	
Jbp923	Glycylsarcosine	2,550	
Jbp923	Jbp485	770	
Latamoxef (moxalactam)	Glycylsarcosine	12,300	
Losartan	Cefadroxil		
Losartan	Glycylsarcosine	24	
Losartan	Glycylsarcosine	52	
Metampicillin	Glycylsarcosine	12,700	
Nifedipine	Cefixime		0.007
NP-647	Glycylsarcosine		
Oseltamivir	Glycylsarcosine		27,400
Oxacillin	Glycylsarcosine	12,000	
Piroxicam	Glycylsarcosine		
Sertraline	Glycylsarcosine		
Temocapril	Glycylsarcosine		
Valacyclovir	Glycylsarcosine		
Valsartan	Cefadroxil		
Valsartan	Glycylsarcosine	390	
Valsartan	Glycylsarcosine	0.56	
Verapamil	Cefixime		0.0283

7.2.2.4 PepT1 SAR The 3D structure of the substrate-binding site of PepT1 is not yet known, but its template has been proposed by the large variety of substrates. The latest model is illustrated in Figure 7.11. The peptide bond is not required for substrate-binding specificity of PepT1 transporter. Just the two oppositely charged head groups separated by a four spacer carbon unit with a distance >500 and <635 pm were identified as a minimal

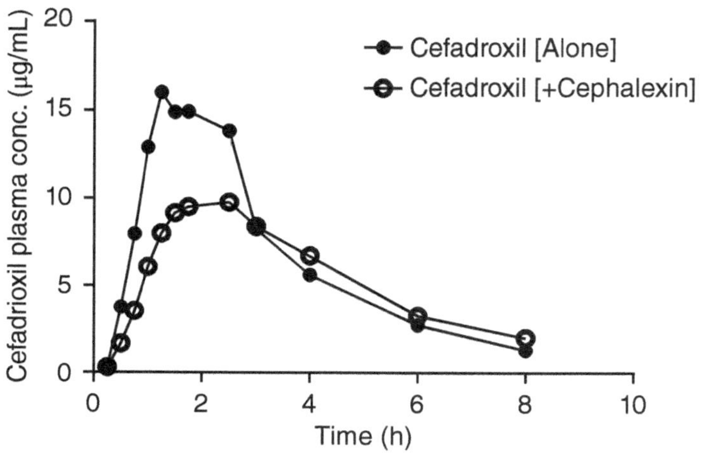

FIGURE 7.10 The mean plasma concentration of cefadroxil in healthy volunteers in absence (-•-) and presence (-o-) of 5 mg/kg cephalexin (45 mg/kg). Garrigues et al. 1991 [142]. Reproduced with permission of Springer.

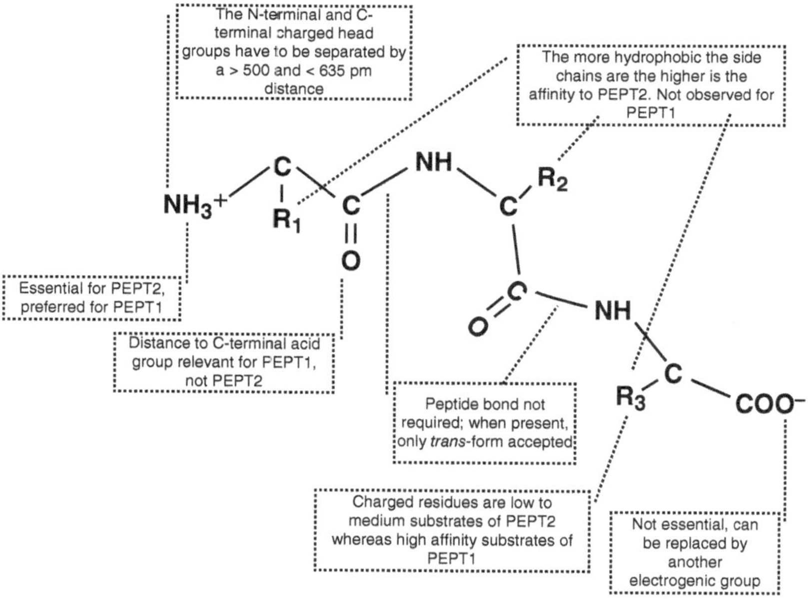

FIGURE 7.11 Key features in compounds determining recognition as a peptide transporter substrate. The structural components that critically control affinity of key elements are presented on the backbone of a tripeptide as model compound. Rubio-Aliaga and Daniel 2008 [112]. Reproduced with permission of Taylor & Francis.

requirement. However, if a peptide bond is present, only the backbone carbonyl function and not the nitrogen increases substrate-binding affinity significantly. This minimal configuration also explains the efficient transport of δ-aminolevulinic acid that serves as a precursor for the endogenous porphyrin accumulation on which photodynamic therapy of tumors is based. The side chains provided in di- and tripeptide and in xenobiotics with charge polarity and conformation are important in affecting the binding affinities. Besides a preferred free N-terminal amino group, there is a high electron density around the terminal carboxylic group in dipeptide or alternatively around the carbonyl group of the second amino acid in a tripeptide structure is warranted [112].

High electron densities at the first and third side chains domains as well as the presence of hydrophobic side chains significantly contribute to overall affinity. These models allow good prediction of substrate affinities and a final proof that the compound is indeed transporter and not only serve as competitors or even inhibitors is mandatory.

7.2.2.5 Influence of Genetic Polymorphism on* In Vitro *and* In Vivo *Activity of PEPT1 In 2004, Zhang et al. identified nine nonsynonymous and four synonymous coding region SNPs in PEPT1 transporter gene [143]. They used a DNA polymorphism discovery panel of 44 ethnically diverse individuals. The group transiently transfected HeLa cells with plasmids constructed by site-directed mutagenesis for each of the nine nonsynonymous variants for functional characterization. Gly-Sar uptake was used for transporter kinetics, while western plot and immunocytochemical analyses were utilized to measure protein expression. The mRNA transcription level of all of the mutants was comparable with the level of the reference sequence in transfected HeLa cells. Interestingly, from the nonsynonymous SNPs investigated, P586L was the only SNPs that had 10-fold lower value for V_{max}. This is ascribed to its lower immunoactive protein level and plasma membrane expression suggesting that P586L has a significant impact on PEPT1 degradation, membrane insertion, and/or translation.

Another PEPT1 genetic variant that should be considered is PEPT1-F28Y. It has low allelic frequency and is associated with significantly lower cephalexin uptake. This is attributed to higher K_m value and unlike P586L is not associated with changes in protein expression [144]. Interestingly, the expression of F28Y SNPs is exclusively expressed within the African American population.

Overall, nonsynonymous SNPs commonly affect the activity of transporter protein by altering its expression or binding affinity. In the case of PEPT1, only two SNPs demonstrated this trend, that is, F28Y and P586L. Other SNPs did not affect the activity of PEPT1.

7.2.3 Monocarboxylate Transporters

The bidirectional movement of monocarboxylic acids across the plasma membrane is catalyzed by a family of proton-linked monocarboxylate transporters (MCTs). MCTs are encoded by the *SLC16A* gene family, of which there are 14 known members that were identified through screening genomic and expressed sequence tag databases. To date, seven MCT family members have been functionally characterized: MCTs 1–4 catalyze proton-coupled transport of monocarboxylates such as lactate and pyruvate [145], MCT6 has been shown to transport drugs such as bumetanide and nateglinide [146], MCT8 is a high-affinity thyroid hormone transporter, and MCT10 (TAT1) is an aromatic amino acid transporter. The other members of the MCT family remain to be characterized [145]. This chapter focuses on MCT1 transporter due to its high expression in the human GIT as discussed below.

7.2.3.1 MCT1 Structure and Distribution In 1992, Kim et al. was the first group to identify MCT1 in Chinese hamster ovary cells when altered mevalonate transport was observed as a result of a single-point mutation in a gene of previously unknown putative function [147]. When the cDNA encoding the wild-type version of this gene was expressed in human cells, it catalyzed inhibitor-sensitive monocarboxylate transport and was named MCT1 [148]. cDNA sequences from human, mouse, and rat were later cloned, all translating into a protein of similar weight (~54 kDa). *SLC16A1*, the human MCT1 gene, spans around 44 kb and has been mapped to the chromosomal region 1p13.2-p12. It has four introns and five exons and encodes a 500 amino acid protein (molecular weight of 55 kDa). Human, rat, and mouse MCT1 genes share 87% amino acid sequence identity with CHO MCT1 [149, 150].

Hydropathy plots predict most members of the MCT family, including MCT1, to have 12 transmembrane (TM) α-helical domains with a large intracellular loop between TM segments 6 and 7 and intracellular C- and N-termini [151, 152]. For correct trafficking to the plasma membrane and for functional activity, MCT1 (also MCT 2, 3, and 4) requires an accessory protein, CD147. This is a member of the immunoglobulin superfamily that was identified as an accessory protein essential for MCT1. The proposed interaction between MCT1 and CD147 involves an arginine residue within TM 8 of MCT1 and a glutamic acid residue in the single TM of CD147 [153].

Based on the reported crystal structures of two members of the major facilitator superfamily, the *Escherichia coli* glycerol-3-phosphate transporter (G1pT) and lactose permease (Lac Y) [154, 155], the structure of MCT1 has been modeled [156]. Although the 12 TM domains are not exactly where predicted from the hydrophobicity analysis, the overall structural features are as predicted. The model is consistent with results of studies involving site-directed mutagenesis that have identified several key residues [156].

MCT1 is expressed in most tissues and is especially prominent in heart and red skeletal muscle where it facilitates lactic acid uptake for oxidation. MCT1 is also important in erythrocytes and all cells under hypoxic conditions, where it is involved in the efflux of glycolytically produced lactic acid.

In the GIT, the major expression of MCT1 is in the distal colon, followed by proximal colon, ileum, and jejunum. This is a key difference from the trends reported with PepT1 and OATP2B1 transporters, where expression is only limited to small intestine. This is a key attribute that drove Xenoport scientists to target the use of MCT1 to enable absorption of compounds across the colon and once daily dosing using traditional controlled release formulation targeting colon (see below) [157–159]. Immunohistochemical and densitometric studies suggested that MCT1 expression is located at the apical membranes of these organs [160]. MCT1 is responsible for the absorption of short chain fatty acids such as acetate, propionate, and butyrate, derived from microbial fermentation by the colonic microflora [161].

7.2.3.2 Mechanism of Drug Uptake MCT1 transports monocarboxylates by a facilitative diffusion mechanism, which requires the cotransport of protons. The driving force behind this diffusion is provided by both the substrate and proton concentration gradients, with the pH gradient determining the extent of transport activity [151, 162]. Thus, the low-pH microenvironment found in the lumen of the small intestine provides the driving force for MCT1-mediated oral absorption, which is similar to the mechanisms of drug uptake mediated by PEPT1 and OATP2B1 (Figure 7.1) [163].

Wilson et al. used site-directed mutagenesis to identify key substrate-binding residues and structural modeling. The group suggested that translocation cycle is the mechanism of transport for MCT1 [153]. This model implies that MCT1 exists in an open and closed conformation, consistent with "rocker switch" mechanism, as seen with other members of the major facilitator superfamily [164]. Both the N- and C-terminal halves of MCT1 tilt back and forth against each other along an axis that separates these two domains, allowing the substrate-binding site alternating access to either side of the membrane. The key step is the protonation of the buried Lys^{38}

(K) (driven by lower pH) causing the protein to switch to the open conformation, and allowing the binding of lactate to the protonated Lys^{38}. The lactate and proton then pass through the channel via the Asp^{302-} (D) and Arg^{306+} (F) ion pair, which are in close proximity with Phe^{360} (R^+). It has been suggested that these three putative binding sites control channel (substrate) selectivity [156]. Once lactic acid passes through this selectivity "filter," Lys^{38} is deprotonated (after passing the lactic acid) and MCT1 returns to a closed conformation, opening the Asp^{302}/Arg^{306} site to the opposite side of the membrane and releasing lactic acid into the intracellular space.

7.2.3.3 MCT1 Substrates and Inhibitors Studies on MCT1 expressed in *Xenopus laevis* oocytes [165] and the endogenous MCT1 of erythrocytes [166, 167] have enabled detailed characterization of its properties. MCT1 transports compounds that generally consist of weak organic acids with the carboxyl group attached to a relatively small R group containing lipophilic or hydrophilic properties [168]. Table 7.7 lists known endogenous and exogenous substrates for MCT1 with reported *in vitro* kinetics parameters. MCT1 was shown to transport short chain, unbranched aliphatic monocarboxylates such as acetate and propionate and substituted monocarboxylates pyruvate, lactate, acetoacetate, and β-hydroxybutyrate, with the K_m values for pyruvate and lactate about 0.7 and 3–5 mM, respectively [145]. Other MCT1 monocarboxylate substrates include the branched chain keto-acids (formed from the transamination of leucine, isoleucine, and valine) and the ketone bodies acetoacetate, β-hydroxybutyrate and acetate [167], and exogenous acids *p*-aminohippuric acid, benzoic

TABLE 7.7 A Partial List of Known Substrates for MCT1 Transporter [158, 159, 169–180]

Object	K_m (μM)
Acetoacetate	5,500
α-Ketobutyrate	200
α-Ketoisocaproate	700
α-Ketoisovalerate	1,300
Arbaclofen placarbil	
Atorvastatin	
Carindacillin	
Cefdinir	
δ-β-Hydroxybutyrate	10,100
δ-Lactate	27,500
Ferulic acid	
Fluorescein	
Gabapentin enacarbil	220
γ-Hydroxybutyrate	7,700
Ketoprofen	
Lactate	2,600
L-Lactate	450,000
Naproxen (*S*-Naproxen)	
Pravastatin	
Pyruvate	1,000
Salicylic acid	
Simvastatin	
Valproic acid	

acid, γ-hydroxy butyrate, foscarnet, mevolonic acid, and salicylic acid [168, 181]. MCT1 is also thought to be responsible for the intestinal absorption of the β-lactam antibiotics such as carbenicillin indanyl sodium as well as phenethicillin and propicillin [169].

Some drugs can be specifically designed to target uptake via MCT1 to improve oral absorption and extending their half-life by enabling once daily dosing using traditional controlled release formulation targeting colon. An interesting example is XP13512, also known as gabapentin enacarbil, which is rapidly absorbed along the entire length of the GIT via MCT1 [158, 159]. XP13512 is an anionic compound produced by the reversible modification of the amine group of gabapentin (which has limited oral absorption confined to upper small intestine and mediated by absorption via paracellular pathway and system L transporters), with an acyloxyalkylcarbamate promoiety [182]. Unlike gabapentin, increase in the oral dose of XP13512 is associated with proportional increase in the systemic exposure of gabapentin (Figure 7.12) [157]. The less than proportional increase observed with gabapentin is attributed to its affinity to system L transporter. This is a high-affinity and low-capacity transporter that is saturated at high luminal concentrations. Targeting MCT1 transporters that are predominantly expressed in the colon allowed Xenoport scientists to develop extended release formulation of XP13512. This enabled once daily dosing of gabapentin, which reduces dosing frequency and increases patient compliance [183]. Similar approach was adapted in the design of arbaclofen placarbil a novel R-baclofen prodrug that is a good substrate for MCT1 [184].

Nonspecific small molecule inhibitors of MCT1 have been identified, and fall into three broad categories (Table 7.8): (i) competitive inhibitors that are aromatic or bulky monocarboxylates such as 2-oxo-4-methylpentanoate, phenylpyruvate, and α-cyano-4-hydroxycinnamate; (ii) a more potent and diverse range of amphiphilic compounds such as quercetin, phloretin, 5-nitro-2-(3-phenyl-propylamino)-benzoate (NPPB), and niflumic acid; and (iii) reversible such as 4,4′-di-isothiocyanatostilbene-2,2′-disulfonate (DIDS). Interestingly, prolonged incubation with DIDS causes irreversible inhibition of MCT1 through covalent modification of the protein [187]. Other examples of irreversible MCT1 inhibitors include a range of thiol and amino reagents, particularly the organomercurial thiol reagent *p*-chloromercuribenzene sulfonate (*p*CMBS) [167]. It should be noted that all the above inhibitors are nonselective, with potencies in the micrometer range.

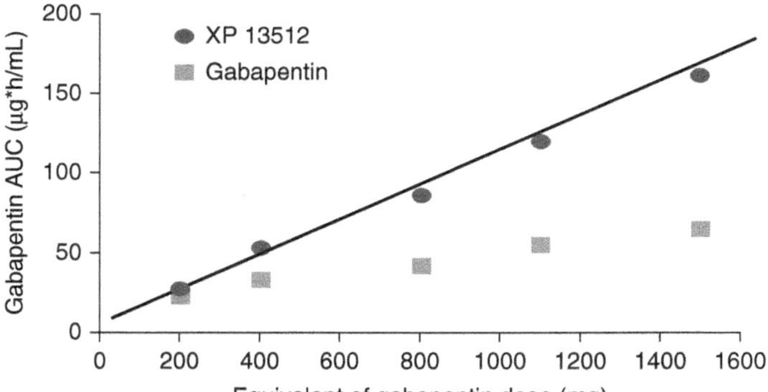

FIGURE 7.12 Dose proportionality of gabapentin blood $AUC_{(0\rightarrow\infty)}$ following single oral doses of XP13512 immediate release capsules or oral gabapentin.

TABLE 7.8 A Partial List of Known Inhibitors for MCT1 Transporter [170, 171, 184–186]

Precipitant	Object	K_i (μm)	IC_{50} (μm)	Percent of Inhibition (%)
Arbaclofen placarbil	Lactate		1100	
Benzoic acid	Rosmarinic acid			
Diclofenac	Benzoic acid		100	
Diflunisal	Benzoic acid		50	
Gabapentin enacarbil	Lactate		620	
Ketoprofen	Benzoic acid	380	440	
Lactate	Lactate			60
L-Lactate	Lactate			60
Naproxen (*S*-Naproxen)	Benzoic acid	220	250	
Rosmarinic acid, luteolin, and apigenin mixture	Benzoic acid			

TIP:

Unlike PepT1 and OATP2B1, MCT1 transporter has a high expression in the colon. In addition, it is a low-affinity/high-capacity transporter. These attributes enabled its use to develop controlled release formulation that target the colon and can be effectively used to extend the half-life of compounds such as gabapentin enacarbil. It should be emphasized that since this transporter is of high capacity, it provides a key advantage, which is the ability to absorb high oral doses with low risk of saturation. If your NME is a substrate for MCT1 transporter, keep these attributes in mind; they may come handy!

7.2.3.4 Influence of Genetic Polymorphism on In Vitro ***and*** In Vivo ***Activity of MCT1*** Lean and Lee were the first to comprehensively analyze *SLC16A1*, MCT1 gene, and identify its genetic variations [188]. The assessment was conducted in the ethnic Chinese group of the Singaporean population ($n = 95$). Using DNA sequencing, the group identified seven genetic mutations of *SLC16A1* gene. A total of four novel nonsynonymous mutations were reported, two in the promoter region and two in the coding exons. In addition, one mutation in the intron and two mutations in the 3′ untranslated region were identified. The -363-855T>C is a novel genetic mutation detected in the promoter region. Another novel SNP is 1282G>A (Val428Ile), which was reported as heterozygotic ($N = 4$). A conversion from valine to isoleucine amino acid at 428 position in MCT1 protein is associated with this mutation. Since both amino acids are nonpolar, this mutation is predicted to be benign and not to affect the functional activity of the MCT1 protein. The 1470T>A (Asp490Glu) mutation was identified as a common polymorphism, where the T and A allele frequency are 34.7% and 65.3%, respectively. Interestingly, this allele frequency is comparable to that shown with the Caucasian population [189]. Furthermore, Poly-Phen program predicted this SNP to be benign and is unlikely to be associated with a significant impact on the functional activity of the MCT1 protein. IVS3-17A>C in intron 3 and 2258 (755)A>G in 3′ untranslated region are novel mutations, which are found to be common polymorphisms in the local Chinese population.

7.3 DRUG ABSORPTION AND CONCENTRATION GRADIENT

Absorption rates of most drugs can be described by Fick's law of simple diffusion. The flux rate of drug absorbed is determined by the product of the coefficient of diffusion, the surface area of the mucosa, and the concentration gradient of the drug [190]. It should be emphasized that the concentration gradient across the unstirred water layer and epithelial plasma membrane leads to a reduced drug concentration in the enterocytes, creating a favorable sink condition. Furthermore, the drug in the intracellular space will continue to diffuse along a concentration gradient into capillary blood, and the mucosal capillary blood drains into the mesenteric veins, which congregate to form the hepatic portal vein. In addition to the mucosal blood flow, there is significant contribution from the splanchnic area, such as the splenic vein and gateric vein, to the portal blood flow. Again, this leads to a significant reduction in the drug concentration in the portal vein before entering the liver. The drug concentration in the portal vein will be diluted further by intimate mixing with hepatic arterial blood before it passes evenly through the hepatocytes. Collectively, it is not unexpected that the drug concentration in hepatocytes would be significantly lower than that in enterocytes [3].

7.3.1 Saturable Intestinal Uptake

The overall transport of a drug across the intestinal enterocytes can be obtained by fitting a model incorporating saturable and nonsaturable components to the drug transport data at 37 °C [191]:

$$J = \frac{J_{\mathrm{max}} \cdot C}{K_{\mathrm{m}} + C} + K_{\mathrm{D}}C \tag{7.1}$$

The total flux (J) of a compound across intestinal membrane is determined by four variables: J_{max} is the maximal uptake rate, K_{m} the transporter substrate-binding affinity, K_{D} the kinetic constant for nonsaturable transport, and C the luminal drug concentration. The impact of intestinal transporters on the overall absorption of drug across the intestine is determined by the ratio of the active process ($J_{\mathrm{max}}C/(K_{\mathrm{m}} + C)$) to the total flux, J, of the drug molecule. Therefore, for a drug molecule that is high solubility/high permeability, that is, *BDDCS* Class I, transcellular passive diffusion is the main route of absorption. In general, for this class of compounds, dissolution rates generally exceed gastric emptying rates when formulated as immediate release products. As a result, these compounds are usually well absorbed and due to their high solubility, efflux and influx transporters will be saturated and their contribution to the overall J will be minor. For high-permeability/low-solubility drugs "*BDDCS* Class II," the oral absorption of these compounds tend to be dissolution rate limited, affecting the value of variable C, which leads to erratic and incomplete absorption from the GIT of animals and humans. This phenomenon is partially attributed to the fact that the compound luminal concentration is not high enough to saturate the efflux transporters. Therefore, to increase the oral bioavailability of *BDDCS* Class II drugs, medicinal chemists should strive to increase drug solubility or target high-affinity uptake transporters, for example, "OATP2B1 transporters" with K_{m} values higher than the luminal drug concentration to minimize potential saturation. They should also avoid interactions with high-affinity efflux transporters to maximize the net positive flux [15–17].

For low-permeability, high-solubility compounds "*BDDCS* Class III," the oral absorption of these compounds is permeability rate limited and dissolution usually occurs rapidly. Therefore, to increase the oral bioavailability of this class of molecules, influx transporter located at the intestinal brush membrane should be targeted. As shown above, influx transporters are divided into two categories: low-affinity/high-capacity transporters (PepT1 and

MCT1), and high-affinity/low-capacity transporters (OATP2B1). When targeted, medicinal chemists should pay close attention to the compound-projected oral dose. Compounds with an oral dose <50 mg are associated with a luminal concentration of about <0.4 mM. At this concentration, low-affinity/high-capacity transporters will lead to a comparable flux profile to that shown by high-affinity/low-capacity transporters. When simulations of high-affinity/low-capacity (OATP2B1) and low-affinity/high-capacity (PepT1) transporter flux were modeled across clinically relevant human oral doses using Equation 7.1. In the case of high-affinity/low-capacity (OATP2B1), K_m was set to 0.27 mM and J_{max} was set to 300 pmol/min/mg. In the case of low-affinity/high-capacity (PepT1), K_m was set to 5.4 mM and J_{max} was set to 3000 pmol/min/mg [192]. The calculated intestinal flux for each transporter intersects at ~50 mg demonstrating that high-affinity (low-capacity) transporters such as OATP2B1 exhibit maximal increases in flux at low oral doses but saturate beyond doses of ~50 mg, whereas low-affinity (high-capacity) transporters such as PepT1 and MCT1 do not saturate until oral doses exceed 1000 mg.

PepT1 transporter is believed to play a better role in increasing the overall oral exposure of compounds with high oral dose as compared to OATP2B1. It is considered as an attractive target for *BDDCS* Class III drugs with projected efficacious doses greater than 50 mg, which is relevant to antibacterial, antiviral, cardiovascular, and metabolic disease areas. It is interesting to note that known substrates for PepT1 transporters that include cephalosporins, penicillins, ACE inhibitors, and antivirals tend to have high oral doses (>50 mg), whereas known substrates for OATP2B1 transporters such as various statins and antidiabetics tend to have low oral doses (<50 mg).

BDDCS Class IV compounds (e.g., acetazolamide, amphotericin, chlorothiazide, ciprofloxacin, and furosemide) have both low-solubility and low-permeability profiles [12, 14]. Therefore, these compounds usually have poor oral bioavailability and are associated with significant inter- and intraindividual variability. For these set of compounds, approaches applicable to Class II and III compounds may be helpful in improving their oral absorption.

7.4 CONCLUSION

It is well established that many transporter proteins are expressed in the GIT. While the physiological impact of many of these proteins can be easily understood, their impact on drug molecules is described to varying degrees across the spectrum of known transporters. For example, the impact of MCT-1 on the absorption of drug molecules has not been studied extensively, while the impact of the efflux transporter P-gp is more widely understood. It is interesting to consider that most of the transporters known to impact drug molecules are expressed on the apical membrane of the enterocyte. It is expected that many compounds would need a "partner" transporter on the basolateral membrane to secrete compounds across the basolateral cell membrane into blood; these "basolateral" transporters and therefore their impact on drug design is not well established. That said, current information on transporter proteins shows that the absorption of several drug molecules can be affected by transporters and this should be considered in rational design process. The impact of the SAR highlighted here should aid in the potential to leverage transporters in facilitating the absorption of drug molecules appropriately for a particular drug series; however, for most transport proteins, the lack of crystal structure currently limits the level to which transporters can be leveraged currently in drug design. This can be a very beneficial strategy for compounds that have low membrane permeability *per se*, that is, compounds in BDDCS Class III and IV. However, as described above, consideration of the likely oral dose size is critical. There is a lesser impact of transporter proteins on compounds with reasonable passive permeability, that is, BDDCS

Class I and II. The scholarship described and the translation of its impact on the rate and extent of absorption should lead to the inclusion of the measured kinetics in *in vitro* systems, that is, K_m, V_{max}, and K_D values, against these transporters in system biology models to predict and fully understand transporter effects in the process of absorption across the human GIT of drug in particular those that are BDDCS Class III or IV.

7.5 CHAPTER 7 TUTORIALS

1. What are the differences between intestinal low-capacity/high-affinity and high-capacity/low-affinity uptake transporters? Provide examples on these two categories.
2. Why are Pept1 transporter substrates not appropriate to target for colonic controlled release formulation?
3. What is the significance of dose and permeability when you consider using intestinal uptake transporters to improve the oral absorption (f_a) of your molecule?

7.6 CHAPTER 7 TUTORIALS ANSWERS KEY

1. What are the differences between intestinal low-capacity/high-affinity and high-capacity/low-affinity uptake transporters? Provide examples on these two categories.

   Answer: These are SLC uptake transporters that are located in the gastrointestinal tract. Low-capacity/high-affinity transporters are those proteins that have low intestinal transporters expression leading to low capacity, that is, low V_{max}. Furthermore, their substrates have high affinity, that is, K_m, and therefore get saturated at low doses. High-capacity/low-affinity transporters are transporters with high intestinal expression and their substrates have low affinity (high K_m values). Therefore, unlike low-capacity/high-affinity transporters they are saturated at high doses. An example on the first transporter class is OATP2B1. Examples on the second class are MCT1 and PEPT1 transporters.

2. Why are Pept1 transporter substrates not appropriate to target for colonic controlled release formulation?

   Answer: Unlike MCT1, Pept1 transporter has poor expression in the colon. Therefore, if Pept1 transporter is the sole driver for compound absorption across the intestine, its substrate is predicted to have poor permeability in the colon. Overall, this leads to poor absorption and the failure of extending the absorption of the molecule. Instead, you may consider formulation such as gastric retention technology to extend the absorption of the molecule. It is pivotal when you consider a particular intestinal transporter for you to be aware of its expression and where it is located, that is, is limited to small intestine with poor expression in colon and vice versa.

3. What is the significance of dose and permeability when you consider using intestinal uptake transporters to improve the oral absorption (f_a) of your molecule?

   Answer: The significance is oral dose is driven in part by the capacity of the transporter. For example, low-capacity transporters are typically saturated at low oral dose. It is interesting to note that many OATP2B1 (low-capacity transporter) substrates that are available in the market tend to have low oral doses (<100 mg). On the other hand, PEPT1 substrates (high-capacity transporter) tend to have high oral doses (>1 g). Example on the former is talinolol and examples on the latter are many of the large dose penicillins and cephalosporins. Uptake intestinal transporters' role in the drug absorption is partly driven by the permeability of its substrate. For example, while they are good substrates for OATP2B1, many of the statins such as simvastatin/atorvastatin, you note an increase in their overall oral exposure in the presence of grapefruit juice (known OATP2B1/P-gp/CYP3A4 inhibitor). This indicates that OATP2B1 role in their absorption is limited due to their high intestinal permeability leading to an f_a almost complete.

REFERENCES

1. Hughes, J.D., et al., *Physiochemical drug properties associated with in vivo toxicological outcomes.* Bioorg Med Chem Lett, 2008. 18(17): p. 4872–4875.
2. Lipinski, C.A., *Drug-like properties and the causes of poor solubility and poor permeability.* J Pharmacol Toxicol Methods, 2000. 44(1): p. 235–249.
3. Hurst, S., Loi, C.M., Brodfuehrer, J. and El-Kattan, A., *Impact of physiological, physicochemical and biopharmaceutical factors in absorption and metabolism mechanisms on the drug oral bioavailability of rats and humans.* Expert Opin Drug Metab Toxicol, 2007. 3(4): p. 469–489.
4. Thomas, V.H., Bhattachar, S., Hitchingham, L., Zocharski, P., Naath, M., Surendran, N., Stoner, C.L. and El-Kattan, A., *The road map to oral bioavailability: an industrial perspective.* Expert Opin Drug Metab Toxicol, 2006. 2(4): p. 591–608.
5. Comess, K.M. and Schurdak, M.E., *Affinity-based screening techniques for enhancing lead discovery.* Curr Opin Drug Disc Devel, 2004. 7(4): p. 411–416.
6. Kola, I. and Landis, J., *Can the pharmaceutical industry reduce attrition rates?* Nat Rev Drug Discov, 2004. 3(8): p. 711–715.
7. Stevens, J.L. and Baker, T.K., *The future of drug safety testing: expanding the view and narrowing the focus.* Drug Discov Today, 2009. 14(3–4): p. 162–167.
8. Krejsa, C.M., Horvath, D., Rogalski, S.L., Penzotti, J.E., Mao, B., Barbosa, F. and Migeon, J.C., *Predicting ADME properties and side effects: the BioPrint approach.* Curr Opin Drug Discov Devel, 2003. 6(4): p. 470–480.
9. Kramer, J.A., Sagartz, J.E. and Morris, D.L., *The application of discovery toxicology and pathology towards the design of safer pharmaceutical lead candidates.* Nat Rev Drug Discov, 2007. 6(8): p. 636–649.
10. Price, D.A., Blagg, J., Jones, L., Greene, N. and Wager, T., *Physicochemical drug properties associated with* in vivo *toxicological outcomes: a review.* Expert Opin Drug Metab Toxicol, 2009. 5(8): p. 921–931.
11. Jenkins, K.M., Angeles, R., Quintos, M.T., Xu, R., Kassel, D.B. and Rourick, R.A., *Automated high throughput ADME assays for metabolic stability and cytochrome P450 inhibition profiling of combinatorial libraries.* J Pharm Biomed Anal, 2004. 34(5): p. 989–1004.
12. Kerns, E.H., *High throughput physicochemical profiling for drug discovery.* J Pharm Sci, 2001. 90(11): p. 1838–1858.
13. Masimirembwa, C.M., Bredberg, U. and Andersson, T.B., *Metabolic stability for drug discovery and development: pharmacokinetic and biochemical challenges.* Clin Pharmacokinet, 2003. 42(6): p. 515–528.
14. Amidon, G.L., Lennernas, H., Shah, V.P. and Crison, J.R., *A theoretical basis for a biopharmaceutic drug classification: the correlation of* in vitro *drug product dissolution and* in vivo *bioavailability.* Pharm Res, 1995. 12(3): p. 413–420.
15. Benet, L.Z., *The drug transporter–metabolism alliance: uncovering and defining the interplay.* Mol Pharm, 2009. 6(6): p. 1631–1643.
16. Custodio, J.M., Wu, C.Y. and Benet, L.Z., *Predicting drug disposition, absorption/elimination/transporter interplay and the role of food on drug absorption.* Adv Drug Deliv Rev, 2008. 60(6): p. 717–733.

17. Wu, C.Y. and Benet, L.Z., *Predicting drug disposition via application of BCS: transport/absorption/ elimination interplay and development of a biopharmaceutics drug disposition classification system.* Pharm Res, 2005. 22(1): p. 11–23.
18. Larsen, M., Holm, R., Jensen, K.G., Brodin, B. and Nielsen, C.U., *Intestinal gaboxadol absorption via PAT1 (SLC36A1): modified absorption* in vivo *following co-administration of L-tryptophan.* Br J Pharmacol, 2009. 157(8): p. 1380–1389.
19. Larsen, M., Larsen, B.B., Frolund, B. and Nielsen, C.U., *Transport of amino acids and GABA analogues via the human proton-coupled amino acid transporter, hPAT1: characterization of conditions for affinity and transport experiments in Caco-2 cells.* Eur J Pharm Sci, 2008. 35(1–2): p. 86–95.
20. Yamamoto, A., Akanuma, S., Tachikawa, M. and Hosoya, K., *Involvement of LAT1 and LAT2 in the high- and low-affinity transport of L-leucine in human retinal pigment epithelial cells (ARPE-19 cells).* J Pharm Sci, 2010. 99(5): p. 2475–2482.
21. Cao, X., et al., *Why is it challenging to predict intestinal drug absorption and oral bioavailability in human using rat model.* Pharm Res, 2006. 23(8): p. 1675–1686.
22. Englund, G., Rorsman, F., Ronnblom, A., Karlbom, U., Lazorova, L., Grasjo, J., Kindmark, A. and Artursson, P., *Regional levels of drug transporters along the human intestinal tract: co-expression of ABC and SLC transporters and comparison with Caco-2 cells.* Eur J Pharm Sci, 2006. 29(3–4): p. 269–277.
23. Sun, D., Lennernas, H., Welage, L.S., Barnett, J.L., Landowski, C.P., Foster, D., Fleisher, D., Lee, K.D. and Amidon, G.L., *Comparison of human duodenum and Caco-2 gene expression profiles for 12,000 gene sequences tags and correlation with permeability of 26 drugs.* Pharm Res, 2002. 19(10): p. 1400–1416.
24. Anderson, C.M. and Thwaites, D.T., *Hijacking solute carriers for proton-coupled drug transport.* Physiology (Bethesda), 2010. 25(6): p. 364–377.
25. Giacomini, K.M., et al., *Membrane transporters in drug development.* Nat Rev Drug Discov. 2010. 9(3): p. 215–236.
26. Benet, L.Z., *Predicting drug disposition via application of a biopharmaceutics drug disposition classification system.* Basic Clin Pharmacol Toxicol. 2010. 106(3): p. 162–167.
27. Shugarts, S. and Benet, L.Z., *The role of transporters in the pharmacokinetics of orally administered drugs.* Pharm Res, 2009. 26(9): p. 2039–2054.
28. Kullak-Ublick, G.A., Ismair, M.G., Stieger, B., Landmann, L., Huber, R., Pizzagalli, F., Fattinger, K., Meier, P.J. and Hagenbuch, B., *Organic anion-transporting polypeptide B (OATP-B) and its functional comparison with three other OATPs of human liver.* Gastroenterology, 2001. 120(2): p. 525–533.
29. Tamai, I., Nezu, J., Uchino, H., Sai, Y., Oku, A., Shimane, M. and Tsuji, A., *Molecular identification and characterization of novel members of the human organic anion transporter (OATP) family.* Biochem Biophys Res Commun, 2000. 273(1): p. 251–260.
30. Hagenbuch, B. and Meier, P.J., *The superfamily of organic anion transporting polypeptides.* Biochim Biophys Acta, 2003. 1609(1): p. 1–18.
31. Sai, Y., Kaneko, Y., Ito, S., Mitsuoka, K., Kato, Y., Tamai, I., Artursson, P. and Tsuji, A., *Predominant contribution of organic anion transporting polypeptide OATP-B (OATP2B1) to apical uptake of estrone-3-sulfate by human intestinal Caco-2 cells.* Drug Metab Dispos, 2006. 34(8): p. 1423–1431.
32. Kobayashi, D., Nozawa, T., Imai, K., Nezu, J., Tsuji, A. and Tamai, I., *Involvement of human organic anion transporting polypeptide OATP-B (SLC21A9) in pH-dependent transport across intestinal apical membrane.* J Pharmacol Exp Ther, 2003. 306(2): p. 703–708.
33. Bronger, H., Konig, J., Kopplow, K., Steiner, H.H., Ahmadi, R., Herold-Mende, C., Keppler, D. and Nies, A.T., *ABCC drug efflux pumps and organic anion uptake transporters in human gliomas and the blood–tumor barrier.* Cancer Res, 2005. 65(24): p. 11419–11428.
34. St-Pierre, M.V., Hagenbuch, B., Ugele, B., Meier, P.J. and Stallmach, T., *Characterization of an organic anion-transporting polypeptide (OATP-B) in human placenta.* J Clin Endocrinol Metab, 2002. 87(4): p. 1856–1863.

35. Gao, B., Huber, R.D., Wenzel, A., Vavricka, S.R., Ismair, M.G., Reme, C. and Meier, P.J., *Localization of organic anion transporting polypeptides in the rat and human ciliary body epithelium.* Exp Eye Res, 2005. 80(1): p. 61–72.

36. Pizzagalli, F., Varga, Z., Huber, R.D., Folkers, G., Meier, P.J. and St-Pierre, M.V., *Identification of steroid sulfate transport processes in the human mammary gland.* J Clin Endocrinol Metab, 2003. 88(8): p. 3902–3912.

37. Oswald, S., Groer, C., Drozdzik, M. and Siegmund, W., *Mass spectrometry-based targeted proteomics as a tool to elucidate the expression and function of intestinal drug transporters.* AAPS J, 2013. 15(4): p. 1128–1140.

38. Nozawa, T., Imai, K., Nezu, J., Tsuji, A. and Tamai, I., *Functional characterization of pH-sensitive organic anion transporting polypeptide OATP-B in human.* J Pharmacol Exp Ther, 2004. 308(2): p. 438–445.

39. Vijay, N. and Morris, M.E., *Role of monocarboxylate transporters in drug delivery to the brain.* Curr Pharm Des, 2013. 20(10): p. 1487–1498.

40. Halestrap, A.P. and Wilson, M.C., *The monocarboxylate transporter family-role and regulation.* IUBMB Life, 2012. 64(2): p. 109–119.

41. Foley, D.W., Rajamanickam, J., Bailey, P.D. and Meredith, D., *Bioavailability through PepT1: the role of computer modelling in intelligent drug design.* Curr Comput Aided Drug Des, 2010. 6(1): p. 68–78.

42. Varma, M.V., Rotter, C.J., Chupka, J., Whalen, K.M., Duignan, D.B., Feng, B., Litchfield, J., Goosen, T.C. and El-Kattan, A.F., *pH-Sensitive interaction of HMG-CoA reductase inhibitors (statins) with organic anion transporting polypeptide 2B1.* Mol Pharm, 2011. 8(4): p. 1303–1313.

43. Fukazawa, I., Uchida, N., Uchida, E. and Yasuhara, H., *Effects of grapefruit juice on pharmacokinetics of atorvastatin and pravastatin in Japanese.* Br J Clin Pharmacol, 2004. 57(4): p. 448–455.

44. Zaidenstein, R., Soback, S., Gips, M., Avni, B., Dishi, V., Weissgarten, Y., Golik, A. and Scapa, E., *Effect of grapefruit juice on the pharmacokinetics of losartan and its active metabolite E3174 in healthy volunteers.* Ther Drug Monit, 2001. 23(4): p. 369–373.

45. Rogers, J.D., et al., *Grapefruit juice has minimal effects on plasma concentrations of lovastatin-derived 3-hydroxy-3-methylglutaryl coenzyme A reductase inhibitors.* Clin Pharmacol Ther, 1999. 66(4): p. 358–366.

46. Lilja, J.J., Kivisto, K.T. and Neuvonen, P.J., *Duration of effect of grapefruit juice on the pharmacokinetics of the CYP3A4 substrate simvastatin.* Clin Pharmacol Ther, 2000. 68(4): p. 384–390.

47. Lilja, J.J., Neuvonen, M. and Neuvonen, P.J., *Effects of regular consumption of grapefruit juice on the pharmacokinetics of simvastatin.* Br J Clin Pharmacol, 2004. 58(1): p. 56–60.

48. Lilja, J.J., Kivisto, K.T. and Neuvonen, P.J., *Grapefruit juice–simvastatin interaction: effect on serum concentrations of simvastatin, simvastatin acid, and HMG-CoA reductase inhibitors.* Clin Pharmacol Ther, 1998. 64(5): p. 477–483.

49. Ando, H., Tsuruoka, S., Yanagihara, H., Sugimoto, K., Miyata, M., Yamazoe, Y., Takamura, T., Kaneko, S. and Fujimura, A., *Effects of grapefruit juice on the pharmacokinetics of pitavastatin and atorvastatin.* Br J Clin Pharmacol, 2005. 60(5): p. 494–497.

50. Kantola, T., Kivisto, K.T. and Neuvonen, P.J., *Grapefruit juice greatly increases serum concentrations of lovastatin and lovastatin acid.* Clin Pharmacol Ther, 1998. 63(4): p. 397–402.

51. Uno, T. and Yasui-Furukori, N., *Effect of grapefruit juice in relation to human pharmacokinetic study.* Curr Clin Pharmacol, 2006. 1(2): p. 157–161.

52. Lemieux, M.J., Huang, Y. and Wang, D.N., *The structural basis of substrate translocation by the* Escherichia coli *glycerol-3-phosphate transporter: a member of the major facilitator superfamily.* Curr Opin Struct Biol, 2004. 14(4): p. 405–412.

53. Meier-Abt, F., Mokrab, Y. and Mizuguchi, K., *Organic anion transporting polypeptides of the OATP/SLCO superfamily: identification of new members in nonmammalian species, comparative modeling and a potential transport mode.* J Membr Biol, 2005. 208(3): p. 213–227.

54. Tamai, I. and Nakanishi, T., *OATP transporter-mediated drug absorption and interaction*. Curr Opin Pharmacol, 2013. 13(6): p. 859–863.

55. Ieiri, I., et al., *Microdosing clinical study: pharmacokinetic, pharmacogenomic (SLCO2B1), and interaction (grapefruit juice) profiles of celiprolol following the oral microdose and therapeutic dose*. J Clin Pharmacol, 2012. 52(7): p. 1078–1089.

56. Ohura, K., Nozawa, T., Murakami, K. and Imai, T., *Evaluation of transport mechanism of prodrugs and parent drugs formed by intracellular metabolism in Caco-2 cells with modified carboxylesterase activity: temocapril as a model case*. J Pharm Sci, 2011. 100(9): p. 3985–3994.

57. Vaidyanathan, S., Camenisch, G., Schuetz, H., Reynolds, C., Yeh, C.M., Bizot, M.N., Dieterich, H.A., Howard, D. and Dole, W.P., *Pharmacokinetics of the oral direct renin inhibitor aliskiren in combination with digoxin, atorvastatin, and ketoconazole in healthy subjects: the role of P-glycoprotein in the disposition of aliskiren*. J Clin Pharmacol, 2008. 48(11): p. 1323–1338.

58. Karlgren, M., Vildhede, A., Norinder, U., Wisniewski, J.R., Kimoto, E., Lai, Y., Haglund, U. and Artursson, P., *Classification of inhibitors of hepatic organic anion transporting polypeptides (OATPs): influence of protein expression on drug–drug interactions*. J Med Chem, 2012. 55(10): p. 4740–4763.

59. Treiber, A., Schneiter, R., Hausler, S. and Stieger, B., *Bosentan is a substrate of human OATP1B1 and OATP1B3: inhibition of hepatic uptake as the common mechanism of its interactions with cyclosporin A, rifampicin, and sildenafil*. Drug Metab Dispos, 2007. 35(8): p. 1400–1407.

60. Kalgutkar, A.S., et al., *Role of transporters in the disposition of the selective phosphodiesterase-4 inhibitor (+)-2-[4-({[2-(benzo[1,3]dioxol-5-yloxy)-pyridine-3-carbonyl]-amino}-methyl)-3-fl uorophenoxy]-propionic acid in rat and human*. Drug Metab Dispos, 2007. 35(11): p. 2111–2118.

61. Takeuchi, K., Sugiura, T., Umeda, S., Matsubara, K., Horikawa, M., Nakamichi, N., Silver, D.L., Ishiwata, N. and Kato, Y., *Pharmacokinetics and hepatic uptake of eltrombopag, a novel platelet-increasing agent*. Drug Metab Dispos, 2011. 39(6): p. 1088–1096.

62. Shen, H., et al., *Cynomolgus monkey as a potential model to assess drug interactions involving hepatic organic anion transporting polypeptides:* in vitro, in vivo, *and* in vitro-*to*-in vivo *extrapolation*. J Pharmacol Exp Ther, 2013. 344(3): p. 673–685.

63. Grube, M., Kock, K., Karner, S., Reuther, S., Ritter, C.A., Jedlitschky, G. and Kroemer, H.K., *Modification of OATP2B1-mediated transport by steroid hormones*. Mol Pharmacol, 2006. 70(5): p. 1735–1741.

64. Fahrmayr, C., Konig, J., Auge, D., Mieth, M. and Fromm, M.F., *Identification of drugs and drug metabolites as substrates of multidrug resistance protein 2 (MRP2) using triple-transfected MDCK-OATP1B1-UGT1A1-MRP2 cells*. Br J Pharmacol, 2012. 165(6): p. 1836–1847.

65. Oswald, S., Konig, J., Lutjohann, D., Giessmann, T., Kroemer, H.K., Rimmbach, C., Rosskopf, D., Fromm, M.F. and Siegmund, W., *Disposition of ezetimibe is influenced by polymorphisms of the hepatic uptake carrier OATP1B1*. Pharmacogenet Genomics, 2008. 18(7): p. 559–568.

66. Ming, X., Knight, B.M. and Thakker, D.R., *Vectorial transport of fexofenadine across Caco-2 cells: involvement of apical uptake and basolateral efflux transporters*. Mol Pharm, 2011. 8(5): p. 1677–1686.

67. Noe, J., Portmann, R., Brun, M.E. and Funk, C., *Substrate-dependent drug–drug interactions between gemfibrozil, fluvastatin and other organic anion-transporting peptide (OATP) substrates on OATP1B1, OATP2B1, and OATP1B3*. Drug Metab Dispos, 2007. 35(8): p. 1308–1314.

68. Satoh, H., Yamashita, F., Tsujimoto, M., Murakami, H., Koyabu, N., Ohtani, H. and Sawada, Y., *Citrus juices inhibit the function of human organic anion-transporting polypeptide OATP-B*. Drug Metab Dispos, 2005. 33(4): p. 518–523.

69. Gao, C., Zhang, H., Guo, Z., You, T., Chen, X. and Zhong, D., *Mechanistic studies on the absorption and disposition of scutellarin in humans: selective OATP2B1-mediated hepatic uptake is a likely key determinant for its unique pharmacokinetic characteristics*. Drug Metab Dispos, 2012. 40(10): p. 2009–2020.

70. Konig, J., Glaeser, H., Keiser, M., Mandery, K., Klotz, U. and Fromm, M.F., *Role of organic anion-transporting polypeptides for cellular mesalazine (5-aminosalicylic acid) uptake*. Drug Metab Dispos, 2011. 39(6): p. 1097–1102.

71. Visentin, M., Chang, M.H., Romero, M.F., Zhao, R. and Goldman, I.D., *Substrate- and pH-specific antifolate transport mediated by organic anion-transporting polypeptide 2B1 (OATP2B1-SLCO2B1)*. Mol Pharmacol, 2012. 81(2): p. 134–142.

72. Hirano, M., Maeda, K., Shitara, Y. and Sugiyama, Y., *Drug–drug interaction between pitavastatin and various drugs via OATP1B1*. Drug Metab Dispos, 2006. 34(7): p. 1229–1236.

73. Kusuhara, H., et al., *Pharmacokinetic interaction study of sulphasalazine in healthy subjects and the impact of curcumin as an* in vivo *inhibitor of BCRP*. Br J Pharmacol, 2012. 166(6): p. 1793–1803.

74. Shirasaka, Y., Kuraoka, E., Spahn-Langguth, H., Nakanishi, T., Langguth, P. and Tamai, I., *Species difference in the effect of grapefruit juice on intestinal absorption of talinolol between human and rat*. J Pharmacol Exp Ther, 2010. 332(1): p. 181–189.

75. Ishiguro, N., Maeda, K., Saito, A., Kishimoto, W., Matsushima, S., Ebner, T., Roth, W., Igarashi, T. and Sugiyama, Y., *Establishment of a set of double transfectants coexpressing organic anion transporting polypeptide 1B3 and hepatic efflux transporters for the characterization of the hepatobiliary transport of telmisartan acylglucuronide*. Drug Metab Dispos, 2008. 36(4): p. 796–805.

76. Ho, R.H., Tirona, R.G., Leake, B.F., Glaeser, H., Lee, W., Lemke, C.J., Wang, Y. and Kim, R.B., *Drug and bile acid transporters in rosuvastatin hepatic uptake: function, expression, and pharmacogenetics*. Gastroenterology, 2006. 130(6): p. 1793–1806.

77. Letschert, K., Faulstich, H., Keller, D. and Keppler, D., *Molecular characterization and inhibition of amanitin uptake into human hepatocytes*. Toxicol Sci, 2006. 91(1): p. 140–149.

78. Roth, M., Timmermann, B.N. and Hagenbuch, B., *Interactions of green tea catechins with organic anion-transporting polypeptides*. Drug Metab Dispos, 2011. 39(5): p. 920–926.

79. Koenen, A., Kock, K., Keiser, M., Siegmund, W., Kroemer, H.K. and Grube, M., *Steroid hormones specifically modify the activity of organic anion transporting polypeptides*. Eur J Pharm Sci, 2012. 47(4): p. 774–780.

80. Shirasaka, Y., Shichiri, M., Mori, T., Nakanishi, T. and Tamai, I., *Major active components in grapefruit, orange, and apple juices responsible for OATP2B1-mediated drug interactions*. J Pharm Sci, 2013. 102(9): p. 3418–3426.

81. Grube, M., et al., *Organic anion transporting polypeptide 2B1 is a high-affinity transporter for atorvastatin and is expressed in the human heart*. Clin Pharmacol Ther, 2006. 80(6): p. 607–620.

82. Vavricka, S.R., Van Montfoort, J., Ha, H.R., Meier, P.J. and Fattinger, K., *Interactions of rifamycin SV and rifampicin with organic anion uptake systems of human liver*. Hepatology, 2002. 36(1): p. 164–172.

83. Sahi, J., et al., *Metabolism and transporter-mediated drug–drug interactions of the endothelin-A receptor antagonist CI-1034*. Chem Biol Interact, 2006. 159(2): p. 156–168.

84. Lu, W.J., Huang, J.D. and Lai, M.L., *The effects of ergoloid mesylates and ginkgo biloba on the pharmacokinetics of ticlopidine*. J Clin Pharmacol, 2006. 46(6): p. 628–634.

85. Bachmakov, I., Glaeser, H., Fromm, M.F. and Konig, J., *Interaction of oral antidiabetic drugs with hepatic uptake transporters: focus on organic anion transporting polypeptides and organic cation transporter 1*. Diabetes, 2008. 57(6): p. 1463–1469.

86. Annaert, P., Ye, Z.W., Stieger, B. and Augustijns, P., *Interaction of HIV protease inhibitors with OATP1B1, 1B3, and 2B1*. Xenobiotica, 2010. 40(3): p. 163–176.

87. Kis, O., Zastre, J.A., Ramaswamy, M. and Bendayan, R., *pH dependence of organic anion-transporting polypeptide 2B1 in Caco-2 cells: potential role in antiretroviral drug oral bioavailability and drug–drug interactions*. J Pharmacol Exp Ther, 2010. 334(3): p. 1009–1022.

88. Mandery, K., Bujok, K., Schmidt, I., Keiser, M., Siegmund, W., Balk, B., Konig, J., Fromm, M.F. and Glaeser, H., *Influence of the flavonoids apigenin, kaempferol, and quercetin on the function*

of organic anion transporting polypeptides 1A2 and 2B1. Biochem Pharmacol, 2010. 80(11): p. 1746–1753.

89. Taub, M.E., et al., *Digoxin is not a substrate for organic anion-transporting polypeptide transporters OATP1A2, OATP1B1, OATP1B3, and OATP2B1 but is a substrate for a sodium-dependent transporter expressed in HEK293 cells*. Drug Metab Dispos, 2011. 39(11): p. 2093–2102.
90. Klatt, S., Fromm, M.F. and Konig, J., *The influence of oral antidiabetic drugs on cellular drug uptake mediated by hepatic OATP family members*. Basic Clin Pharmacol Toxicol, 2013. 112(4): p. 244–250.
91. Kock, K., Xie, Y., Hawke, R.L., Oberlies, N.H. and Brouwer, K.L., *Interaction of silymarin flavonolignans with organic anion-transporting polypeptides*. Drug Metab Dispos, 2013. 41(5): p. 958–965.
92. Kalgutkar, A.S., et al., *Elucidation of the biochemical basis for a clinical drug–drug interaction between atorvastatin and 5-(N-(4-((4-ethylbenzyl)thio)phenyl)sulfamoyl)-2-methyl benzoic acid (CP-778875), a subtype selective agonist of the peroxisome proliferator-activated receptor alpha*. Xenobiotica, 2013. 43(11): p. 963–972.
93. Wlcek, K., Koller, F., Ferenci, P. and Stieger, B., *Hepatocellular organic anion-transporting polypeptides (OATPs) and multidrug resistance-associated protein 2 (MRP2) are inhibited by silibinin*. Drug Metab Dispos, 2013. 41(8): p. 1522–1528.
94. Xu, F., Li, Z., Zheng, J., Gee Cheung, F.S., Chan, T., Zhu, L., Zhuge, H. and Zhou, F., *The inhibitory effects of the bioactive components isolated from* Scutellaria baicalensis *on the cellular uptake mediated by the essential solute carrier transporters*. J Pharm Sci, 2013. 102(11): p. 4205–4211.
95. Di, L., et al., *Development of a new permeability assay using low-efflux MDCKII cells*. J Pharm Sci, 2011. 100(11): p. 4974–4985.
96. Varma, M.V., et al., *pH-Dependent solubility and permeability criteria for provisional biopharmaceutics classification (BCS and BDDCS) in early drug discovery*. Mol Pharm, 2012. 9(5): p. 1199–1212.
97. Varma, M.V., Chang, G., Lai, Y., Feng, B., El-Kattan, A.F., Litchfield, J. and Goosen, T.C., *Physicochemical property space of hepatobiliary transport and computational models for predicting rat biliary excretion*. Drug Metab Dispos, 2012. 40(8): p. 1527–1537.
98. Hanggi, E., Grundschober, A.F., Leuthold, S., Meier, P.J. and St-Pierre, M.V., *Functional analysis of the extracellular cysteine residues in the human organic anion transporting polypeptide, OATP2B1*. Mol Pharmacol, 2006. 70(3): p. 806–817.
99. Miyagawa, M., Maeda, K., Aoyama, A. and Sugiyama, Y., *The eighth and ninth transmembrane domains in organic anion transporting polypeptide 1B1 affect the transport kinetics of estrone-3-sulfate and estradiol-17beta-D-glucuronide*. J Pharmacol Exp Ther, 2009. 329(2): p. 551–557.
100. Gui, C. and Hagenbuch, B., *Amino acid residues in transmembrane domain 10 of organic anion transporting polypeptide 1B3 are critical for cholecystokinin octapeptide transport*. Biochemistry, 2008. 47(35): p. 9090–9097.
101. Gui, C., Miao, Y., Thompson, L., Wahlgren, B., Mock, M., Stieger, B. and Hagenbuch, B., *Effect of pregnane X receptor ligands on transport mediated by human OATP1B1 and OATP1B3*. Eur J Pharmacol, 2008. 584(1): p. 57–65.
102. Kim, K.A., Lee, H.M., Joo, H.J., Park, I.B. and Park, J.Y., *Effects of polymorphisms of the SLCO2B1 transporter gene on the pharmacokinetics of montelukast in humans*. J Clin Pharmacol, 2013. 53(11): p. 1186–1193.
103. Gong, I.Y. and Kim, R.B., *Impact of genetic variation in OATP transporters to drug disposition and response*. Drug Metab Pharmacokinet, 2013. 28(1): p. 4–18.
104. Nakanishi, T. and Tamai, I., *Genetic polymorphisms of OATP transporters and their impact on intestinal absorption and hepatic disposition of drugs*. Drug Metab Pharmacokinet, 2012. 27(1): p. 106–121.
105. Kim, K.A., Joo, H.J., Lee, H.M. and Park, J.Y., *SLCO2B1 genetic polymorphisms in a Korean population: pyrosequencing analyses and comprehensive comparison with other populations*. Mol Biol Rep, 2013. 40(7): p. 4211–4217.

106. Nozawa, T., Nakajima, M., Tamai, I., Noda, K., Nezu, J., Sai, Y., Tsuji, A. and Yokoi, T., *Genetic polymorphisms of human organic anion transporters OATP-C (SLC21A6) and OATP-B (SLC21A9): allele frequencies in the Japanese population and functional analysis*. J Pharmacol Exp Ther, 2002. 302(2): p. 804–813.

107. Imanaga, J., et al., *The effects of the SLCO2B1 c.1457C>T polymorphism and apple juice on the pharmacokinetics of fexofenadine and midazolam in humans*. Pharmacogenet Genomics, 2011. 21(2): p. 84–93.

108. Bailey, D.G., *Fruit juice inhibition of uptake transport: a new type of food–drug interaction*. Br J Clin Pharmacol, 2010. 70(5): p. 645–655.

109. Mougey, E.B., Lang, J.E., Wen, X. and Lima, J.J., *Effect of citrus juice and SLCO2B1 genotype on the pharmacokinetics of montelukast*. J Clin Pharmacol, 2011. 51(5): p. 751–760.

110. Tapaninen, T., Karonen, T., Backman, J.T., Neuvonen, P.J. and Niemi, M., *SLCO2B1 c.935G>A single nucleotide polymorphism has no effect on the pharmacokinetics of montelukast and aliskiren*. Pharmacogenet Genomics, 2013. 23(1): p. 19–24.

111. Liang, R., Fei, Y.J., Prasad, P.D., Ramamoorthy, S., Han, H., Yang-Feng, T.L., Hediger, M.A., Ganapathy, V. and Leibach, F.H., *Human intestinal H^+/peptide cotransporter. Cloning, functional expression, and chromosomal localization*. J Biol Chem, 1995. 270(12): p. 6456–6463.

112. Rubio-Aliaga, I. and Daniel, H., *Peptide transporters and their roles in physiological processes and drug disposition*. Xenobiotica, 2008. 38(7–8): p. 1022–1042.

113. Herrera-Ruiz, D., Wang, Q., Gudmundsson, O.S., Cook, T.J., Smith, R.L., Faria, T.N. and Knipp, G.T., *Spatial expression patterns of peptide transporters in the human and rat gastrointestinal tracts, Caco-2* in vitro *cell culture model, and multiple human tissues*. AAPS PharmSci, 2001. 3(1): p. E9.

114. Barr, W.H., Zola, E.M., Candler, E.L., Hwang, S.M., Tendolkar, A.V., Shamburek, R., Parker, B. and Hilty, M.D., *Differential absorption of amoxicillin from the human small and large intestine*. Clin Pharmacol Ther, 1994. 56(3): p. 279–285.

115. Chan, K.K., Buch, A., Glazer, R.D., John, V.A. and Barr, W.H., *Site-differential gastrointestinal absorption of benazepril hydrochloride in healthy volunteers*. Pharm Res, 1994. 11(3): p. 432–437.

116. Jappar, D., Wu, S.P., Hu, Y. and Smith, D.E., *Significance and regional dependency of peptide transporter (PEPT) 1 in the intestinal permeability of glycylsarcosine:* in situ *single-pass perfusion studies in wild-type and Pept1 knockout mice*. Drug Metab Dispos, 2010. 38(10): p. 1740–1746.

117. Posada, M.M. and Smith, D.E., *Relevance of PepT1 in the intestinal permeability and oral absorption of cefadroxil*. Pharm Res, 2013. 30(4): p. 1017–1025.

118. Liu, W., Liang, R., Ramamoorthy, S., Fei, Y.J., Ganapathy, M.E., Hediger, M.A., Ganapathy, V. and Leibach, F.H., *Molecular cloning of PEPT 2, a new member of the H^+/peptide cotransporter family, from human kidney*. Biochim Biophys Acta, 1995. 1235(2): p. 461–466.

119. Shen, H., Smith, D.E., Keep, R.F. and Brosius, F.C., 3rd, *Immunolocalization of the proton-coupled oligopeptide transporter PEPT2 in developing rat brain*. Mol Pharm, 2004. 1(4): p. 248–256.

120. Adibi, S.A., *The oligopeptide transporter (Pept-1) in human intestine: biology and function*. Gastroenterology, 1997. 113(1): p. 332–340.

121. Irie, M., Terada, T., Katsura, T., Matsuoka, S. and Inui, K., *Computational modelling of H^+-coupled peptide transport via human PEPT1*. J Physiol, 2005. 565(Pt 2): p. 429–439.

122. Rubio-Aliaga, I. and Daniel, H., *Mammalian peptide transporters as targets for drug delivery*. Trends Pharmacol Sci, 2002. 23(9): p. 434–440.

123. Luckner, P. and Brandsch, M., *Interaction of 31 beta-lactam antibiotics with the H^+/peptide symporter PEPT2: analysis of affinity constants and comparison with PEPT1*. Eur J Pharm Biopharm, 2005. 59(1): p. 17–24.

124. Lafforgue, G., Arellano, C., Vachoux, C., Woodley, J., Philibert, C., Dupouy, V., Bousquet-Melou, A., Gandia, P. and Houin, G., *Oral absorption of ampicillin: role of paracellular route vs. PepT1 transporter*. Fundam Clin Pharmacol, 2008. 22(2): p. 189–201.

125. Brandsch, M., *Transport of drugs by proton-coupled peptide transporters: pearls and pitfalls*. Expert Opin Drug Metab Toxicol, 2009. 5(8): p. 887–905.

126. Zhang, Q., Liu, Q., Wu, J., Wang, C., Peng, J., Ma, X. and Liu, K., *PEPT1 involved in the uptake and transepithelial transport of cefditoren* in vivo *and* in vitro. Eur J Pharmacol, 2009. 612(1–3): p. 9–14.

127. Ganapathy, M.E., Prasad, P.D., Mackenzie, B., Ganapathy, V. and Leibach, F.H., *Interaction of anionic cephalosporins with the intestinal and renal peptide transporters PEPT 1 and PEPT 2.* Biochim Biophys Acta, 1997. 1324(2): p. 296–308.

128. Liu, Z., Wang, C., Liu, Q., Meng, Q., Cang, J., Mei, L., Kaku, T. and Liu, K., *Uptake, transport and regulation of JBP485 by PEPT1* in vitro *and* in vivo. Peptides, 2011. 32(4): p. 747–754.

129. Li, F., Hong, L., Mau, C.I., Chan, R., Hendricks, T., Dvorak, C., Yee, C., Harris, J. and Alfredson, T., *Transport of levovirin prodrugs in the human intestinal Caco-2 cell line.* J Pharm Sci, 2006. 95(6): p. 1318–1325.

130. Varma, M.V., Eriksson, A.H., Sawada, G., Pak, Y.A., Perkins, E.J. and Zimmerman, C.L., *Transepithelial transport of the group II metabotropic glutamate 2/3 receptor agonist (1S,2S,5R,6S)-2-aminobicyclo[3.1.0]hexane-2,6-dicarboxylate (LY354740) and its prodrug (1S,2S,5R,6S)-2-[(2'S)-(2'-amino)propionyl]aminobicyclo[3.1.0]hexane-2,6-dicarboxylate (LY544344).* Drug Metab Dispos, 2009. 37(1): p. 211–220.

131. Hu, Y., Chen, X. and Smith, D.E., *Species-dependent uptake of glycylsarcosine but not oseltamivir in* Pichia pastoris *expressing the rat, mouse, and human intestinal peptide transporter PEPT1.* Drug Metab Dispos, 2012. 40(7): p. 1328–1335.

132. Ganapathy, M.E., Huang, W., Wang, H., Ganapathy, V. and Leibach, F.H., *Valacyclovir: a substrate for the intestinal and renal peptide transporters PEPT1 and PEPT2.* Biochem Biophys Res Commun, 1998. 246(2): p. 470–475.

133. Bretschneider, B., Brandsch, M. and Neubert, R., *Intestinal transport of beta-lactam antibiotics: analysis of the affinity at the H^+/peptide symporter (PEPT1), the uptake into Caco-2 cell monolayers and the transepithelial flux.* Pharm Res, 1999. 16(1): p. 55–61.

134. Boldhane, S.P. and Kuchekar, B.S., *Gastroretentive drug delivery of metformin hydrochloride: formulation and* in vitro *evaluation using 3(2) full factorial design.* Curr Drug Deliv, 2009. 6(5): p. 477–485.

135. Hoffman, A., Stepensky, D., Lavy, E., Eyal, S., Klausner, E. and Friedman, M., *Pharmacokinetic and pharmacodynamic aspects of gastroretentive dosage forms.* Int J Pharm, 2004. 277(1–2): p. 141–153.

136. Li, M., Anderson, G.D., Phillips, B.R., Kong, W., Shen, D.D. and Wang, J., *Interactions of amoxicillin and cefaclor with human renal organic anion and peptide transporters.* Drug Metab Dispos, 2006. 34(4): p. 547–555.

137. Wenzel, U., Kuntz, S., Diestel, S. and Daniel, H., *PEPT1-mediated cefixime uptake into human intestinal epithelial cells is increased by Ca^{2+} channel blockers.* Antimicrob Agents Chemother, 2002. 46(5): p. 1375–1380.

138. Knutter, I., Kottra, G., Fischer, W., Daniel, H. and Brandsch, M., *High-affinity interaction of sartans with H^+/peptide transporters.* Drug Metab Dispos, 2009. 37(1): p. 143–149.

139. Omkvist, D.H., Brodin, B. and Nielsen, C.U., *Ibuprofen is a non-competitive inhibitor of the peptide transporter hPEPT1 (SLC15A1): possible interactions between hPEPT1 substrates and ibuprofen.* Br J Pharmacol, 2010. 161(8): p. 1793–1805.

140. Khomane, K.S., et al., *Mechanistic insights into PEPT1-mediated transport of a novel antiepileptic, NP-647.* Mol Pharm, 2012. 9(9): p. 2458–2468.

141. Nielsen, C.U., Frolund, S., Abdulhadi, S., Sari, H., Langthaler, L., Nohr, M.K., Kall, M.A., Brodin, B. and Holm, R., *Sertraline inhibits the transport of PAT1 substrates* in vivo *and* in vitro. Br J Pharmacol, 2013. 170(5): p. 1041–1052.

142. Garrigues, T.M., Martin, U., Peris-Ribera, J.E. and Prescott, L.F., *Dose-dependent absorption and elimination of cefadroxil in man.* Eur J Clin Pharmacol, 1991. 41(2): p. 179–183.

143. Zhang, E.Y., Fu, D.J., Pak, Y.A., Stewart, T., Mukhopadhyay, N., Wrighton, S.A. and Hillgren, K.M., *Genetic polymorphisms in human proton-dependent dipeptide transporter PEPT1: implications for the functional role of Pro586.* J Pharmacol Exp Ther, 2004. 310(2): p. 437–445.

144. Anderle, P., Nielsen, C.U., Pinsonneault, J., Krog, P.L., Brodin, B. and Sadee, W., *Genetic variants of the human dipeptide transporter PEPT1*. J Pharmacol Exp Ther, 2006. 316(2): p. 636–646.

145. Halestrap, A.P. and Meredith, D., *The SLC16 gene family – from monocarboxylate transporters (MCTs) to aromatic amino acid transporters and beyond*. Pflugers Arch, 2004. 447(5): p. 619–628.

146. Murakami, Y., Kohyama, N., Kobayashi, Y., Ohbayashi, M., Ohtani, H., Sawada, Y. and Yamamoto, T., *Functional characterization of human monocarboxylate transporter 6 (SLC16A5)*. Drug Metab Dispos, 2005. 33(12): p. 1845–1851.

147. Kim, C.M., Goldstein, J.L. and Brown, M.S., *cDNA cloning of MEV, a mutant protein that facilitates cellular uptake of mevalonate, and identification of a point mutation responsible for its gain in function*. J Biol Chem, 1992. 267: p. 23113–23121.

148. Kim-Garcia, C., Li, X., Luna, J. and Francke, U., *cDNA cloning of the human monocarboxylate transporter 1 and chromosomal location of the SLC16A1 locus to 1p13.2-p12*. Genomics, 1994. 23: p. 500–503.

149. Jackson, V.N., Price, N.T. and Halestrap, A.P., *cDNA cloning of MCT1, a monocarboxylate transporter from rat skeletal muscle*. Biochim Biophys Acta, 1995. 1238(2): p. 193–196.

150. Carpenter, L. and Halestrap, A.P., *Cloning and sequencing of the monocarboxylate transporter from mouse Ehrlich Lettre tumour cell confirms its identity as MCT1 and demonstrates that glycosylation is not required for MCT1 function*. Biochim Biophys Acta, 1996. 1279: p. 157–163.

151. Halestrap, A.P. and Price, N.T., *The proton-linked monocarboxylate transporter (MCT) family: structure, function and regulation*. Biochem J, 1999. 343(Pt 2): p. 281–299.

152. Poole, R.C., Sansom, C.E. and Halestrap, A.P., *Studies of the membrane topology of the rat erythrocyte H^+/lactate cotransporter (MCT1)*. Biochem J, 1996. 320: p. 817–824.

153. Wilson, M.C., Meredith, D., Bunnun, C., Sessions, R.B. and Halestrap, A.P., *Studies on the DIDS-binding site of monocarboxylate transporter 1 suggest a homology model of the open conformation and a plausible translocation cycle*. J Biol Chem, 2009. 284(30): p. 20011–20021.

154. Abramson, J., Smirnova, I., Kasho, V., Verner, G., Kaback, H.R. and Iwata, S., *Structure and mechanism of the lactose permease of* Escherichia coli. Science, 2003. 301: p. 610–615.

155. Huang, Y., Lemieux, M.J., Song, J., Auer, M. and Wang, D.-N., *Structure and mechanism of the glycerol-3-phosphate transporter from* Escherichia coli. Science, 2003. 301(5633): p. 616–620.

156. Manoharan, C., Wilson, M.C., Sessions, R.B. and Halestrap, A.P., *The role of charged residues in the transmembrane helices of monocarboxylate transporter 1 and its ancillary protein basigin in determining plasma membrane expression and catalytic activity*. Mol Membr Biol, 2006. 23(6): p. 486–498.

157. Cundy, K.C., Sastry, S., Luo, W., Zou, J., Moors, T.L. and Canafax, D.M., *Clinical pharmacokinetics of XP13512, a novel transported prodrug of gabapentin*. J Clin Pharmacol, 2008. 48(12): p. 1378–1388.

158. Cundy, K.C., et al., *XP13512 [(+/−)-1-([(alpha-isobutanoyloxyethoxy)carbonyl] aminomethyl)-1-cyclohexane acetic acid], a novel gabapentin prodrug: I. Design, synthesis, enzymatic conversion to gabapentin, and transport by intestinal solute transporters*. J Pharmacol Exp Ther, 2004. 311(1): p. 315–323.

159. Cundy, K.C., et al., *XP13512 [(+/−)-1-([(alpha-isobutanoyloxyethoxy)carbonyl] aminomethyl)-1-cyclohexane acetic acid], a novel gabapentin prodrug: II. Improved oral bioavailability, dose proportionality, and colonic absorption compared with gabapentin in rats and monkeys*. J Pharmacol Exp Ther, 2004. 311(1): p. 324–333.

160. Gill, R.K., Saksena, S., Alrefai, W.A., Sarwar, Z., Goldstein, J.L., Carroll, R.E., Ramaswamy, K. and Dudeja, P.K., *Expression and membrane localization of MCT isoforms along the length of the human intestine*. Am J Physiol Cell Physiol, 2005. 289(4): p. C846–C852.

161. Cummings, J.H. and Macfarlane, G.T., *The control and consequences of bacterial fermentation in the human colon*. J Appl Bacteriol, 1991. 70(6): p. 443–459.

162. Juel, C., *Lactate–proton cotransport in skeletal muscle*. Physiol Rev, 1997. 77(2): p. 321–358.

163. Groer, C., Bruck, S., Lai, Y., Paulick, A., Busemann, A., Heidecke, C.D., Siegmund, W. and Oswald, S., *LC–MS/MS-based quantification of clinically relevant intestinal uptake and efflux transporter proteins*. J Pharm Biomed Anal, 2013. 85: p. 253–261.

164. Law, C.J., Maloney, P.C. and Wang, D.-N., *Ins and outs of major facilitator superfamily antiporters*. Annu Rev Microbiol, 2008. 62(1): p. 289–305.

165. Bröer, S., Schneider, H.P., Broer, A., Rahman, B., Hamprecht, B. and Deitmer, J.W., *Characterization of the monocarboxylate transporter 1 expressed in* Xenopus laevis *oocytes by changes in cytosolic pH*. Biochem J, 1998. 333: p. 167–174.

166. Deuticke, B., *Monocarboxylate transport in erythrocytes*. J Membr Biol, 1982. 70(2): p. 89–103.

167. Poole, R.C. and Halestrap, A.P., *Transport of lactate and other monocarboxylates across mammalian plasma membranes*. Am J Physiol 1993. 264(4): p. C761–C782.

168. Enerson, B.E. and Drewes, L.R., *Molecular features, regulation, and function of monocarboxylate transporters: implications for drug delivery*. J Pharm Sci, 2003. 92(8): p. 1531–1544.

169. Li, Y.-H., Ito, K., Tsuda, Y., Kohda, R., Yamada, H. and Itoh, T., *Mechanism of intestinal absorption of an orally active beta-lactam prodrug: uptake and transport of carindacillin in Caco-2 cells*. J Pharmacol Exp Ther 1999. 290(3): p. 958–964.

170. Jenkinson, S.E., Chung, G.W., van Loon, E., Bakar, N.S., Dalzell, A.M. and Brown, C.D., *The limitations of renal epithelial cell line HK-2 as a model of drug transporter expression and function in the proximal tubule*. Pflugers Arch, 2012. 464(6): p. 601–611.

171. Choi, J.S., Jin, M.J. and Han, H.K., *Role of monocarboxylic acid transporters in the cellular uptake of NSAIDs*. J Pharm Pharmacol, 2005. 57(9): p. 1185–1189.

172. Li, Y.H., Tanno, M., Itoh, T. and Yamada, H., *Role of the monocarboxylic acid transport system in the intestinal absorption of an orally active beta-lactam prodrug: carindacillin as a model*. Int J Pharm, 1999. 191(2): p. 151–159.

173. Tsuji, A., Tamai, I., Nakanishi, M., Terasaki, T. and Hamano, S., *Intestinal brush-border transport of the oral cephalosporin antibiotic, cefdinir, mediated by dipeptide and monocarboxylic acid transport systems in rabbits*. J Pharm Pharmacol, 1993. 45(11): p. 996–998.

174. Utoguchi, N. and Audus, K.L., *Carrier-mediated transport of valproic acid in BeWo cells, a human trophoblast cell line*. Int J Pharm, 2000. 195(1–2): p. 115–124.

175. Hosoya, K., Kondo, T., Tomi, M., Takanaga, H., Ohtsuki, S. and Terasaki, T., *MCT1-mediated transport of L-lactic acid at the inner blood–retinal barrier: a possible route for delivery of monocarboxylic acid drugs to the retina*. Pharm Res, 2001. 18(12): p. 1669–1676.

176. Neuhoff, S., Ungell, A.L., Zamora, I. and Artursson, P., *pH-Dependent passive and active transport of acidic drugs across Caco-2 cell monolayers*. Eur J Pharm Sci, 2005. 25(2–3): p. 211–220.

177. Konishi, Y., Hagiwara, K. and Shimizu, M., *Transepithelial transport of fluorescein in Caco-2 cell monolayers and use of such transport in* in vitro *evaluation of phenolic acid availability*. Biosci Biotechnol Biochem, 2002. 66(11): p. 2449–2457.

178. Konishi, Y. and Shimizu, M., *Transepithelial transport of ferulic acid by monocarboxylic acid transporter in Caco-2 cell monolayers*. Biosci Biotechnol Biochem, 2003. 67(4): p. 856–862.

179. Tsuji, A., Saheki, A., Tamai, I. and Terasaki, T., *Transport mechanism of 3-hydroxy-3-methylglutaryl coenzyme A reductase inhibitors at the blood–brain barrier*. J Pharmacol Exp Ther, 1993. 267(3): p. 1085–1090.

180. Wu, X., Whitfield, L.R. and Stewart, B.H., *Atorvastatin transport in the Caco-2 cell model: contributions of P-glycoprotein and the proton–monocarboxylic acid co-transporter*. Pharm Res, 2000. 17(2): p. 209–215.

181. Lam, W.K., Felmlee, M.A. and Morris, M.E., *Monocarboxylate transporter-mediated transport of gamma-hydroxybutyric acid in human intestinal Caco-2 cells*. Drug Metab Disposition, 2010. 38(3): p. 441–447.

182. Cundy, K.C., et al., *XP13512 [(±)-1-([(α-isobutanoyloxyethoxy)carbonyl]aminomethyl)-1-cyclohexane acetic acid], a novel gabapentin prodrug: I. Design, synthesis, enzymatic conversion*

to gabapentin, and transport by intestinal solute transporters. J Pharmacol Exp Ther, 2004. 311(1): p. 315–323.

183. Lal, R., Sukbuntherng, J., Luo, W., Chen, D., Blumenthal, R., Ho, J. and Cundy, K.C., *Clinical pharmacokinetics of gabapentin after administration of gabapentin enacarbil extended-release tablets in patients with varying degrees of renal function using data from an open-label, single-dose pharmacokinetic study*. Clin Ther, 2012. 34(1): p. 201–213.
184. Lal, R., et al., *Arbaclofen placarbil, a novel R-baclofen prodrug: improved absorption, distribution, metabolism, and elimination properties compared with R-baclofen*. J Pharmacol Exp Ther, 2009. 330(3): p. 911–921.
185. Kobayashi, M., Otsuka, Y., Itagaki, S., Hirano, T. and Iseki, K., *Inhibitory effects of statins on human monocarboxylate transporter 4*. Int J Pharm, 2006. 317(1): p. 19–25.
186. Fale, P.L., Ascensao, L. and Serralheiro, M.L., *Effect of luteolin and apigenin on rosmarinic acid bioavailability in Caco-2 cell monolayers*. Food Funct, 2013. 4(3): p. 426–431.
187. Poole, R.C. and Halestrap, A.P., *Reversible and irreversible inhibition, by stilbenedisulphonates, of lactate transport into rat erythrocytes – identification of some new high-affinity inhibitors*. Biochem J, 1991. 275: p. 307–312.
188. Lean, C.B. and Lee, E.J., *Genetic variations in the MCT1 (SLC16A1) gene in the Chinese population of Singapore*. Drug Metab Pharmacokinet, 2009. 24(5): p. 469–474.
189. Merezhinskaya, N., Fishbein, W.N., Davis, J.I. and Foellmer, J.W., *Mutations in MCT1 cDNA in patients with symptomatic deficiency in lactate transport*. Muscle Nerve, 2000. 23(1): p. 90–97.
190. Garanero, G., Ramanathan, S. and Amidon, G., Gastrointestinal Dissolution and Absorption of Drugs, in *Drug Bioavailability: Estimation of Solubility, Permeability, Absorption and Bioavailability*, Waterbeemd, H.v.d., Lennernas, H. and Artursson, P., Editors. 2005, Wiley-VCH: Zurich. p. 191–210.
191. Varma, M.V.S., et al., *Biopharmaceutic classification system: a scientific framework for pharmacokinetic optimization in drug research*. Curr Drug Metab, 2004. 5(5): p. 375–388.
192. Balimane, P. and Sinko, P., *Effect of ionization on the variable uptake of valacyclovir via the human intestinal peptide transporter (hPepT1) in CHO cells*. Biopharm Drug Dispos, 2000. 21(5): p. 165–174.

8

EXTENDED CLEARANCE CLASSIFICATION SYSTEM (ECCS) AND ITS UTILITY IN PREDICTING CLEARANCE RATE-DETERMINING STEP IN DRUG DISCOVERY

8.1 INTRODUCTION

In drug discovery, the optimization of clearance within a chemical series is a key step to identify a new molecular entity (NME). This optimization process can be used to achieve optimum plasma exposure necessary to effectively modulate new drug targets. Drug molecules are eliminated by renal, biliary, active uptake mediated by hepatoselective transporters (e.g., organic anion transporting polypeptides 1B1 (OATP1B1) and 1B3 (OATP1B3)), and/or metabolism. To predict human clearance mechanism and rate, significant investment in human *in vitro* reagents and preclinical experiments is warranted. However, various tools are available to predict human clearance that include but not limited to scaling clearance from human liver microsomes (HLM) and hepatocytes, sandwich culture human hepatocytes (SCHH), suspension hepatocytes, renal proximal tubules cells, transfected cell lines, and single species scaling from preclinical species (e.g., rat, dog) [1–4]. Successful and timely implementation of the proper tool would enable medicinal chemists to identify an NME that is capable of achieving the systemic and target exposure required. It also allows drug metabolism scientists to successfully predict potential drug–drug interactions (DDI) or drug–food interactions and impact of enzymes/transporters pharmacogenomics on drug exposure. Failure to identify the correct mechanism of elimination would waste significant resources and lead to the optimization of parameters that are not relevant to the systemic clearance of NMEs. Therefore, a framework that provides a prospective guidance on the rate-determining step of drug clearance is a prerequisite for successful and timely optimization of the medicinal chemistry design in bringing forward best-in-class compounds in efficient resource-sparing manner.

Oral Bioavailability Assessment: Basics and Strategies for Drug Discovery and Development, First Edition.
Ayman F. El-Kattan.

In this chapter, the physiological basis of clearance and the associated drug-related physicochemical descriptors are discussed. In addition, the new extended clearance classification system (ECCS) is presented and its utility in drug discovery to enable prospective prediction of clearance rate-determining step using available *in vitro* and preclinical *in vivo* models.

8.2 HEPATIC CLEARANCE

Based on physiologically based pharmacokinetics principles (PBPK), factors that affect hepatic clearance are determined by system (drug-independent) and drug-dependent parameters. Examples on system-dependent parameters include hepatic blood flow, protein content of cells, and hepatocyte volume, whereas drug-dependent parameters include drug plasma protein binding, red blood cells partition, intrinsic clearance, which is the intrinsic ability of the liver to clear the drug. Intrinsic hepatic clearance is driven by the drug affinity to hepatic metabolizing enzymes, basolateral and canalicular transporters [5–7]. It is important to characterize the transporter(s) metabolizing enzyme(s) interplay to successfully predict the hepatic clearance of NMEs (for reviews, [4, 6, 8, 9]). The mathematical expression defining this interplay is termed extended clearance concept and shown in Equation 8.1 [5]:

$$CL_{int,h} = \frac{(PS_{influx} + PS_{pd}) \cdot CL_{int}}{(PS_{efflux} + PS_{pd} + CL_{int})} \tag{8.1}$$

Herein, total hepatic clearance (CL_h) is calculated by assuming well-stirred conditions (Equation 8.2) [7]:

$$CL_h = Q_h \cdot \frac{f_b \cdot (PS_{influx} + PS_{pd}) \cdot CL_{int}}{Q_h \cdot (PS_{efflux} + PS_{pd} + CL_{int}) + f_b \cdot (PS_{influx} + PS_{pd}) \cdot CL_{int}} = Q_h \cdot E_h \tag{8.2}$$

PS_{influx} and PS_{efflux} are the active (transporter-mediated) sinusoidal influx and basolateral efflux clearances. PS_{pd} represents passive diffusion clearance (Figure 8.1). CL_{int} is the sum of the metabolic and biliary intrinsic clearances ($CL_{int,met} + CL_{int,bile}$), Q_h is the liver blood flow,

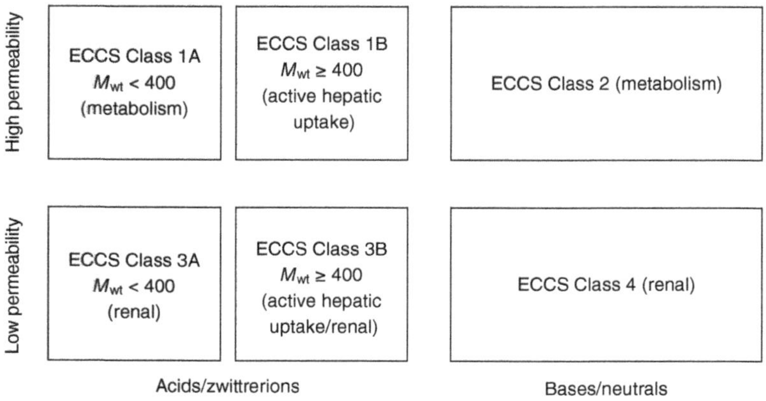

FIGURE 8.1 The framework of extended clearance classification system (ECCS) for identifying the predominant mechanism that determines systemic clearance of drugs [10].

E_h is the hepatic extraction ratio, and $f_{u,b}$ is the unbound fraction in blood, which is calculated using Equation 8.3.

$$f_{u,b} = \frac{f_{u,plasma}}{B/P} \tag{8.3}$$

where $f_{u,plasma}$ is the unbound fraction in plasma and B/P is the blood to plasma ratio.

It should be emphasized that prediction of the hepatic clearance for drugs that are substrates for a combination of metabolizing enzymes and influx or efflux transporters would be biased if only one process is considered [4, 5, 9, 11–13]. However, Equation 8.2 is simplified when "rapid-equilibrium" between liver and blood compartments is achieved. This equilibrium is typically realized when the drug molecule is not a substrate for active hepatic uptake transporters (e.g., organic anion transporting polypeptides 1B1/1B3) and the PS_{pd} is significantly higher than CL_{int}. Equation 8.2 is simplified to Equation 8.4 [7]:

$$CL_h = Q_h \cdot \frac{f_b \cdot CL_{int}}{Q_h + f_b \cdot CL_{int}} = Q_h \cdot E_h \tag{8.4}$$

This equation is usually used in drug discovery to predict the hepatic clearance for rapid equilibrium compounds that are only substrates for cytochrome P-450 (CYP) and other metabolizing enzymes [14–16].

For rapid equilibrium compounds, hepatic clearance can be predicted using HLM [1–3]. This is used mainly for drug molecules that are substrates for CYP enzymes. However, to predict the human hepatic clearance for drug molecules that are substrates for UDP glucuronosyltransferases (UGTs), sulfotransferases, aldehyde oxidase, and glutathione S-transferase (GST), among others, human hepatocytes is the method of choice [1–3, 17, 18].

Hepatic uptake can be the rate-determining step of the systemic clearance of the drug molecule, if PS_{pass} is significantly lower than CL_{int} and the compound is a substrate for uptake transporters (e.g., OATP1B1 and 1B3 transporters) leading to PS_{influx} significantly higher than PS_{pass}. Equation 8.2 can be reduced to Equation 8.5.

$$CL_h = Q_h \cdot \frac{f_b \cdot PS_{influx}}{Q_h + f_b \cdot PS_{influx}} = Q_h \cdot E_h \tag{8.5}$$

Examples on drug molecules that have hepatic uptake as a rate-determining step for their systemic clearance include HMG-CoA reductase inhibitors (statins) and angiotensin II antagonists (sartans) [11, 19–22]. Indeed, a number of studies suggested that active uptake mediated by OATP1B1 and 1B3 transporters expressed at the basolateral membrane of liver hepatocytes could be the rate-determining step in hepatic clearance, not only for compounds that are metabolically stable such as rosuvastatin [23] and pravastatin [24–26], but also for compounds that are extensively metabolized such as atorvastatin [27, 28], glyburide [29, 30], and repaglinide [31, 32]. This result is substantiated by clinically relevant DDIs, leading to changes in systemic exposure of these drugs that are primarily attributed to the inhibition of hepatic uptake mediated by members of the OATP family rather than the inhibition of hepatic metabolism or biliary efflux [9, 27, 33]. Furthermore, polymorphism in *SLCO1B1* (encoding OATP1B1) was reported to lead to major changes in systemic exposure of statins, which, in turn, regulates the relative exposure in peripheral tissue such as muscle and the risk of toxicity such as rhabdomyolysis [34–37]. These examples clearly demonstrate the significant role of active hepatic uptake mediated by OATP transporter family in determining the systemic exposure of these molecules. It should be emphasized that affinity of drug molecules with active hepatic uptake as a rate-determining step to enzymatic metabolism and biliary efflux would determine the hepatic exposure and excretion from the body. Therefore, inhibition or pharmacogenomics variations in these efflux transporters and/or metabolizing enzymes would lead to major changes in the liver exposure and minimal change in the plasma. This example is valid provided that no alteration in activity of these transporters and enzymes is realized along the intestine.

8.2.1 Physicochemical Determinants of Active Hepatic Uptake

OATPs are the key transporters involved in active hepatic uptake [38]. The main physicochemical drivers for OATP substrates are not conclusive considering the limited data sets available. In 2012, Varma et al. analyzed the physicochemical properties of 219 diverse commercial and proprietary that are substrates for human OATP transporters. It is interesting to note that drug ionization state was reported to play a crucial role in the substrate interaction with OATPs [39]. Indeed, mainly acids and zwitterions are substrates for OATP transporters. Almost no bases were shown to be substrates for these transporters. Furthermore, OATP substrates tend to have a molecular weight that is larger than 400 Da. They also have high polarity shown with high polar surface area. In addition, lipophilicity ($\log D$) did not appear to play such a key role. It should be emphasized that OATP substrates tend to have a wide permeability range. For example, active uptake could be the rate-determining step in the hepatic clearance of compounds with low permeability (e.g., pravastatin, rosuvastatin, valsartan) and high permeability (e.g., atorvastatin, cerivastatin, glyburide, repaglinide, and telmisartan) drugs [38, 40, 41]. Collectively, the group indicated that hepatic active uptake could be the predominant clearance mechanism for acids and zwitterions with a molecular weight >400 Da, irrespective of their membrane permeability [39].

TIP:

You continue to read and listen for leading scientists emphasizing that changes in metabolizing enzymes or efflux transporters activity that are mediated by drug inhibition or genetic variations of these proteins will not influence the drug systemic exposure of a molecule with a rate-determining step as active hepatic uptake such as statins or sartans. This assumption is only valid if no alteration is observed in these enzymes/transporters along the intestine. For example, genetic variations in BCRP and MRP2 have a significant impact on the systemic exposure of rosuvastatin and pravastatin, respectively [42–44]. These changes are mainly attributed to alteration in the fraction that is absorbed. It is clear that there is a tendency to underestimate the potential contribution of the intestine to the overall disposition of the molecule. For compounds that have low permeability, keep a close eye on the potential impact of efflux transporters, for example, P-gp, MRP2, BCRP on their absorption. The potential impact is typically underestimated due to the general belief that these transporters can be easily saturated for doses higher than 100 mg. For compounds that are extensively metabolized, keep a close eye on ones that are substrate for CYP3A4, since higher expression for CYP3A4 in the intestine relative to other metabolizing enzymes is reported. This is associated with the observation that CYP3A4 substrates tend to have a lower average of fraction that escapes intestinal metabolism (f_g) [45].

8.2.2 Physicochemical Determinants of Hepatobiliary Transport

Drug secretion into the bile is an active process that is driven by efflux transporters lining the bile canaliculi (e.g., MRP2, BCRP, and P-gp transporters) [46, 47]. Drug molecules that are substrates for MRP2 and BCRP transporters are acidic in nature and include sulfates and glucuronide conjugates. However, P-gp substrates are typically basic hydrophobic drug molecules. Examples of MRP2 and BCRP substrates are pravastatin and rosuvastatin, respectively

[11, 38]. It is pivotal to emphasize that there is an overlap in the chemical space of the substrates of hepatic uptake and hepatobiliary transport. For example, the dominant physicochemical properties of drugs excreted in bile are similar to the previous section: ionization state, molecular weight, lipophilicity, and polarity [39, 48–50].

Overall, polar acidic drugs with high molecular weight and low permeability are taken up by hepatoselective OATP transporters and consequently eliminated in bile [39, 51]. While substrate affinity toward canalicular efflux transporters is a prerequisite for active secretion into bile, it is evident that hepatic sinusoidal uptake is the rate-determining step in the hepatic clearance of compounds eliminated in bile (Equation 8.2).

8.2.3 Physicochemical Determinants of Metabolism

Generally, drug molecules that are metabolized in the liver are lipophilic in nature with high passive permeability. In 2005, Benet and colleagues proposed biopharmaceutics drug disposition classification system (BDDCS) that indicated that compounds with high passive permeability tend to have a high extent of metabolism (>70%) and vice versa [52–54]. Following metabolism, these highly permeable compounds are excreted/eliminated from the body as phase I and/or phase II metabolites. For these drug molecules that are not substrates for OATP transporters, hepatic clearance can be predicted only from metabolic clearance measured using HLM and/or hepatocytes (Equation 8.4).

Lipophilicity ($\log P$ or $\log D_{7.4}$) has been closely correlated with affinity to metabolizing enzymes [55–57]. Varma et al. noted that gut and hepatic extraction are significantly influenced by drug lipophilicity, in which drug molecules with $c \log D_{7.4} > 3$ demonstrate high intestinal and hepatic extraction and vice versa [58]. The charge of the chemical substrate is also an important determinant, with bases tending to be more readily metabolized by the majority of enzymes. There are exceptions based on affinity for specific drug metabolizing enzymes. For example, CYP2C9 substrates are relatively acidic, while CYP3A and CYP2D6 substrates are either basic or neutral [59, 60].

8.3 RENAL CLEARANCE

Renal clearance of a drug molecule is an outcome of the glomerular filtration, tubular secretion, and reabsorption processes. It is mathematically shown as

$$\mathrm{CL}_{\mathrm{renal}} = (f_{\mathrm{b}} \cdot \mathrm{GFR} + \mathrm{CL}_{\mathrm{sec}}) \cdot (1 - F_{\mathrm{reabs}}) \tag{8.6}$$

where f_{b} is the unbound drug fraction in blood, GFR is the glomerular filtration rate, $\mathrm{CL}_{\mathrm{sec}}$ is the renal secretory clearance, and F_{reabs} is the fraction of filtered and secreted drug that is reabsorbed in renal proximal tubules. $\mathrm{CL}_{\mathrm{sec}}$ can be expressed as Equation 8.7 assuming a well-stirred model:

$$\mathrm{CL}_{\mathrm{sec}} = Q_{\mathrm{r}} \cdot \frac{f_{\mathrm{b}} \cdot \mathrm{CL}_{\mathrm{int,sec}}}{Q_{\mathrm{r}} + f_{\mathrm{b}} \cdot \mathrm{CL}_{\mathrm{int,sec}}} \tag{8.7}$$

where Q_{r} is the renal blood flow and $\mathrm{CL}_{\mathrm{int,sec}}$ is the intrinsic secretory clearance that can be described by Equation 8.8:

$$\mathrm{CL}_{\mathrm{int,sec}} = \frac{\mathrm{PS}_{\mathrm{influx,b}} \cdot \mathrm{PS}_{\mathrm{efflux,a}}}{(\mathrm{PS}_{\mathrm{efflux,b}} + \mathrm{PS}_{\mathrm{efflux,a}})} \tag{8.8}$$

$\mathrm{PS}_{\mathrm{influx,b}}$, $\mathrm{PS}_{\mathrm{efflux,b}}$, $\mathrm{PS}_{\mathrm{influx,a}}$, and $\mathrm{PS}_{\mathrm{efflux,a}}$ are influx and efflux intrinsic transport clearances across the basolateral and apical membranes of proximal tubule cells, respectively. GFR is the filtration of about 10% of total renal blood flow at the glomerulus of the nephron and is defined by the blood flow rate and f_b. While the filtration process occurs with all drug molecules, glomerular filtration contribution to total clearance is typically low due to low human GFR value of 1.78 mL/min/kg. It is interesting to note that tubular secretion allows the transport of drug molecules from blood to proximal tubular lumen. This process is governed by uptake transporters lining the basolateral side of the proximal tubular cells, $f_{b,}$ and blood flow rate [61, 62]. The overall contribution of renal clearance to the overall drug clearance is also determined by the extent of renal tubular reabsorption that is usually proportional to the passive permeability of drug molecule and is influenced by urine flow and pH [63, 64].

Renal secretion is an active process that is mediated by polyspecific members of the organic ion transporter family (*SLC22*). These transporters are expressed at the basolateral membrane of proximal tubules cells, for example, organic cation transporter 2 (OCT2) and organic anion transporters 1, 2, and 3 (OAT1, OAT2, and OAT3) [65–67]. Substrates for these transporters have apparent low passive permeability across the basolateral membrane compared to the apical membrane [68], and a high counter concentration-gradient created by water reabsorption makes the contribution of passive tubular secretion negligible, if any. It should be emphasized that while active secretion involves passage across two membranes of proximal tubule cells, drug uptake from the blood compartment to cell will be the rate-determining process toward systemic clearance [69], especially for hydrophilic drugs with minimum passive and active back flux. The efflux transporters at the urine side would determine the kidney tissue concentrations.

8.3.1 Physicochemical Determinants of Renal Clearance

In 2012, our group reported that drugs with high renal recovery tend to have low passive permeability [70]. They are typically ionized at physiological pH and have affinity for uptake renal transporters such as OAT1/OAT3/OCT2 [70, 71].

8.4 PROPOSAL OF EXTENDED CLEARANCE CLASSIFICATION SYSTEM (ECCS)

As discussed earlier, three physicochemical properties appear to play a key role in defining the rate-determining step of drug clearance and they are ionization state, molecular weight, and membrane permeability [39, 51, 70, 72]. Our group hypothesized that these key parameters could be used to prospectively predict a molecule's predominant clearance mechanism, or rate-determining process toward its systemic clearance. These principal parameters were used to propose a new classification system, the ECCS. As shown in Figure 8.1, drug molecules are classified into six categories [10]:

- Class 1A: low-molecular-weight (M_{wt} < 400 Da), high-permeability acids and zwitterions for which metabolism is the predominant clearance mechanism.
- Class 1B: high-molecular-weight (M_{wt} > 400 Da), high-permeability acids and zwitterions for which hepatic uptake is the predominant clearance mechanism.
- Class 2: high-permeability bases and neutrals for which metabolism is the predominant clearance mechanism.
- Class 3A: low-molecular-weight (M_{wt} < 400 Da), low-permeability acids and zwitterions, for which renal elimination is the predominant clearance mechanism.
- Class 3B: high-molecular-weight (M_{wt} > 400 Da), low-permeability acids and zwitterions, for which hepatic uptake or renal elimination is the predominant clearance mechanism.
- Class 4: low-permeability bases and neutrals, for which renal clearance is the predominant clearance mechanism.

A cutoff value of 5×10^{-6} cm/s for the membrane permeability is used to define high- and low-permeability classes, based on a previous analysis conducted by our group [51]. Using a smaller set of compounds with apparent membrane permeability measured across a clonal cell line of MDCK-LE, this statistically derived a cutoff value distinguished between high and low intestinal absorption (f_a) and renal clearance with high sensitivity and specificity. The molecular weight cutoff (400 Da) enables distinction between OATP substrates (Class 1B/3B) and nonsubstrates (Class 1A/3A) within the subset of high-permeability acids and zwitterions. Class 1A and 1B once in the liver, they are metabolized and excreted in the urine/bile as phase I and II metabolites. The overall extent of metabolism is >70% of dose (Figure 8.1). The rate-determining step for Class 1A is metabolism, while the rate-determining step for Class 1B is active hepatic uptake mediated by OATP transporters. Class 3 is divided into two subclasses Class 3A and Class 3B. For Class 3A compounds, their elimination is usually driven by renal elimination. However, Class 3B compounds are eliminated by a mix of renal and biliary elimination following active uptake mediated by OATP transporters. For Class 4 compounds, the rate-determining step is renal elimination. The ionization state of the molecule can be obtained as a measured value or using *in silico* programs such as MoKa (version 2.5.4, Molecular discovery). This classification was successfully validated using a database that was published by Obach et al. [72], Varma et al. [58, 70], Berellini et al. [73], and Lombardo et al. [74]. For more information, refer to Varma et al. [10].

8.4.1 General Characteristics of the ECCS Classes

8.4.1.1 Class 1A These are acidic or zwitterionic compounds with high permeability (MDCK-LE $P_{app} \geq 5 \times 10^{-6}$ cm/s) and low molecular weight <400 Da. Hepatic metabolism is the rate-determining step of the clearance of Class 1A compounds with an extent of metabolism that is greater than 70%. Transporters have no impact on the systemic clearance and exposure of these molecules. Examples of Class 1A drugs include nonsteroidal anti-inflammatory drugs such as ibuprofen and ketoprofen (Table 8.1). These compounds are typically metabolized by CYP2C enzymes [131], although other metabolic pathways such as glucuronidation prevail for certain drugs (e.g., valproic acid). Human *in vitro* tools such as HLM (CYP substrates) or human hepatocytes (UGT or CYP substrates) can be effectively used to predict the systemic clearance of these molecules [3].

8.4.1.2 Class 1B These are acidic or zwitterionic compounds with high permeability (MDCK-LE $P_{app} \geq 5 \times 10^{-6}$ cm/s) and high molecular weight (≥400 Da). While these molecules have a high extent of metabolism (>70%), Class 1B compounds are predominantly systemically cleared by active hepatic uptake mediated by OATPs (Equation 8.5). Once cleared from the blood compartment to the liver compartment via these active uptake transporters, Class 1B compounds are metabolized and excreted in the bile and/or urine as phase I and phase II metabolites with an extent of metabolism higher than 70%. Class 1B drug molecules are substrates to OATP1B1 and/or OATP1B3 transporters and are predominantly metabolized by CYP 3A4/CYP2C/UGT enzymes [38, 40, 41]. To predict the hepatic clearance of these molecules, the use of human *in vitro* tools such as suspension hepatocytes and SCHH is warranted [38, 132–138]. It should be emphasized that relying only on HLM/human hepatocytes will underestimate systemic clearance, although they can be critical for modeling liver concentrations [132, 139].

8.4.1.3 Class 2 These are neutral or basic compounds with high permeability (MDCK-LE $P_{app} \geq 5 \times 10^{-6}$ cm/s). Class 2 compounds are predominantly cleared by metabolism. They cross the basolateral membrane of the hepatocytes via passive diffusion and are metabolized and excreted in the urine/bile as phase I or II metabolites. Inhibitors for metabolizing enzymes have a significant impact on the plasma exposure of Class 2 molecules, given the predominance of this clearance mechanism. For example, the CYP3A substrate midazolam increased

TABLE 8.1 Representative Examples per ECCS Class and the Major Transporters and Metabolizing Enzymes Involved in Their Clearance

ECCS Class	Drug	Metabolic Enzymes	Primary Transporters	References
Class 1A	Acetyl salicylic acid	Carboxylesterase 1/2		[75]
	Diclofenac	CYP2C9, UGT2B7		[76, 77]
	Ibuprofen	CYP2C19, CYP2C9		[78, 79]
	Rosiglitazone	CYP2C8		[80]
	Valproic acid	UGT2B7, UGT1A6 and UGT1A9.		[81]
	Warfarin	CYP2C9		[82]
Class 1B	Atorvastatin	CYP3A4	OATPs	[83, 84]
	Bosentan	CYP2C9, CYP3A4	OATPs	[85]
	Cerivastatin	CYP2C8, CYP3A4	OATPs	[86]
	Fluvastatin	CYP2C9	OATPs	[87]
	Glyburide	CYP2C9, CYP3A4	OATPs	[29]
	Pitavastatin	UGT1A3 and UGT2B7	OATPs	[88, 89]
	Repaglinide	CYP2C8, CYP3A	OATPs	[13, 32]
	Telmisartan	UGTs	OATPs	[90]
Class 2	Abacavir	Alcohol dehydrogenase		[91]
	Alfentanil	CYP3A4		[92, 93]
	Amitriptyline	CYP2D6 and CYP2C19		[94]
	Amlodipine	CYP3A4		[95]
	Budesonide	CYP3A4		[96]
	Caffeine	CYP1A2		[97]
	Diazepam	CYP3A4, CYP2C19		[98]
	Diltiazem	CYP3A4		[99]
	Erlotinib	CYP3A4		[100]
	Felodipine	CYP3A4		[101]
	Imipramine	CYP2D6		[102]
	Labetalol	UGT1A1		[103]
	Midazolam	CYP3A4		[104]
	Morphine	UGT2B7	*p*-Glycoprotein	[105, 106]
	Nifedipine	CYP3A4		[107]
	Omeprazole	CYP2C19		[108]
	Propranolol	CYP2D6		[109]
	Sildenafil	CYP3A4		[110]
	Triazolam	CYP3A4		[93]
	Verapamil	CYP3A4		[101]
	Voriconazole	CYP2C19, 2C9, 3A4		[111, 112]
	Ziprasidone	Aldehyde Oxidase, CYP		[113]

TABLE 8.1 (*Continued*)

ECCS Class	Drug	Metabolic Enzymes	Primary Transporters	References
Class 3A	Amoxicillin		OAT1	[114]
	Cefotaxime		OAT3	[115]
	Ceftizoxime		OAT1, OAT3	[116]
	Furosemide		OAT1, OAT3	[117]
	Methotrexate		OAT1, OAT3	[118, 119]
	Penicillin_G		OAT1, OAT3	[120]
	Amoxicillin		OAT1	[114]
	Cefotaxime		OAT3	[115]
	Ceftizoxime		OAT1, OAT3	[116]
	Irbesartan		OATPs	[121]
Class 3B	Furosemide		OAT1, OAT3	[117]
Class 3A	Amoxicillin		OAT1	[114]
	Cefotaxime		OAT3	[115]
	Cefoperazone		OATP1B3, MRP2	[122]
	Ceftizoxime		OAT1, OAT3	[116]
	Irbesartan		OATPs	[121]
	Furosemide		OAT1, OAT3	[117]
	Methotrexate		OAT1, OAT3	[118, 119]
	Penicillin_G		OAT1, OAT3	[120]
	Rosuvastatin		OATPs, BCRP	[123]
	Valsartan		OATPs, MRP2	[124]
	Methotrexate		OAT1, OAT3	[118, 119]
	Penicillin_G		OAT1, OAT3	[120]
	Rosuvastatin		OATPs, BCRP	[123]
	Valsartan		OATPs, MRP2	[124]
Class 4	Acyclovir		OAT1, OAT3	[125]
	Famotidine		OCT2	[126]
	Lamivudine		OCT2, MATEs	[127, 128]
	Ranitidine		OCT2	[129]
	Sitagliptin		OAT3	[130]

Source: Varma et al. 2015 [10]. Reproduced with permission of Springer.

plasma exposure by 2541% when coadministered with the potent CYP3A4 inhibitor ritonavir [140]. Representative examples of Class 2 include amitriptyline, diazepam, imipramine, midazolam, nifedipine, sildenafil, and voriconazole (Table 8.1). Due to the expected rapid equilibrium between blood and liver compartments, systemic clearance of Class 2 compounds can be described by Equation 8.4; and HLM, human hepatocytes, or other human *in vitro* systems aligned with the underlying metabolic process, are effective in predicting the clearance of this class of compounds.

8.4.1.4 Class 3A These are acids or zwitterions with low permeability (MDCK-LE $P_{app} < 5 \times 10^{-6}$ cm/s) and small molecular weight (<400 Da). Class 3A compounds are predominantly cleared by renal clearance as unchanged drug in urine [39, 141, 142]. Renal uptake transporters such as OAT1, OAT3, and OCT2 transporters play a key role in the clearance of Class 3A compounds. Amoxicillin and furosemide are representative examples.

The prediction of renal clearance can be achieved via single species scaling or by using PBPK modeling [90, 133], whereas metabolic stability studies would underestimate the clearance of these molecules.

8.4.1.5 ***Class 3B*** These are acids or zwitterions with low permeability (MDCK-LE $P_{app} < 5 \times 10^{-6}$ cm/s) and large molecular weight (≥400 Da). Class 3B compounds are predominantly cleared by hepatic active uptake and/or renal clearance, followed by elimination as unchanged drug in bile or urine [39, 141, 142]. Hepatic and renal uptake transporters such as OATP1B1, OATP1B3, OATP2B1, OAT1, OAT3, and OCT2 play a key role in the clearance of Class 3B compounds. Therefore, substrate specificity toward hepatic transporters versus renal transporters dictates the clearance pathway to be via hepatic or renal (Table 8.1). Pravastatin, rosuvastatin, and valsartan are representative examples. Similar to Class 1B compounds, the prediction of active hepatic uptake clearance can be achieved via *in vitro* hepatocyte uptake studies [90, 133], whereas metabolic stability studies would underestimate the clearance of these molecules. Renal clearance can be predicted by single species scaling or physiologically based pharmacokinetic models.

8.4.1.6 ***Class 4*** These are neutral or basic compounds with low permeability (MDCK-LE $P_{app} < 5 \times 10^{-6}$ cm/s). Class 4 compounds are predominantly systemically cleared by glomerular filtration and active renal secretion that is mediated by basolateral transporters such as OCT2, OAT1, OAT2, and/or OAT3 (Table 8.1). While OCT2 generally prefer basic compounds, OATs are also capable of transporting bases along with acids and zwitterions [70, 143]. Acyclovir, famotidine, and sitagliptin are some examples of this class. Renal clearance can be predicted by single species scaling or physiologically based pharmacokinetic models [144].

8.5 CONCLUSIONS

In the drug design phase, an earlier understanding of the predominant clearance mechanism in a chemical series increases efficiency by focusing on characterizing the appropriate ADME parameters to enable the rapid identification of the best-in-class clinical candidate that is devoid of drug interaction risk and capable of robustly testing the mechanism in humans. Understanding the clearance rate-determining step of a drug candidate is a prerequisite for the successful prediction of clinical pharmacokinetics and DDIs, as well as for determining the impact of genetic variations of metabolizing enzymes and transporters on drug exposure. The proposed ECCS framework enables the early identification of clearance mechanism, based on membrane permeability and compound ionization, obtained using high-throughput *in vitro* or *in silico* tools. Overall, ECCS increases efficiency by providing the basis to determine the right *in vitro* and *in vivo* experimental approach(es) to be conducted for timely and reliable prediction of clearance and pharmacokinetics in humans and hence guidance of medicinal chemistry design to achieve best-in-class molecules.

8.6 CHAPTER 8 TUTORIALS

Determine the ECCS classification, human renal recovery, extent of metabolism, and rate-determining step of the compounds below.

Compound No.	**1**	**2**	**3**	**4**	**5**	**6**	**7**	**8**	**9**	**10**	**11**	**12**
Ionization	Zwitterion	Acidic	Acidic	Acidic	Zwitterion	Zwitterion	Basic	Basic	Neutral	Basic	Basic	Neutral
M_{wt}	171	180	481	551	501	452	474	408	346	252	259	225
PSA	63	64	149	154	81	79	122	100	107	114	41	119
$c\log P$	−0.7	1	1.9	4.2	2	5.3	2	3.4	3.1	0.2	2.75	−2.4
$c\log D$ (pH 7.4)	−1.4	−1.4	−2.2	2.4	1.23	2	2.5	1.5	3.6	−0.4	0.8	−1.5
MDCK-LE permeability (10^{-6} cm/s)	0.5	22.9	0.9	7.3	0.4	15.6	19.3	1.9	60	0.7	13.9	0.96
f_a	0.9	1	0.3	1	0.1	1	1	1	1	0.7	1	0.1
cHLM (μL/min/mg)	7.1	7.8	7.1	11.6	7.1	33.7	80	14.8	160	7.2	42	6.65
ECCS classification												
Human renal recovery (%)												
Extent of metabolism (%)												
Rate determining clearance step												

Compound	Gabapentin	Aspirin	Rosuvastatin	Bosentan	Fexofenadine	Repaglinide	Sildenafil	Amlodipine	Nifedipine	Cimetidine	Propranolol	Acyclovir
No.	**1**	**2**	**3**	**4**	**5**	**6**	**7**	**8**	**9**	**10**	**11**	**12**
Ionization	Zwitterion	Acidic	Acidic	Acidic	Zwitterion	Zwitterion	Basic	Basic	Neutral	Basic	Basic	Neutral
M_{wt}	171	180	481	551	501	452	474	408	346	252	259	225
PSA	63	64	149	154.	81	79	122	100	107	114	41	119
$c \log P$	−0.7	1.0	1.9	4.2	2.0	5.3	2.0	3.4	3.1	0.2	2.75	−2.4
$c \log D$ (pH 7.4)	−1.4	−1.4	−2.2	2.4	1.23	2	2.5	1.5	3.6	−0.4	0.8	−1.5
MDCK-LE permeability (10^{-6} cm/s)	0.5	22.9	0.9	7.3	0.4	15.6	19.3	1.9	60	0.7	13.9	0.96
f_a	0.9	1	0.3	1	0.1	1	1	1	1	0.7	1	0.1
cHLM (μL/min/mg)	7.1	7.8	7.1	11.6	7.1	33.7	80	14.8	160	7.2	42	6.65
ECCS classification	3A	1a	3B	1b	3B	1b	2	2	2	4	2	4
Human renal recovery (%)	100	1.4	5	2	25	1.5	7.5	10	0.01	62	0.3	75
Extent of metabolism (%)	0	98	5	98	<30	>90	>90	>90	>90		>90	>90
Rate determining clearance step	Renal	Metabolism	Active uptake	Active uptake	Active uptake	Active uptake	Metabolism	Metabolism	Metabolism	Renal	Metabolism	Renal

8.7 CHAPTER 8 TUTORIAL ANSWER KEY

Answer: The ECCS classification, human renal recovery, extent of metabolism, and rate-determining step of compounds are shown on page 320. Be aware of amlodipine ECCS classification. Based on MDCK-LE value and ionization, the compound was supposed to be classified as ECCS Class 4. However, the compound has high f_a in humans, suggesting that there might be an issue with its permeability determination. Since this compound has high lipophilicity, it has a high likelihood to have high nonspecific binding that can limit its permeability and bias the assessment. It is important to triangulate your assessment of permeability by relating it to observed absorption in preclinical models and also to understand its renal recovery in preclinical species and rate of metabolism. Overall, these inputs can provide a more rational prediction of your ECCS classification. Also, check out cimetidine: while its permeability is low, its oral absorption is almost complete. We did not classify cimetidine as ECCS Class 2 similar to amlodipine, since the compound has high renal recovery and its absorption is driven by active uptake along the intestine, rendering its f_a high. Again, it is prudent to triangulate your assessment to come up with a better ECCS classification.

REFERENCES

1. Hosea, N.A. and Jones, H.M., *Predicting pharmacokinetic profiles using* in silico *derived parameters*. Mol Pharm, 2013. 10(4): p. 1207–1215.
2. Hosea, N.A., *Drug design tools* – in silico, in vitro *and* in vivo *ADME/PK prediction and interpretation: is PK in monkey an essential part of a good human PK prediction?* Curr Top Med Chem, 2011. 11(4): p. 351–357.
3. Hosea, N.A., et al., *Prediction of human pharmacokinetics from preclinical information: comparative accuracy of quantitative prediction approaches*. J Clin Pharmacol, 2009. 49(5): p. 513–533.
4. Barton, H.A., Lai, Y., Goosen, T.C., Jones, H.M., El-Kattan, A.F., Gosset, J.R., Lin, J. and Varma, M.V., *Model-based approaches to predict drug–drug interactions associated with hepatic uptake transporters: preclinical, clinical and beyond*. Expert Opin Drug Metab Toxicol, 2013. 9(4): p. 459–472.
5. Liu, L. and Pang, K.S., *The roles of transporters and enzymes in hepatic drug processing*. Drug Metab Dispos, 2005. 33(1): p. 1–9.
6. Pang, K.S., Maeng, H.J. and Fan, J., *Interplay of transporters and enzymes in drug and metabolite processing*. Mol Pharm, 2009. 6(6): p. 1734–1755.
7. Pang, K.S. and Rowland, M., *Hepatic clearance of drugs. I. Theoretical considerations of a "well-stirred" model and a "parallel tube" model. Influence of hepatic blood flow, plasma and blood cell binding, and the hepatocellular enzymatic activity on hepatic drug clearance*. J Pharmacokinet Biopharm, 1977. 5(6): p. 625–653.
8. Yoshida, K., Maeda, K. and Sugiyama, Y., *Hepatic and intestinal drug transporters: prediction of pharmacokinetic effects caused by drug–drug interactions and genetic polymorphisms*. Annu Rev Pharmacol Toxicol, 2013. 53: p. 581–612.
9. Shitara, Y., Horie, T. and Sugiyama, Y., *Transporters as a determinant of drug clearance and tissue distribution*. Eur J Pharm Sci, 2006. 27(5): p. 425–446.
10. Varma, M.V., Steyn, S.J., Allerton, C. and El-Kattan, A.F., *Predicting clearance mechanism in drug discovery: extended clearance classification system (ECCS)*. Pharm Res, 2015. 32(12): p. 3785–3802.
11. Shitara, Y. and Sugiyama, Y., *Pharmacokinetic and pharmacodynamic alterations of 3-hydroxy-3-methylglutaryl coenzyme A (HMG-CoA) reductase inhibitors: drug–drug interactions and interindividual differences in transporter and metabolic enzyme functions*. Pharmacol Ther, 2006. 112(1): p. 71–105.

12. Camenisch, G. and Umehara, K., *Predicting human hepatic clearance from* in vitro *drug metabolism and transport data: a scientific and pharmaceutical perspective for assessing drug–drug interactions*. Biopharm Drug Dispos, 2012. 33(4): p. 179–194.
13. Varma, M.V., Lin, J., Bi, Y.A., Rotter, C.J., Fahmi, O.A., Lam, J.L., El-Kattan, A.F., Goosen, T.C. and Lai, Y., *Quantitative prediction of repaglinide-rifampicin complex drug interactions using dynamic and static mechanistic models: delineating differential CYP3A4 induction and OATP1B1 inhibition potential of rifampicin*. Drug Metab Dispos, 2013. 41(5): p. 966–974.
14. De Buck, S.S., Sinha, V.K., Fenu, L.A., Nijsen, M.J., Mackie, C.E. and Gilissen, R.A., *Prediction of human pharmacokinetics using physiologically based modeling: a retrospective analysis of 26 clinically tested drugs*. Drug Metab Dispos, 2007. 35(10): p. 1766–1780.
15. Jones, H.M., et al., *Simulation of human intravenous and oral pharmacokinetics of 21 diverse compounds using physiologically based pharmacokinetic modelling*. Clin Pharmacokinet, 2011. 50(5): p. 331–347.
16. Obach, R.S., Baxter, J.G., Liston, T.E., Silber, B.M., Jones, B.C., Macintyre, F., Rance, D.J. and Wastall, P., *The prediction of human pharmacokinetic parameters from preclinical and* in vitro *metabolism data*. J Pharmacol Exp Ther, 1997. 283(1): p. 46–58.
17. Williams, J.A., Hyland, R., Jones, B.C., Smith, D.A., Hurst, S., Goosen, T.C., Peterkin, V., Koup, J.R. and Ball, S.E., *Drug–drug interactions for UDP-glucuronosyltransferase substrates: a pharmacokinetic explanation for typically observed low exposure (AUCi/AUC) ratios*. Drug Metab Dispos, 2004. 32(11): p. 1201–1208.
18. Di, L., Feng, B., Goosen, T.C., Lai, Y., Steyn, S.J., Varma, M.V. and Obach, R.S., *A perspective on the prediction of drug pharmacokinetics and disposition in drug research and development*. Drug Metab Dispos, 2013. 41(12): p. 1975–1993.
19. Fenner, K.S., Jones, H.M., Ullah, M., Kempshall, S., Dickins, M., Lai, Y., Morgan, P. and Barton, H.A., *The evolution of the OATP hepatic uptake transport protein family in DMPK sciences: from obscure liver transporters to key determinants of hepatobiliary clearance*. Xenobiotica, 2012. 42(1): p. 28–45.
20. International Transporter Consortium, et al., *Membrane transporters in drug development*. Nat Rev Drug Discov, 2010. 9(3): p. 215–236.
21. Kalliokoski, A. and Niemi, M., *Impact of OATP transporters on pharmacokinetics*. Br J Pharmacol, 2009. 158(3): p. 693–705.
22. Shimizu, K., et al., *Whole-body distribution and radiation dosimetry of [11C]telmisartan as a biomarker for hepatic organic anion transporting polypeptide (OATP) 1B3*. Nucl Med Biol, 2012. 39(6): p. 847–853.
23. Bergman, E., Forsell, P., Tevell, A., Persson, E.M., Hedeland, M., Bondesson, U., Knutson, L. and Lennernas, H., *Biliary secretion of rosuvastatin and bile acids in humans during the absorption phase*. Eur J Pharm Sci, 2006. 29(3–4): p. 205–214.
24. Yamazaki, M., Kobayashi, K. and Sugiyama, Y., *Primary active transport of pravastatin across the liver canalicular membrane in normal and mutant Eisai hyperbilirubinemic rats*. Biopharm Drug Dispos, 1996. 17(7): p. 607–621.
25. Varma, M.V., Lai, Y., Feng, B., Litchfield, J., Goosen, T.C. and Bergman, A., *Physiologically based modeling of pravastatin transporter-mediated hepatobiliary disposition and drug–drug interactions*. Pharm Res, 2012. 29(10): p. 2860–2873.
26. Watanabe, T., Kusuhara, H., Maeda, K., Shitara, Y. and Sugiyama, Y., *Physiologically based pharmacokinetic modeling to predict transporter-mediated clearance and distribution of pravastatin in humans*. J Pharmacol Exp Ther, 2009. 328(2): p. 652–662.
27. Maeda, K., Ikeda, Y., Fujita, T., Yoshida, K., Azuma, Y., Haruyama, Y., Yamane, N., Kumagai, Y. and Sugiyama, Y., *Identification of the rate-determining process in the hepatic clearance of atorvastatin in a clinical cassette microdosing study*. Clin Pharmacol Ther, 2011. 90(4): p. 575–581.
28. Lau, Y., Huang, Y., Frassetto, L. and Benet, L., *Effect of OATP1B transporter inhibition on the pharmacokinetics of atorvastatin in healthy volunteers*. Clin Pharmacol Ther (St Louis, MO, U S), 2006. 81(2): p. 194–204.

29. Varma, M.V., Scialis, R.J., Lin, J., Bi, Y.A., Rotter, C.J., Goosen, T.C. and Yang, X., *Mechanism-based pharmacokinetic modeling to evaluate transporter–enzyme interplay in drug interactions and pharmacogenetics of glyburide*. AAPS J, 2014. 16(4): p. 736–748.

30. Zheng, H., Huang, Y., Frassetto, L. and Benet, L., *Elucidating rifampin's inducing and inhibiting effects on glyburide pharmacokinetics and blood glucose in healthy volunteers: unmasking the differential effects of enzyme induction and transporter inhibition for a drug and its primary metabolite*. Clin Pharmacol Ther (St Louis, MO, USA), 2008. 85(1): p. 78–85.

31. Kalliokoski, A., Backman, J.T., Kurkinen, K.J., Neuvonen, P.J. and Niemi, M., *Effects of gemfibrozil and atorvastatin on the pharmacokinetics of repaglinide in relation to SLCO1B1 polymorphism*. Clin Pharmacol Ther, 2008. 84(4): p. 488–496.

32. Varma, M.V., Lai, Y., Kimoto, E., Goosen, T.C., El-Kattan, A.F. and Kumar, V., *Mechanistic modeling to predict the transporter- and enzyme-mediated drug–drug interactions of repaglinide*. Pharm Res, 2013. 30(4): p. 1188–1199.

33. Prueksaritanont, T., et al., *Pitavastatin is a more sensitive and selective OATP1B clinical probe than rosuvastatin*. Br J Clin Pharmacol, 2014. 78(3): p. 587–598.

34. Group, S.C., et al., *SLCO1B1 variants and statin-induced myopathy – a genomewide study*. N Engl J Med, 2008. 359(8): p. 789–799.

35. Ieiri, I., Higuchi, S. and Sugiyama, Y., *Genetic polymorphisms of uptake (OATP1B1, 1B3) and efflux (MRP2, BCRP) transporters: implications for inter-individual differences in the pharmacokinetics and pharmacodynamics of statins and other clinically relevant drugs*. Expert Opin Drug Metab Toxicol, 2009. 5(7): p. 703–729.

36. Niemi, M., Neuvonen, P.J., Hofmann, U., Backman, J.T., Schwab, M., Lutjohann, D., von Bergmann, K., Eichelbaum, M. and Kivisto, K.T., *Acute effects of pravastatin on cholesterol synthesis are associated with SLCO1B1 (encoding OATP1B1) haplotype *17*. Pharmacogenet Genomics, 2005. 15(5): p. 303–309.

37. Nishizato, Y., et al., *Polymorphisms of OATP-C (SLC21A6) and OAT3 (SLC22A8) genes: consequences for pravastatin pharmacokinetics*. Clin Pharmacol Ther, 2003. 73(6): p. 554–565.

38. Shitara, Y., Maeda, K., Ikejiri, K., Yoshida, K., Horie, T. and Sugiyama, Y., *Clinical significance of organic anion transporting polypeptides (OATPs) in drug disposition: their roles in hepatic clearance and intestinal absorption*. Biopharm Drug Dispos, 2013. 34(1): p. 45–78.

39. Varma, M.V., Chang, G., Lai, Y., Feng, B., El-Kattan, A.F., Litchfield, J. and Goosen, T.C., *Physicochemical property space of hepatobiliary transport and computational models for predicting rat biliary excretion*. Drug Metab Dispos, 2012. 40(8): p. 1527–1537.

40. Elsby, R., Hilgendorf, C. and Fenner, K., *Understanding the critical disposition pathways of statins to assess drug–drug interaction risk during drug development: it's not just about OATP1B1*. Clin Pharmacol Ther, 2012. 92(5): p. 584–598.

41. Li, R., Barton, H.A. and Varma, M.V., *Prediction of pharmacokinetics and drug–drug interactions when hepatic transporters are involved*. Clin Pharmacokinet, 2014. 53(8): p. 659–678.

42. Hua, W.J., Hua, W.X. and Fang, H.J., *The role of OATP1B1 and BCRP in pharmacokinetics and DDI of novel statins*. Cardiovasc Ther, 2012. 30(5): p. e234–e241.

43. Kivisto, K.T. and Niemi, M., *Influence of drug transporter polymorphisms on pravastatin pharmacokinetics in humans*. Pharm Res, 2007. 24(2): p. 239–247.

44. Niemi, M., et al., *Association of genetic polymorphism in ABCC2 with hepatic multidrug resistance-associated protein 2 expression and pravastatin pharmacokinetics*. Pharmacogenet Genomics, 2006. 16(11): p. 801–808.

45. Hisaka, A., Nakamura, M., Tsukihashi, A., Koh, S. and Suzuki, H., *Assessment of intestinal availability (FG) of substrate drugs of cytochrome p450s by analyzing changes in pharmacokinetic properties caused by drug–drug interactions*. Drug Metab Dispos, 2014. 42(10): p. 1640–1645.

46. Kusuhara, H. and Sugiyama, Y., *Pharmacokinetic modeling of the hepatobiliary transport mediated by cooperation of uptake and efflux transporters*. Drug Metab Rev, 2010. 42(3): p. 539–550.

47. Lai, Y., *Identification of interspecies difference in hepatobiliary transporters to improve extrapolation of human biliary secretion.* Expert Opin Drug Metab Toxicol, 2009. 5(10): p. 1175–1187.
48. Yang, X., Gandhi, Y.A., Duignan, D.B. and Morris, M.E., *Prediction of biliary excretion in rats and humans using molecular weight and quantitative structure–pharmacokinetic relationships.* AAPS J, 2009. 11(3): p. 511–525.
49. Fagerholm, U., *Prediction of human pharmacokinetics–biliary and intestinal clearance and enterohepatic circulation.* J Pharm Pharmacol, 2008. 60(5): p. 535–542.
50. Luo, G., Johnson, S., Hsueh, M.M., Zheng, J., Cai, H., Xin, B., Chong, S., He, K. and Harper, T.W., In silico *prediction of biliary excretion of drugs in rats based on physicochemical properties.* Drug Metab Dispos, 2010. 38(3): p. 422–430.
51. Varma, M.V., et al., *pH-dependent solubility and permeability criteria for provisional biopharmaceutics classification (BCS and BDDCS) in early drug discovery.* Mol Pharm, 2012. 9(5): p. 1199–1212.
52. Benet, L.Z., Broccatelli, F. and Oprea, T.I., *BDDCS applied to over 900 drugs.* AAPS J, 2011. 13(4): p. 519–547.
53. Custodio, J.M., Wu, C.-Y. and Benet, L.Z., *Predicting drug disposition, absorption/elimination/transporter interplay and the role of food on drug absorption.* Adv Drug Delivery Rev, 2008. 60(6): p. 717–733.
54. Wu, C.-Y. and Benet, L.Z., *Predicting drug disposition via application of BCS: transport/absorption/elimination interplay and development of a biopharmaceutics drug disposition classification system.* Pharm Res, 2005. 22(1): p. 11–23.
55. Lipinski, C.A., Lombardo, F., Dominy, B.W. and Feeney, P.J., *Experimental and computational approaches to estimate solubility and permeability in drug discovery and development settings.* Adv Drug Delivery Rev, 2012. 64: p. 4–17.
56. Veber, D.F., Johnson, S.R., Cheng, H.-Y., Smith, B.R., Ward, K.W. and Kopple, K.D., *Molecular properties that influence the oral bioavailability of drug candidates.* J Med Chem, 2002. 45(12): p. 2615–2623.
57. Camenisch, G., Alsenz, J., van de Waterbeemd, H. and Folkers, G., *Estimation of permeability by passive diffusion through Caco-2 cell monolayers using the drugs' lipophilicity and molecular weight.* Eur J Pharm Sci, 1998. 6(4): p. 313–319.
58. Varma, M.V., Obach, R.S., Rotter, C., Miller, H.R., Chang, G., Steyn, S.J., El-Kattan, A. and Troutman, M.D., *Physicochemical space for optimum oral bioavailability: contribution of human intestinal absorption and first-pass elimination.* J Med Chem, 2010. 53(3): p. 1098–1108.
59. Lewis, D.F., *Homology modelling of human cytochromes P450 involved in xenobiotic metabolism and rationalization of substrate selectivity.* Exp Toxicol Pathol, 1999. 51(4): p. 369–374.
60. Lewis, D.F., Modi, S. and Dickins, M., *Structure–activity relationship for human cytochrome P450 substrates and inhibitors.* Drug Metab Rev, 2002. 34(1–2): p. 69–82.
61. Ho, R.H. and Kim, R.B., *Transporters and drug therapy: implications for drug disposition and disease.* Clin Pharmacol Ther, 2005. 78(3): p. 260–277.
62. Shitara, Y., Sato, H. and Sugiyama, Y., *Evaluation of drug–drug interaction in the hepatobiliary and renal transport of drugs.* Annu Rev Pharmacol Toxicol, 2005. 45: p. 689–723.
63. Sun, H., Frassetto, L. and Benet, L.Z., *Effects of renal failure on drug transport and metabolism.* Pharmacol Ther, 2006. 109: p. 1–11.
64. Jappar, D., Hu, Y., Keep, R.F. and Smith, D.E., *Transport mechanisms of carnosine in SKPT cells: contribution of apical and basolateral membrane transporters.* Pharm Res, 2009. 26(1): p. 172–181.
65. Ahlin, G., Karlsson, J., Pedersen, J.M., Gustavsson, L., Larsson, R., Matsson, P., Norinder, U., Bergstrom, C.A. and Artursson, P., *Structural requirements for drug inhibition of the liver specific human organic cation transport protein 1.* J Med Chem, 2008. 51(19): p. 5932–5942.
66. Inui, K.I., Masuda, S. and Saito, H., *Cellular and molecular aspects of drug transport in the kidney.* Kidney Int, 2000. 58(3): p. 944–958.
67. Launay-Vacher, V., Izzedine, H., Karie, S., Hulot, J.S., Baumelou, A. and Deray, G., *Renal tubular drug transporters.* Nephron Physiol, 2006. 103(3): p. p97–p106.

68. Masereeuw, R. and Russel, F.G., *Mechanisms and clinical implications of renal drug excretion*. Drug Metab Rev, 2001. 33(3–4): p. 299–351.

69. Sirianni, G.L. and Pang, K.S., *Organ clearance concepts: new perspectives on old principles*. J Pharmacokinet Biopharm, 1997. 25(4): p. 449–470.

70. Varma, M.V., Feng, B., Obach, R.S., Troutman, M.D., Chupka, J., Miller, H.R. and El-Kattan, A., *Physicochemical determinants of human renal clearance*. J Med Chem, 2009. 52(15): p. 4844–4852.

71. Feng, B., LaPerle, J.L., Chang, G. and Varma, M.V., *Renal clearance in drug discovery and development: molecular descriptors, drug transporters and disease state*. Expert Opin Drug Metab Toxicol, 2010. 6(8): p. 939–952.

72. Obach, R.S., Lombardo, F. and Waters, N.J., *Trend analysis of a database of intravenous pharmacokinetic parameters in humans for 670 drug compounds*. Drug Metab Dispos, 2008. 36(7): p. 1385–1405.

73. Berellini, G., Waters, N.J. and Lombardo, F., In silico *prediction of total human plasma clearance*. J Chem Inf Model, 2012. 52(8): p. 2069–2078.

74. Lombardo, F., Obach, R.S., Varma, M.V., Stringer, R. and Berellini, G., *Clearance mechanism assignment and total clearance prediction in human based upon* in silico *models*. J Med Chem, 2014. 57(10): p. 4397–4405.

75. Tang, M., Mukundan, M., Yang, J., Charpentier, N., LeCluyse, E.L., Black, C., Yang, D., Shi, D. and Yan, B., *Antiplatelet agents aspirin and clopidogrel are hydrolyzed by distinct carboxylesterases, and clopidogrel is transesterificated in the presence of ethyl alcohol*. J Pharmacol Exp Ther, 2006. 319(3): p. 1467–1476.

76. Davies, N.M. and Anderson, K.E., *Clinical pharmacokinetics of diclofenac. Therapeutic insights and pitfalls*. Clin Pharmacokinet, 1997. 33(3): p. 184–213.

77. King, C., Tang, W., Ngui, J., Tephly, T. and Braun, M., *Characterization of rat and human UDP-glucuronosyltransferases responsible for the* in vitro *glucuronidation of diclofenac*. Toxicol Sci, 2001. 61(1): p. 49–53.

78. Andersson, T.B., Bredberg, E., Ericsson, H. and Sjoberg, H., *An evaluation of the* in vitro *metabolism data for predicting the clearance and drug–drug interaction potential of CYP2C9 substrates*. Drug Metab Dispos, 2004. 32(7): p. 715–721.

79. McGinnity, D.F., Parker, A.J., Soars, M. and Riley, R.J., *Automated definition of the enzymology of drug oxidation by the major human drug metabolizing cytochrome P450s*. Drug Metab Dispos, 2000. 28(11): p. 1327–1334.

80. Kirchheiner, J., et al., *Pharmacokinetics and pharmacodynamics of rosiglitazone in relation to CYP2C8 genotype*. Clin Pharmacol Ther, 2006. 80(6): p. 657–667.

81. Chu, X.M., Zhang, L.F., Wang, G.J., Zhang, S.N., Zhou, J.H. and Hao, H.P., *Influence of UDP-glucuronosyltransferase polymorphisms on valproic acid pharmacokinetics in Chinese epilepsy patients*. Eur J Clin Pharmacol, 2012. 68(10): p. 1395–1401.

82. Takahashi, H. and Echizen, H., *Pharmacogenetics of warfarin elimination and its clinical implications*. Clin Pharmacokinet, 2001. 40(8): p. 587–603.

83. Lennernas, H., *Clinical pharmacokinetics of atorvastatin*. Clin Pharmacokinet, 2003. 42(13): p. 1141–1160.

84. Lau, Y.Y., Okochi, H., Huang, Y. and Benet, L.Z., *Multiple transporters affect the disposition of atorvastatin and its two active hydroxy metabolites: application of* in vitro *and* ex situ *systems*. J Pharmacol Exp Ther, 2006. 316(2): p. 762–771.

85. Dingemanse, J. and van Giersbergen, P.L., *Clinical pharmacology of bosentan, a dual endothelin receptor antagonist*. Clin Pharmacokinet, 2004. 43(15): p. 1089–1115.

86. Shitara, Y., Hirano, M., Sato, H. and Sugiyama, Y., *Gemfibrozil and its glucuronide inhibit the organic anion transporting polypeptide 2 (OATP2/OATP1B1:SLC21A6)-mediated hepatic uptake and CYP2C8-mediated metabolism of cerivastatin: analysis of the mechanism of the clinically relevant drug–drug interaction between cerivastatin and gemfibrozil*. J Pharmacol Exp Ther, 2004. 311(1): p. 228–236.

87. Neuvonen, P.J., Niemi, M. and Backman, J.T., *Drug interactions with lipid-lowering drugs: mechanisms and clinical relevance*. Clin Pharmacol Ther, 2006. 80(6): p. 565–581.

88. Hirano, M., Maeda, K., Shitara, Y. and Sugiyama, Y., *Contribution of OATP2 (OATP1B1) and OATP8 (OATP1B3) to the hepatic uptake of pitavastatin in humans*. J Pharmacol Exp Ther, 2004. 311(1): p. 139–146.

89. Fujino, H., Saito, T., Tsunenari, Y. and Kojima, J., *Effect of gemfibrozil on the metabolism of pitavastatin–determining the best animal model for human CYP and UGT activities*. Drug Metabol Drug Interact, 2004. 20(1–2): p. 25–42.

90. Li, R., Ghosh, A., Maurer, T.S., Kimoto, E. and Barton, H.A., *Physiologically based pharmacokinetic prediction of telmisartan in human*. Drug Metab Dispos, 2014. 42(10): p. 1646–1655.

91. McDowell, J.A., Chittick, G.E., Stevens, C.P., Edwards, K.D. and Stein, D.S., *Pharmacokinetic interaction of abacavir (1592U89) and ethanol in human immunodeficiency virus-infected adults*. Antimicrob Agents Chemother, 2000. 44(6): p. 1686–1690.

92. Pea, F. and Furlanut, M., *Pharmacokinetic aspects of treating infections in the intensive care unit: focus on drug interactions*. Clin Pharmacokinet, 2001. 40(11): p. 833–868.

93. Baneyx, G., Parrott, N., Meille, C., Iliadis, A. and Lave, T., *Physiologically based pharmacokinetic modeling of CYP3A4 induction by rifampicin in human: influence of time between substrate and inducer administration*. Eur J Pharm Sci, 2014. 56: p. 1–15.

94. Samer, C.F., Lorenzini, K.I., Rollason, V., Daali, Y. and Desmeules, J.A., *Applications of CYP450 testing in the clinical setting*. Mol Diagn Ther, 2013. 17(3): p. 165–184.

95. Bailey, D.G. and Dresser, G.K., *Interactions between grapefruit juice and cardiovascular drugs*. Am J Cardiovasc Drugs, 2004. 4(5): p. 281–297.

96. Edsbacker, S. and Andersson, T., *Pharmacokinetics of budesonide (Entocort EC) capsules for Crohn's disease*. Clin Pharmacokinet, 2004. 43(12): p. 803–821.

97. Brosen, K., *Drug interactions and the cytochrome P450 system. The role of cytochrome P450 1A2*. Clin Pharmacokinet, 1995. 29(Suppl 1): p. 20–25.

98. Niwa, T., Shiraga, T., Ishii, I., Kagayama, A. and Takagi, A., *Contribution of human hepatic cytochrome p450 isoforms to the metabolism of psychotropic drugs*. Biol Pharm Bull, 2005. 28(9): p. 1711–1716.

99. Sutton, D., Butler, A.M., Nadin, L. and Murray, M., *Role of CYP3A4 in human hepatic diltiazem N-demethylation: inhibition of CYP3A4 activity by oxidized diltiazem metabolites*. J Pharmacol Exp Ther, 1997. 282(1): p. 294–300.

100. van Erp, N.P., Gelderblom, H. and Guchelaar, H.J., *Clinical pharmacokinetics of tyrosine kinase inhibitors*. Cancer Treat Rev, 2009. 35(8): p. 692–706.

101. Ohno, Y., Hisaka, A. and Suzuki, H., *General framework for the quantitative prediction of CYP3A4-mediated oral drug interactions based on the AUC increase by coadministration of standard drugs*. Clin Pharmacokinet, 2007. 46(8): p. 681–696.

102. Bogni, A., Monshouwer, M., Moscone, A., Hidestrand, M., Ingelman-Sundberg, M., Hartung, T. and Coecke, S., *Substrate specific metabolism by polymorphic cytochrome P450 2D6 alleles*. Toxicol In Vitro, 2005. 19(5): p. 621–629.

103. Jeong, H., Choi, S., Song, J.W., Chen, H. and Fischer, J.H., *Regulation of UDP-glucuronosyltransferase (UGT) 1A1 by progesterone and its impact on labetalol elimination*. Xenobiotica, 2008. 38(1): p. 62–75.

104. Paine, M.F., Shen, D.D., Kunze, K.L., Perkins, J.D., Marsh, C.L., McVicar, J.P., Barr, D.M., Gillies, B.S. and Thummel, K.E., *First-pass metabolism of midazolam by the human intestine*. Clin Pharmacol Ther, 1996. 60(1): p. 14–24.

105. Crowe, A., *The influence of P-glycoprotein on morphine transport in Caco-2 cells. Comparison with paclitaxel*. Eur J Pharmacol, 2002. 440(1): p. 7–16.

106. Chau, N., Elliot, D.J., Lewis, B.C., Burns, K., Johnston, M.R., Mackenzie, P.I. and Miners, J.O., *Morphine glucuronidation and glucosidation represent complementary metabolic pathways that are*

both catalyzed by UDP-glucuronosyltransferase 2B7: kinetic, inhibition, and molecular modeling studies. J Pharmacol Exp Ther, 2014. 349(1): p. 126–137.

107. Foti, R.S., Rock, D.A., Wienkers, L.C. and Wahlstrom, J.L., *Selection of alternative CYP3A4 probe substrates for clinical drug interaction studies using* in vitro *data and* in vivo *simulation*. Drug Metab Dispos, 2010. 38(6): p. 981–987.

108. Kanazawa, H., Okada, A., Higaki, M., Yokota, H., Mashige, F. and Nakahara, K., *Stereospecific analysis of omeprazole in human plasma as a probe for CYP2C19 phenotype*. J Pharm Biomed Anal, 2003. 30(6): p. 1817–1824.

109. Narimatsu, S., et al., *Regio- and stereoselective oxidation of propranolol enantiomers by human CYP2D6, cynomolgus monkey CYP2D17 and marmoset CYP2D19*. Chem Biol Interact, 2011. 189(3): p. 146–152.

110. Gupta, M., Kovar, A. and Meibohm, B., *The clinical pharmacokinetics of phosphodiesterase-5 inhibitors for erectile dysfunction*. J Clin Pharmacol, 2005. 45(9): p. 987–1003.

111. Leveque, D., Nivoix, Y., Jehl, F. and Herbrecht, R., *Clinical pharmacokinetics of voriconazole*. Int J Antimicrob Agents, 2006. 27(4): p. 274–284.

112. Jeu, L., Piacenti, F.J., Lyakhovetskiy, A.G. and Fung, H.B., *Voriconazole*. Clin Ther, 2003. 25(5): p. 1321–1381.

113. Beedham, C., Miceli, J.J. and Obach, R.S., *Ziprasidone metabolism, aldehyde oxidase, and clinical implications*. J Clin Psychopharmacol, 2003. 23(3): p. 229–232.

114. Li, M., Anderson, G.D., Phillips, B.R., Kong, W., Shen, D.D. and Wang, J., *Interactions of amoxicillin and cefaclor with human renal organic anion and peptide transporters*. Drug Metab Dispos, 2006. 34(4): p. 547–555.

115. Yee, S.W., et al., *Reduced renal clearance of cefotaxime in Asians with a low-frequency polymorphism of OAT3 (SLC22A8)*. J Pharm Sci, 2013. 102(9): p. 3451–3457.

116. Ueo, H., Motohashi, H., Katsura, T. and Inui, K., *Human organic anion transporter hOAT3 is a potent transporter of cephalosporin antibiotics, in comparison with hOAT1*. Biochem Pharmacol, 2005. 70(7): p. 1104–1113.

117. Hasannejad, H., et al., *Interactions of human organic anion transporters with diuretics*. J Pharmacol Exp Ther, 2004. 308(3): p. 1021–1029.

118. Liu, Q., Wang, C., Meng, Q., Huo, X., Sun, H., Peng, J., Ma, X., Sun, P. and Liu, K., *MDR1 and OAT1/OAT3 mediate the drug–drug interaction between puerarin and methotrexate*. Pharm Res, 2014. 31(5): p. 1120–1132.

119. El-Sheikh, A.A., Greupink, R., Wortelboer, H.M., van den Heuvel, J.J., Schreurs, M., Koenderink, J.B., Masereeuw, R. and Russel, F.G., *Interaction of immunosuppressive drugs with human organic anion transporter (OAT) 1 and OAT3, and multidrug resistance-associated protein (MRP) 2 and MRP4*. Transl Res, 2013. 162(6): p. 398–409.

120. Ye, J., Liu, Q., Wang, C., Meng, Q., Sun, H., Peng, J., Ma, X. and Liu, K., *Benzylpenicillin inhibits the renal excretion of acyclovir by OAT1 and OAT3*. Pharmacol Rep, 2013. 65(2): p. 505–512.

121. Werner, D., et al., *Determinants of steady-state torasemide pharmacokinetics: impact of pharmacogenetic factors, gender and angiotensin II receptor blockers*. Clin Pharmacokinet, 2008. 47(5): p. 323–332.

122. Nakakariya, M., Shimada, T., Irokawa, M., Koibuchi, H., Iwanaga, T., Yabuuchi, H., Maeda, T. and Tamai, I., *Predominant contribution of rat organic anion transporting polypeptide-2 (Oatp2) to hepatic uptake of beta-lactam antibiotics*. Pharm Res, 2008. 25(3): p. 578–585.

123. Kitamura, S., Maeda, K., Wang, Y. and Sugiyama, Y., *Involvement of multiple transporters in the hepatobiliary transport of rosuvastatin*. Drug Metab Dispos, 2008. 36(10): p. 2014–2023.

124. Yamashiro, W., Maeda, K., Hirouchi, M., Adachi, Y., Hu, Z. and Sugiyama, Y., *Involvement of transporters in the hepatic uptake and biliary excretion of valsartan, a selective antagonist of the angiotensin II AT1-receptor, in humans*. Drug Metab Dispos, 2006. 34(7): p. 1247–1254.

125. Ye, J., Liu, Q., Wang, C., Meng, Q., Peng, J., Sun, H., Kaku, T. and Liu, K., *Inhibitory effect of JBP485 on renal excretion of acyclovir by the inhibition of OAT1 and OAT3*. Eur J Pharm Sci, 2012. 47(2): p. 341–346.

126. Tahara, H., Kusuhara, H., Endou, H., Koepsell, H., Imaoka, T., Fuse, E. and Sugiyama, Y., *A species difference in the transport activities of H2 receptor antagonists by rat and human renal organic anion and cation transporters*. J Pharmacol Exp Ther, 2005. 315(1): p. 337–345.

127. Muller, F., Konig, J., Hoier, E., Mandery, K. and Fromm, M.F., *Role of organic cation transporter OCT2 and multidrug and toxin extrusion proteins MATE1 and MATE2-K for transport and drug interactions of the antiviral lamivudine*. Biochem Pharmacol, 2013. 86(6): p. 808–815.

128. Choi, C.I., Bae, J.W., Keum, S.K., Lee, Y.J., Lee, H.I., Jang, C.G. and Lee, S.Y., *Effects of OCT2 c.602C > T genetic variant on the pharmacokinetics of lamivudine*. Xenobiotica, 2013. 43(7): p. 636–640.

129. Bourdet, D.L., Pritchard, J.B. and Thakker, D.R., *Differential substrate and inhibitory activities of ranitidine and famotidine toward human organic cation transporter 1 (hOCT1; SLC22A1), hOCT2 (SLC22A2), and hOCT3 (SLC22A3)*. J Pharmacol Exp Ther, 2005. 315(3): p. 1288–1297.

130. Chu, X.Y., et al., *Transport of the dipeptidyl peptidase-4 inhibitor sitagliptin by human organic anion transporter 3, organic anion transporting polypeptide 4C1, and multidrug resistance P-glycoprotein*. J Pharmacol Exp Ther, 2007. 321(2): p. 673–683.

131. Lewis, D.F., *Modelling human cytochromes P450 for evaluating drug metabolism: an update*. Drug Metabol Drug Interact, 2000. 16(4): p. 307–324.

132. Varma, M.V., Bi, Y.A., Kimoto, E. and Lin, J., *Quantitative prediction of transporter- and enzyme-mediated clinical drug–drug interactions of organic anion-transporting polypeptide 1B1 substrates using a mechanistic net-effect model*. J Pharmacol Exp Ther, 2014. 351(1): p. 214–223.

133. Li, R., Barton, H.A., Yates, P.D., Ghosh, A., Wolford, A.C., Riccardi, K.A. and Maurer, T.S., *A "middle-out" approach to human pharmacokinetic predictions for OATP substrates using physiologically-based pharmacokinetic modeling*. J Pharmacokinet Pharmacodyn, 2014. 41(3): p. 197–209.

134. Jones, H.M., et al., *Mechanistic pharmacokinetic modeling for the prediction of transporter-mediated disposition in humans from sandwich culture human hepatocyte data*. Drug Metab Dispos, 2012. 40(5): p. 1007–1017.

135. Poirier, A., Cascais, A.-C., Funk, C. and Lavé, T., *Prediction of pharmacokinetic profile of valsartan in human based on* in vitro *uptake transport data*. J Pharmacokinet Pharmacodyn, 2009. 36(6): p. 585–611.

136. Soars, M.G., Grime, K., Sproston, J.L., Webborn, P.J. and Riley, R.J., *Use of hepatocytes to assess the contribution of hepatic uptake to clearance in vivo*. Drug Metab Dispos, 2007. 35(6): p. 859–865.

137. Ménochet, K., Kenworthy, K.E., Houston, J.B. and Galetin, A., *Use of mechanistic modeling to assess interindividual variability and interspecies differences in active uptake in human and rat hepatocytes*. Drug Metab Dispos, 2012. 40(9): p. 1744–1756.

138. Swift, B., Pfeifer, N.D. and Brouwer, K.L., *Sandwich-cultured hepatocytes: an* in vitro *model to evaluate hepatobiliary transporter-based drug interactions and hepatotoxicity*. Drug Metab Rev, 2010. 42(3): p. 446–471.

139. Watanabe, T., Kusuhara, H., Maeda, K., Kanamaru, H., Saito, Y., Hu, Z. and Sugiyama, Y., *Investigation of the rate-determining process in the hepatic elimination of HMG-CoA reductase inhibitors in rats and humans*. Drug Metab Dispos, 2010. 38(2): p. 215–222.

140. Greenblatt, D.J., Peters, D.E., Oleson, L.E., Harmatz, J.S., MacNab, M.W., Berkowitz, N., Zinny, M.A. and Court, M.H., *Inhibition of oral midazolam clearance by boosting doses of ritonavir, and by 4,4-dimethyl-benziso-(2H)-selenazine (ALT-2074), an experimental catalytic mimic of glutathione oxidase*. Br J Clin Pharmacol, 2009. 68(6): p. 920–927.

141. Tomita, Y., Maeda, K. and Sugiyama, Y., *Ethnic variability in the plasma exposures of OATP1B1 substrates such as HMG-CoA reductase inhibitors: a kinetic consideration of its mechanism*. Clin Pharmacol Ther, 2013. 94(1): p. 37–51.

142. Maeda, K. and Sugiyama, Y., *Transporter biology in drug approval: regulatory aspects*. Mol Aspects Med, 2013. 34(2–3): p. 711–718.

143. Feng, B., Varma, M.V., Costales, C., Zhang, H. and Tremaine, L., In vitro *and* in vivo *approaches to characterize transporter-mediated disposition in drug discovery*. Expert Opin Drug Discov, 2014. 9(8): p. 873–890.

144. Paine, S.W., Ménochet, K., Denton, R., McGinnity, D.F. and Riley, R.J., *Prediction of human renal clearance from preclinical species for a diverse set of drugs that exhibit both active secretion and net reabsorption*. Drug Metab Dispos, 2011. 39(6): p. 1008–1013.

9

IN VITRO AND *IN SITU* APPROACHES TO MEASURE INTESTINAL PERMEABILITY AND EFFLUX TRANSPORTERS

9.1 INTRODUCTION

In drug discovery, understanding the absorption of the drug molecule is a prerequisite for successful drug development. Scientists aim at identifying a new molecular entity (NME) that dissolves and demonstrates optimum solubility at various physiological pHs that range from 1.2 to 7.4. These molecules are also chemically stable under these conditions, and once in solution, the compound is absorbed and crosses the apical membrane of intestinal enterocytes. Absorption is determined by the interplay of all these factors (i.e., solubility, dissolution, and permeability) and is mediated by passive and/or active transport [1–5]. In this chapter, we assume that the NME is not solubility/dissolution limited.

In vitro permeability is measured as a flux or the rate at which a molecule passes through a membrane chosen to mimic the apical membrane of the gastrointestinal tract (GIT) barrier. The physicochemical properties that influence the compound permeability are discussed elsewhere (Chapter 4) and are not reviewed herein. The intestinal permeability measuring unit is centimeters per second. Interestingly, the actual values are on the order of micrometers per second typically expressed as $0.01–10 \times 10^{-6}$ cm/s. The root numbers can be <0.01 or as high as several hundred (10^{-4} cm/s).

Drug molecules may cross the intestinal wall via passive or active transport. Passive diffusion is the most common route of absorption and occurs via two pathways: the paracellular pathway where the drug diffuses through the tight junctions between enterocytes or a transcellular pathway, which requires drug diffusion across the lipophilic cell membrane of intestinal enterocytes [5]. Furthermore, as discussed in Chapter 6, efflux transporters such as P-glycoprotein (P-gp), MRP2, and BCRP have a significant impact on the absorption of drug molecules, in particular those with low permeability. As a result, understanding the role of

Oral Bioavailability Assessment: Basics and Strategies for Drug Discovery and Development, First Edition.
Ayman F. El-Kattan.

these transporters and characterizing their impact on drug molecule absorption and disposition represented by the determination of the drug K_m and V_{max} is a key for successful drug development [3, 6–9].

Artificial membranes and few *in silico* permeability models are based exclusively on passive diffusion, and several have been shown to correlate with fraction absorbed. In contrast, cell-based membrane permeability models (such as Caco-2 or MDCK, discussed subsequently) have active or carrier-mediated components that operate in combination with passive diffusion. Intestinal influx transporters may improve overall transcellular permeability across intestinal enterocytes. However, efflux transporters such as P-gp, MRP2, or BCRP are effective in pumping drug molecules back into the intestinal lumen, which would reduce overall intestinal permeability and lower NME absorption and oral bioavailability.

In the following sections, known *in vitro* and *in situ* tools for the determination of drug permeability and the impact of efflux transporters on drug absorption and disposition are discussed with emphasis on system design, strength and shortcomings, metrics, and calculation methodologies.

9.2 PARALLEL ARTIFICIAL MEMBRANE PERMEABILITY ASSAY

In 1998, the parallel artificial membrane permeability assay (PAMPA) was first introduced as a novel approach for the prediction of *in vivo* absorption of NMEs [10]. This tool determines the intrinsic intestinal permeability (P_{eff}) of an NME across an artificial membrane that mimics the composition of the intestinal biological membrane. Specifically, PAMPA is used to model passive transport of NMEs [10, 11]. The method is widely used in the drug industry. It provides a simple, cost-effective, and high-throughput alternative to *in vivo* oral bioavailability studies [12, 13].

9.2.1 System Design

PAMPA features the use of a 96-well plate filled with a donor solution covered with another 96-well plate that contains a receiver solution. The two chambers are divided by a porous hydrophobic filter that is embedded in a lipid-infused artificial membrane arranged in a form of monolayer (Figure 9.1). Phospholipids are the main components of PAMPA monolayer. The phospholipids are composed of polar phosphate group attached to two lipophilic fatty acyl side chains. Examples of phospholipids commonly used in PAMPA are phosphotidylcholine, phosphotidylserine, phosphotidylethanolamine, phosphatidylinositol, and cholesterol [14]. The phospholipids formulate a biologically relevant monolayer. Both phosphotidylcholine and phosphotidylethanolamine are the main phospholipids used as artificial membrane components of PAMPA since they are neutral. Unlike other charged phospholipids, they tend

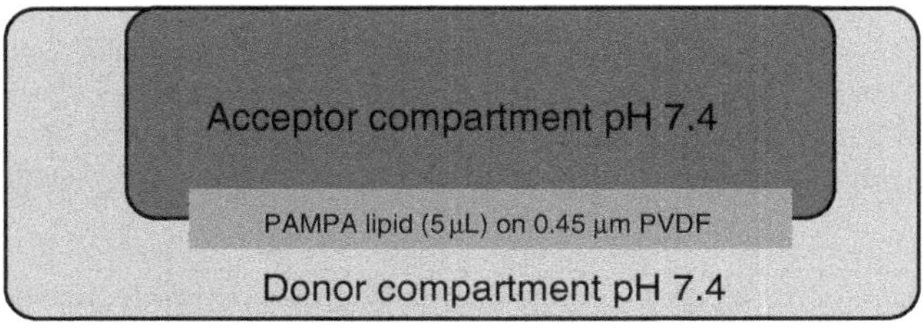

FIGURE 9.1 Schematic diagram of a PAMPA donor/acceptor compartments assembly.

to have less retention, which would increase confidence in the measured permeability and its relevance [12].

The solution of the donor and receptor chambers are typically made from buffered saline solution at physiologically relevant pHs (e.g., pH = 6.5 or 7.4). It is well established in the literature that solution pH has a significant impact on the permeability of charged molecules [5]. For example, Kerns et al. reported a better permeability for acidic molecules when measured at low pH, which is not unexpected considering their lower extent of ionization at low pH [15]. Indeed, the drug transport across PAMPA membrane is affected by drug solubility, pH gradient, phospholipid content, surfactant composition, and secondary chemical stabilizers such as bile salts [10, 14, 16]. It is also influenced by factors such as hydrogen bonding, molecular weight, lipophilicity, and rotational bonds [14, 15]. A PAMPA monolayer membrane is typically composed of 4% (w/v) dioleyoylphosphatidylcholine (DOPC, charge neutral) and 2% (w/v) stearic acid (negative charge) in dodecane. The monolayer supported on a polyvinylidenefluoride (PVDF) filter. Relative to other systems that contain 10% or more lecithin content, the DOPC monolayer has relatively low membrane retention. Furthermore, there are many commonalities between the DOPC/stearic acid and intestinal cell membranes. Overall, this makes DOPC/stearic acid combination the formula of choice for use in PAMPA.

9.2.2 PAMPA Strengths and Shortcomings

Compared to other *in vitro* permeability assays, PAMPA offers a simple, cost-effective, and high-throughput alternative. It can be easily automated with the aid of a liquid handling system. PAMPA is effective in rank ordering the permeability of NMEs within a chemical series. However, it is unable to rank permeability for compounds from different chemical series. PAMPA produces reproducible results in particular when 2% DOPC monolayer is used as reported by Avdeef et al. and Troutman [11, 14].

There are a few disadvantages with PAMPA that may limit its utility. For example, it does not predict the role of active transport mediated by intestinal efflux or influx transporters [10, 13, 15, 17, 18]. PAMPA has been reported to produce false negative results for compounds that are classified as well-absorbed molecules [18]. This technique is mainly effective in assessing the risk of poor permeability where it can be successfully used to rank order molecules; however, it is not effective in quantifying permeability and use as a basis to predict oral absorption in humans. In addition, the difference in the lipid monolayer composition and preparation among different laboratories lead to a major interlab variability that would limit our ability to compare obtained results.

9.2.3 PAMPA Metrics

An effective permeability (P_{eff}) value that is higher than 10×10^{-6} cm/s indicates that oral absorption is not limited by poor intestinal permeability, whereas $P_{eff} < 2.5 \times 10^{-6}$ cm/s indicates that absorption is permeability limited. Retention factor (R_f) is an important parameter that should be considered when assessing the permeability of NMEs, where high R_f value ($R_f > 0.9$) indicates drug/membrane retention and/or high nonspecific drug binding to labware. It also reduces confidence in the obtained permeability measurement [13, 19–21].

9.3 Caco-2 AND MDCK IMPLEMENTATION IN DRUG DISCOVERY AND DEVELOPMENT

Caco-2 cells are derived from human epithelial colorectal adenocarcinoma cells and were first developed by Kettering Institute for Cancer Research (Figure 9.2) [22]. Drug metabolism and

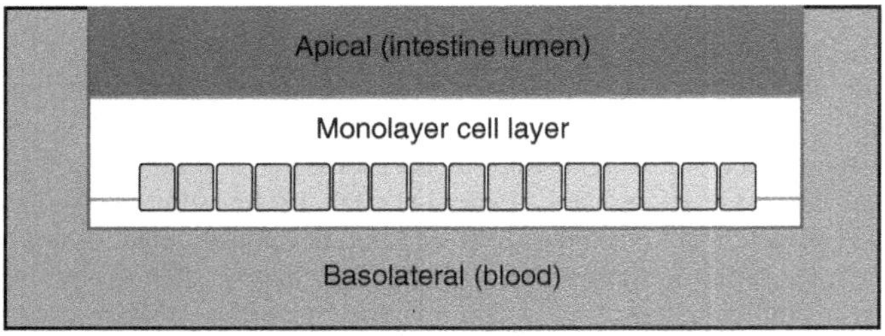

FIGURE 9.2 Systematic design of the *in vitro* cell culture monolayer experiment.

pharmacokinetics (DMPK) scientists have a major interest in the utility of Caco-2 as an effective tool to predict the extent of oral absorption in humans [22]. This interest is ascribed to the similarities between the Caco-2 monolayer and normal intestinal enterocytes [23]. For example, Caco-2 cell line has tight junctions and brush border membrane that is lined with high-density microvilli that is similar to the small intestine apical membrane. It also contains desmosomes, which are specialized structures of cell membrane and provide cell-to-cell adhesion. Furthermore, similar to the small intestine, known efflux (i.e., P-gp, BCRP, MRP2) and influx (i.e., Pept1, OATP2B1, MCT1) transporters are expressed at its apical membranes. Collectively, these commonalities enabled DMPK and formulation scientists to use Caco-2 as an effective *in vitro* tool to assess intestinal permeability, the mechanism of drug absorption, test various formulations, and investigate potential drug–drug interactions [22, 24–27]. Indeed, Caco-2 is the gold standard that is approved by the FDA for the assessment of permeability of NMEs in support of biopharmaceutics classification system (BCS) biowaivers [22, 24–27].

The enterocytes that line the villi of the intestinal tract pass through two transitional steps, which are proliferation and differentiation [23]. The intestinal enterocytes originate as stem cells in the villi crypts. They differentiate into mature functional absorptive enterocytes that later translocate to the tip of the villi. Over time, the functional enterocytes eventually shed into the intestinal lumen and are replaced by new stem cells. This proliferation cycle from the crypts to the villi tips is associated with constant renewal of newly differentiated intestinal epithelial cells and maintain the integrity of the intestinal tract. Differentiation happens when the stem cells translocate away from the crypts toward the villi tip and become fully functional absorptive enterocytes. It should be emphasized that the development and differentiation of the Caco-2 cell line mimics that reported with intestinal enterocyte. Caco-2 cells were the only human colon carcinoma cell lines to spontaneously differentiate into fully functional intestinal enterocytes [23]. This feature continues to be a major driver for the interest of DMPK and formulation scientists in the utility of this cell line in determining intestinal permeability.

9.3.1 System Design

Caco-2 cells and other cell lines such as MDCK (Madin–Darby Canine Kidney) cell lines are cultured in flasks, and they grow to form a monolayer in a process that is referred to as confluency [23]. The confluent cells are then separated and seeded on a porous permeable filter. The process is termed passage when this is repeated [23]. The cells sitting on the filter and arranged in a monolayer separates the apical chamber (lumen side also known as donor well) from the basolateral chamber (blood side also known as acceptor well). The lumen side vehicle pH is typically set at a pH of 7.4 or 6.5. The same applies to the basolateral chamber. Apparent permeability (P_{app}) is calculated using the following equation and by measuring the NME concentration in the receiving compartment (basolateral chamber) at multiple time points:

$$P_{\text{app}} = \frac{V_{\text{acc}}}{A_t} \cdot \frac{C_{\text{acc}}}{C_{\text{don}}} \tag{9.1}$$

where V_{acc} is the volume of the acceptor well, A the surface area of the monolayer, t time, C_{acc} the concentration in the acceptor well at time t, and C_{don} the NME concentration in the donor well (often assumed consistent with time).

9.3.2 Caco-2 Strengths and Shortcomings

A drug may cross the intestinal enterocytes by the following mechanisms: carrier-mediated transport, transcytosis, and passive transcellular and paracellular routes. Caco-2 has been reported to be effective in determining the predominant mechanism of absorption [28]. If the NME is actively transported, then it is possible to determine the transporter involved in the absorption of the drug molecule. Several research reports indicated that Caco-2 cell line is effective in assessing the permeability of compounds absorbed by passive transcellular pathway. For example, Lennernaes et al. reported that Caco-2 model yields a close approximation of an NME's permeability if the transport system is via the passive transcellular pathway [29]. Interestingly, Caco-2 is not a viable model for assessing compounds that are hydrophilic with small molecular weight ($M_{\text{wt}} < 350$ Da) and absorbed through the paracellular pathway such as pregabalin and gabapentin [30]. Unlike lipophilic molecules that are predominantly absorbed via the transcellular pathway, paracellularly absorbed molecules have a longer intestinal retention time before being absorbed [31]. It is generally accepted that hydrophilic molecules diffuse down the length of the villi. This process should result in an increase of overall area available for absorption and enable diffusion through the paracellular pathway located in the crypt region [31]. Caco-2 is not a good model for assessing the intestinal permeability and absorption of paracellular absorbed drugs. For example, pregabalin and gabapentin are both absorbed via paracellular pathway. They have poor permeability in Caco-2 cell lines. However, in the clinic, these molecules have almost complete oral absorption [30]. It should be noted that paracellular pathway can be identified using Caco-2 by assessing the impact of presence and absence of calcium on the permeability of tested articles. Calcium is needed to ensure the integrity of the tight junction. In its absence, the integrity of the tight junction is compromised, and hence, it opens up leading to a significant increase in the permeability of compounds absorbed by paracellular pathway. However, the absence of calcium is not associated with an increase in the permeability for compounds absorbed via transcellular pathway [32]. Indeed, opening the tight junction by eliminating calcium from the surrounding media was used by researchers to improve the overall absorption of paracellular absorbed molecules using chelating agents such as EDTA or by the use of chitosan [32]. Similar to other cell lines, there is an inherent significant interlaboratory variability in the data reported with Caco-2, which renders comparison of data obtained from different laboratories challenging.

9.3.3 MDCK Strengths and Shortcomings

The MDCK cell line is originally derived from a normal female cocker spaniel kidney by S. Madin and N. Darby [33]. These are epithelial cells that are not comparable to human intestinal enterocytes or Caco-2 cells. However, they are not as expensive, over five times faster to grow, and less borne to contamination [34, 35]. Both Caco-2 and MDCK cell lines are effective in assessing the permeability of compounds that are absorbed via passive transcellular pathway. However, MDCK falls short in characterizing the absorption that is influenced by intestinal transporter. This result is mainly attributed to the inherent low expression of known efflux and influx transporter. To overcome this limitation, scientists tend to transfect this cell line with a transporter of choice such as P-gp or BCRP. This approach would allow a better understanding of the potential impact of these transporters on the absorption of NMEs [36, 37].

In 2011, Pfizer scientists developed new MDCK cell lines that have low expression of efflux transporters (MDCK-LE) such as P-gp, MRP2, and BCRP [1]. The goal is to have a cell line that only assesses the inherent passive permeability of the molecule without the impact of efflux transporters. Similar to MDCK, MDCK-LE can grow at a fast rate and less porne to contamination. MDCK-LE is also effectively used to predict the oral absorption of NMEs and was recently used as a basis for identifying clearance rate-determining step [38]. This prediction was established based on the observation that compounds that have high extent of metabolism tend to have high oral absorption (high f_a) provided that solubility/dissolution is a not limiting factor. As a result, Varma et al. used receiver operating characteristic technique to identify an MDCK-LE permeability value that can differentiate between high and low absorbed drug molecules. Compounds with a $P_{app} > 5 \times 10^{-6}$ cm/s tend to have high oral absorption ($f_a \geq 85\%$) as well as an extent of metabolism that is >70%, whereas compounds with low permeability ($P_{app} < 5 \times 10^{-6}$ cm/s) typically have low absorption ($f_a < 85\%$) and extent of metabolism (<70%) [39].

9.3.4 Metrics

As mentioned earlier, both Caco-2 and MDCK correlate well to each other, only when the NME is diffusing via passive transcellular pathway. When considering carrier-mediated processes, MDCK/MDR1 should be used for comparison with *in vivo* values. In general, apparent permeability values exceeding 10×10^{-6} cm/s indicate that permeability will not be the limiting step to oral absorption. Studies have indicated that the fraction absorbed nears 100% in this regime [40]. Values below 1×10^{-6} cm/s usually suggest that permeability may be an issue; the oral absorption of the compounds that reside in this permeability category may fall between 0% and 20% [40].

For MDCK-LE, Varma et al. established a relationship between apparent permeability measured in MDCK-LE and oral absorption for a set of compounds. The group reported that a cutoff value of $\geq 5 \times 10^{-6}$ cm/s is associated with fraction absorbed ≥85%, indicating complete oral absorption. It should be emphasized that this is valid provided that the absorption is not dissolution/solubility limited [39].

9.3.5 Calculation Methodologies to Assess $CL_{int,P\text{-}glycoprotein}$

In 2003, Sugiyama et al. discussed approaches that can be used to measure the intrinsic transport clearance mediated by P-gp-mediated efflux activity ($CL_{int,P\text{-}gp}$) using Caco-2. It is derived using the following equation [41]:

$$PS_{app,A\text{-}to\text{-}B} = PS_{app,A\text{-}to\text{-}B} \times \frac{PS_{basolateral,passive}}{PS_{apical,passive} + PS_{basolateral,passive} + CL_{int,P\text{-}gp}} \quad (9.2)$$

where $PS_{app,A\text{-}to\text{-}B}$ is the apparent apical-to-basolateral permeability–surface area product (PS). $PS_{apical,passive}$ and $PS_{basolateral,passive}$ represent PS of apical and basolateral membrane passive transport, respectively. If we assume that $PS_{apical,passive}$ and $PS_{basolateral,passive}$ are equal and are represented as $PS_{passive}$, then Equation 9.2 can be rearranged as Equation 9.3:

$$PS_{app,A\text{-}to\text{-}B} = \frac{(PS_{passive})^2}{(2 \cdot PS_{passive} + CL_{int,P\text{-}gp})}$$

$$PS_{passive} = 2 \cdot PS_{app,A\text{-}to\text{-}B,inh} \tag{9.3}$$

where $PS_{app,A\text{-}to\text{-}B,inh}$ represents PS measured in the presence of P-gp inhibitors. $PS_{passive}$ represents passive PS across each membrane, assuming influx and efflux passive permeability are similar across both apical and basolateral membrane and no other active transport is involved. P-gp-mediated flux is calculated using Equation 9.4:

$$\text{P-gp mediated flux} = CL_{int,P\text{-}gp} \cdot C_{donor} \tag{9.4}$$

Compound P-gp affinity (K_m) and maximum flux rate (V_{max}) can be calculated using GraphPad Prism software and by plotting a relationship between compound P-gp flux calculated using Equation 9.4 and donor concentration.

9.3.6 Caco-2 Use in Assessing the Impact of P-gp and Other Efflux Transporters on Intestinal Absorption

To assess if the compound is a P-gp, Caco-2 and MDCK–MDR1 transfected cell lines are used by assessing the efflux ratio of the molecule. This ratio is calculated using Equation 9.5:

$$\text{Efflux ratio} = \frac{P_{app\ B\rightarrow A}}{P_{app\ B\rightarrow A}} \tag{9.5}$$

where $P_{app,B\rightarrow A}$ is the apparent permeability measured from basal to apical and $P_{app,B\rightarrow A}$ the apparent permeability measured from apical to basal. Typically, compounds with an efflux ratio >2 are considered substrate for efflux transporter. In addition, K_m and V_{max} values can be calculated using statistical software such as GraphPad Prism by setting a relationship between efflux ratio and tested article concentration. K_m is the drug affinity and V_{max} is the maximum efflux ratio achieved.

Furthermore, the potential impact of efflux transporter inhibitors on the cellular uptake (intracellular accumulation) of NME can be examined by assessing the impact of known efflux transporter inhibitors on the compounds profile. For example, impact of efflux transporters on cellular uptake of tested articles can be measured by assessing the impact of P-gp inhibitors such as verapamil (100 μM), nifedipine (100 μM), and quinidine (50 mM), MRP1/2 inhibitors, for example, probenecid (200 μM), MK-571 (100 μM), MRP3 (etoposide (100 μM)), MRP4 (celecoxib (100 μM)), and MRP5 (sulfinpyrazone (100 μM)). To assess impact on cellular uptake, the following procedure is used:

1. All inhibitors are freshly prepared by dissolving in DMSO and diluted by phosphate buffer. The final concentration of DMSO in the incubations is 0.2% (v/v). The inhibitors listed above at the concentration recommended were reported to show little cytotoxicity (≈10%) to the cells tested when incubated for 4 h [42].
2. All inhibitors are preincubated with cells for 2 h and coincubated for a further 30 min in the presence of NME at 0.1 or 1.0 μM.

3. To halt any transporter-mediated uptake, the cells are then washed with cold PBS buffer five times.
4. Thereafter, the cells are harvested, lysed by sonication, and extracted using ice-cold acetonitrile:methanol mixture (1:1, v/v, with 0.01 N HCl).
5. The supernatant is dried using a SpeedVac and the residue reconstituted with the mobile phase and 10–20 μL and injected into LC–MS for the determination of NME concentrations.
6. All uptake assays in the absence and presence of various inhibitors are conducted in at least three independent experiments on different days.

9.4 SINGLE-PASS INTESTINAL PERFUSION

Single-pass intestinal perfusion (SPIP) is an *in situ* technique that is used in the evaluation of intestinal permeability of NMEs across the rat intestinal membrane. It is also effectively used to assess the role of efflux and influx transporters and their impact on oral absorption of NMEs.

9.4.1 System Design

Rats are anesthetized using a 0.5 mL mixture of ketamine (75 mg/kg) and xylazine (5 mg/kg) administered by intraperitoneal injection [42]. During surgery and throughout the experiment, the anesthetized rats' temperature is maintained by placing them on a 37 °C heating pad and under a heating lamp [43]. Then, three surgical procedures are conducted: First, the right jugular vein is cannulated with a heparinized (100 IU/mL) polyethylene cannula with a 0.8 mm o.d. Second, a semicircular incision is made and an 8- to 12-cm-long segment of either ileum or jejunum is isolated and cut at both sides of the intestinal segment. The length of the isolated segment is then recorded. The intestinal content is then gently flushed 3–5 times from the isolated segment using 15 mL of blank perfusion buffer maintained at 37 °C. Two glass cannulas (4.0 mm o.d.) are inserted into both ends of the intestinal segment. A polyethylene tube (5.5 mm o.d.) is connected between the proximal end of the intestinal segment glass cannula and external syringe pump. The perfusion buffer that has the tested article is perfused at a constant flow rate of 0.2 mL/min. The distal end is ligated and kept *in vivo*. Around 10 min prior to mesenteric artery cannulation, 1 mL of heparin saline solution (100 IU/mL) is injected in the jugular vein. The mesenteric vein draining the selected intestinal segment is cannulated using a heparinized (100 IU/mL) 22G polyethylene catheter, which is secured using a drop of cyanoacrylate adhesive. Based on the blood flow from the cannulated mesenteric artery, the infusion rate using the peristaltic pump connected with the jugular vein is adjusted; the rate is typically 0.25 mL/min.

The intestinal perfusion buffer is composed of 138 mM NaCl, 2.7 mM KCl, 8.1 mM Na_2HPO_4, and 1.5 mM KH_2PO_4. Phenol red (50 μg/mL) is added to the solution as a nonabsorbable marker. Antipyrine (1.05 mM) is also included in each perfusion as a reference probe for membrane integrity and passive permeability (see below). The pH is adjusted to the targeted value of 6.5 or 7.4. Tested article is perfused with phenol red and antipyrine [43].

The perfusion is initiated by infusing the tested article solution from the syringe pump at a rate of 1 mL/min for 5 min. This procedure is then followed throughout the experiment that lasts for 90 min with an infusion rate of 0.2 mL/min. It should be emphasized that the blood from the mesenteric artery is collected in EDTA lined 1.5 mL tubes at 5 min intervals for 90 min. Similarly, the intestinal perfusate samples are collected from the distal segment glass cannula every 5 min for 60 min. The blood samples are centrifuged at 6000*g* for 10 min and the obtained plasma are transferred to tubes and stored at −20 °C. The perfusate and plasma samples are analyzed using LC/MS. In addition to the tested article, antipyrine is typically used as a marker for the integrity of the experiment, and as a reference standard for the permeability class boundary, and phenol red is used as a nonabsorbable marker.

The intestinal permeability can be calculated based on compound disappearance from the lumen (P_{lumen}) and the appearance of drug in the blood using the following equations (P_{blood}) [44]:

$$P_{\text{lumen}} = -\frac{Q}{2\pi rl} \ln \frac{C_{\text{lumen}}}{C_{\text{o}}} \tag{9.6}$$

$$P_{\text{blood}} = \frac{dX/dt}{A \times C_{\text{o}}} \tag{9.7}$$

where C_{lumen} and C_{o} correspond to the inlet and outlet compound concentrations, respectively, r the intestinal radius (cm), and l the length (cm) of the isolated intestinal segment. Q is the flow rate (0.2 mL/min), determined by the external syringe pump, and A the surface area of the intestinal segment ($=2\pi rl$, cm^2).

Net water flux (NWF) values were estimated based on inlet (C_{in}) and outlet (C_{out}) concentrations of phenol red using Equation 9.8:

$$\text{NWF} = \frac{\left(1 - \left[\frac{C_{\text{outphenolred}}}{C_{\text{inphenolred}}}\right]\right)}{L} \times Q_{\text{in}} \tag{9.8}$$

A negative NWF indicates loss of fluid from the mucosal side (lumen) to the serosal side (blood). A positive NWF indicates secretion of fluid into the segment.

9.4.2 SPIP Use in Assessing the Impact of P-gp and Other Efflux Transporters on Intestinal Absorption

To test the potential impact of efflux transporters on the intestinal permeability of NMEs, scientists usually consider using chemical inhibitors of efflux transporters to assess the impact of efflux transporters on intestinal permeability using rat intestinal perfusion methodology, or to determine intestinal permeability using efflux transporter knock out animals and compare to that measured in wild type animals using intestinal perfusion methodology.

9.4.2.1 SPIP Use in Assessing the Impact of P-gp and Other Efflux Transporters on Intestinal Absorption To assess impact of efflux transporters on the intestinal transporter of NMEs, known chemical inhibitors are used. Yu et al. used known efflux transporters to characterize their contribution to the intestinal permeability of tanshinone IIB (TIB), which is the primary active ingredient from the root of *Salvia miltiorrhiza* [42]. The group used inhibitors such as verapamil (P-gp inhibitor infused using a concentration of 100 μM), quinidine at 50 μM (a P-gp inhibitor), probenecid at 200 μM (an MRP1/2 inhibitor), MK-571 at 100 μM (an MRP1/2 inhibitor), etoposide (20 μM, an MRP3 inhibitor), celecoxib at 100 μM (an MRP4 inhibitor), sulfinpyrazone (100 μM, an MRP5 inhibitor), and indomethacin (100 μM, an MRP6 and MRP1/2 inhibitor) to assess efflux transporters impact on the intestinal transport of TIB [45–47]. The permeability of TIB based on the P_{lumen} was concentration dependent and significantly increased with increasing concentration from 0.1, 0.5, and 2.0 μM ($p < 0.05$). This low-to-moderate permeability suggests the potential involvement of saturable efflux mechanism in the intestinal permeability of TIB. There was a significant ($p < 0.01$ or 0.05) increase (about 2- to 3.5-fold) in the P_{blood} measured in the presence of verapamil (100 μM), nifedipine (100 μM), or quinidine (50 μM) relative to TIB given alone. However, this trend was not observed in P_{lumen} of TIB measured in the presence of these P-gp inhibitors. The group attributed these findings of small impact of P-gp efflux on the disappearance of

TIB in the lumen to potential significant tissue uptake and binding, which may reduce the efflux inhibition activities of these inhibitors. In addition, for these agents to work, the tested article should either have low solubility/dissolution or given at a low luminal concentration that is lower than its K_m. The low efflux inhibition is most likely attributed to one or more of these factors not being satisfied. It also suggests that assessing the potential role of P-gp to the intestinal transport of tested article may be compromised if the assessment is only limited for measurement of P_{lumen} and P_{blood} is not determined.

TIP:

When considering SPIP as a technique to assess potential contribution of efflux transporter to the intestinal permeability of NME, it is pivotal to measure both P_{lumen} and P_{blood} since P-gp inhibitors are not always effective in inhibiting the P-gp-mediated efflux due to the above-mentioned reasons. While it is expensive and resource intensive, obtaining both values is key for putting P-gp impact in perspective.

9.4.3 SPIP Strengths and Shortcomings

SPIP is considered as one of the key tools available for accurate prediction of NME intestinal permeability that is at the discretion of DMPK and formulation scientists. Interestingly, while accurate, human intestinal perfusion technique is too expensive to use. Salphati et al. reported a good correlation between permeability measured using rat SPIP versus that reported in humans ($R^2 = 0.88$). In this report, the correlation coefficient, also referred to as Spearman rank (indicated by r_s), was used to compare the two sets of acquired data [43]. Similar to R^2, when r_s is close to one, a strong correlation exists between the two observations. P_{lumen} values for 14 compounds were compared with the corresponding literature human f_a values. A strong significant correlation was reported with an r_s of 0.92. The relationship was not as significant between the same literature human f_a values and the apparent permeability (P_{app}) measured in cell-based assays, where r_s was only 0.61 and 0.59 for Caco-2 and MDCK cell lines, respectively [43]. This further substantiates the superiority of SPIP as a technique to measure effective intestinal permeability of NMEs relative to *in vitro* cell lines. This result is not unexpected when we consider that SPIP provides an environment that more closely resembles a normal physiological state (i.e., intact blood supply, functional intestinal barrier, presence of drug metabolizing enzymes, and numerous transporters) [43]. Table 9.1 contains a list of the experimental rat P_{eff} obtained from the SPIP experiments in comparison to the literature-based human values. It should be emphasized that the rank order of the compound effective permeability using SPIP was comparable to that obtained from human intestinal perfusion studies. Furthermore, rat P_{eff} (P_{lumen}) is commonly used in predicting human oral absorption as depicted in Figure 9.3 [48, 49].

The main limitation of the SPIP method is that it is resource intensive and time-consuming. It is also not a high-throughput screen and expensive relative to studies conducted using *in vitro* cell lines such as MDCK/MDR1 or Caco-2. Furthermore, the compound tested should be in solution to accurately measure the permeability and minimize the potential impact of poor solubility on permeability measurement, which may not be feasible for compounds that are BDDCS Class 2 or 4. As indicated earlier, P_{blood} should be calculated along with P_{lumen} to enable appropriate assessment of the potential contribution of P-gp to the intestinal transport of NMEs.

TABLE 9.1 Similar Correlation Between P_{eff} Values in Literature-Based Human and Rat Perfusion Values

Compound	P_{eff} (10^{-4} cm/s) Rat, Exp	Man, Literature	Rank Order (Rat/Man)
Antipyrine	0.73 ± 0.02	5.6 ± 1.6	7/7
Atenolol	0.18 ± 0.09	0.15 ± 0.2	3/2
Furosamide	0.19 ± 0.03	0.3 ± 0.3	4/3
Hydrochlorothiazide	0.07 ± 0.03	0.04 ± 0.05	1/1
Ketoprofen	1.09 ± 0.71	8.5 ± 3.9	8/9
Metoprolol	0.59 ± 0.13	1.5 ± 0.9	5/5
Naproxen	1.67 ± 0.82	8.0 ± 4.2	9/8
Propranolol	0.66 ± 0.29	2.8 ± 1.3	6/9
Terbutaline	0.12 ± 0.09	0.3 ± 0.3	2/3

Source: Salphati et al. 2001 [43]. Reproduced with permission of John Wiley & Sons.

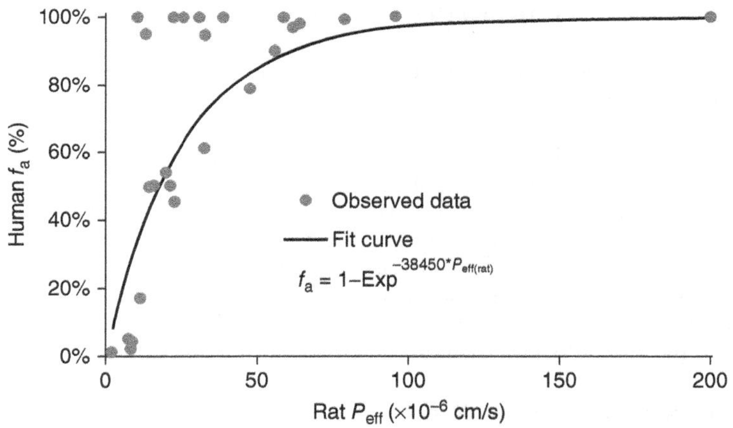

FIGURE 9.3 The relationship between rat P_{eff} versus human f_a [48, 49].

9.5 MISCELLANEOUS PERMEABILITY MODELS AND TECHNOLOGIES REVIEW

There is a range of *in vitro* tools outside the typical methodologies (PAMPA, Caco-2, and MDCK) that are used for permeability determination. The following section briefly discusses the more commonly found cell and noncell lines that are frequently considered.

9.5.1 Living Cell Line Permeability Models

9.5.1.1 Lewis Lung Carcinoma Porcine Kidney Cells (LLC-PK1) LLC-PK1 is a cell line that stands for Lewis Lung Carcinoma Porcine Kidney Cells (LLC-PK1). This is a cell line that originated from the distal tubules of porcine kidney cells. It is used in drug industry to assess compounds permeability and commonly transfected with known transporters (e.g., P-gp,

BCRP, MRP2) to assess their impact on drug permeability and gain a better understanding of their role in drug absorption and disposition [50–53].

In 2002, Li et al. published a report on the investigation of the intestinal absorption and renal function of pyrimidine nucleobase uracil using Caco-2 and LLC-PK1 cell lines [54]. The group measured the apical to basolateral (A–B) transport of radiolabeled uracil using the two cell lines. Liquid scintillation counting (LSC) was used as an analytical tool to measure uracil levels at the basolateral side of the monolayer. The group also calculated the apparent linear initial rate (pmol/cm^2/min), the maximum uptake rate, the Michaelis–Menten constant (K_m), and the linear clearance of radiolabeled uracil. The authors concluded that the transport mechanism of uracil significantly differed based on the cell line investigated [54]. In part, this conclusion is attributed to differences in transporters' expression between the two cell lines.

Adachi et al. investigated the extent to which intestinal absorption of known P-gp substrate drugs is affected by interindividual differences in the expression level of P-gp, and simultaneous administration of P-gp substrates or inhibitors. They used both rat SPIP and human P-gp transfected LLC-PK1 cell lines. The group reported a significant correlation in the product ratio of the apparent membrane permeability coefficient and the surface area when the *in situ* and *in vitro* experiments were compared [55]. Based on their findings, they recommended the use of the *in vitro* transcellular transport across LLC-PK1 monolayers to assess and predict the impact of P-gp on the intestinal absorption of drug molecules [55].

9.5.1.2 Rat Fetal Intestinal Epithelial Cells (2/4/A1) In search of an *in vitro* cell line that mimics the permeability across human jejunum of compounds absorbed via passive pathway, Tavelin et al. developed Rat Fetal Intestinal Epithelial Cells (2/4/A1) [56, 57]. This is a conditionally immortalized cell line that is a temperature-sensitive mutant of the growth-supporting oncogene Simian virus. In 2/4/A1 monolayers, the transport of low permeability drugs was comparable to human jejunum and up to 300 times faster than that in Caco-2 monolayers. Furthermore, a better relationship was reported between 2/4/A1 cell line permeability and human fraction absorbed for 13 assessed radiolabeled drugs in comparison with Caco-2 cell line. It is interesting to note that 2/4/A1 required only 6–8 days to achieve confluence compared with 21–35 days cultivated at 37 °C needed with Caco-2 cell line [58]. This is the key advantage for 2/4/A1 rendering it less resource intensive and would lead to a faster turnaround time.

9.5.1.3 Rat Small Intestine Cells (IEC-18) Ma et al. introduced the utility of microporous filter-grown IEC-18 epithelial cells, derived from rat ileal crypts as an *in vitro* tool to assess intestinal permeability [59]. IEC-18 cells become confluent when grown on hydrated collagen-coated Millicell-CM permeable inserts. The group reported the presence of tight junctions, desmosomes, and the development of microvilli at the apical surface using both scanning and transmission electron microscopy. It is interesting to note that the net electrical resistance of IEC-18 epithelium to that of small intestinal tissue, with or without muscularis or serosal layers was comparable. The group concluded that the IEC-18 cell line is a good model for assessing the intestinal permeability for compounds absorbed via transcellular and paracellular pathways.

9.5.2 Noncell Line Permeability Models

9.5.2.1 Chromatographic Methods In 2002, Meng et al. assessed the membrane permeability of drug molecules by modeling the relative diffusion rate of a drug within a membrane using a chromatographic approach [60]. The group calculated the permeability of drugs into the stationary phase by providing membrane partition coefficient for a series of analogs. They attributed the correlation between the diffusion coefficients and hydrophobicity ($\log P_{oct}$) of drug molecules tested to a demonstration that lipophilic molecules penetrate deeper into the stationary phase. It should be noted that the group used C_{18} derivatized HPLC column where the slopes of the theoretical plate height versus the flow rate were found to increase quantitatively. This increase indicates that the mass transfer resistance slows down the penetration of molecules and was used to rank order drug molecules tested.

9.5.2.2 Blood–Brain Barrier Model While permeability assessment across the blood–brain barrier (BBB) is outside the scope of intestinal permeability, many of the techniques used in its modeling are shared with that used with intestinal permeability. Deli et al. reviewed many of the *in vitro* BBB models, which include bovine, human, porcine, and rodent examples [61]. In a separate study, Gulyaeva et al. compared octanol-buffer partitioning and high-performance liquid chromatography as noncell permeability models to predict BBB distribution [62].

9.6 CONCLUSION

Determination of intestinal permeability to predict human oral absorption of NME is a cornerstone of developing viable orally available clinical candidates. Various *in vitro* and *in situ* approaches are available for DMPK and formulation scientists to consider. The selection of any of these tools should be based on prior understanding of the compound physiochemical properties and the pros and cons of each tool. In addition, determining the potential contribution of efflux transporter to the intestinal transport of NMEs is pivotal in particular if the compound has low intestinal permeability and/or will be coadministered with compounds that are P-gp substrate. For example, talinolol is a known P-gp substrate that saturates P-gp transporter at therapeutically relevant doses. Therefore, coadministration with digoxin, a known P-gp substrate with narrow therapeutic index, leads to an increase in the overall plasma exposure of digoxin and result in potential side effect. Therefore, understanding this phenomenon is an important aspect of compound clinical development and will be part of the drug label [63].

REFERENCES

1. Di, L., et al., *Development of a new permeability assay using low-efflux MDCKII cells*. J Pharm Sci, 2011. 100(11): p. 4974-4985.
2. Varma, M.V., Obach, R.S., Rotter, C., Miller, H.R., Chang, G., Steyn, S.J., El-Kattan, A. and Troutman, M.D., *Physicochemical space for optimum oral bioavailability: contribution of human intestinal absorption and first-pass elimination*. J Med Chem, 2010. 53(3): p. 1098–1108.
3. Varma, M.V., Ambler, C.M., Ullah, M., Rotter, C.J., Sun, H., Litchfield, J., Fenner, K.S. and El-Kattan, A.F., *Targeting intestinal transporters for optimizing oral drug absorption*. Curr Drug Metab, 2010. 11(9): p. 730–742.
4. Hurst, S., Loi, C.M., Brodfuehrer, J. and El-Kattan, A., *Impact of physiological, physicochemical and biopharmaceutical factors in absorption and metabolism mechanisms on the drug oral bioavailability of rats and humans*. Expert Opin Drug Metab Toxicol, 2007. 3(4): p. 469–489.
5. Thomas, V.H., Bhattachar, S., Hitchingham, L., Zocharski, P., Naath, M., Surendran, N., Stoner, C.L. and El-Kattan, A., *The road map to oral bioavailability: an industrial perspective*. Expert Opin Drug Metab Toxicol, 2006. 2(4): p. 591–608.
6. Darwich, A.S., Neuhoff, S., Jamei, M. and Rostami-Hodjegan, A., *Interplay of metabolism and transport in determining oral drug absorption and gut wall metabolism: a simulation assessment using the "Advanced Dissolution, Absorption, Metabolism (ADAM)" model*. Curr Drug Metab, 2010. 11(9): p. 716–729.

7. Zakeri-Milani, P. and Valizadeh, H., *Intestinal transporters: enhanced absorption through P-glycoprotein-related drug interactions*. Expert Opin Drug Metab Toxicol, 2014. 10(6): p. 859–871.
8. Shi, J.G., Zhang, Y. and Yeleswaram, S., *The relevance of assessment of intestinal P-gp inhibition using digoxin as an* in vivo *probe substrate*. Nat Rev Drug Discov, 2011. 10(1): p. 75; author reply 75.
9. Tachibana, T., Kato, M., Takano, J. and Sugiyama, Y., *Predicting drug–drug interactions involving the inhibition of intestinal CYP3A4 and P-glycoprotein*. Curr Drug Metab, 2010. 11(9): p. 762–777.
10. Kansy, M., Senner, F. and Gubernator, K., *Physicochemical high throughput screening: parallel artificial membrane permeation assay in the description of passive absorption processes*. J Med Chem, 1998. 41(7): p. 1007–1010.
11. Troutman, M., PAMPA – The Frontline Screen for Intrinsic (Membrane) Permeability. 2006, N/A: Groton, CT. p. PowerPoint presentation.
12. Seo, P.R., Teksin, Z.S., Kao, J.P.Y. and Polli, J.E., *Lipid composition effect on permeability across PAMPA*. Eur J Pharm Sci, 2006. 29(3–4): p. 259–268.
13. Avdeef, A., *The rise of PAMPA*. Expert Opin Drug Metab Toxicol, 2005. 1(2): p. 325–342.
14. Avdeef, A., *Absorption and Drug Development*. 1st ed. 2003, John Wiley & Sons, Inc.: Hoboken, NJ. p. 287.
15. Kerns, E.H., Di, L., Petusky, S., Farris, M., Ley, R. and Jupp, P., *Combined application of parallel artificial membrane permeability assay and Caco-2 permeability assays in drug discovery*. J Pharm Sci, 2004. 93(6): p. 1440–1453.
16. Willmann, S., Schmitt, W., Keldenich, J., Lippert, J. and Dressman, J.B., *A physiological model for the estimation of the fraction dose absorbed in humans*. J Med Chem, 2004. 47(16): p. 4022–4031.
17. Liu, H., Sabus, C., Carter, G.T., Du, C., Avdeef, A. and Tischler, M., In vitro *permeability of poorly aqueous soluble compounds using different solubilizers in the PAMPA assay with liquid chromatography/mass spectrometry detection*. Pharm Res, 2003. 20(11): p. 1820–1826.
18. Avdeef, A. and Tsinman, O., *PAMPA – A drug absorption* in vitro *model. 13. Chemical selectivity due to membrane hydrogen bonding: in combo comparisons of HDM-, DOPC-, and DS-PAMPA models*. Eur J Pharm Sci, 2006. 28(1–2): p. 43–50.
19. Rositano, F. and Ferraro, D.O., *Ecosystem services provided by agroecosystems: a qualitative and quantitative assessment of this relationship in the Pampa region, Argentina*. Environ Manage, 2014. 53(3): p. 606–619.
20. Reis, J.M., Sinko, B. and Serra, C.H., *Parallel artificial membrane permeability assay (PAMPA) – Is it better than Caco-2 for human passive permeability prediction?* Mini Rev Med Chem, 2010. 10(11): p. 1071–1076.
21. Kansy, M., Avdeef, A. and Fischer, H., *Advances in screening for membrane permeability: high-resolution PAMPA for medicinal chemists*. Drug Discov Today Technol, 2004. 1(4): p. 349–355.
22. Hidalgo, I.J., *Assessing the absorption of new pharmaceuticals*. Curr Top Med Chem, 2001. 1(5): p. 385–401.
23. Delie, F. and Rubas, W., *A human colonic cell line sharing similarities with enterocytes as a model to examine oral absorption: advantages and limitations of the Caco-2 model*. Crit Rev Ther Drug Carrier Syst, 1997. 14(3): p. 221–286.
24. Nkansah, P., et al., *Development and evaluation of novel solid nanodispersion system for oral delivery of poorly water-soluble drugs*. J Control Release, 2013. 169(1–2): p. 150–161.
25. Cook, J.A., et al., *Refining the* in vitro *and* in vivo *critical parameters for P-glycoprotein, [I]/IC50 and [I2]/IC50, that allow for the exclusion of drug candidates from clinical digoxin interaction studies*. Mol Pharm, 2010. 7(2): p. 398–411.
26. Fenner, K.S., Troutman, M.D., Kempshall, S., Cook, J.A., Ware, J.A., Smith, D.A. and Lee, C.A., *Drug–drug interactions mediated through P-glycoprotein: clinical relevance and* in vitro-in vivo *correlation using digoxin as a probe drug*. Clin Pharmacol Ther, 2009. 85(2): p. 173–181.
27. Wils, P., Phung-Ba, V., Warnery, A., Lechardeur, D., Raeissi, S., Hidalgo, I.J. and Scherman, D., *Polarized transport of docetaxel and vinblastine mediated by P-glycoprotein in human intestinal epithelial cell monolayers*. Biochem Pharmacol, 1994. 48(7): p. 1528–1530.

28. Artursson, P., Palm, K. and Luthman, K., *Caco-2 monolayers in experimental and theoretical predictions of drug transport*. Adv Drug Delivery Rev, 2001. 46(1–3): p. 27–43.

29. Lennernaes, H., Palm, K., Fagerholm, U. and Artursson, P., *Comparison between active and passive drug transport in human intestinal epithelial (Caco-2) cells* in vitro *and human jejunum* in vivo. Int J Pharm, 1996. 127(1): p. 103–107.

30. Jezyk, N., Li, C., Stewart, B.H., Wu, X., Bockbrader, H.N. and Fleisher, D., *Transport of pregabalin in rat intestine and Caco-2 monolayers*. Pharm Res, 1999. 16(4): p. 519–526.

31. Shah, P., Jogani, V., Bagchi, T. and Misra, A., *Role of Caco-2 cell monolayers in prediction of intestinal drug absorption*. Biotechnol Prog, 2006. 22(1): p. 186–198.

32. Merzlikine, A., Rotter, C., Rago, B., Poe, J., Christoffersen, C., Thomas, V.H., Troutman, M. and El-Kattan, A., *Effect of chitosan glutamate, carbomer 974P, and EDTA on the* in vitro *Caco-2 permeability and oral pharmacokinetic profile of acyclovir in rats*. Drug Dev Ind Pharm, 2009. 35(9): p. 1082–1091.

33. Bohets, H., Annaert, P., Mannens, G., Van Beijsterveldt, L., Anciaux, K., Verboven, P., Meuldermans, W. and Lavrijsen, K., *Strategies for absorption screening in drug discovery and development*. Curr Top Med Chem, 2001. 1(5): p. 367–383.

34. Irvine, J.D., Takahashi, L., Lockhart, K., Cheong, J., Tolan, J.W., Selick, H.E. and Grove, J.R., *MDCK (Madin–Darby canine kidney) cells: a tool for membrane permeability screening*. J Pharm Sci, 1999. 88(1): p. 28–33.

35. Ungell, A.-L.B., *Caco-2 replace or refine?* Drug Discov Today Technol, 2004. 1(4): p. 423–430.

36. Soldner, A., Benet, L.Z., Mutschler, E. and Christians, U., *Active transport of the angiotensin-II antagonist losartan and its main metabolite EXP 3174 across MDCK–MDR1 and caco-2 cell monolayers*. Br J Pharmacol, 2000. 129(6): p. 1235–1243.

37. Soldner, A., Christians, U., Susanto, M., Wacher, V.J., Silverman, J.A. and Benet, L.Z., *Grapefruit juice activates P-glycoprotein-mediated drug transport*. Pharm Res, 1999. 16(4): p. 478–485.

38. Varma, M.V., Steyn, S.J., Allerton, C. and El-Kattan, A.F., *Predicting clearance mechanism in drug discovery: extended clearance classification system (ECCS)*. Pharm Res, 2015. 32(12): p. 3785–3802.

39. Varma, M.V., et al., *pH-Dependent solubility and permeability criteria for provisional biopharmaceutics classification (BCS and BDDCS) in early drug discovery*. Mol Pharm, 2012. 9(5): p. 1199–1212.

40. Fiese, E.F., *General pharmaceutics – the new physical pharmacy*. J Pharm Sci, 2003. 92(7): p. 1331–1342.

41. Adachi, Y., Suzuki, H. and Sugiyama, Y., *Quantitative evaluation of the function of small intestinal P-glycoprotein: comparative studies between* in situ *and* in vitro. Pharm Res, 2003. 20(8): p. 1163–1169.

42. Yu, X.Y., et al., *Role of ATP-binding cassette drug transporters in the intestinal absorption of tanshinone IIB, one of the major active diterpenoids from the root of* Salvia miltiorrhiza. Xenobiotica, 2007. 37(4): p. 375–415.

43. Salphati, L., Childers, K., Pan, L., Tsutsui, K. and Takahashi, L., *Evaluation of a single-pass intestinal-perfusion method in rat for the prediction of absorption in man*. J Pharm Pharmacol, 2001. 53(7): p. 1007–1013.

44. Zhang, J., et al., *A mechanistic study of the intestinal absorption of cryptotanshinone, the major active constituent of* Salvia miltiorrhiza. J Pharmacol Exp Ther, 2006. 317(3): p. 1285–1294.

45. Tian, Q., et al., *Human multidrug resistance associated protein 4 confers resistance to camptothecins*. Pharm Res, 2005. 22(11): p. 1837–1853.

46. Tian, R., Koyabu, N., Morimoto, S., Shoyama, Y., Ohtani, H. and Sawada, Y., *Functional induction and de-induction of P-glycoprotein by St. John's wort and its ingredients in a human colon adenocarcinoma cell line*. Drug Metab Dispos, 2005. 33(4): p. 547–554.

47. Borst, P. and Elferink, R.O., *Mammalian ABC transporters in health and disease*. Annu Rev Biochem, 2002. 71: p. 537–592.

48. Zakeri-Milani, P., Valizadeh, H., Tajerzadeh, H., Azarmi, Y., Islambolchilar, Z., Barzegar, S. and Barzegar-Jalali, M., *Predicting human intestinal permeability using single-pass intestinal perfusion in rat*. J Pharm Pharm Sci, 2007. 10(3): p. 368–379.

49. Levet-Trafit, B., Gruyer, M.S., Marjanovic, M. and Chou, R.C., *Estimation of oral drug absorption in man based on intestine permeability in rats*. Life Sci, 1996. 58(24): p. PL359–PL363.
50. Takada, T., Suzuki, H. and Sugiyama, Y., *Characterization of polarized expression of point- or deletion-mutated human BCRP/ABCG2 in LLC-PK1 cells*. Pharm Res, 2005. 22(3): p. 458–464.
51. Mita, S., Suzuki, H., Akita, H., Hayashi, H., Onuki, R., Hofmann, A.F. and Sugiyama, Y., *Vectorial transport of unconjugated and conjugated bile salts by monolayers of LLC-PK1 cells doubly transfected with human NTCP and BSEP or with rat Ntcp and Bsep*. Am J Physiol Gastrointest Liver Physiol, 2006. 290(3): p. G550–G556.
52. Salama, N.N., Yang, Z., Bui, T. and Ho, R.J., *MDR1 haplotypes significantly minimize intracellular uptake and transcellular P-gp substrate transport in recombinant LLC-PK1 cells*. J Pharm Sci, 2006. 95(10): p. 2293–2308.
53. Takeuchi, T., Yoshitomi, S., Higuchi, T., Ikemoto, K., Niwa, S., Ebihara, T., Katoh, M., Yokoi, T. and Asahi, S., *Establishment and characterization of the transformants stably-expressing MDR1 derived from various animal species in LLC-PK1*. Pharm Res, 2006. 23(7): p. 1460–1472.
54. Li, H., Chung, S.-J. and Shim, C.-K., *Characterization of the transport of uracil across Caco-2 and LLC-PK1 cell monolayers*. Pharm Res, 2002. 19(10): p. 1495–1501.
55. Adachi, Y., Suzuki, H. and Sugiyama, Y., *Quantitative evaluation of the function of small intestinal P-glycoprotein: comparative studies between* in situ *and* in vitro. Pharm Res, 2003. 20(8): p. 1163–1169.
56. Tavelin, S., Milovic, V., Ocklind, G., Olsson, S. and Artursson, P., *A conditionally immortalized epithelial cell line for studies of intestinal drug transport*. J Pharmacol Exp Ther, 1999. 290(3): p. 1212–1221.
57. Tavelin, S., Milovic, V., Ocklind, G., Olsson, S. and Artursson, P., *A conditionally immortalized epithelial cell line for studies of intestinal drug transport*. J Pharmacol Exp Ther, 1999. 290(3): p. 1212–1221.
58. Tavelin, S., Taipalensuu, J., Soederberg, L., Morrison, R., Chong, S. and Artursson, P., *Prediction of the oral absorption of low-permeability drugs using small intestine-like 2/4/A1 cell monolayers*. Pharm Res, 2003. 20(3): p. 397–405.
59. Ma, T.Y., Hollander, D., Bhalla, D., Nguyen, H. and Krugliak, P., *IEC-18, a nontransformed small intestinal cell line for studying epithelial permeability*. J Lab Clin Med, 1992. 120(2): p. 329–341.
60. Meng, Q.C., Johansson, J.S. and Eckenhoff, R.G., *Chromatographic approach for determining the relative membrane permeability of drugs*. J Chromatogr, B: Anal Technol Biomed Life Sci, 2002. 774(1): p. 89–95.
61. Deli, M.A., Abraham, C.S., Kataoka, Y. and Niwa, M., *Permeability studies on* in vitro *blood–brain barrier models: physiology, pathology, and pharmacology*. Cell Mol Neurobiol, 2005. 25(1): p. 59–127.
62. Gulyaeva, N., Zaslavsky, A., Lechner, P., Chlenov, M., McConnell, O., Chait, A., Kipnis, V. and Zaslavsky, B., *Relative hydrophobicity and lipophilicity of drugs measured by aqueous two-phase partitioning, octanol-buffer partitioning and HPLC. A simple model for predicting blood–brain distribution*. Eur J Med Chem, 2003. 38(4): p. 391–396.
63. Westphal, K., et al., *Oral bioavailability of digoxin is enhanced by talinolol: evidence for involvement of intestinal P-glycoprotein*. Clin Pharmacol Ther, 2000. 68(1): p. 6–12.

10

IN SILICO APPROACHES TO PREDICT INTESTINAL PERMEABILITY

10.1 INTRODUCTION

The call to use *in silico* models to accurately predict drug absorption and disposition following dosing in humans will continue to be a major driver for scientists in drug metabolism and pharmaceutical science departments. These prediction models enable efficient use of resources and proactive identification of liabilities ahead of new molecular entities (NMEs) synthesis. Absorption is an outcome that is dependent on the solubility and permeability of drug molecules [1, 2]. *In silico* approaches to accurately predict these key variables is pivotal to identify NMEs that have good absorption profiles in humans or to proactively identify liabilities. The vision is to screen large libraries of compounds *in silico* first to predict NMEs permeability and absorption profiles. This information can then guide and direct subsequent synthesis efforts performed by medicinal chemists. However, there are limitations and caveats to current simple and complex permeability prediction methods that should be considered [3]. A general overview of how most permeability prediction models are derived as well as some literature models and commercial models available are discussed herein.

10.2 PREDICTION OF HUMAN INTESTINAL PERMEABILITY AND ABSORPTION BASED ON In vitro PERMEABILITY INPUTS

Current permeability prediction models are successfully used to measure transcellular passive intestinal human permeability. This model is accomplished by the measurement of permeability using validated *in vitro* cellular monolayer models such as Caco-2, MDCK/MDR1, and MDCK low efflux cell lines permeability, for which there is a large amount of experimental

Oral Bioavailability Assessment: Basics and Strategies for Drug Discovery and Development, First Edition. Ayman F. El-Kattan.

data published (Chapter 9) [4, 5]. Subsequent application of the standard known correlations of measured permeability in *in vitro* settings to human absorption (f_a) provides the basis for the human permeability extrapolations [4, 6–13].

In silico predictive permeability models are built based on experimental data obtained *in vitro* using Caco-2 and other relevant monolayer cellular models. Due to similarity in morphological structure and enzyme and transporter expression to human intestine, Caco-2 provides by far the most comprehensive data set [14, 15]. It should be emphasized that this data is rather biased toward compounds with high permeability that is commonly available. Statistical models are then used to fit the experimental permeability data to molecules' physicochemical properties also known as molecular descriptors. The value of these predictive models is driven by the quality of the experimental data used in the training. Typically, better prediction outcome is realized when the permeability range spans low, moderate, and high permeability compounds. Furthermore, the complexity of available models ranges from simple experimental partition models to more complex physiologically based models. While complex models have better accuracy, they are generally more resource intensive and require time to run and build. Therefore, it is pivotal for drug scientists to consider the questions to be answered to decide on the model of choice [16].

10.2.1 Statistical Models

Contemporary permeability prediction models are built by first identifying molecular descriptors of the training set. Then a statistical model is used to fit these descriptors to the experimental permeability data. Once the designated descriptors are demonstrated to fit the experimental data, the model is then used to predict the permeability of NMEs.

10.2.1.1 Molecular Descriptors Molecular descriptors have various categories. The simplest category includes physicochemical properties such as molecular size, atom counts, and molecular weight [17]. These basic descriptors are often termed 1D descriptors. Another category termed 2D descriptors, which define complex aspects of drug molecules and include drug flexibility, lipophilicity/hydrophilicity, electronic charge, counts of hydrogen bonding donor (HBD), and hydrogen bonding acceptor (HBA) atoms. These are the attributes associated with the rule of five [18]. More complex descriptors (3D descriptors) include polarizability, a variety of surface-area properties, side chain folding interactions, solvation properties, and electrotopologic physicochemical properties, which consider compound local topology, electronegativity, and neighbor perturbation of individual atoms [19–21].

Advanced models usually involve conformational analysis of the drug molecule to determine its lowest energy configuration. Appropriate molecular descriptors associated with those minimized conformations are then derived. These models are more accurate and superior to simple models [19]. Various statistical programs are used to generate molecular descriptors. Table 10.1 outlines the main programs available to run these calculations and include but not limited to Molconn-Z 3.50, HYBOT, MolSurf (Qemist AB), VolSurf, Cerius^2, GRID, and the Chemistry Development Kit (CDK).

As outlined earlier, once the set of molecular descriptors are defined and generated, the data can be correlated to the existing experimental permeability data. For example, molecular descriptors such as polar surface area and log P were correlated with *in vitro* permeability/*in vivo* permeability and usually classified permeability into low, moderate, and high permeability bins [20, 24–26]. The permeability categorization is determined by utilizing discriminant analysis-partial least squares (DA-PLS). The simple models are usually used as the first screens to assess permeability for large libraries [27]. However, definitive quantitative complex models use multivariant analysis (MVA) and either multiple linear regressions (MLR) or multiple

TABLE 10.1 Common Molecular Descriptor Programs and Their Relevant Information [22, 23]

Name	Associated Molecular Descriptors	Literature and Website References
Chemistry Development Kit (CDK)	Provides descriptors on constitutional, topological, geometric, electric, and hybrid	www.cdk.com
GRID	Used to study interactions of molecules (with water, enzymes, etc.), provide descriptors therefrom	www.moldiscovery.com
HYBOT	Hydrogen bond-related descriptors	www.timtec.net/software/hybot-plus.htm
VolSurf	Provides a variety of proprietary VolSurf descriptors	www.moldiscovery.com
MolSurf (Qemist AB)	Surface, log *P*, polarity, HBD, HBA	www.3.interscience.wiley.com/cgi-bin/abstract/85006687/abstract
Molconn-Z 3.50	Provides a variety of proprietary Molconn-Z descriptors	www.tripos.com
Cerius	Provides a wide range of molecular descriptors; predicts ADME properties from structure	www.accelrys.com

nonlinear regression (MNLR) to correlate combinations of several molecular descriptors to permeability [17, 27, 28]. It should be emphasized that MVA is recommended when the number of experimental observations (i.e., experimental permeability measurements) is 5–10 times more than the number of molecular descriptors being considered, whereas partial least squares (PLS) modeling, a more advanced approach, works well if there are fewer experimental observations (experimental permeability measurements) and many molecular descriptors are being fit [29, 30]. Modeling of experimental data and molecular descriptors by complex neural networks is also reported [31–35]. It is interesting to note that these programs model specific data sets well but can be easily "overparameterized" on particular data sets. As a result, the prediction of permeability for NMEs outside of the original data physicochemical space is generally not well predicted.

There are many research reports citing models that are developed by academic and industrial labs. However, the most useful models are the standardized, commercial permeability, and *in vivo* prediction models summarized in Table 10.2.

10.2.1.2 ADMET™ Predictor and GastroPlus® ADMET™ predictor is complex computer software that uses NMEs' structural features and calculated 2D and 3D descriptors to predict various inputs that can be incorporated into mechanistic PBPK models (e.g., GastroPlus®) to

TABLE 10.2 Commercial Permeability and *In Silico* Models and Their Relevant Information [36–40]

Model Name	Description	Comments
iDEA® structure-based models (SBM)	*Inputs:* SMILES[a] string *Output:* Caco-2 apparent permeability (P_{eff}), predicted human fraction of dose absorbed (f_a)	www.lionbioscience.com
ADMET™ predictor (formerly QMPRPlus™)	*Inputs:* 2D or 3D descriptors or SMILES string; rat *in situ* permeability[a] *Output:* MDCK effective permeability is predicted from structure alone; *in vitro* or rodent *in vivo* permeability inputs will give direct human P_{eff}	www.simulations-plus.com
GastroPlus™	*Inputs:* pK_a, P_{eff} from *in vitro* models, solubility, dose/formulation, and diffusivity *Output*: f_a, rate/extent of absorption, f_a versus time	www.simulations-plus.com
SimCYP®	*Inputs:* pK_a, P_{eff} from *in vitro* models, solubility, dose/formulation, and diffusivity *Output*: f_a, rate/extent of absorption, f_a versus time	http://www.certara.com/software/pbpk-modeling/simcyp-pbpk/

[a]SMILES: simple molecular input line entry system.

predict and simulate the absorption and disposition of NMEs. It is efficient since it is reported to process around 200,000 molecules/h on a personal computer. Examples of key 2D and 3D molecular descriptors that the program generates include but not limited to molecular weight, number of (total, aromatic, aliphatic) rings, number of specific functional groups, geometric descriptors (moments of inertia, radii of gyration, surface areas), atomic partial charges, number of heteroatoms, fraction of single (double, triple) bonds, and charge descriptors [41]. Various research groups demonstrated the utility of ADMET predictor in predicting drug molecules pK_a, human effective permeability in jejunum, MDCK apparent permeability, solubility in various matrices, log P and log D, blood–brain barrier permeability, human plasma protein binding, volume of distribution, blood to plasma ratio, likelihood for P-glycoprotein efflux, and many relevant inputs that are used in building reliable PBPK models using GastroPlus [42–46].

GastroPlus is physiologically based pharmacokinetics (PBPK) simulation commercial software [36, 37]. It has physiological gut models for various preclinical models such as mice, rats, minipigs, rabbits, cats, beagle dogs, cynomolgus monkeys, rhesus monkeys, and humans in both fasted and fed conditions. It simulates the plasma exposure following intravenous, oral, ocular, inhalation, dermal, or subcutaneous dosing of NMEs to preclinical species or humans [46–49]. GastroPlus implements the Advanced Compartmental Absorption and Transit (ACAT) model to predict oral drug absorption [36, 37]. ACAT was established based on the work published by Yu and Amidon [50]. This comprehensive model uses drug- and system-dependent parameters to accurately predict the absorption of NMEs following oral dosing in preclinical models and humans. For drug-dependent parameters, it includes six states of drug components (e.g., unreleased, undissolved, dissolved, degraded, metabolized, and absorbed) and three states of excreted drugs (e.g., unreleased, undissolved, and dissolved). The model also considers drug physicochemical properties such as pK_a, particle size/density, solubility at various intestinal pHs and states, dissolution rates, and intestinal permeability at various gastrointestinal tract regions. Furthermore, the model includes dosage forms and doses in relevant calculations and simulations. For system-dependent parameters, the gastrointestinal tract is divided into nine compartments (stomach, seven segments of small intestine and colon). Intestinal physiological conditions such as gastric emptying, intestinal and colon transit rates, drug transport, and diffusion across the lumen are also taken into account. ACAT assesses the impact of drug ionization on dissolution and absorption by using relevant pH-dependent solubility and $\log D$ models. In addition, it determines the potential impact of bile salts on the *in vivo* drug solubility and permeability and provides mechanistic models to investigate *in vivo* precipitation by considering nucleation kinetics of investigated drug molecule. The ACAT model is also effectively used to calculate paracellular permeability and pathways. ACAT is considered the tool of choice to predict the utility of new formulations in improving overall oral absorption of NMEs by using advanced mechanistic *in vitro–in vivo* correlation module.

ACAT was successful in integrating intestinal metabolizing enzymes and transporter expression and activity to model nonlinear kinetics and absorption of known substrates of enzymes and transporters, for example, digoxin and propranolol [51, 52]. This model is also effective in predicting food–drug interactions and drug–drug interactions.

10.2.1.3 SimCYP® and ADAM Model SimCYP® is another mechanistic PBPK simulation model that was originally established to predict CYP-mediated drug–drug interactions [53]. Over the years, SimCYP utility has been expanded to enable simulation of various aspects of drug absorption and disposition as well as prediction of impact of interindividual variability, enzymes, and transporters pharmacogenomics on drug pharmacokinetics and dynamics.

SimCYP uses advanced dissolution, absorption, and metabolism (ADAM) simulator, a mechanistic physiologically based model, to predict the extent and rate of absorption of drug molecules from physicochemical and *in vitro* data. The ADAM model incorporates relevant system-dependent parameters related to human gastrointestinal tract such as intestinal surface area available for absorption, intestinal region-specific gut wall permeability and luminal pH, enterocytic blood flow, enterohepatic recirculation, gastric emptying, and small intestine and colonic transit times. The food effect on the oral absorption of drug molecules is also considered.

In addition to drug physiochemical properties and *in vitro* inputs such as protein binding, blood to plasma ratio, metabolic stability assessed using microsomes, hepatocyte, or recombinant enzymes, SimCYP model takes into consideration the interaction of the drug with various influx and efflux hepatic, renal, and intestinal transporters. This approach would allow accurate prediction of the impact of these proteins on the absorption and disposition of the drug in humans. It also enables successful prediction of the variability of drug pharmacokinetics, drug–drug interactions, food–drug interactions and impact of transporters pharmacogenomics on drug absorption and disposition [54–59].

10.3 CONCLUSION

To date, most permeability models predict only passive transcellular processes. More recent SimCYP and GastroPlus models attempt to address intestinal efflux transporters and their impact on drug absorption [45, 47, 51, 58, 60–62]. In addition, models for specific transporters and intestinal metabolism prediction are being developed. Interestingly, some permeability models have been combined with activity models in an effort to guide researchers to the most active molecules with the best permeability properties [3, 63]. New models are also being developed to predict Caco-2 and human intestinal effective permeability from PAMPA and MDCK-low efflux models, which are often faster and more straightforward to operate in a high-throughput mode [64].

REFERENCES

1. Varma, M.V., Obach, R.S., Rotter, C., Miller, H.R., Chang, G., Steyn, S.J., El-Kattan, A. and Troutman, M.D., *Physicochemical space for optimum oral bioavailability: contribution of human intestinal absorption and first-pass elimination*. J Med Chem, 2010. 53(3): p. 1098–1108.
2. Hurst, S., Loi, C.M., Brodfuehrer, J. and El-Kattan, A., *Impact of physiological, physicochemical and biopharmaceutical factors in absorption and metabolism mechanisms on the drug oral bioavailability of rats and humans*. Expert Opin Drug Metab Toxicol, 2007. 3(4): p. 469–489.
3. Egan, W.J. and Lauri, G., *Prediction of intestinal permeability*. Adv Drug Delivery Rev, 2002. 54(3): p. 273–289.
4. Artursson, P., Palm, K. and Luthman, K., *Caco-2 monolayers in experimental and theoretical predictions of drug transport*. Adv Drug Delivery Rev, 1996. 22: p. 67–84.
5. Yamashita, F., Wanchana, S. and Hashida, M., *Quantitative structure/property relationship analysis of Caco-2 permeability using a genetic algorithm*. J Pharm Sci, 2002. 91: p. 2230–2239.
6. Artursson, P. and Karlsson, J., *Correlation between oral drug absorption in humans and apparent drug permeability coefficients in human intestinal epithelial (Caco-2) cells*. Biochem Biophys Res Commun, 1991. 175: p. 880–885.
7. Chiou, W.L. and Barve, A., *Linear correlation of the fraction of oral dose absorbed of 64 drugs between human and rats*. Pharm Res, 1998. 15: p. 1792–1795.
8. Cogburn, J.N., Donovan, M.G. and Schasteen, C.S., *A model of human small intestine absorptive cells. 1. Transport barrier*. Pharm Res, 1991. 8: p. 210–216.
9. Rubas, W., Jezyk, N. and Grass, G.M., *Comparison of the permeability characteristics of a human colonic epithelial (Caco-2) cell line to colon of rabbit, monkey and dog intestine and human drug absorption*. Pharm Res, 1993. 10(1): p. 113–118.
10. Stewart, B.H., Chan, O.H., Lu, R.H., Reyner, E.L., Schmid, H.L., Hamilton, H.W., Steinbaugh, B.A. and Taylor, M.D., *Comparison of intestinal permeabilities determined in multiple* in vitro *and* in situ *models: relationship to absorption in humans*. Pharm Res, 1995. 12: p. 693–699.
11. Yazdanian, M., Glynn, S.L., Wright, J.L. and Hawi, A., *Correlating partitioning and Caco-2 cell permeability of structurally diverse small molecular weight compounds*. Pharm Res, 1998. 15: p. 1490–1494.
12. Yee, S., In vitro *permeability across Caco-2 cells (colonic) can predict* in vivo *(small intestine) absorption in man-fact or myth*. Pharm Res, 1997. 14: p. 763–766.
13. Varma, M.V., et al., *pH-Dependent solubility and permeability criteria for provisional biopharmaceutics classification (BCS and BDDCS) in early drug discovery*. Mol Pharm, 2012. 9(5): p. 1199–1212.
14. Oswald, S., Groer, C., Drozdzik, M. and Siegmund, W., *Mass spectrometry-based targeted proteomics as a tool to elucidate the expression and function of intestinal drug transporters*. AAPS J, 2013. 15(4): p. 1128–1140.

15. Groer, C., Bruck, S., Lai, Y., Paulick, A., Busemann, A., Heidecke, C.D., Siegmund, W. and Oswald, S., *LC–MS/MS-based quantification of clinically relevant intestinal uptake and efflux transporter proteins*. J Pharm Biomed Anal, 2013. 85: p. 253–261.
16. Stenberg, P., Norinder, U., Luthman, K. and Artursson, P., *Experimental and computational screening models for the prediction of intestinal drug absorption*. J Med Chem, 2001. 44(12): p. 1927–1937.
17. Hou, T.J., Zhang, W., Xia, K., Qiao, X.B. and Xu, X.J., *ADME evaluation in drug discovery. 5. Correlation of Caco-2 permeation with simple molecular properties*. J Chem Inf Comput Sci, 2004. 44: p. 585–1600.
18. Lipinski, C.A., Lombardo, F., Dominy, B.W. and Feeney, P.J., *Experimental and computational approaches to estimate solubility and permeability in drug discovery and development settings*. Adv Drug Delivery Rev, 1997. 23: p. 3–25.
19. Stenberg, P., Bergstroem, C.A.S., Luthman, K. and Artursson, P., *Theoretical predictions of drug absorption in drug discovery and development*. Clin Pharmacokinet, 2002. 41(11): p. 877–899.
20. Stenberg, P., Luthman, K., Ellens, H., Lee, C.P., Smith, P.L., Lago, A., Elliott, J.D. and Artursson, P., *Prediction of the intestinal absorption of endothelin receptor antagonists using three theoretical methods of increasing complexity*. Pharm Res, 1999. 16(10): p. 1520–1526.
21. Ungell, A.-L.B., *Caco-2 replace or refine?* Drug Discov Today Technol, 2004. 1(4): p. 423–430.
22. Guangli, M. and Yiyu, C., *Predicting Caco-2 permeability using support vector machine and chemistry development kit*. J Pharm Pharm Sci, 2006. 9(2): p. 210–221.
23. Pastor, M., Cruciani, G. and Watson, K.A., *A strategy for the incorporation of water molecules present in a ligand binding site into a three-dimensional quantitative structure–activity relationship analysis*. J Med Chem, 1997. 40(25): p. 4089–4102.
24. Oprea, T. and Gottfries, J., *Toward minimalistic modeling of oral drug absorption*. J Mol Graph Model, 1999. 17: p. 261–274.
25. Osterberg, T. and Norinder, U., *Prediction of polar surface area and drug transport processes using simple parameters and PLS statistics*. J Chem Inf Comput Sci, 2000. 40: p. 1408–1411.
26. Osterberg, T. and Norinder, U., *Prediction of drug transport processes using simple parameters and PLS statistics – the use of ACD/logP and ACD/ChemSketch descriptors*. Eur J Pharm Sci, 2001. 12(3): p. 327–337.
27. Ponce, Y.M., Perez, M.A.C., Zaldivar, V.R. and Ofori, E., *Total and local quadratic indices of the "molecular pseudograph's atom adjacency matrix" application to prediction of Caco-2 permeability of drugs*. Int J Mol Sci, 2003. 4(8): p. 512–536.
28. Camenisch, G., Alsenz, J., van de Waterbeemd, H. and Folkers, G., *Estimation of permeability by passive diffusion through Caco-2 cell monolayers using the drugs' lipophilicity and molecular weight*. Eur J Pharm Sci, 1998. 6: p. 317–324.
29. Cruciani, G., Carosati, E., De Boeck, B., Ethirajulu, K., Mackie, C., Howe, T. and Vianello, R., *Metasite: understanding metabolism in human cytochromes from the perspective of the chemist*. J Med Chem, 2005. 48: p. 6970–6979.
30. Norinder, U., Osterberg, T. and Artursson, P., *Theoretical calculation and prediction of Caco-2 cell permeability using MolSurf parameterization and PLS statistics*. Pharm Res, 1996. 14: p. 1786–1791.
31. Bergstroem, C.A.S., *Computational models to predict aqueous drug solubility, permeability and intestinal absorption*. Expert Opin Drug Metab Toxicol, 2005. 1(4): p. 613–627.
32. Clark, D.E. and Grootenhuis, P.D.J., *Progress in computational methods for the prediction of ADMET properties*. Curr Opin Drug Discovery Dev, 2002. 5(3): p. 382–390.
33. Dorronsoro, I., Chana, A., Abasolo, I., Castro, A., Gil, C., Stud, M. and Martinez, A., *CODES/neural network model: a useful tool for* in silico *prediction of oral absorption and blood–brain barrier permeability of structurally diverse drugs*. QSAR Comb Sci, 2004. 23: p. 89–98.
34. Fujiwara, S., Yamashita, F. and Hashida, M., *Prediction of Caco-2 cell permeability using a combination of MO-calculation and neural network*. Int J Pharm, 2002. 237: p. 95–105.
35. Subramanian, G. and Kitchen, D.B., *Computational approaches for modeling human intestinal absorption and permeability*. J Mol Model, 2006. 12(5): p. 577–589.

36. Huang, W., Lee, S.L. and Yu, L.X., *Mechanistic approaches to predicting oral drug absorption*. AAPS J, 2009. 11(2): p. 217–224.

37. Agoram, B., Woltosz, W.S. and Bolger, M.B., *Predicting the impact of physiological and biochemical processes on oral drug bioavailability*. Adv Drug Deliv Rev, 2001. 50(Suppl 1): p. S41–S67.

38. Parrott, N. and Lave, T., *Prediction of intestinal absorption: comparative assessment of GASTRO-PLUS and IDEA*. Eur J Pharm Sci, 2002. 17(1–2): p. 51–61.

39. Geerts, T. and Vander Heyden, Y., In silico *predictions of ADME-Tox properties: drug absorption*. Comb Chem High Throughput Screen, 2011. 14(5): p. 339–361.

40. Darwich, A.S., Aslam, U., Ashcroft, D.M. and Rostami-Hodjegan, A., *Meta-analysis of the turnover of intestinal epithelia in preclinical animal species and humans*. Drug Metab Dispos, 2014. 42(12): p. 2016–2022.

41. Lennernas, H., *Human intestinal permeability*. J Pharm Sci, 1998. 87(4): p. 403–410.

42. Fraczkiewicz, R., Lobell, M., Goller, A.H., Krenz, U., Schoenneis, R., Clark, R.D. and Hillisch, A., *Best of both worlds: combining pharma data and state of the art modeling technology to improve* in Silico *pKa prediction*. J Chem Inf Model, 2015. 55(2): p. 389–397.

43. Hosey, C.M. and Benet, L.Z., *Predicting the extent of metabolism using* in vitro *permeability rate measurements and* in silico *permeability rate predictions*. Mol Pharm, 2015. 12(5): p. 1456–1466.

44. Vucicevic, J., Nikolic, K., Dobricic, V. and Agbaba, D., *Prediction of blood–brain barrier permeation of alpha-adrenergic and imidazoline receptor ligands using PAMPA technique and quantitative-structure permeability relationship analysis*. Eur J Pharm Sci, 2015. 68: p. 94–105.

45. Abuasal, B.S., Bolger, M.B., Walker, D.K. and Kaddoumi, A., In silico *modeling for the non-linear absorption kinetics of UK-343,664: a P-gp and CYP3A4 substrate*. Mol Pharm, 2012. 9(3): p. 492–504.

46. Hosea, N.A. and Jones, H.M., *Predicting pharmacokinetic profiles using* in silico *derived parameters*. Mol Pharm, 2013. 10(4): p. 1207–1215.

47. Watson, K.J., Davis, J. and Jones, H.M., *Application of physiologically based pharmacokinetic modeling to understanding the clinical pharmacokinetics of UK-369,003*. Drug Metab Dispos, 2011. 39(7): p. 1203–1213.

48. Jones, H.M., et al., *Simulation of human intravenous and oral pharmacokinetics of 21 diverse compounds using physiologically based pharmacokinetic modelling*. Clin Pharmacokinet, 2011. 50(5): p. 331–347.

49. Allan, G., Davis, J., Dickins, M., Gardner, I., Jenkins, T., Jones, H., Webster, R. and Westgate, H., *Pre-clinical pharmacokinetics of UK-453,061, a novel non-nucleoside reverse transcriptase inhibitor (NNRTI), and use of* in silico *physiologically based prediction tools to predict the oral pharmacokinetics of UK-453,061 in man*. Xenobiotica, 2008. 38(6): p. 620–640.

50. Yu, L.X. and Amidon, G.L., *A compartmental absorption and transit model for estimating oral drug absorption*. Int J Pharm, 1999. 186(2): p. 119–125.

51. Reyner, E.L., Sevidal, S., West, M.A., Clouser-Roche, A., Freiwald, S., Fenner, K., Ullah, M., Lee, C.A. and Smith, B.J., In vitro *characterization of axitinib interactions with human efflux and hepatic uptake transporters: implications for disposition and drug interactions*. Drug Metab Dispos, 2013. 41(8): p. 1575–1583.

52. Wang, Y., Wang, Z., Zuo, Z., Tomlinson, B., Lee, B.T., Bolger, M.B. and Chow, M.S., *Clinical pharmacokinetics of buffered propranolol sublingual tablet (Promptol)–application of a new "physiologically based" model to assess absorption and disposition*. AAPS J, 2013. 15(3): p. 787–796.

53. Rostami-Hodjegan, A. and Tucker, G., 'In silico' *simulations to assess the* 'in vivo' *consequences of* 'in vitro' *metabolic drug–drug interactions*. Drug Discov Today Technol, 2004. 1(4): p. 441–448.

54. Sjogren, E., et al., In vivo *methods for drug absorption – comparative physiologies, model selection, correlations with* in vitro *methods (IVIVC), and applications for formulation/API/excipient characterization including food effects*. Eur J Pharm Sci, 2014. 57: p. 99–151.

55. Kostewicz, E.S., et al., *PBPK models for the prediction of* in vivo *performance of oral dosage forms*. Eur J Pharm Sci, 2014. 57: p. 300–321.

56. Jamei, M., Bajot, F., Neuhoff, S., Barter, Z., Yang, J., Rostami-Hodjegan, A. and Rowland-Yeo, K., *A mechanistic framework for* in vitro–in vivo *extrapolation of liver membrane transporters: prediction of drug–drug interaction between rosuvastatin and cyclosporine*. Clin Pharmacokinet, 2014. 53(1): p. 73–87.

57. Idkaidek, N.M. and Najib, N., In silico *vs.* in vivo *human intestinal permeability*. Drug Res (Stuttg), 2014. 64(12): p. 693–694.

58. Neuhoff, S., Yeo, K.R., Barter, Z., Jamei, M., Turner, D.B. and Rostami-Hodjegan, A., *Application of permeability-limited physiologically-based pharmacokinetic models: part I-digoxin pharmacokinetics incorporating P-glycoprotein-mediated efflux*. J Pharm Sci, 2013. 102(9): p. 3145–3160.

59. Varma, M.V., Lai, Y., Feng, B., Litchfield, J., Goosen, T.C. and Bergman, A., *Physiologically based modeling of pravastatin transporter-mediated hepatobiliary disposition and drug–drug interactions*. Pharm Res, 2012. 29(10): p. 2860–2873.

60. Johnson, S.R. and Zheng, W., *Recent progress in the computational prediction of aqueous solubility and absorption*. AAPS J, 2006. 8(1): p. E27–E40.

61. Prasad, B., Evers, R., Gupta, A., Hop, C.E., Salphati, L., Shukla, S., Ambudkar, S.V. and Unadkat, J.D., *Interindividual variability in hepatic organic anion-transporting polypeptides and P-glycoprotein (ABCB1) protein expression: quantification by liquid chromatography tandem mass spectroscopy and influence of genotype, age, and sex*. Drug Metab Dispos, 2014. 42(1): p. 78–88.

62. Neuhoff, S., Yeo, K.R., Barter, Z., Jamei, M., Turner, D.B. and Rostami-Hodjegan, A., *Application of permeability-limited physiologically-based pharmacokinetic models: part II – prediction of P-glycoprotein mediated drug–drug interactions with digoxin*. J Pharm Sci, 2013. 102(9): p. 3161–3173.

63. Penzotti, J.E., Landrum, G.A. and Putta, S., *Building predictive ADMET models for early decisions in drug discovery*. Curr Opin Drug Discovery Dev, 2004. 7(1): p. 49–61.

64. Fujikawa, M., Ano, R., Nakao, K., Shimizu, R. and Akamatsu, M., *Relationships between structure and high-throughput screening permeability of diverse drugs with artificial membranes: application to prediction of Caco-2 cell permeability*. Bioorg Med Chem, 2005. 13(15): p. 4721–4732.

11

IN VIVO PRECLINICAL APPROACHES TO DECONVOLUTE THE CONTRIBUTION OF FIRST-PASS EFFECT FROM ORAL ABSORPTION

11.1 INTRODUCTION

Drug discovery teams have goals of identifying new molecular entities (NMEs) that are safe, orally bioavailable, and biologically active. They usually conduct *in vivo* studies using preclinical species and typically start with rodents such as mice and rats to investigate NMEs' safety, efficacy, and pharmacokinetics profile. Poor bioavailability is troublesome and may impede biological assessment if the route of administration is oral. Therefore, determining the cause of poor bioavailability is pivotal in deciding on the changes in the molecular structure of NMEs needed to overcome poor oral bioavailability and assess biological activity. It is well established in the literature that there are major species differences in the expression and activity of metabolizing enzymes and transporters (Chapter 2) that are usually associated with lower clearance and high bioavailability in upper species compared to rodents such as mice and rats. As a result, assessing biological activity using oral route is not always feasible in rodents.

This chapter has three objectives: First, to review *in vivo* preclinical approaches that are commonly used in drug discovery and development to deconvolute the contribution of first-pass effects from poor oral absorption of NMEs with poor oral bioavailability. Second, to provide specific examples on known metabolizing enzymes and efflux transporters inhibitors that can be used to identify the mechanism leading to NMEs poor oral bioavailability. Third, to discuss various routes of administration that can be considered if NMEs has poor oral bioavailability that impedes its biological assessment in rodents.

Oral Bioavailability Assessment: Basics and Strategies for Drug Discovery and Development, First Edition.
Ayman F. El-Kattan.

11.2 *IN VIVO* ESTIMATION OF ORAL BIOAVAILABILITY AND ITS COMPONENTS

Oral bioavailability (F) is a product of three parameters: f_a, which is the fraction of the drug that is absorbed from the intestinal lumen to the intestinal enterocytes, f_g, which is the fraction that escapes intestinal metabolism, and f_h, which is the fraction that escapes hepatic metabolism. F is calculated using this equation:

$$F = f_a \cdot f_g \cdot f_h \tag{11.1}$$

The methods for estimating f_a, f_g, and f_h are discussed as follows.

11.2.1 f_h Calculation Following Intravenous Dosing

f_h can be calculated following intravenous dosing in preclinical animal models and humans using this equation [1]:

$$f_h = 1 - E_h = 1 - \frac{CL_h}{Q} = 1 - \frac{CL_{total}}{Q} \tag{11.2}$$

where E_h is the hepatic extraction ratio, CL_h the total hepatic clearance, and Q the hepatic blood flow. In general, most of the drug molecules are predominantly eliminated in the liver [2]. As a result, it is generally accepted that CL_h is equal to CL_{total}. In drug discovery, CL_{total} obtained following intravenous (IV) dosing is initially used to calculate f_h and F. Hepatic extraction ratio is a key determinant of the drug f_h. Drug molecules with high hepatic extraction ($E_h \geq 0.7$) will have low bioavailability when given orally ($F \leq 0.3$). On the other hand, drug molecules with low hepatic extraction ($E_h \leq 0.3$) will have high oral bioavailability. It should be stressed that these assumptions are valid provided that the drug is completely absorbed, not significantly metabolized in the intestine, and clearance is not different following both intravenous and oral dosing.

$f_a{\cdot}f_g$ can be calculated following oral and IV dosing by

$$f_a \cdot f_g = \frac{F}{1 - \frac{CL_h}{Q}} \tag{11.3}$$

11.3 FRACTION OF THE DOSE ABSORBED INTO THE PORTAL BLOOD AFTER ORAL ADMINISTRATION ($F_A{\cdot}F_G$)

The main methods that are used to estimate $f_a{\cdot}f_g$ in preclinical species are as follows [3].

11.3.1 $f_a{\cdot}f_g$ Calculation Using Dose-Normalized AUC Ratio Between Oral and Intraportal Vein/Intraperitoneal Administration

Following intraportal and oral administration, $f_a{\cdot}f_g$ can be calculated using the following equation:

$$f_a \cdot f_g = \frac{AUC_{po,0-\infty} \cdot D_{port}}{AUC_{port,0-\infty} \cdot D_{po}} \tag{11.4}$$

where D_{port} and D_{po} are doses after intraportal vein administration and oral dosing, respectively. $AUC_{port,0\to\infty}$ and $AUC_{po,0\to\infty}$ are the systemic plasma AUC values following intraportal vein and oral administration, respectively. It should be emphasized that if obtaining portal vein cannulated animals is technically and/or economically not feasible, scientists may opt to use intraperitoneal administration instead and use Equation 11.4 to calculate $f_a{\cdot}f_g$. The main disadvantage of dosing the tested article as an injection bolus in either portal vein or peritoneal cavity is the higher likelihood of first hepatic metabolism saturation using these techniques. This is attributed to the rapid rate of absorption and the corresponding high drug concentrations achieved in the liver, which increases likelihood of hepatic first-pass saturation. To overcome this liability, it is recommended to give the dose as an infusion in the portal vein at a rate that is comparable to the oral absorption rate. For example, Nickerson and Toler used droloxifene as a probe substrate to assess the utility of intraperitoneal injection bolus and intraportal infusion (at a rate and duration that is comparable to that shown after oral drug delivery) as procedures to assess the factors that determine drug oral bioavailability [4]. The group reported that following intraperitoneal bolus of droloxifene, systemic plasma exposure was double that observed following portal vein infusion at a rate comparable to oral absorption rate and duration of 5 h. This is attributed to the saturation of hepatic first-pass enzymes mediated by high absorption rate and portal vein concentration. The group recommended using Equation 11.4. However, instead of giving the compound as intraperitoneal bolus injection, it is administered in the portal vein at an infusion rate that is comparable to that determined following oral administration. This should reduce the likelihood of saturation of hepatic first pass, improve the quality of the results, and accurately determine the factors that lead to compound poor bioavailability.

11.3.2 Mass Balance

Following oral administration, Equation 11.5 can be used to calculate the amount absorbed in the portal vein using the following equation (A_a) [5]:

$$A_a = Q_{pv} \cdot R_b \cdot (\mathrm{AUC}_{po,pv} - \mathrm{AUC}_{po,sys}) \tag{11.5}$$

where Q_{pv} is the portal vein blood flow (in rats, Q_{pv} is assumed to be 32.9 mL/min/kg); R_b, the blood to plasma ratio; and $\mathrm{AUC}_{po,pv}$ and $\mathrm{AUC}_{po,sys}$ are the drug blood exposure in the portal vein and systemic circulation following oral dosing, respectively. $f_a{\cdot}f_g$ can be estimated using this equation [5]:

$$f_a \cdot f_g = \frac{Q_{pv} \cdot R_b \cdot (\mathrm{AUC}_{po,pv} - \mathrm{AUC}_{po,sys})}{D_{po}} \tag{11.6}$$

where D_{po} is the drug oral dose [5]. Blood samples can be collected from portal vein in either portal-vein-cannulated animals by serial bleeding or individual animals by terminal bleeding. A key advantage of this methodology over the clearance method (see Section 11.3.3) is its ability to assess the impact of selective inhibitor for either efflux transporter/uptake transporter/metabolizing enzyme on the $f_a{\cdot}f_g$ without the need to determine the effect of these inhibitors on the systemic clearance using intravenous administration [5].

This procedure has its own limitations that should be taken into consideration when designing the portal-vein-cannulated rat experiments. For example, portal vein cannulation may have an impact on portal blood flow as well as drug protein binding for highly bound drug molecules by alteration of albumin amount. This may affect drug distribution and the final estimate. In addition, portal vein cannulation is resource intensive, technically challenging, and expensive. The use of individual animals is typically associated with high interanimal variability, which may compromise the quality of the assessment and require an increase in the number of animals per time point to overcome this limitation.

Mass balance principle obeys Fick's first law and assumes that the total drug amount in the portal vein is the result of drug absorbed into the portal vein from mesenteric vein combined with that coming from systemic circulation via mesenteric artery. Therefore, this methodology may underestimate the total amount of drug absorbed, when the drug has significant systemic intestinal metabolism prior to its absorption. This is one of the key factors that favor the use of clearance method (see below) to calculate $f_a \cdot f_g$ for compounds with significant systemic intestinal metabolism [6]. However, this method is appropriate for compounds with low intestinal first-pass metabolism or those primarily influenced by uptake and/or efflux transporters lining the apical surface of the intestinal epithelium. Of note, $f_a \cdot f_g$ calculated using mass balance methodology is influenced by the accuracy of portal blood flow rate, which is also influenced by surgery and food consumption.

11.3.3 Clearance Method

A_a and $f_a \cdot f_g$ can be also estimated using clearance method. The equations are as follows:

$$A_a = \mathrm{CL_b} \cdot \mathrm{AUC_{po,pv}} \quad (11.7)$$

$$f_a \cdot f_g = \frac{\mathrm{CL_b} \cdot \mathrm{AUC_{po,pv}}}{D_{po}} \quad (11.8)$$

where $\mathrm{CL_b}$ is the systemic blood clearance (or systemic plasma clearance when blood and plasma concentration are the same or blood to plasma ratio is estimated).

This system describes the absorption process using a model of three compartments, which are systemic blood, intestine, and liver compartment. It assumes that the drug follows first-order linear kinetics and distribution in each compartment is perfusion limited, that is, homogenous and instantaneous. Furthermore, drug elimination is limited to intestinal and/or hepatic clearance. It should be emphasized that if inhibitors are used to assess the impact of metabolizing enzymes/efflux transporters on the oral bioavailability of NMEs, it is prudent to determine the role of these inhibitors in the clearance of the tested article using Equation 11.8. This will enable scientists to measure $f_a \cdot f_g$ changes using clearance methodology.

11.4 APPROACHES TO DIFFERENTIATE THE POOR ABSORPTION FROM FIRST-PASS EFFECT CONTRIBUTION TO LOW ORAL BIOAVAILABILITY

In drug discovery and development, timely understanding of key drivers and mechanisms responsible for the poor oral bioavailability of NMEs is pivotal in addressing the cause and identifying potential solution. The factors leading to poor bioavailability can range from significant first pass that is mediated by high hepatic and/or intestinal metabolism or poor absorption driven by efflux transporter, low permeability, and/or poor solubility. The following text gives an overview of various tools that can be used to identify the cause of NMEs poor bioavailability using known modulators of metabolizing enzymes, efflux transporters, or intestinal pH.

11.4.1 Aminobenzotriazole

In a series of articles, Balani et al. studied the utility of 1-aminobenzotriazole (ABT) as a tool to differentiate the contribution of intestinal and hepatic first-pass effect mediated by cytochrome P450 enzymes to the poor bioavailability of NMEs and used antipyrine as a probe cytochrome

P450 substrate (Figure 11.1) [7, 8]. *In vitro*, the group demonstrated that ABT is an irreversible pan inhibitor of cytochrome P450 enzymes. Following 30-min preincubation, ABT inhibited CYP1A2, 2B6, 2C9, 2C19, 2D6, and 3A4 in a concentration-dependent manner at a concentration range of 0–500 μM. With 2-h ABT pretreatment, a single oral dose of 100 mg/kg in rats and 20 mg/kg in dogs and monkeys was effective in inhibiting the intravenous plasma clearance of antipyrine by 93%, 96%, and 83%, respectively.

In 2006, Strelevitz et al. further expanded the utility of ABT by investigating the impact of ABT different routes of administration, namely, intravenous and oral dosing on the plasma exposure of midazolam, a known CYP3A4 probe substrate. They proposed a hypothesis that ABT given intravenously would only inhibit cytochrome P450 enzymes located in the liver, while ABT dosed orally would inhibit cytochrome P450 located in both intestine and liver. This would allow them to determine the cause of the poor oral bioavailability of midazolam whether it is high first pass in the intestine and/or liver. They compared the impact of ABT route of administration on the oral exposure of midazolam. As illustrated in Figure 11.2, ABT only increased the plasma exposure of midazolam when dosed orally. However, intravenous dosing was associated with no increase in midazolam systemic exposure. This suggests that the poor oral bioavailability of midazolam is primarily attributed to significant intestinal first pass, that is, low f_g rather than low f_h and/or f_a.

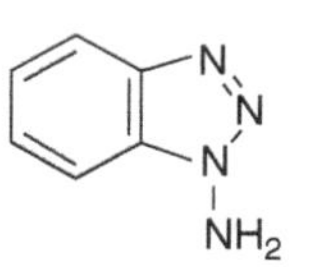

FIGURE 11.1 The chemical structure of 1-aminobenzotriazole.

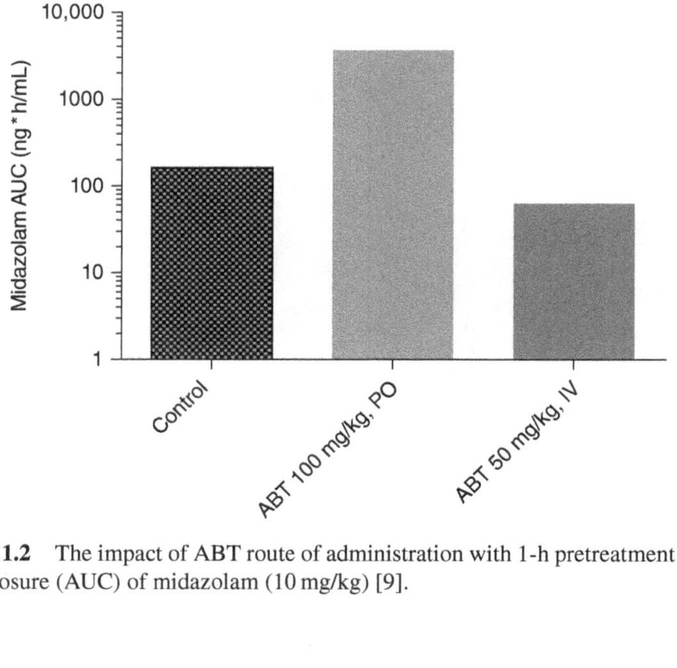

FIGURE 11.2 The impact of ABT route of administration with 1-h pretreatment on the oral plasma exposure (AUC) of midazolam (10 mg/kg) [9].

11.4.2 GrapeFruit Juice

The use of grapefruit juice in understanding the contribution of CYP3A4, P-glycoportein, and OATP2B1 to the absorption and intestinal first pass effect of drug molecules in preclinical species and humans is reviewed in Chapter 7.

11.4.3 CP-100,356

In 2009, Kalgutkar et al. evaluated the utility of the diaminoquinazoline derivative CP-100,356 as an *in vivo* probe to selectively assess MDR1-/BCRP-mediated drug efflux and its impact on overall oral exposure of NME using rat as a preclinical model (Figure 11.3) [10]. *In vitro*, the group reported that CP-100,356 was not an inhibitor for major human cytochrome P450 enzymes including CYP3A4 ($IC_{50} > 50\,\mu M$) or MRP2 ($IC_{50} > 15\,\mu M$). It was a weak inhibitor of OATP1B1 ($IC_{50} \sim 66\,\mu M$). However, it inhibited acetoxymethyl calcein (calcein-AM) uptake with an IC_{50} of 0.5 μM and digoxin transport ($IC_{50} \sim 1.2\,\mu M$) using MDR1-transfected MDCKII cells. Furthermore, CP-100,356 inhibited prazosin transport with an IC_{50} of 1.5 μM using human BCRP-transfected MDCKII cells. Collectively, this suggested that CP-100,356 is a potent dual inhibitor for MDR1 and BCRP transporters *in vitro*.

To assess the P-glycoprotein and BCRP *in vivo* inhibition in rats, CP-100,356 was coadministered with P-glycoprotein substrate fexofenadine and dual P-glycoprotein/BCRP substrate prazosin. At a dose of 24 mg/kg, CP-100,356 increased fexofenadine rate (C_{max}) and extent of systemic exposure (AUC) 36- and 80-fold increase, respectively. CP-100,356 increased prazosin oral AUC 2.6-fold increase. Interestingly, coadministration of CP-100,356 and cytochrome P4503A substrate midazolam did not result in increase in the systemic exposure of midazolam in rat. In conclusion, CP-100,356 is a viable approach that can be used to understand the potential contribution of P-glycoprotein/BCRP to the overall disposition of a new chemical entity of interest.

11.4.4 Elacridar

Elacridar, also known as GF-120198, is widely used as a dual inhibitor of P-glycoprotein and BCRP transporter. It is used preclinically and in humans to assess the contribution of P-glycoprotein and BCRP efflux to the absorption and disposition of drug molecules (Figure 11.4).

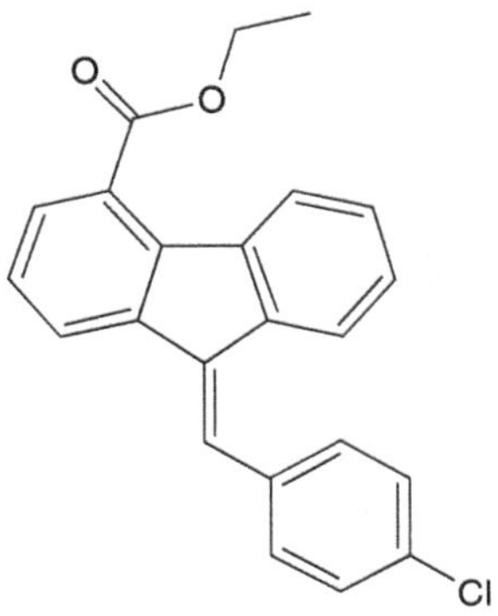

FIGURE 11.3 The chemical structure of CP-100,356 [10].

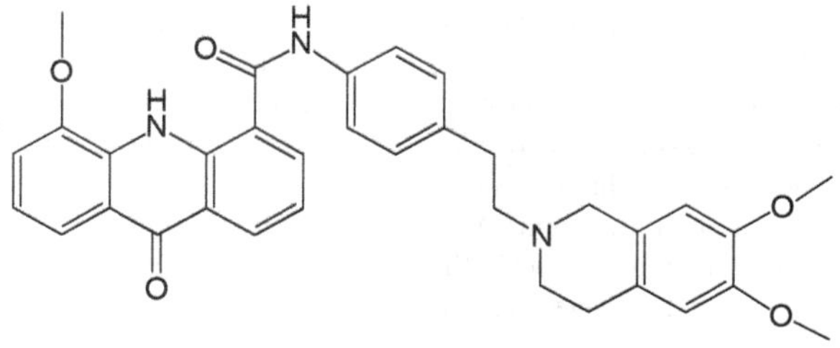

FIGURE 11.4 The chemical structure of elacridar (GF120198).

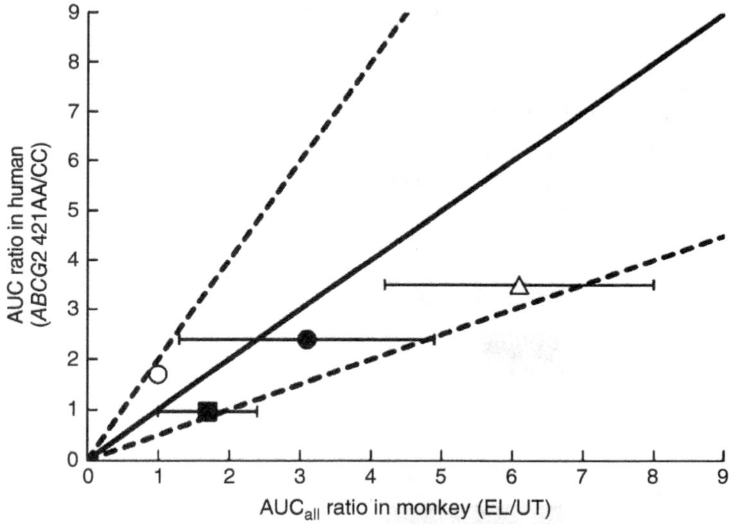

FIGURE 11.5 The relationship of the AUC ratio between monkeys (with elacridar (EL)/without elacridar (UT)) and human (ABCG2 421AA/CC) after oral administration of rosuvastatin (•), pitavastatin (■), fluvastatin (○), and sulfasalazine (Δ). The solid lines represent a 1:1 correspondence and the dotted lines represent the twofold difference. Karibe et al. 2015 [11]. Reproduced with permission of Springer.

It is well established in the literature that ABCG2 421C>A variant has a significant impact on the oral exposure (AUC and C_{max}) of known BCRP substrates such as rosuvastatin, sulfasalazine, and fluvastatin in humans with minimal impact on the elimination terminal half-life ($t_{1/2}$) of these molecules. These findings suggest that this genetic variant impact is mainly limited to the absorption rather than the elimination of these molecules. To project and estimate the AUC changes in ABCG2 421C>A genetic variants for BCRP substrate drugs in humans, Karibe et al. proposed the use of cynomolgus monkeys pretreated with a BCRP inhibitor, GF-120198, as useful animal model. At a 20 mg/kg PO dose, GF-120198 pretreatment had an impact on the monkey oral exposure of rosuvastatin, sulfasalazine, pitavastatin, and fluvastatin represented by the AUC ratio of treated to untreated animals that was similar to that observed to the oral AUC ratio of human (ABCG2 421AA/CC) (Figure 11.5) [11]. This suggests the potential utility of this model as an approach to predict the impact of human (ABCG2 421AA/CC) on the oral exposure of BCRP substrate.

Topotecan is topoisomerase I inhibitor that is used in the treatment of ovarian cancer and lung cancer, as well as other cancer types [12]. Kruijtzer et al. assessed the impact of elacridar on the intravenous and oral exposure of topotecan in cancer patients. After oral dose of topotecan (1 mg/m^2), the AUC and renal recovery (Xu%) increased significantly when coadministered with single oral dose of elacridar (1000 mg) (Figure 11.6) ($P < 0.05$). However, small impact was observed on the intravenous AUC and half-life of topotecan. These findings suggest that P-glycoprotein plays a key role in limiting the oral absorption of topotecan. However, it has a minimal impact on the disposition and elimination of topotecan.

11.4.5 Curcumin

The use of curcumin to investigate the potential role of BCRP to the efflux of NMEs is reviewed in Chapter 7.

11.4.6 Zosuquidar and Ko143

Zosuquidar (ZSQ) is antineoplastic drug candidate, also known as LY335979. *In vitro*, this compound is a potent P-glycoprotein inhibitor with an IC_{50} range of 0.02–0.07 μM [14, 15].

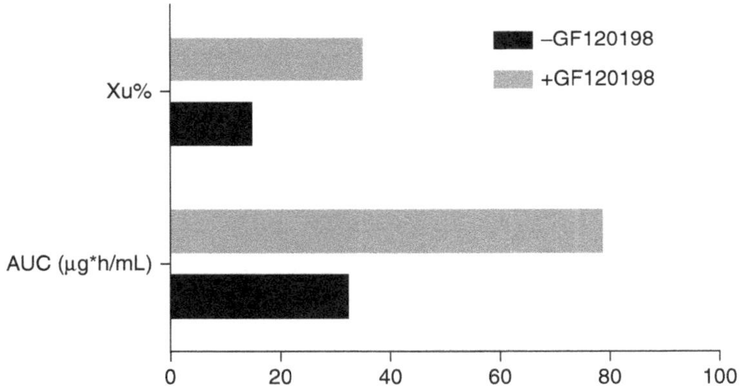

FIGURE 11.6 The impact of 1000 mg of GF-120198 PO dose on the renal recovery and exposure of Topotecan (1 mg/m^2 PO dose). Adapted from Kruijtzer et al. 2002 [12].

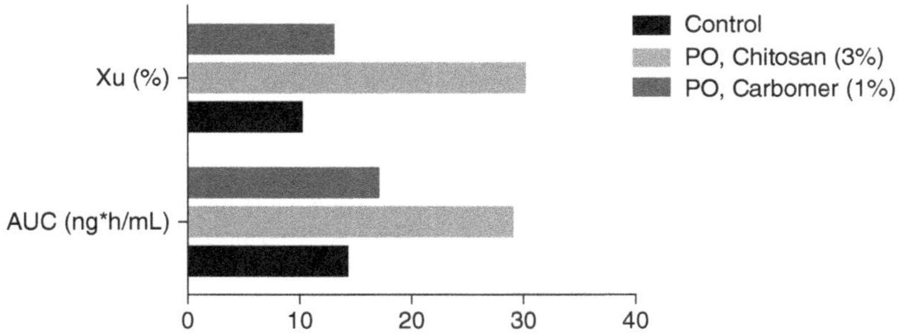

FIGURE 11.7 The impact of formulation on the renal recovery (Xu%) and AUC following acyclovir 2.5 mg/kg PO dose [13].

It is used with Ko143, a potent BCRP inhibitor (IC_{50} = 0.01–3.79 μM), to assess the effect of P-glycoprotein and BCRP on the oral absorption of NMEs (Figure 11.8) [16, 17]. Matsuda et al. demonstrated the utility of these efflux transporter inhibitors using three model drugs, fexofenadine, sulfasalazine, and topotecan, which are P-glycoprotein, BCRP, and P-glycoprotein and BCRP substrates, respectively [5]. The study was conducted using portal-vein-cannulated rats after pretreatment with ZSQ (30 mg/kg) and/or Ko143 (10 mg/kg) 40 min prior to oral administration of model drugs. Intestinal availability ($f_a \cdot f_g$) was determined using mass balance methodology (Section 11.3.2). Mass balance was an effective approach in dissecting the impacts of various efflux transporters on the absorption of model drugs. For example, around 71% of fexofenadine was effluxed by P-glycoprotein. On the other hand, 79% of sulfasalazine was effluxed by BCRP. Topotecan was effluxed by both transporters (Figure 11.9).

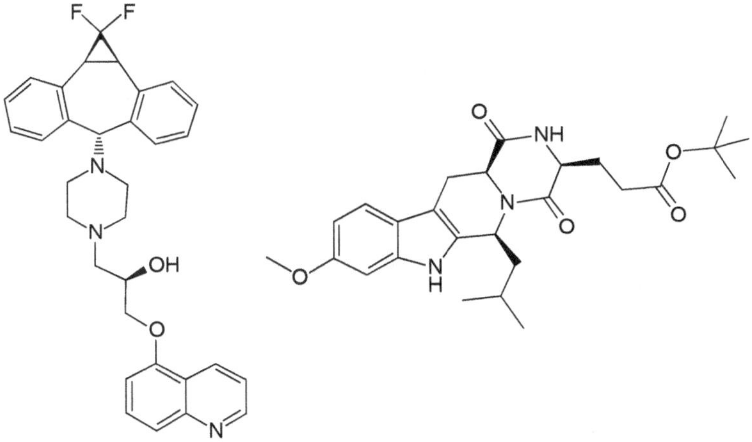

FIGURE 11.8 The chemical structures of zosuquidar and Ko143, respectively.

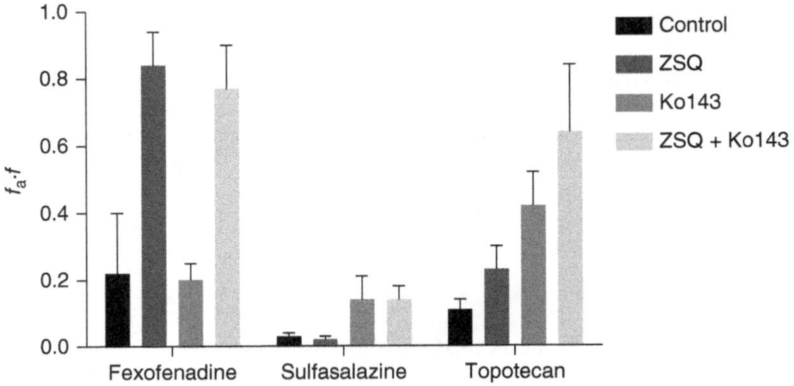

FIGURE 11.9 The $f_a \cdot f_g$ of fexofenadine (5 mg/kg), sulfasalazine (5 mg/kg), and topotecan (0.3 mg/kg) following ZSQ (30 mg/kg) and/or Ko143 (10 mg/kg) in portal vein cannulated rats. Adapted from Matsuda et al. 2013 [5].

It should be reiterated that when mass balance methodology is used, there is no need to investigate the impact of efflux transporter inhibitor on the systemic clearance of tested article. This is a key advantage of this methodology that clearance methodology lacks.

TIP:

Matsuda et al. approach is a good example of not only determining the cause of poor bioavailability of NMEs, that is, is it first pass and/or poor absorption, but also defining the biological mechanism leading to poor bioavailability, whether it is driven by efflux transporter(s) or metabolizing enzyme(s) [5].

11.4.7 Famotidine and Pentagastrin

Many of the drugs available in the market are weak acids or bases [18]. In humans, changes in the pH along the gastrointestinal tract would have a substantial impact on the extent of ionization of these molecules, which affects their overall absorption and systemic exposure. Indeed, understanding the impact of gastric pH on solubility, dissolution rate, and absorption of such molecules is critical in enabling the successful optimization and selection of new formulations. It reduces the impact of gastrointestinal pH on oral absorption of promising drug candidates and variability in oral exposure in humans.

It is neither feasible nor ethical to run these studies in humans early in drug discovery. Instead, *in vitro* and *in vivo* preclinical models were successfully utilized to investigate these trends. For example, Zhou et al. assessed the impact of pH on the *in vitro* dissolution rate and *in vivo* pharmacokinetics in pH-dependent canine absorption model of weak bases using ketoconazole and dipyridamole as probe molecules. The group studied the *in vitro* pH dissolution profiles of ketoconazole and dipyridamole using USP apparatus II and an Opt-Diss Fiber Optic UV System. *In vivo* studies were performed with crossover design in three groups including no treatment, pentagastrin treatment, and famotidine treatment. Pentagastrin treatment group maintained the gastric pH at ~2–3 by giving pentagastrin (6 μg/kg) intramuscularly 30 min before dosing of probe molecules. Famotidine treatment group maintained the gastric pH at ~5–7.5 by administering famotidine at an oral dose of 40 mg/dog, 3 h before dosing. Ketoconazole (200 mg/dog) or dipyridamole (75 mg/dog) was administered orally [19]. Consistent with reports in the literature, both ketoconazole and dipyridamole demonstrated pH-dependent dissolution profiles *in vitro*. Indeed, the increase in the dissolution medium pH from 1.2 to 6.8 was associated with a reduction in the extent of dissolution of the two compounds tested. These trends were in concordance with the *in vivo* studies, where the area under plasma time curve (AUC) increased nearly 30-fold for ketoconazole and ninefold for dipyridamole with pentagastrin (gastric pH at 2–3) as compared to famotidine (gastric pH 5–7.5) treatments, respectively. Taken collectively, changes in pH along gastrointestinal tract should be investigated to understand the impact on the dissolution and absorption of weak bases and acids. This is pivotal and can be achieved using a combination of *in vitro* dissolution studies and *in vivo* pH-dependent canine absorption model.

11.4.8 The Use of Renal Recovery in Measuring Oral Absorption

If a compound is predominantly excreted as unchanged drug in the urine, renal recovery can be used as a tool to investigate its absorption and the impact of formulations on its overall oral exposure. In 2009, Merzlikine et al. investigated the impact of chitosan glutamate and carbomer 974P formulations on the oral absorption of paracellularly absorbed model drug, acyclovir, in rats by investigating their impact on its oral systemic exposure and amount excreted unchanged into urine (X_u%) [13]. In rats, chitosan glutamate increased acyclovir oral AUC by twofold and the X_u% by threefold. In contrast, the effect of carbomer 974P was not statistically significant at 1% level (Figure 11.7). Chitosan is an effective permeability enhancer that can increase plasma exposure of molecules absorbed via paracellular pathway. This is ascribed to its ability to loosen and open the tight junction leading to higher absorption through this pathway. In summary, renal recovery can be effectively used to assess the absorption and impact of formulations on the pharmacokinetics of NMEs predominantly eliminated by renal excretion.

11.5 APPROACHES TO ASSESS THE PHARMACODYNAMICS ACTIVITY OF NME WITH POOR ORAL BIOAVAILABILITY

Drug molecules with good oral absorption, high rat hepatic/intestinal extraction ratios, and poor oral bioavailability tend to have better oral bioavailability in higher species such as dogs, monkeys, and humans [18]. This trend is supported by multiple reports in the literature and primarily attributed to species differences in physiology, metabolizing enzymes expression and activity as discussed in Chapter 2. Figure 11.10 illustrates the relationship between rat and human plasma clearance. Drugs that have high rat clearance (>49 mL/min/kg, assuming total rat hepatic blood flow of 70 mL/min/kg) would not necessarily have high clearance in humans (>14 mL/min/kg, assuming total human hepatic blood flow of 20 mL/min/kg). Indeed, in this set of data, 61% of drug molecules with high rat clearance had a human clearance that was smaller than 14 mL/min/kg [20]. Widman et al. investigated the disposition and pharmacokinetic profiles of remoxipride in rodents, dogs, and humans [21]. Consistent with its high permeability, remoxipride was completely absorbed following oral dosing in all species investigated. Unlike dogs and humans, the oral bioavailability was low in rodents, which was attributed to its high blood clearance that was equal to or higher than liver blood flow. This led to significant first-pass effect in rodents, which limited remoxipride oral exposure. However, in

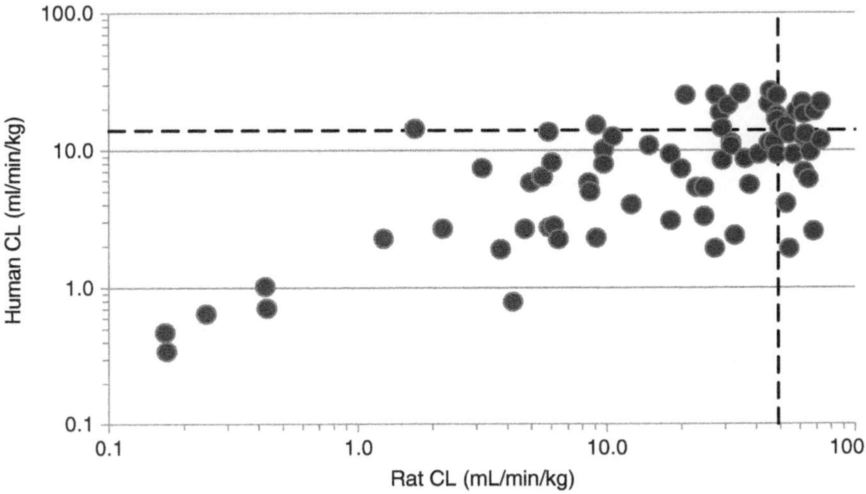

FIGURE 11.10 The relationship between rat and human clearance. Lombardo et al. 2013 [20]. Reproduced with permission of John Wiley & Sons.

both dogs and humans, clearance values were low rendering the compound orally bioavailable (>90%). Similar examples on species differences in first-pass effect were reported with reboxetine, atomoxetine, and indinavir [22–25]. These successful medications would not reach the market if the discovery team solely depended on assessing the biological activity of these molecules in rats using oral dosing, where poor exposure would limit the plasma exposure of these molecules at the site of action. This may lead scientists to jump to the conclusion that these molecules are not potent enough to warrant further investment. To overcome this liability, discovery scientists usually opt to assess the biological activity in rodent species using other routes of administration that have the ability to avoid intestinal and/or hepatic first-pass effects typically associated with oral dosing. Following is a discussion of these routes and dosing procedures that are considered when assessing the biological activity for compounds with high first-pass effect.

11.5.1 Intraperitoneal (IP) Dosing

The peritoneal cavity is considered the largest serosal pouch located between the parietal and visceral peritoneum. These are two membranes that isolate visceral organs from abdominal wall. It secretes a fluid that functions as a lubricant and has anti-inflammatory properties. Injection in the intraperitoneal sac known as intraperitoneal dosing (IP) is widely used in preclinical species and to a lesser extent in humans. Once administered, drug crosses the visceral peritoneum, which is highly innervated with blood vessels and pool into portal vein ending into the liver (Figure 11.11).

In drug discovery, IP is an effective procedure that is used to assess biological activities of NMEs, where they are dosed in a form of solution or suspension. This procedure is

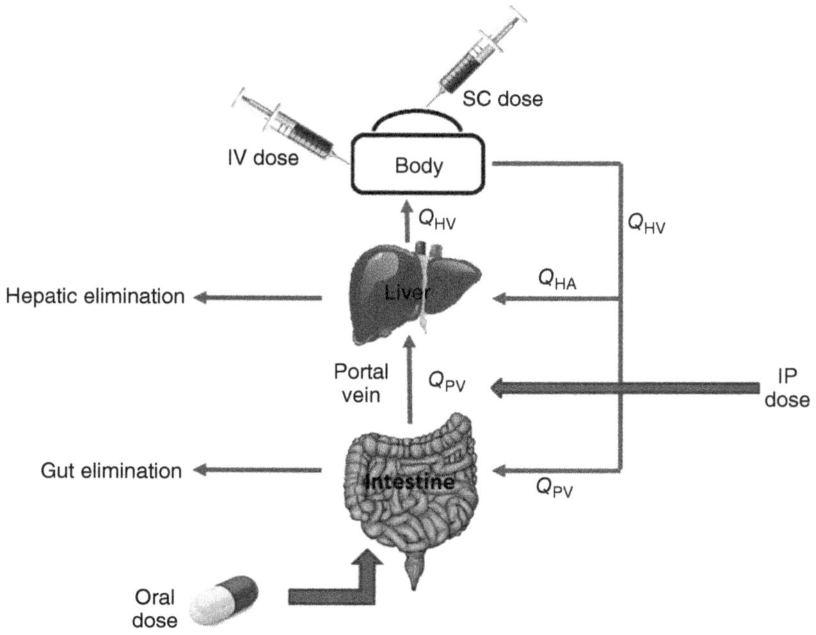

FIGURE 11.11 Routes of administration to assess biological activity in preclinical species.

not technically challenging when compared with oral or intravenous dosing. It overcomes intestinal, but not hepatic, first-pass effect. Nonetheless, IP dosing is associated with a higher likelihood of saturation of liver first-pass effect since the rate of entry in the liver is faster than the rate of absorption following oral dosing [4].

11.5.2 Subcutaneous (SC) Dosing

Subcutaneous dosing is an injection in the hypodermis, also known as subcutaneous tissue, which is located between skin and muscle. Hypodermis is composed of loose connective tissue and fat lobules. It is also highly innervated with blood vessels and nerves. Tested articles are given in a form of solution or suspension. Absorption following subcutaneous (SC) dosing is slow, yet if the compound is not crashing out of solution, it will have a plasma exposure that is comparable to that given following intravenous dosing. This is attributed to the fact that this route of administration avoids hepatic and intestinal first-pass effect, rendering the compound systemically bioavailable (Figure 11.11).

While this route of administration is superior to the IP dosing due to its ability to achieve higher plasma exposure, the maximum dosing volume that can be given is smaller relative to IP (Table 11.1). This is a pivotal factor that should be considered when selecting the route of administration. It dictates the maximum dose that can be administered (Table 11.1) [26].

11.5.3 Minipumps

Osmotic pumps such as that offered by Alzaet® are implantable pumps that have been successfully used in research to deliver small molecules, peptides, and protein drugs at continuous and controlled release rates in preclinical animal models, for example, mice and rats [27, 28]. The pharmacokinetic principles that govern this methodology are similar to that reported with intravenous infusion. Herein, it will be critical to measure tested article clearance to determine the rate of release that is needed to achieve targeted efficacious plasma concentration. In addition, close attention should be exercised to ensure that tested article dosed is in solution. Therefore, measuring drug solubility in osmotic pump vehicle is another factor that should be evaluated and will dictate vehicle selection.

The release rate that is needed to achieve a targeted efficacious plasma concentration is calculated using this equation:

$$K_o = CL_b \cdot C_{ss} \tag{11.9}$$

where K_o is the pump release rate, CL_b the tested article blood clearance, and C_{ss} the total targeted efficacious blood concentration. K_o unit is mass per unit time, for example, µg/h.

TABLE 11.1 Administration Volumes Considered Good Practice (and Possible Maximum Dose Volume)

Species	mL/kg Oral	S.C.	I.P.	I.M.	I.V. (Bolus)	I.V. (Slow Injection)
Mouse	10 (50)	10 (40)	20 (80)	0.05 (0.1)	5	(25)
Rat	10 (40)	5 (10)	10 (20)	0.1 (0.2)	5	(20)
Dog	10 (15)	1 (2)	5 (20)	0.25 (0.5)	2.5	(5)
Marmoset	10 (15)	2 (5)	(20)	0.25 (0.5)	2.5	(10)
Minipig	10 (15)	1 (2)	(20)	0.25 (0.5)	2.5	(5)

Source: Diehl et al. 2001 [26]. Reproduced with permission of John Wiley & Sons.

The solubility in pump vehicle that is needed to achieve the targeted plasma concentration is calculated using the following equation:

$$C_{\mathrm{d}} = \frac{K_{\mathrm{o}}}{Q} \tag{11.10}$$

where C_{d} is the solubility of the tested article that needs to be achieved to attain the targeted efficacious plasma exposure. Q is the pump infusion rate, which is determined by pump design and varies depending on pump specifications and targeted study length.

Osmotic pumps can be installed intraperitoneally or subcutaneously. This will determine the plasma concentration obtained. As discussed earlier, the site of drug administration determines if the compound would avoid either intestinal and/or hepatic first-pass effect.

TIP:

As shown in Equations 11.9 and 11.10, the use of osmotic pump is limited by three factors: (1) the targeted efficacious exposure, (2) the clearance of the molecule, and (3) the solubility of the tested article in the pump vehicle that has a limited volume. Therefore, calculating these parameters will determine the feasibility of using osmotic pumps to assess biological activities of NMEs. In practice, the use of osmotic pumps is not feasible when the tested articles have high clearance combined with high protein binding and efficacious plasma concentration.

As a technique, osmotic pump has multiple advantages. For example, it is a cost-effective and practical approach that reduces costs and avoids the need for nighttime or weekend dosing by lab technicians. It enables an extended period of treatment, reduces likelihood of contamination, and allows group housing. It successfully minimizes animal stress related to recurrent handling and multiple dosing, which is another benefit that improves the quality of reported findings.

11.5.4 The Impact of Intravenous, Intraperitoneal, Subcutaneous, and Oral Dosing on Reboxetine Plasma Exposure: Case Study

As discussed earlier, route of administration has a significant impact on the systemic exposure of compounds that have significant first-pass effect. [*S*,*S*]-Reboxetine is a norepinephrine reuptake inhibitor drug that is used outside the United States in the treatment of unipolar depression [29]. In rats, it has high clearance (81 mL min/kg) that is primarily mediated by hepatic metabolism. Figure 11.12 depicts plasma concentration–time profiles following IV, SC, IP, and PO administration of 5 mg/kg of [*S*,*S*]-reboxetine to Sprague Dawley rats. Table 11.2 summarizes relevant pharmacokinetic parameters following IV, SC, IP, and PO administration (5 mg/kg) of [*S*,*S*]-reboxetine to Sprague Dawley rats (Pfizer Internal Data). IV dosing had an AUC of 1032 ng*h/mL. SC administration was associated with an AUC and *F* of 798% and 798 ng*h/mL, respectively, which was significantly higher than IP and PO dosing (Table 11.2). Overall, the obtained findings are consistent with the high hepatic and intestinal first-pass effects of [*S*,*S*]-reboxetine. Both SC and IV dosing avoids hepatic and intestinal first-pass effect; therefore, they had the highest plasma exposure. On the other hand, IP avoids only intestinal first-pass effect but not hepatic first-pass effect. As a result, IP AUC and *F*% was lower than both IV and SC, yet higher than PO, since the latter neither avoids hepatic nor intestinal first-pass effects (Figure 11.11).

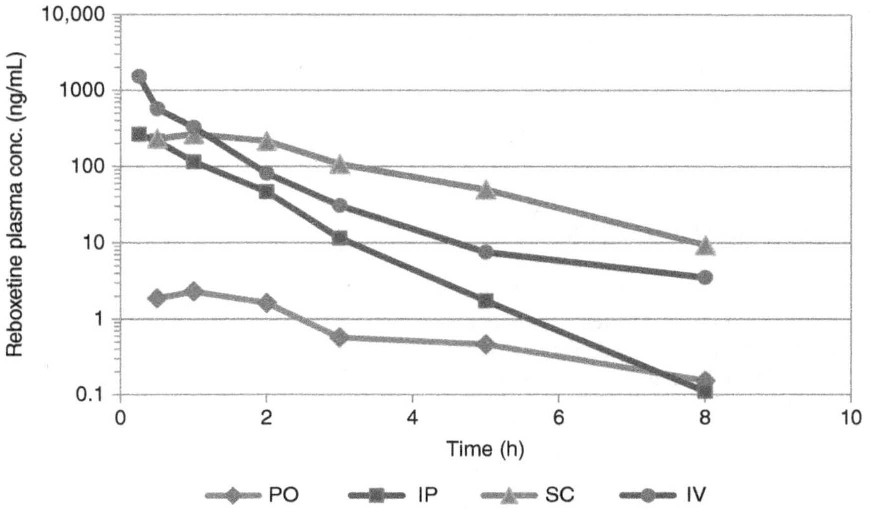

FIGURE 11.12 Plasma [*S*,*S*]-reboxetine concentration–time profiles after intravenous, subcutaneous, intraperitoneal, and oral administration (5 mg/kg) of [*S*,*S*]-reboxetine to Sprague Dawley rats (Pfizer internal data).

TABLE 11.2 The Pharmacokinetic Parameters of [*S*,*S*]-Reboxetine Following 5 mg/kg IV, SC, PO, and IP Dosing to Sprague Dawley Rats (Pfizer Internal Data)

Dosing Route	$AUC_{(0\to\infty)}$ (ng*h/mL)	CL (mL/min/kg)	V_{dss} (L/kg)	F (%)
IV	1032	81	5.5	–
SC	798	–	–	77
IP	284	–	–	28
PO	0.01	–	–	0.001

TIP:

It should be emphasized that SC and IP dosing may be associated with poor systemic exposure although the tested article has good hepatic and intestinal stability. This may result from poor NME solubility, where compound would crash out of solution once dosed. As a result, it is a good practice to consider formulation that would ensure the compound solubility and reduce its precipitation.

11.6 CONCLUSION

Medicinal chemists supporting discovery projects are keen to understand the causes of NMEs poor bioavailability. In this chapter, we reviewed various approaches that can be utilized to deconvolute the causes of poor oral bioavailability. Special emphasis was placed on the use of various inhibitors that selectively inhibit known transporters/metabolizing enzymes. This would enable scientists to identify the mechanism contributing to the poor bioavailability. We also provided an overview of alternative routes of administration that can be considered to assess the biological activity of molecules that have poor first pass and bioavailability in rodents yet are projected to have good pharmacokinetics profiles in upper preclinical species and humans.

11.7 TUTORIAL CHAPTER 11

1. Drug X is in advanced preclinical development. It has poor bioavailability following PO dosing in SPD rats. The team is seeking your input on the assessment of the causes of poor bioavailability and whether you would expect similar trends in humans. You have performed both IV and PO dosing in portal-vein-cannulated rats. Following are the obtained PK parameters:

PK Parameters	IV Route	PO Route
Dose (mg/kg)	3	3
C_{max} (ng/mL)	–	5.7
T_{max} (h)	–	6.7
AUC_{sys} (ng*h/mL)	1000	100
AUC_{portal} (ng*h/mL)	–	800
CL (mL/min/kg)	50	–
V_{dss} (L/kg)	13	–
$T_{1/2}$ (h)	3	5
Q_{pv} (mL/min/kg)	39	–

Mass balance equation:

$$F_a \cdot F_g = \frac{Q_{pv} \cdot (AUC_{po,pv} - AUC_{po,sys})}{D_{po}}$$

Clearance method equation:

$$F_a \cdot F_g = \frac{CI_b \cdot AUC_{po \cdot pv}}{D_{po}}$$

Using the data and equations above, assess the main factors contributing to the poor oral exposure of Drug X and the potential of observing it in upper species.

If there is a difference between the values generated using mass balance versus clearance method, what does that mean?

2. How would you identify the main enzyme class that is responsible for the high first-pass effect observed with Drug X?

3. True or false

Famotidine is an important agent that can be used to assess the potential contribution of CYP-mediated metabolism to the overall clearance of the molecules in preclinical species.

ABT is an important agent that can be used in preclinical species to understand the impact of alteration in the extent of ionization, dissolution rate, and solubility on the absorption of new tested articles.

The difference in the $f_a \cdot f_g$ calculated using mass balance or clearance methods indicates that the molecule undergoes significant systemic intestinal metabolism.

Similar to SC, when you dose your compound IP, you avoid both hepatic and intestinal first-pass effect.

11.8 CHAPTER 11 TUTORIAL ANSWER KEY

1. Drug X is in advanced preclinical development. It has poor bioavailability following PO dosing in SPD rats. The team is seeking your input on the assessment of the causes of poor bioavailability and whether you would expect similar trends in humans. You have performed both IV and PO dosing in portal-vein-cannulated rats. Following are the obtained PK parameters:

PK Parameters	IV Route	PO Route
Dose (mg/kg)	3	3
C_{max} (ng/mL)	–	5.7
T_{max} (h)	–	6.7
AUC_{sys} (ng*h/mL)	1000	100
AUC_{portal} (ng*h/mL)	–	800
CL (mL/min/kg)	50	–
V_{dss} (L/kg)	13	–
$T_{1/2}$ (h)	3	5
Q_{pv} (mL/min/kg)	39	–

Mass balance equation:

$$F_a \cdot F_g = \frac{Q_{pv} \cdot (AUC_{po,pv} - AUC_{po,sys})}{D_{po}}$$

Clearance method equation:

$$F_a \cdot F_g = \frac{CI_b \cdot AUC_{po \cdot pv}}{D_{po}}$$

Using the data and equations above, assess the main factors contributing to the poor oral exposure of Drug X and the potential of observing it in upper species.

Answer:
Mass balance $f_a{\cdot}f_g = 0.55$
Clearance method $f_a{\cdot}f_g = 0.80$

If there is a difference between the values generated using mass balance versus clearance method, what does that mean?

Answer: As discussed in Section 11.3.2, mass balance principle obeys Fick's first law and assumes that the total drug amount in the portal vein is the result of drug absorbed into the portal vein from mesenteric vein combined with that coming from systemic circulation via mesenteric artery. Therefore, this methodology may underestimate the total amount of drug absorbed, when the drug has significant systemic intestinal metabolism prior to its absorption. This is one of the key factors that favors the use of clearance method to calculate $f_a{\cdot}f_g$ for compounds with significant systemic intestinal metabolism [6]. Collectively, this may explain the difference in the reported $f_a{\cdot}f_g$ calculated using these two methodologies.

2. How would you identify the main enzyme class that is responsible for the high first-pass effect observed with Drug X?

   Answer: Scientists typically consider the use of ABT giving following IV (50 mpk) versus PO (100 mpk) dosing. Depending on the outcome of impact of ABT, they will be able identify the following points:

   If you observe a significant increase in the plasma exposure of your tested article in the presence of ABT, this suggests that the compound elimination is mainly driven by CYP-mediated metabolism.

   If you note a higher increase in plasma exposure following PO rather than IV dosing of ABT, this may suggest that the compound undergoes higher intestinal first-pass relative to hepatic first-pass effect.

3. True or false
   Famotidine is an important agent that can be used to assess the potential contribution of CYP-mediated metabolism to the overall clearance of the molecules in preclinical species.
   (F)
   ABT is an important agent that can be used in preclinical species to understand the impact of alteration in the extent of ionization, dissolution rate, and solubility on the absorption of new tested articles.
   (F)
   The difference in the $f_a{\cdot}f_g$ calculated using mass balance or clearance methods indicates that the molecule undergoes significant systemic intestinal metabolism.
   (T)
   Similar to SC, when you dose your compound IP, you avoid both hepatic and intestinal first-pass effect.
   (F)

REFERENCES

1. El-Kattan, A., Holliman, C. and Cohen, L., *Quantitative Bioanalysis in Drug Discovery and Development: Principles and Applications*, in *Mass Spectrometry in Drug Metabolism and Pharmacokinetics*, Ramanathan, R., Editor. 2008, Wiley: Hoboken. p. 87–122.

2. Williams, J.A., Hyland, R., Jones, B.C., Smith, D.A., Hurst, S., Goosen, T.C., Peterkin, V., Koup, J.R. and Ball, S.E., *Drug–drug interactions for UDP-glucuronosyltransferase substrates: a pharmacokinetic explanation for typically observed low exposure (AUCi/AUC) ratios*. Drug Metab Dispos, 2004. 32(11): p. 1201–1208.

3. Kwon, Y., *Handbook of Essential Pharmacokinetics, Pharmacodynamics, and Drug Metabolism for Industrial Scientists*. 2001, Kluwer Academic/Plenum Publishers: New York.

4. Nickerson, D.F. and Toler, S.M., *Intraperitoneal and intraportal administration of droloxifene to the Sprague-Dawley rat: assessing the first-pass effect*. Xenobiotica, 1997. 27(6): p. 627–632.

5. Matsuda, Y., Konno, Y., Hashimoto, T., Nagai, M., Taguchi, T., Satsukawa, M. and Yamashita, S., In vivo *assessment of the impact of efflux transporter on oral drug absorption using portal vein-cannulated rats*. Drug Metab Dispos, 2013. 41(8): p. 1514–1521.

6. Kwon, Y. and Inskeep, P.B., *Theoretical considerations on two equations for estimating the extent of absorption after oral administration of drugs*. Pharm Res, 1996. 13(4): p. 566–569.

7. Balani, S.K., Li, P., Nguyen, J., Cardoza, K., Zeng, H., Mu, D.X., Wu, J.T., Gan, L.S. and Lee, F.W., *Effective dosing regimen of 1-aminobenzotriazole for inhibition of antipyrine clearance in guinea pigs and mice using serial sampling*. Drug Metab Dispos, 2004. 32(10): p. 1092–1095.

8. Balani, S.K., Zhu, T., Yang, T.J., Liu, Z., He, B. and Lee, F.W., *Effective dosing regimen of 1-aminobenzotriazole for inhibition of antipyrine clearance in rats, dogs, and monkeys*. Drug Metab Dispos, 2002. 30(10): p. 1059–1062.

9. Strelevitz, T.J., Foti, R.S. and Fisher, M.B., In vivo *use of the P450 inactivator 1-aminobenzotriazole in the rat: varied dosing route to elucidate gut and liver contributions to first-pass and systemic clearance*. J Pharm Sci, 2006. 95(6): p. 1334–1341.

10. Kalgutkar, A.S., Frederick, K.S., Chupka, J., Feng, B., Kempshall, S., Mireles, R.J., Fenner, K.S. and Troutman, M.D., *N-(3,4-dimethoxyphenethyl)-4-(6,7-dimethoxy-3,4-dihydroisoquinolin-2[1H]-yl)-6,7-dimethoxyquinazolin-2-amine (CP-100,356) as a "chemical knock-out equivalent" to assess the impact of efflux transporters on oral drug absorption in the rat*. J Pharm Sci, 2009. 98(12): p. 4914–4927.

11. Karibe, T., et al., *Evaluation of the usefulness of breast cancer resistance protein (BCRP) knockout mice and BCRP inhibitor-treated monkeys to estimate the clinical impact of BCRP modulation on the pharmacokinetics of BCRP substrates*. Pharm Res, 2014. 32 (5): p. 1634–1647.

12. Kruijtzer, C.M., Beijnen, J.H., Rosing, H., ten Bokkel Huinink, W.W., Schot, M., Jewell, R.C., Paul, E.M. and Schellens, J.H., *Increased oral bioavailability of topotecan in combination with the breast cancer resistance protein and P-glycoprotein inhibitor GF120918*. J Clin Oncol, 2002. 20(13): p. 2943–2950.

13. Merzlikine, A., Rotter, C., Rago, B., Poe, J., Christoffersen, C., Thomas, V.H., Troutman, M. and El-Kattan, A., *Effect of chitosan glutamate, carbomer 974P, and EDTA on the* in vitro *Caco-2 permeability and oral pharmacokinetic profile of acyclovir in rats*. Drug Dev Ind Pharm, 2009. 35(9): p. 1082–1091.

14. Shaik, N., Giri, N., Pan, G. and Elmquist, W.F., *P-glycoprotein-mediated active efflux of the anti-HIV1 nucleoside abacavir limits cellular accumulation and brain distribution*. Drug Metab Dispos, 2007. 35(11): p. 2076–2085.

15. Heredi-Szabo, K., et al., *A P-gp vesicular transport inhibition assay – optimization and validation for drug–drug interaction testing*. Eur J Pharm Sci, 2013. 49(4): p. 773–781.

16. Weiss, J., Rose, J., Storch, C.H., Ketabi-Kiyanvash, N., Sauer, A., Haefeli, W.E. and Efferth, T., *Modulation of human BCRP (ABCG2) activity by anti-HIV drugs*. J Antimicrob Chemother, 2007. 59(2): p. 238–245.

17. Jin, H.E., Song, B., Kim, S.B., Shim, W.S., Kim, D.D., Chong, S., Chung, S.J. and Shim, C.K., *Transport of gemifloxacin, a 4th generation quinolone antibiotic, in the Caco-2 and engineered MDCKII cells, and potential involvement of efflux transporters in the intestinal absorption of the drug*. Xenobiotica, 2013. 43(4): p. 355–367.

18. Thomas, V.H., Bhattachar, S., Hitchingham, L., Zocharski, P., Naath, M., Surendran, N., Stoner, C.L. and El-Kattan, A., *The road map to oral bioavailability: an industrial perspective.* Expert Opin Drug Metab Toxicol, 2006. 2(4): p. 591–608.

19. Zhou, R., et al., *pH-Dependent dissolution* in vitro *and absorption* in vivo *of weakly basic drugs: development of a canine model.* Pharm Res, 2005. 22(2): p. 188–192.

20. Lombardo, F., et al., *Comprehensive assessment of human pharmacokinetic prediction based on* in vivo *animal pharmacokinetic data, part 2: clearance.* J Clin Pharmacol, 2013. 53(2): p. 178–191.

21. Widman, M., Nilsson, L.B., Bryske, B. and Lundstrom, J., *Disposition of remoxipride in different species. Species differences in metabolism.* Arzneimittelforschung, 1993. 43(3): p. 287–297.

22. Cocchiara, G., Battaglia, R., Pevarello, P. and Strolin Benedetti, M., *Comparison of the disposition and of the metabolic pattern of Reboxetine, a new antidepressant, in the rat, dog, monkey and man.* Eur J Drug Metab Pharmacokinet, 1991. 16(3): p. 231–239.

23. Sauer, J.M., Ponsler, G.D., Mattiuz, E.L., Long, A.J., Witcher, J.W., Thomasson, H.R. and Desante, K.A., *Disposition and metabolic fate of atomoxetine hydrochloride: the role of CYP2D6 in human disposition and metabolism.* Drug Metab Dispos, 2003. 31(1): p. 98–107.

24. Mattiuz, E.L., et al., *Disposition and metabolic fate of atomoxetine hydrochloride: pharmacokinetics, metabolism, and excretion in the Fischer 344 rat and beagle dog.* Drug Metab Dispos, 2003. 31(1): p. 88–97.

25. Lin, J.H., Chiba, M., Balani, S.K., Chen, I.W., Kwei, G.Y., Vastag, K.J. and Nishime, J.A., *Species differences in the pharmacokinetics and metabolism of indinavir, a potent human immunodeficiency virus protease inhibitor.* Drug Metab Dispos, 1996. 24(10): p. 1111–1120.

26. Diehl, K.H., et al., *A good practice guide to the administration of substances and removal of blood, including routes and volumes.* J Appl Toxicol, 2001. 21(1): p. 15–23.

27. Whittlesey, K.J. and Shea, L.D., *Delivery systems for small molecule drugs, proteins, and DNA: the neuroscience/biomaterial interface.* Exp Neurol, 2004. 190(1): p. 1–16.

28. Ranade, V.V., *Drug delivery systems. 4. Implants in drug delivery.* J Clin Pharmacol, 1990. 30(10): p. 871–889.

29. Sepede, G., Corbo, M., Fiori, F. and Martinotti, G., *Reboxetine in clinical practice: a review.* Clin Ter, 2012. 163(4): p. e255–e262.

12

IN VITRO APPROACHES TO ASSESS HEPATIC METABOLISM AND FIRST-PASS EFFECT

12.1 INTRODUCTION

As discussed in Chapter 8, prospective identification for the clearance rate-determining step for new molecular entities (NMEs) based on measured and calculated *in vitro* inputs (e.g., permeability, ionization, and molecular weight) is a prerequisite for using the right *in vitro* tools early on in drug discovery to successfully predict NMEs human clearance. Prospective assessments increase the accuracy of human dose calculation. This approach ensures that the drug will have a systemic exposure in clinical development that is appropriate to test the mechanism at the site of action. Extended clearance classification system (ECCS) is a system that was designed by Pfizer scientists and can proactively assign the right class for NMEs to focus on the relevant *in vitro* tools by identifying/predicting clearance rate-determining step [1]. Compounds that are ECCS Class 1a and 2 have metabolism as a rate-determining step. ECCS Class 1a are compounds that are either acids/zwitterions with high permeability and low molecular weight <400 Da. ECCS Class 2 molecules are either bases or neutral with high permeability [1]. Clearance as a rate can be calculated using *in vitro* tools such as microsomes, cytosol, S9 fraction, and hepatocyte [2]. Accurate prediction of hepatic metabolic clearance is important since metabolism is the predominant route of elimination of many drugs in the market as noted by many research groups [1, 3]. Furthermore, the selection of the *in vitro* tools will depend on the enzyme(s) driving the elimination of the molecule.

This chapter has two objectives, first to review known *in vitro* tools available to measure the metabolic clearance for NMEs; second to discuss the steps that are recommended to enable accurate prediction of clearance for compounds with metabolism as the rate-determining step.

Oral Bioavailability Assessment: Basics and Strategies for Drug Discovery and Development, First Edition. Ayman F. El-Kattan.

12.2 *IN VITRO* TOOLS AVAILABLE TO MEASURE METABOLIC CLEARANCE

Various *in vitro* tools are available at the discretion of drug metabolism and pharmacokinetics (DMPK) scientists to calculate and predict human hepatic clearance. The selection of the tool is determined by the type of enzyme that is responsible for the elimination of the molecule. Following is a discussion of these tools and enzymes expression reported with these systems.

12.2.1 Liver Microsomes

Liver microsomes preparations are considered the golden standard for weeding out NMEs with poor metabolic stability in a high-throughput mode. They are reliably prepared using differential centrifugation [4]. A key driver for liver microsomes use is attributed to their wide availability at competitive affordable prices from multiple commercial sources. This enables reliable and easy screen of significant number of NMEs. Furthermore, microsomes can be obtained either from single individuals or as a pool of subjects from either same sex or different genders. This has been used as a basis to investigate potential interindividual variability in different populations with different genetic variations backgrounds. This allows the assessment of the impact of age and sex on drug metabolism [5–7]. It should be emphasized that these preparations require the use of appropriate cofactors such as NADPH-regenerating system to ensure that the availability of these cofactors is not a limiting factor for ensuring the success of the evaluation. It is prudent to be aware of the disadvantages of these preparations to maximize the value of their utility. For example, microsomes only express metabolizing enzymes that are endoplasmic reticulum bound and these include cytochrome P450 (CYPs), flavin monooxygenases (FMO), and glucuronosyltransferases (UGTs). Therefore, metabolism that is mediated by other enzymes that are not endoplasmic reticulum bound is not appropriately assessed using these preparations. It is prudent to have good understanding of the structure–activity relationship (SAR) for various enzymes to ensure that the right *in vitro* tools are used to determine metabolic stability liabilities early on in drug discovery [4].

12.2.2 Liver S9 Fractions

Liver S9 fraction is a preparation that contains a combination of microsomal and cytosolic preparations of liver cells. It expresses most of the Phase I and Phase II enzymes. This allows a more comprehensive assessment of the metabolic profile of NMEs. Similar to other *in vitro* enzyme-based preparations, it is pivotal to include relevant cofactors at levels that allow the completion of enzymatic reactions. For example, for Phase I enzymes, NADPH-regenerating system is needed. While other cofactors are required to ensure full functioning activities of Phase II enzymes, these include but are not limited to 3′-phosphoadenosine 5′-phosphosulfate or UDP-glucuronic acid, that are used in sulfation and glucuronidation reactions, respectively. Collectively, the selection of the cofactor is determined by the question that is asked. Compared to microsomes and hepatocytes, the enzyme expression per unit basis of S9 fractions is lower. This would reduce the sensitivity of this system and its utility in defining the metabolic profile of NMEs, which is a key inherent disadvantage of this system [8, 9].

12.2.3 Supersomes

Supersomes are microsomes that are prepared from baculovirus-based systems that express functional recombinant proteins, that is, metabolizing enzymes in insect cells. These systems successfully express single metabolizing enzymes such as CYPs and UGTs. This is a key attribute of the system that is used to understand the contribution of a single enzyme to the overall metabolism of an NME [10]. Typically, they are used later in drug discovery and development when we are interested in deconvoluting the contribution of a particular enzyme to determine the potential impact of enzyme polymorphism and/or inhibition on the systemic exposure of the NMEs [11, 12].

12.2.4 Hepatocytes

Primary human hepatocytes is the main tool used in drug discovery to provide a holistic assessment of the metabolism profile of NMEs by determining their rate of metabolism, biotransformation, reaction phenotyping, and induction profiles. Hepatocytes offer a more complete physiological system relative to other liver preparations. They express all relevant Phase I and Phase II metabolizing enzymes, and reducing factors such as NADPH are continuously replenished. Primary hepatocytes are either available as cryopreserved or freshly prepared. In the last decade, major advancements were made in the methods of preparation, which ensured consistency in the quality of the preparation as noted in the high cell viability and reproducibility of the data generated. Similar to microsomes, human hepatocytes are either pooled or available from single donors, this allows DMPK scientists to understand interindividual variability driven by enzyme polymorphism or gender impact on clearance and biotransformation profiles [4].

12.3 STEPS TO MEASURE *IN VIVO* HEPATIC CLEARANCE

Hepatic metabolic clearance can be calculated by adapting the following scheme:

1. *In vitro* incubation using microsomes/S9 fraction/supersomes/hepatocytes to calculate apparent hepatic intrinsic clearance ($CL_{int,app}$)
2. Scaling to apparent *in vivo* intrinsic clearance (CL_{int})
3. Models to calculate *in vivo* hepatic blood clearance (CL_h).

The three steps to calculate human hepatic clearance using *in vitro* tools such as microsomes, S9 fraction, supersomes, and hepatocytes are discussed in the following sections.

12.3.1 *In Vitro* Incubation Using Microsomes/S9 Fraction/Supersomes/Hepatocytes to Determine Apparent Hepatic Intrinsic Clearance

There are two methodologies that are used to calculate intrinsic metabolic clearance. The first is based on measuring the formation of specific drug metabolites and the second is by measuring the disappearance of drug concentration from the incubation medium over the time course of the *in vitro* experiment. The following equation is used to calculate intrinsic metabolic clearance using the two methodologies [2]:

$$CL_{int,app} = \sum_{i}^{n} \frac{V_{max,i}}{K_{m,i}} = \left[\frac{\ln \frac{[S]_2}{[S]_1}}{t_2 - t_1}\right] \times \frac{V}{P} = \frac{0.693 \cdot V}{t_{1/2} \cdot P} \tag{12.1}$$

where V_{max} and K_m are Michaelis–Menten metabolite formation parameters for specific metabolic pathway (i), $[S]_2$ the substrate concentration measured at time t_2, V, the incubation volume; and p, the microsome protein or number of hepatocyte cells. Note that V/P ratio is

TABLE 12.1 *In Vitro* and *In Vivo* Inputs Used to Scale $CL_{int,app,scaled}$ [13, 14]

Species	Body Weight (kg)	Liver Weight per Body Weight (g/kg)	Liver Weight (g)	Microsomal Protein Per Liver Weight (mg/g)	Hepatocytes Per Liver Weight (M Cells/g)	Q_h (mL/min/kg)
Rat	0.25	40	10	45	135	70
Dog	12.5	32	400	45	240	40
Monkey	3.5	32	112	45	120	44
Human	70	20	1470	45	120	20

also defined as the amount of microsomal protein per unit gram of the liver (MPPGL) in the case of microsomes incubation or the number of hepatocytes per gram of liver (HPGL) in the case of hepatocyte incubation (Table 12.1). $t_{1/2}$ is the time that is needed for the substrate concentration to go to half of its original value. It is noteworthy to mention that Equation 12.1 calculates the apparent *in vitro* intrinsic hepatic clearance $CL_{int,app}$, since it does not take into consideration the microsomal protein binding $f_{u,mic}$. In early drug discovery, DMPK scientists do not have a solid grasp of the overall metabolic profile of the NME; therefore, measuring intrinsic hepatic clearance using drug disappearance is the method of choice for this setting.

12.3.2 Scaling to Apparent *In Vivo* Intrinsic Clearance

To scale for the apparent *in vivo* intrinsic clearance $CL_{int,app,scaled}$, which is a measure of the maximum metabolic activity of liver (microsomal protein or hepatocytes) for a drug in the absence of limiting physiological factors such as blood flow or blood protein binding, two scaling factors should be incorporated. First, HPGL/MPPGL scaling factor should be used as discussed earlier. Second, expression of the metabolic clearance per whole liver and body weights should be corrected. Collectively, to calculate $CL_{int,app,scaled}$, from microsomes and hepatocytes, S9 fraction, Equations 12.2–12.4, respectively, are used:

$$CL_{int,app,scaled} = \frac{\ln 2}{T_{1/2}(\text{min})} \cdot \frac{\text{mL incubation}}{\text{mg protein}} \cdot \frac{\text{mg protein}}{\text{g liver}} \bullet \frac{\text{g liver weight}}{\text{kg body weight}} \quad (12.2)$$

$$CL_{int,app,scaled} = \frac{\ln 2}{T_{1/2}(\text{min})} \cdot \frac{\text{mL incubation}}{\text{M cells}} \cdot \frac{\text{M cells}}{\text{g liver}} \cdot \frac{\text{g liver weight}}{\text{kg body weight}} \quad (12.3)$$

$$CL_{int,app,scaled} = \frac{\ln 2}{T_{1/2}(\text{min})} \cdot \frac{\text{mL incubation}}{\text{mg protein}} \cdot \frac{121\,\text{mg S9 protein}}{\text{g liver}} \cdot \frac{\text{g liver weight}}{\text{kg body weight}} \quad (12.4)$$

Equation 12.4 used to calculate metabolic clearance from human S9 fraction [8].

Table 12.1 summarizes MPPGL and HPGL values used in scaling these $CL_{int,app,scaled}$ for various species.

To calculate the $CL_{int,app,scaled}$ from supersomes, the following steps are carried out:

1. Measure the disappearance of the tested article over time and determine half-life using Equation 12.1.

TABLE 12.2 Enzyme Abundance and ISEF Correction Values for Known CYP Enzymes

Enzyme	Abundance (pmol/mg HLM Protein)	ISEF
CYP1A2	52	0.43
CYP2C8	24	0.43
CYP2C9	73	0.78
CYP2C19	14	0.25
CYP2D6	8	0.75
CYP3A4	137	0.24

Source: Li et al. 2014 [15]. Reproduced with permission of John Wiley & Sons.

2. $CL_{int,app}$ is calculated using Equation 12.5:

$$CL_{int,app}\ (\mu L/min/pmol)\frac{\ln 2}{T_{1/2}(min)}\cdot\frac{mL\ \ incubation}{mg\ \ protein} \tag{12.5}$$

3. $CL_{int,app,scaled}$ is calculated using Equation 12.6:

$$CL_{int,\ app,scaled}\ (mL/min/kg) = \left(\sum_i CL_{int,\ app}\cdot \text{Abundance}\ (pmol/mg\ \ HLM)\cdot \text{ISEF}\right)\cdot\frac{mg\ \ protein}{g}\cdot\frac{g}{kg} \tag{12.6}$$

The abundance and ISEF values for known CYP enzymes are depicted in Table 12.2 [15]. As for human MPPGL, it is shown in Table 12.1.

12.3.3 Models to Calculate *In Vivo* CL

There are three models that are used to calculate *in vivo* clearance and these are well-stirred, parallel-tube, and dispersion models and are discussed as follows.

12.3.3.1 Well-Stirred Model In 1971, Gillette was the first to introduce the concept of well-stirred model of hepatic drug clearance [16]. It is the simplest model to calculate human metabolic clearance, where the drug is assumed to instantaneously and homogenously distribute throughout liver interstitial water and the free drug concentration in the blood and liver interstitial water is the same. Herein, we assume 100% mixing where drug distribution in the liver is rather perfusion limited with no impact of active hepatic uptake on the distribution of the drug [17]. The drug clearance is calculated using Equation 12.7:

$$CL_h(mL/min/kg) = \frac{Q_h\cdot f_{u,blood}\cdot CL_{int,app,scaled}/f_{u,mic}}{Q_h + f_{u,blood}\cdot CL_{int,app,scaled}/f_{u,mic}} \tag{12.7}$$

where Q_h is the hepatic blood flow, $f_{u,blood}$ is the drug unbound fraction in blood, and $f_{u,mic}$ is the drug microsomal unbound fraction. $f_{u,blood}$ is calculated using Equation 12.8:

$$f_{u,blood} = \frac{f_{u,plasma}}{B/P\ \ \text{ratio}} \tag{12.8}$$

where $f_{u,plasma}$ is the fraction unbound in plasma and *B*/*P* ratio is the blood to plasma ratio of the tested article.

12.3.3.2 Parallel-Tube Model Another widely used model to calculate hepatic clearance is parallel-tube model that assumes that the liver is a group of similar cylindrical tubes arranged in parallel and that have equal velocity and intrinsic properties. Herein, the enzymes and efflux biliary transporters are evenly distributed. Unlike the well-stirred model, this system produces a concentration gradient within the liver from the portal vein to hepatic vein. The hepatic CL is calculated using Equation 12.9 [18]:

$$CL_h \ (mL/min/kg) = Q_h \left[1 - \exp^{\left(\frac{f_{u,blood} \cdot CL_{int,app,scaled}/f_{u,mic}}{Q_h} \right)} \right] \tag{12.9}$$

12.3.3.3 Dispersion Model The dispersion model is widely used to calculate human hepatic clearance and considered more flexible and representative of the human liver [17, 19, 20]. The solute distribute by the dispersion of the blood that happens by variations in both blood flow and convective forces. CL_h is calculated using Equation 12.10:

$$CL_h \ (mL/min/kg) = Q_h \left[1 - \frac{4a}{(1+a)^2 \exp^{[(a-1)/2Dn]} - (1-a)^2 \exp^{[-(a+1)/2Dn]}} \right] \tag{12.10}$$

where Dn = 0.17, $a = \sqrt{1 + 4RnDn}$ and $Rn = \frac{f_{u,blood} \cdot CL_{in,app,scaled}/f_{u,mic}}{Q_h}$.

It should be emphasized that the dispersion model becomes either the well-stirred model or parallel tube model, when the DN approaches infinity or zero, respectively [17, 19, 20].

12.3.3.4 Hepatic Extraction Ratio and f_h Calculations To calculate hepatic extraction ratio (E_h) and f_h using the three models, the following equations can be used, respectively:

$$E_h = \frac{f_{u,blood} \cdot CL_{int,app,scaled}/f_{u,mic}}{Q_h + f_{u,blood} \bullet CL_{int,app,scaled}/f_{u,mic}} \tag{12.11}$$

$$E_h = 1 - \exp^{-\left(\frac{f_{u,blood} \cdot CL_{int,app,scaled}/f_{u,mic}}{Q_h} \right)} \tag{12.12}$$

$$E_h = 1 - \exp^{-\left(\frac{f_{u,blood} \cdot CL_{int,app,scaled}/f_{u,mic}}{Q_h} + \left(\frac{f_{u,blood} \cdot CL_{int,app,scaled}/f_{u,mic}}{Q_h} \right)^2 Dn \right)} \tag{12.13}$$

The f_h of an NME can be calculated using Equation 12.4.

$$F_h = 1 - E_h \tag{12.14}$$

12.3.4 f_g Calculation

The f_g calculation was discussed in Chapter 5. The reader is encouraged to check to gain better understanding of f_g calculation and assessment.

12.4 CONCLUSION

Prediction for the contribution of metabolism to the overall clearance of NME is a prerequisite for accurate dose prediction. This prediction is particularly relevant for compounds that are ECCS Class 1a and 2, where metabolism is the rate-determining step for their clearance. It is clear that the selection of the tool and method of choice is determined by the driving clearance mechanism. For example, if UGT is the enzyme that is driving the compound elimination in such a case, hepatocyte is the method of choice. Typically, in early drug discovery settings, DMPK scientists escalate their investments, where they start with microsomes, since they are readily available. Once more interest in the lead compound is realized, DMPK scientists may opt to predict the CL using *in vitro* tools that are more resource intensive, such as hepatocyte, supersomes, or S9 fraction. This can shed light into the main enzymes responsible for the elimination of the molecule, predict drug–drug interactions liability, and impact of enzyme polymorphism on the compound disposition and elimination.

12.5 CHAPTER 12 TUTORIALS

1. Calculate rat, dog, monkey and human $CL_{int,app,scaled}$, CL_b, and extraction ratio (E_h) using parallel and well-stirred methodologies for drug x and measured in hepatocyte.

Species	$T_{1/2}$ (min)	g Liver/kg Body Weight	Cells/mL	Cells/g Liver	*B/P*	fu_p	fu_{hep}/fu_{mic}	*Q* (mL/min/kg)	$CL_{int,app}$ (μL/min/Mill)[a]	$CL_{int, app, scaled}$ (mL/min/kg)[b]	CL_b (WS) (mL/min/kg)[c]	CL_b (Parallel) (mL/min/kg)[d]	E_h
Rat	34.4	40	5.00×10^5	1.35×10^8	0.71	0.062	0.80	70					
Dog	282	32	5.00×10^5	2.4×10^8	0.68	0.162	0.83	40					
Monkey	282	32	5.00×10^5	1.2×10^8	0.68	0.162	0.83	44					
Human	100.25	20	5.00×10^5	1.2×10^8	0.7	0.091	0.69	20					

[a]Calculated from Equation 12.1.
[b]Calculated from Equation 12.3.
[c]Calculated from Equation 12.7.
[d]Calculated from Equation 12.9.

12.6 CHAPTER 12 TUTORIAL ANSWER KEY

1. Calculate rat, dog, monkey and human $CL_{int,app,scaled}$, CL_b, and extraction ratio (E_h) using parallel and well-stirred methodologies for drug X and measured in hepatocyte.

Species	$T_{1/2}$ (min)	g Liver/kg Body Weight	Cells/mL	Cells/g Liver	*B/P*	fu_p	fu_{hep}/fu_{mic}	*Q* (mL/min/kg)	$CL_{int,app}$ (μL/min/Mill)[a]	$CL_{int, app, scaled}$ (mL/min/kg)[b]	CL_b (WS) (mL/min/kg)[c]	CL_b (Parallel) (mL/min/kg)[d]	E_h
Rat	34.4	40	5.00×10^5	1.35×10^8	0.71	0.062	0.80	70	40.3	217.6	17.74	20.1	0.25
Dog	282	32	5.00×10^5	2.4×10^8	0.68	0.162	0.83	40	4.9	37.8	8.53	9.5	0.21
Monkey	282	32	5.00×10^5	1.2×10^8	0.68	0.162	0.83	44	4.9	18.9	4.82	5.1	0.11
Human	100.25	20	5.00×10^5	1.2×10^8	0.7	0.091	0.69	20	13.8	33.2	4.76	5.4	0.24

[a]Calculated from Equation 12.1.
[b]Calculated from Equation 12.3.
[c]Calculated from Equation 12.7.
[d]Calculated from Equation 12.9.

REFERENCES

1. Varma, M.V., Steyn, S.J., Allerton, C. and El-Kattan, A.F., *Predicting clearance mechanism in drug discovery: extended clearance classification system (ECCS)*. Pharm Res, 2015. 32(12): p. 3785–3802.
2. Houston, J.B. and Galetin, A., *Methods for predicting* in vivo *pharmacokinetics using data from* in vitro *assays*. Curr Drug Metab, 2008. 9(9): p. 940–951.
3. Williams, J.A., Hyland, R., Jones, B.C., Smith, D.A., Hurst, S., Goosen, T.C., Peterkin, V., Koup, J.R. and Ball, S.E., *Drug–drug interactions for UDP-glucuronosyltransferase substrates: a pharmacokinetic explanation for typically observed low exposure (AUCi/AUC) ratios*. Drug Metab Dispos, 2004. 32(11): p. 1201–1208.
4. Plant, N., *Strategies for using* in vitro *screens in drug metabolism*. Drug Discov Today, 2004. 9(7): p. 328–336.
5. Sorensen, L.B., Sorensen, R.N., Miners, J.O., Somogyi, A.A., Grgurinovich, N. and Birkett, D.J., *Polymorphic hydroxylation of perhexiline* in vitro. Br J Clin Pharmacol, 2003. 55(6): p. 635–638.
6. Court, M.H., Krishnaswamy, S., Hao, Q., Duan, S.X., Patten, C.J., Von Moltke, L.L. and Greenblatt, D.J., *Evaluation of 3'-azido-3'-deoxythymidine, morphine, and codeine as probe substrates for UDP-glucuronosyltransferase 2B7 (UGT2B7) in human liver microsomes: specificity and influence of the UGT2B7*2 polymorphism*. Drug Metab Dispos, 2003. 31(9): p. 1125–1133.
7. Springer, D., Paul, L.D., Staack, R.F., Kraemer, T. and Maurer, H.H., *Identification of cytochrome p450 enzymes involved in the metabolism of 4'-methyl-alpha-pyrrolidinopropiophenone, a novel scheduled designer drug, in human liver microsomes*. Drug Metab Dispos, 2003. 31(8): p. 979–982.
8. Nishimuta, H., Houston, J.B. and Galetin, A., *Hepatic, intestinal, renal, and plasma hydrolysis of prodrugs in human, cynomolgus monkey, dog, and rat: implications for* in vitro–in vivo *extrapolation of clearance of prodrugs*. Drug Metab Dispos, 2014. 42(9): p. 1522–1531.
9. Nagilla, R., Frank, K.A., Jolivette, L.J. and Ward, K.W., *Investigation of the utility of published* in vitro *intrinsic clearance data for prediction of* in vivo *clearance*. J Pharmacol Toxicol Methods, 2006. 53(2): p. 106–116.
10. Roy, P., Yu, L.J., Crespi, C.L. and Waxman, D.J., *Development of a substrate-activity based approach to identify the major human liver P-450 catalysts of cyclophosphamide and ifosfamide activation based on cDNA-expressed activities and liver microsomal P-450 profiles*. Drug Metab Dispos, 1999. 27(6): p. 655–666.
11. Afshar, M. and Thormann, W., *Capillary electrophoretic investigation of the enantioselective metabolism of propafenone by human cytochrome P-450 SUPERSOMES: evidence for atypical kinetics by CYP2D6 and CYP3A4*. Electrophoresis, 2006. 27(8): p. 1526–1536.
12. McGinnity, D.F., Griffin, S.J., Moody, G.C., Voice, M., Hanlon, S., Friedberg, T. and Riley, R.J., *Rapid characterization of the major drug-metabolizing human hepatic cytochrome P-450 enzymes expressed in* Escherichia coli. Drug Metab Dispos, 1999. 27(9): p. 1017–1023.
13. Di, L., *The role of drug metabolizing enzymes in clearance*. Expert Opin Drug Metab Toxicol, 2014. 10(3): p. 379–393.
14. Davies, B. and Morris, T., *Physiological parameters in laboratory animals and humans*. Pharm Res, 1993. 10(7): p. 1093–1095.
15. Li, X., Suhar, T., Glass, L. and Rajaraman, G., *A high-throughput (HTS) assay for enzyme reaction phenotyping in human recombinant P450 enzymes using LC–MS/MS*. Curr Protoc Pharmacol, 2014. 9: p. 9.18.1–9.18.10.
16. Gillette, J.R., *Factors affecting drug metabolism*. Ann N Y Acad Sci, 1971. 179: p. 43–66.
17. Liu, L. and Pang, K.S., *An integrated approach to model hepatic drug clearance*. Eur J Pharm Sci, 2006. 29(3–4): p. 215–230.
18. Pang, K.S. and Rowland, M., *Hepatic clearance of drugs. I. Theoretical considerations of a "well-stirred" model and a "parallel tube" model. Influence of hepatic blood flow, plasma and blood cell binding, and the hepatocellular enzymatic activity on hepatic drug clearance*. J Pharmacokinet Biopharm, 1977. 5(6): p. 625–653.

19. Yamamoto, T., Itoga, H., Kohno, Y., Nagata, K. and Yamazoe, Y., *Prediction of oral clearance from* in vitro *metabolic data using recombinant CYPs: comparison among well-stirred, parallel-tube, distributed and dispersion models*. Xenobiotica, 2005. 35(6): p. 627–646.

20. Ito, K. and Houston, J.B., *Comparison of the use of liver models for predicting drug clearance using* in vitro *kinetic data from hepatic microsomes and isolated hepatocytes*. Pharm Res, 2004. 21(5): p. 785–792.

13

THE UTILITY OF ECCS AS A ROADMAP TO IMPROVE ORAL BIOAVAILABILITY OF NEW MOLECULAR ENTITIES: INDUSTRIAL PERSPECTIVE

13.1 INTRODUCTION

Oral bioavailability is a product of the fraction of the dose that is absorbed following oral dosing (f_a), the fraction that escapes intestinal (f_g), and hepatic first-pass effects (f_h). Drug permeability, solubility, and affinity to influx and efflux transporters affect drug f_a, while drug affinity to uptake transporters, efflux transporters, and metabolizing enzymes lining enterocyte and expressed in the liver determine its f_g and f_h, respectively. Poor oral bioavailability (F) is a leading cause for compound failure in drug discovery and development. New molecular entities (NMEs) with poor F tend to have low plasma exposure and higher inter- as well as intraindividual variability. Combined with short half-life, this would limit the therapeutic usefulness and increase side effects specially for compounds with low therapeutic index [1–3]. To ensure successful discovery and development of viable drug candidates that have optimum drug systemic exposure capable of testing the mechanism at the site of action, it is prudent to optimize the physicochemical properties and pharmacological activity of NMEs in parallel [4]. Indeed, many drug metabolic and pharmacokinetics (DMPK) departments use high-throughput *in vitro* screens to measure physicochemical properties and *in vitro* kinetics parameters of interest such as log $D_{\mathrm{pH}\,7.4}$, polar surface area (PSA), molecular weight (M_{wt}), intrinsic clearance measured in human liver microsomes (HLM) and hepatocytes, active uptake in transporter-transfected cells, permeability, plasma protein, and blood binding, to name a few. Some of these assays are typically run to all compounds synthesized since they are used by medicinal chemists to optimize properties such as solubility, permeability, and metabolic clearance to identify best-in-class molecules capable of testing the mechanism, while sound, without identifying the driver of poor absorption and high intestinal and hepatic first-pass effect, molecular design may be optimized for *in vitro* parameters that are not relevant to the f_a or systemic clearance of a compound. For example, optimizing the compound affinity and interaction with P-glycoprotein transporter

---

Oral Bioavailability Assessment: Basics and Strategies for Drug Discovery and Development, First Edition. Ayman F. El-Kattan.

(ABCB1) (P-gp) while it has high permeability, solubility, and projected human oral dose (>100 mg) may prove inappropriate use of resources. In addition, optimizing the metabolic clearance for compounds that have active hepatic uptake mediated by organic anion transporting polypeptide 1B1/1B3 (OATP1B1 and OATP1B3) as the rate-determining step of their clearance is not consistent with the principles of extended clearance concept [4]. In addition, DMPK departments' research and development budget for many leading companies is shrinking. A strategic approach is capable of streamlining the available resources and prioritizing the implementation of *in vitro* assays and preclinical pharmacokinetics screens that can affect the compound permeability and solubility and define the rate-determining step of compound clearance is needed. As such, an approach such as this would play a key role in increasing efficiency of medicinal chemistry design by focusing on relevant drivers for compound absorption and elimination. This approach would enable precision screening to build structure–activity relationships (SARs) for relevant metabolizing enzymes and transporters to weed out compounds with poor profiles. Once lead compound is identified, more rigor *in vitro* and physicochemical inputs are used to build physiologically based pharmacokinetic (PBPK) models that together with pharmacodynamics inputs obtained from relevant *in vitro* screens or preclinical biology models will enable successful dose and pharmacokinetic profile prediction in human. Furthermore, this will function as a basis for predicting drug–drug interactions, food–drug interactions, and impact metabolizing enzymes and transporters pharmacogenomics on drug exposure and efficacy. In 2015, our group proposed a new classification system, the extended clearance classification system (ECCS), that can be used to identify the rate-determining step of drug clearance. The model was validated using a data set of 315 molecules [4]. The rate-determining step is identified using readily available inputs such as permeability, molecular weight, and ionization and can be used as a basis for prioritizing screens needed to predict clearance for NMEs and optimize the relevant parameters accordingly. Herein, we discuss the utility of ECCS as a roadmap to predict the drivers of drug oral bioavailability.

13.2 ECCS CLASSIFICATION

The rate-determining step of NMEs clearance can be prospectively predicted using ECCS classification. NMEs are classified based on their physiochemical properties (e.g., ionization state, molecular weight, and membrane permeability) and follow six classes:

1. ECCS Class 1A: These are acids/zwitterions NMEs that have high permeability and small molecular weight (<400 Da). ECCS Class 1A rate-determining step of these molecules is metabolism.
2. ECCS Class 1B: These are acids/zwitterions NMEs that have high permeability and large molecular weight (≥400 Da). ECCS Class 1B rate-determining step of these molecules is active hepatic uptake that is mediated by organic anion transporting polypeptide 1B1/1B3. They have high extent of metabolism >70% similar to Class 1A and 2.
3. ECCS Class 2: These molecules are either bases/neutrals. They have high membrane permeability and metabolism is the rate-determining step of their clearance mechanism.
4. ECCS Class 3A: These are acids/zwitterions NMEs that have low permeability and small molecular weight (<400 Da). ECCS Class 3A rate-determining step of these molecules is renal elimination.
5. ECCS Class 3B: These are acids/zwitterions NMEs that have low permeability and large molecular weight (≥400 Da). ECCS Class 3B rate-determining step of these molecules is active hepatic uptake that is mediated by organic anion transporting

polypeptide 1B1/1B3 and/or renal elimination. They have high extent of metabolism <30% similar to Class 3A and 4.

6. ECCS Class 4: These molecules are either bases/neutrals. They have low membrane permeability and renal elimination is the rate-determining step of their clearance mechanism.

Chapter 8 has more detailed description on the ECCS.

13.2.1 f_a *and ECCS*

Absorption is a process that describes the dynamic transfer of a drug molecule from the site of administration following oral dosing, the gastrointestinal tract (GIT), across the enterocyte into mesenteric arteries that pool into the portal vein [2]. Figure 13.1 illustrates the relationship between membrane permeability measured or calculated from LE-MDCK cell lines and human f_a of compounds classified based on their ECCS classification. Drug molecules with membrane permeability $>5\times10^{-6}$ cm/s tend to have an f_a higher than 85% and vice versa [5]. The average f_a for ECCS Classes 1A, 1B, and 2 were 92%, 92%, and 94%, respectively (Figure 13.2). It is interesting to note that from the set of drug molecules investigated and classified as high-permeability molecules (ECCS 1A/1B/2), 18 had reports that indicate that they are substrates for efflux transporters such as P-gp and breast cancer resistance protein (BCRP) transporters. Only 4 out of the 18 molecules had an $f_a<85\%$ and they are methylprednisolone ($f_a<82\%$), losartan ($f_a<80\%$), carvedilol ($f_a<65\%$), and lovastatin ($f_a<31\%$). Intestinal efflux transporters line the apical membrane of enterocytes. They play a role in limiting the oral absorption of drug molecules. Tochibana et al. reported a sensitivity analysis that investigated the key variables that determine the role of P-gp on drug absorption as a function of oral dose [6]. The group indicated that the absorption of P-gp substrate is typically influenced by its effective intestinal permeability, solubility, dose, and affinity to P-gp (K_m) [7]. P-gp would limit the oral absorption of drug molecules if the compound has high oral dose, low solubility, and/or slow dissolution or has a high affinity to P-gp (low K_m value) where the transporter would not be saturated at therapeutically relevant doses. Therefore, ECCS Class 1A/1B/2 molecules that are substrates for P-gp or BCRP may still have a limited absorption if part of the criteria above is met.

Unlike high-permeability drug molecules, those with low permeability such as ECCS Classes 3A, 3B, and 4 have an average f_a of 80%, 52%, and 69%, respectively (Figure 13.2). This result suggests that the impact of intestinal efflux transporters, for example, BCRP, P-gp, and multidrug resistance protein 2 (MRP2), is more substantial [6]. Pharmacogenomics findings for intestinal efflux transporters and intestinal drug interactions for low-permeability molecules substantiated these findings. For example, rosuvastatin and pravastatin are substrates for BCRP and MRP2 transporters, respectively. Pharmacogenomics with these transporters had a significant impact on these compounds systemic exposure [8, 9]. In addition, the absorption of drug molecules that are of ECCS Class 4 tends to be significantly affected by intestinal P-gp drug interaction, since it is associated with changes in plasma exposure with no major alteration in half-life, suggesting minimal impact on clearance and significant role of P-gp in limiting the absorption of these molecules as reported with digoxin [10].

In addition, ECCS Class 3A, 3B, and 4 drug molecules absorption is affected by uptake intestinal transporters such as OATP2B1, PEPT1, and System L transporters. These influx transporters line the apical membrane of intestinal enterocytes and are divided into high-affinity/low-capacity transporters (e.g., system L and OATP2B1) and low-affinity/high-capacity transporters (e.g., PEPT1 and monocarboxylate transporter 1) [3, 11, 12]. High-affinity/low-capacity intestinal uptake transporters would improve the

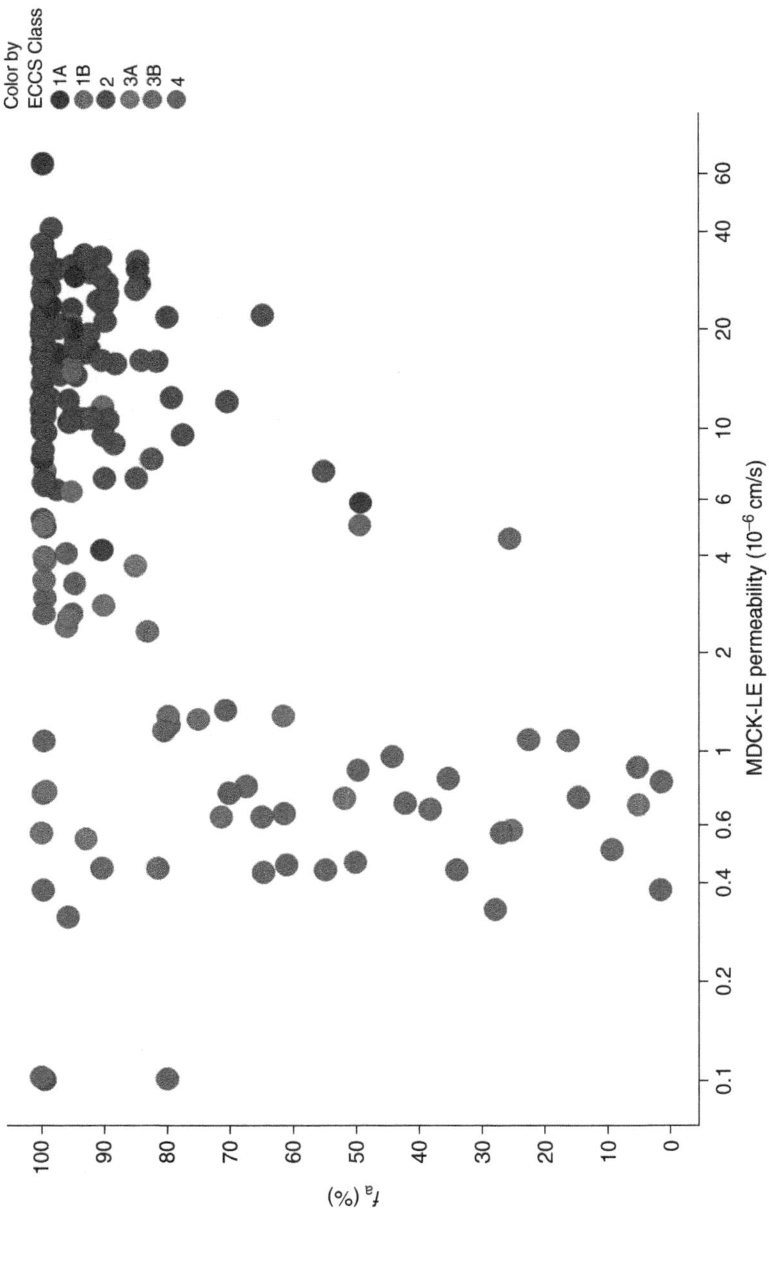

FIGURE 13.1 The relationship between LE-MDCK permeability versus human f_a for set of compounds classified based on ECCS. *(See color plate section for the color representation of this figure.)*

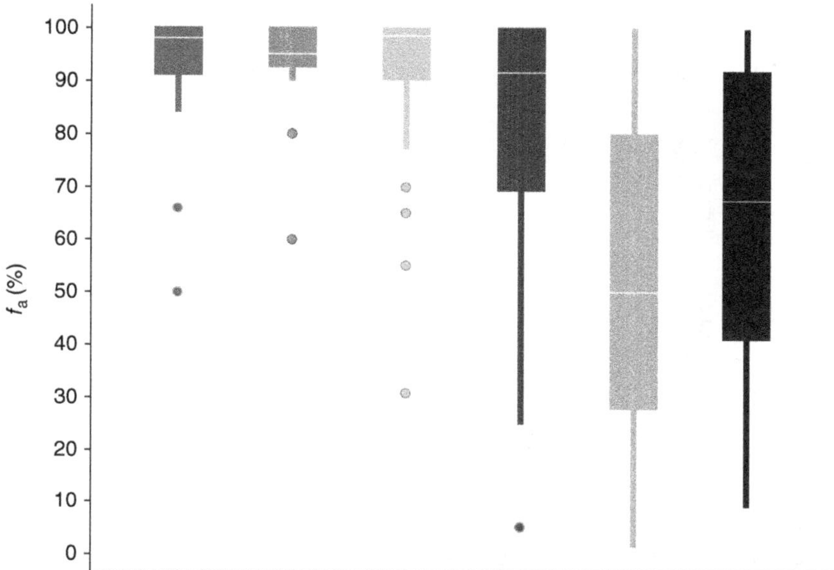

	1A	1B	2	3A	3B	4
Count	19	11	114	20	21	26
Median	98%	95%	99%	92%	50%	68%
Avg	92%	92%	94%	80%	52%	65%

ECCS Class

FIGURE 13.2 The median and average of f_a for compounds classified based on their ECCS.

oral absorption of low-permeability and low oral dose drug molecules (<100 mg) [12]. For example, gabapentin is an ECCS Class 3A molecule that is substrate for System L transporter (high-affinity/low-capacity transporter). Increasing oral dose of gabapentin is associated with lower than proportional increase in systemic exposure in humans [13]. Many penicillins and cephalosporins are substrates for PEPT1 transporter (low-affinity/high-capacity transporter) [12]. These are low-permeability/high oral dose molecules (oral dose > 1 g/day). From the drug molecules investigated, the average f_a for PEPT1 substrates is 70%. This result is not expected based on an average passive membrane permeability of 0.7×10^{-6} cm/s (Table 13.1).

Solubility is also an important factor that should be considered when developing an orally bioavailable molecule. Our group recommended the measurement of the equilibrium solubility of NMEs in either pH 1.2 medium or fasted state simulated intestinal fluid (FaSSIF) medium (pH 6.5) [5], where solubility at pH 1.2 medium is assessed for acidic compounds and FaSSIF medium solubility is used for basic, neutral, and zwitterionic compounds. Using a cutoff of 200 µg/mL, the data set suggested a 93% sensitivity and 86% specificity in predicting both the biopharmaceutics drug disposition classification system (BDDCS) solubility classes [5]. Therefore, NMEs with solubility in relevant matrices higher than 200 µg/mL are considered high solubility molecules and vice versa. These molecules should have low likelihood of solubility-limited absorption.

13.2.2 f_g *and ECCS*

f_g is an important factor that should be investigated when assessing drug bioavailability, since it measures the impact of intestinal first-pass effect on drug bioavailability. For compounds

TABLE 13.1 Molecular Weight, Ionization, ECCS Classification, CL_t, CL_h, f_a, f_g, f_h, f, Main Enzymes, Uptake Transporters, and Efflux Transporters Affecting the Disposition of Molecules [4, 14]

Name	M_{wt}	Ionization	ECCS Class	CL_t (mL/min/kg)	CL_h (mL/min/kg)	f_a	f_g	f_h	f	Main Enzymes	Main Uptake Transporters	Main Efflux Transporters	References
Acenocoumarol	353.3	Acid	1A							CYP2C			[15]
Acetylsalicylic acid	180.2	Acid	1A	9.30	9.2	0.84	1.00	0.54	0.68	Esterase			[16]
Aminosalicylic acid	153.1	Acid	1A	9.30									
Azapropazone	300.4	Acid	1A	0.14	0.1								
Bromfenac	334.2	Acid	1A	1.30	1.3	1.00	0.71	0.94	0.67	UGT			[17]
Clinafloxacin	365.8	Zwitterion	1A	4.70	1.5	1.00	0.96	0.93	0.90				
Dichloroacetic acid	128.9	Acid	1A										
Diclofenac	296.1	Acid	1A	3.50	3.5	1.00	0.64	0.85	0.54			BCRP	[18]
Diflunisal	250.2	Acid	1A	0.10	0.1					UGT			[19]
Enalapril	376	Zwitterion	1A	4.90	4.4	0.66	0.80	0.78	0.41	Esterase			[a]
Entacapone	305.3	Acid	1A	11.70	5.3	0.50	0.65	0.77	0.25	UGT			[a]
Fenoprofen	242.3	Acid	1A							UGT			[19]
Fluorescein	332.3	Acid	1A	1.30									
Gemfibrozil	250.34	Acid	1A	1.70	1.7				0.98	UGT			[a]+[20]
Gliclazide	323.4	Acid	1A			0.97				CYP2C			[21]
Hexobarbital	236.3	Acid	1A	3.40	3.4					CYP2C			[22]
Ibuprofen	206.0	Acid	1A	0.82	0.8	0.95	1.04	0.96	0.95	CYP2C			[a]
Isoxicam	335.3	Acid	1A	0.07	0.1	1.00	0.97	1.00	0.97				
Ketoprofen	254.3	Acid	1A	1.60	1.1	0.92	0.97	0.95	0.85	UGT			[19]
Meloxicam	351.4	Acid	1A	0.11	0.1				0.97	CYP2C			[b]
Milrinone	211.2	Acid	1A	6.20	1.2	1.00	0.97	0.95	0.92				
Naproxen	230.7	Acid	1A	0.13	0.1	0.99	1.00	0.99	0.99	UGT			[19]
Nateglinide	317.4	Acid	1A	1.80	1.7	0.90	0.86	0.93	0.72	CYP2C			[a]
Perindopril	368.48	Zwitterion	1A	4.84	4.5				0.75	Esterase			[a]

Pioglitazone	356.45	Acid	1A	1.43	1.4					CYP2C			[a]
Piroxicam	331.35	Acid	1A							CYP2C			[a]
Pralidoxime	137.2	Acid	1A										
Rosiglitazone	357.4	Zwitterion	1A	0.68	0.7	1.00	1.00	0.97	0.99	CYP2C			[a]
Sinitrodil	238.2	Acid	1A										
Sulfamethoxazole	253.3	Acid	1A	0.36	0.3	1.00	1.01	0.99	1.00				
Thalidomide	258.2	Acid	1A	2.20	2.2								
Thiopental	242.3	Acid	1A	3.90	3.9								
Tiagabine	375.5	Zwitterion	1A	1.60	1.6	0.95	1.02	0.93	0.90	CYP3A4			[a]
Tolbutamide	270.3	Acid	1A	0.21	0.2	0.85	1.01	0.99	0.85				
Tolcapone	273.2	Acid	1A	1.50	1.5				0.60	UGT			[a]
Valproic acid	144.2	Acid	1A	0.16	0.2	1.00	0.91	0.99	0.90	UGT			[a]
Warfarin	308.3	Acid	1A	0.06	0.1	0.98	0.97	1.00	0.95	CYP2C			[b]
Atorvastatin	558.7	Acid	1B						0.12	CYP3A4	OATP	BCRP + P-gp	[a]
Benazepril	424.5	Acid	1B	0.35	0.3			0.98	0.18	Esterase			[a]
Bosentan	551.6	Acid	1B	2.10	2.1					CYP3A4	OATP	MRP2	[23–25]
Candesartan cilexetil	610.68	Acid	1B							Esterase			[b]
Cerivastatin	459.6	Acid	1B	2.90	2.9	1.00	0.69	0.87	0.60	CYP2C	OATP	BCRP	[23, 24]
Dronedarone	556.77	Zwitterion	1B							CYP3A4			[a]
Fluvastatin	411.5	Acid	1B	16.00	16.0	1.00	0.79	0.30	0.24	CYP2C	OATP	BCRP	[26]
Fosinopril	563.68	Acid	1B							Esterase		MRP2	[a]+ [27]
Glimepiride	490.6	Acid	1B	0.50	0.5	1.00	1.02	0.98	1.00	CYP2C	OATP		[a]+ [28, 29]
Glipizide	445.5	Acid	1B	0.56	0.5	1.00	0.97	0.98	0.95				
Glyburide	494.0	Acid	1B	0.82	0.8	1.00	0.99	0.96	0.95	CYP2C	OATP	BCRP + P-gp	[a]+ [30]
Irbesartan	428.5	Acid	1B	2.12	2.1				0.70	UGT	OATP		[a]+ [31]
Levocabastine	420.5	Zwitterion	1B	1.80	0.5				1.00	UGT	OATP		[32]
Losartan	422.9	Acid	1B	8.20	7.3	0.80	0.66	0.68	0.36	CYP2C		P-gp	[a]+ [33]
Olmesartan medoxomil	558.22	Acid	1B							Esterase			[34]
Pitavastatin	421.5	Acid	1B	1.40					0.51		OATP	BCRP	[a]+ [35]
Ramipril	416.5	Zwitterion	1B	23.00	22.7	0.60			0.28	Esterase			[a]
Repaglinide	452.6	Zwitterion	1B	7.80	7.7	0.95	0.99	0.67	0.63	CYP2C	OATP		[a]+ [36, 37]

(*continued*)

TABLE 13.1 (*Continued*)

Name	M_{wt}	Ionization	ECCS Class	CL_t (mL/min/kg)	CL_h (mL/min/kg)	f_a	f_g	f_h	f	Main Enzymes	Main Uptake Transporters	Main Efflux Transporters	References
Sitafloxacin	409.8	Zwitterion	1B	3.70	0.9	0.95	0.97	0.96	0.89	UGT			[38]
Telmisartan	514.6	Acid	1B	8.40	8.3	0.90	0.75	0.64	0.43	UGT	OATP		[39]
Trovafloxacin	416.4	Zwitterion	1B	1.40	1.3	0.95	0.98	0.95	0.88		OATP		[40]
Acebutolol	336.429	Base	2	10.00	7.3	0.90	0.82	0.68	0.50			P-gp	[a]
Acetaminophen	151.2	Neutral	2	5.00	4.9	1.00	1.04	0.79	0.82	UGT			[a]
Adinazolam	351.8	Base	2	6.20	3.6	1.00	0.46	0.84	0.39	CYP3A4			[41]
Alfentanil	416.5	Base	2	3.90	3.9	1.00	0.52	0.83	0.43	CYP3A4			[42]
Alfuzosin	389.4	Base	2	5.90	5.3					CYP3A4			[b]
Alprazolam	308.8	Neutral	2	0.74	0.6	0.92	0.98	0.97	0.88	CYP3A4			[a]
Alprenolol	249.4	Base	2	15.00	14.4	0.93	0.23	0.37	0.08				
Amantadine	151.2	Base	2	4.80	0.0	0.95	0.95	1.00	0.90				[a]
Amiodarone	645.3	Base	2	1.90	1.9	1.00	0.71	0.92	0.65				[b]
Amitriptyline	277.4	Base	2	6.10	6.0	0.95	0.68	0.74	0.48				[a]
Amlodipine	408.9	Base	2	7.00	6.3	0.95	0.93	0.73	0.64	CYP3A4		P-gp	[a]+ [43]
Amphetamine	135.2	Base	2	9.70	5.8								
Antipyrine	188.2	Neutral	2	0.64	0.6	0.97	1.03	0.97	0.97				[44]
Aripiprazole	448.4	Base	2	0.83	0.8				0.87				[a]
Atomoxetine	255.4	Base	2	9.30	9.0	1.00	1.04	0.61	0.63	CYP2D6			[b]
Atropine	289.4	Base	2	7.60	3.3	0.98	0.59	0.86	0.50				
Azelastine	456.5	Base	2	9.00	8.8								
Bepridil	366.5	Base	2										
Betaxolol	307.4	Base	2	3.40	2.7	1.00	1.01	0.88	0.89	CYP2D6			[a]+ [45]
Bevantolol	345.4	Base	2	5.50	5.1								
Biperiden	311.5	Base	2	12.00	12.0	1.00	0.69	0.48	0.33				
Bromazepam	316.2	Neutral	2	0.64	0.64	0.84	0.80	0.97	0.65				[a]
Bromopride	391.2	Base	2	13.00									
Brotizolam	393.7	Neutral	2	2.00						CYP3A4			[46]

Budesonide	430.5	Neutral	2	20.00	19.0	1.00	0.63	0.17	0.11	CYP3A4	P-gp	a
Buflomedil	307.4	Base	2	5.60	4.3	1.00	0.92	0.81	0.75			
Bunazosin	373.4	Base	2	4.80								
Buprenorphine	467.6	Base	2	13.30	13.2				0.28	CYP3A4		a
Butorphanol	327.5	Base	2	40.00	39.2							[44]
Caffeine	194.2	Neutral	2	1.40	1.4	1.00	1.04	0.94	0.98	CYP1A2		[47]
Canagliflozin	453.53	Neutral	2	2.74					0.65	UGT	P-gp	a
Carvedilol	406.5	Base	2	7.80	7.7	0.65	0.58	0.67	0.25	CYP2D6	P-gp	a
Chloramphenicol	323.1	Neutral	2	2.40	2.2	0.90	0.98	0.91	0.80	UGT		[48]
Cibenzoline	262.3	Base	2	8.60	4.4	1.00	1.03	0.81	0.83			[49]
Ciclesonide	540.7	Neutral	2	36.19	36.2				0.01	Esterase		
Cilostazol	369.46	Base	2							CYP3A4		[50]
Clomipramine	314.9	Base	2	8.20								a
Clonazepam	315.7	Neutral	2	0.88	0.9	0.98	1.04	0.96	0.98	CYP3A4		a
Clopidogrel	321.826	Base	2									
Clozapine	326.8	Base	2	2.50	2.5	0.55	0.55	0.89	0.27	CYP1A2		a
Cocaine	303.4	Base	2	32.00	31.7					CYP3A4		[51]
Codeine	299.37	Base	2	11.00	11.0	0.95	1.00	0.45	0.50			a
Conivaptan	498.59	Neutral	2	3.00	3.0	1.00	0.14	0.87	0.12	CYP3A4	P-gp	a
Cotinine	176.2	Neutral	2	0.89	0.8	0.90	1.04	0.97	0.90	UGT		[52, 53]
Dapagliflozin	408.8755	Neutral	2	1.42					0.78	UGT		a
Darifenacin	426.6	Base	2	12.00	10.5							a
Dexamethasone	392.5	Neutral	2	3.30	3.2	0.90	1.03	0.86	0.80	CYP3A4		
Dexfenfluramine	231.3	Base	2	11.00								
Dexmedetomidine	200.3	Base	2	0.51								
Diazepam	284.7	Neutral	2	0.38	0.4	1.00	1.02	0.98	1.00			a
Dilevalol	328.4	Base	2	29.00	28.3							
Diltiazem	414.5	Base	2	13.00	3.8	0.90	0.50	0.84	0.38	CYP3A4	P-gp	a
Diphenhydramine	255.4	Base	2	6.20	6.1				0.72	CYP2D6		a
Dolasetron	324.4	Base	2	11.50	11.5				0.76			a
Domperidone	425.9	Base	2	9.50	9.5	0.93	0.26	0.59	0.14	CYP3A4	P-gp	[50, 54]
Doxazosin	451.5	Base	2	2.10					0.68			

(continued)

TABLE 13.1 (*Continued*)

Name	M_{wt}	Ionization	ECCS Class	CL_t (mL/min/kg)	CL_h (mL/min/kg)	f_a	f_g	f_h	f	Main Enzymes	Main Uptake Transporters	Main Efflux Transporters	References
Doxepin	279.4	Base	2	14.00	14.0				0.30	CYP2C			[a]
Drotaverine	397.5	Neutral	2	3.50	3.5	1.00	0.69	0.85	0.58				
Eletriptan	382.5	Base	2	5.69	5.4				0.50	CYP3A4		P-gp	[a]+ [43]
Encainide	352.5	Base	2	13.00	12.2	0.95	0.67	0.47	0.30				
Enprofylline	194.2	Neutral	2										
Eplerenone	414.5	Neutral	2	3.26	3.2				0.69	CYP3A4			[a]
Erlotinib	393.4	Neutral	2	1.21	1.2				0.60	CYP3A4		BCRP + P-gp	[55, 56]
Esmolol	295.3	Base	2	170.00	169.2					Esterase			[a]
Estradiol	272.4	Neutral	2	30.00	29.9								
Etomidate	244.3	Neutral	2	13.00	12.7								
Etoricoxib	358.8	Neutral	2	0.78	0.8				1.00	CYP3A4			[57]
Ezetimibe	409.44	Base	2	0.60	0.6					UGT		MRP2 + P-gp	[a]
Felodipine	384.3	Neutral	2	11.00	11.0	0.88	0.35	0.52	0.16	CYP3A4			[a]
Fenofibrate	360.835	Neutral	2	0.35	0.4					Esterase			[a]
Fenoximone	248.3	Neutral	2	29.00									
Fentanyl	336.5	Base	2	13.00	12.0				0.50	CYP3A4		P-gp	[a]
Flecainide	414.4	Base	2	4.90	2.8	0.81	0.99	0.88	0.70	CYP2D6			[58]
Fleroxacin	369.3	Base	2	2.40	0.9	1.00	1.04	0.96	1.00				
Fluconazole	306.3	Neutral	2	0.31	0.1	0.95	0.95	1.00	0.90				[a]
Flumazenil	303.3	Neutral	2	16.00	16.0	0.95	0.55	0.31	0.16				
Flunitrazepam	313.3	Neutral	2	1.40	1.4					CYP3A4			[59]
Galanthamine	287.4	Base	2	5.60	4.5								
Gefitinib	446.9	Base	2	7.97	7.8				0.60	CYP3A4		BCRP + P-gp	[60]
Granisetron	312.4	Base	2	9.10	8.1	1.00	0.99	0.65	0.64				[a]
Haloperidol	375.9	Base	2	7.80	7.7	1.00	0.90	0.66	0.60				[a]
Hydrocodone	299.37	Base	2	11.10									[a]
Hydromorphone	285.3	Base	2	14.60	13.7				0.42				[a]

Ibutilide	384.6	Base	2	26.00	24.2							
Ifosfamide	261.1	Base	2	1.10	0.5	1.00	0.94	0.98	0.92	CYP3A4		[a]
Imatinib	493.6	Base	2	3.30	3.1				0.98	CYP3A4	BCRP + P-gp	[a]
Imipramine	280.4	Base	2	13.00	12.8	1.00	1.00	0.36	0.42			[a]
Indinavir	613.8	Neutral	2	18.20					0.65	CYP3A4	P-gp	[a]
Irofulven	246.3	Neutral	2	140.00								
Isomazole	287.3	Base	2									
Isosorbide dinitrate	236.1	Neutral	2	46.00					0.22			
Isosorbide-5-mononitrate	191.1	Neutral	2			1.00						
Isradipine	371.4	Neutral	2	10.00	10.0				0.19			
Ketanserin	395.4	Base	2	6.70	6.6	1.00	0.70	0.71	0.50			
Ketobemidone	247.3	Base	2	24.29						CYP2C + CYP3A4		[61]
Labetalol	328.4	Base	2	25.00	24.4	0.95			0.18	UGT		[a]
Lansoprazole	369	Neutral	2	4.40	4.4	1.00	1.05	0.81	0.85	CYP2C		[a]
Letrozole	285.3	Neutral	2	0.57	0.6	1.00	1.02	0.98	1.00	UGT		[b]
Levorphanol	443.5	Base	2							UGT		[a]
Lidocaine	234.3	Base	2	9.20	8.5				0.35			
Linezolid	337.3	Neutral	2	1.80	1.2	1.00	1.05	0.95	1.00			
Lorazepam	321.2	Neutral	2	1.00	1.0	1.00	0.97	0.96	0.93	UGT		[a]
Lorcainide	370.9	Base	2	16.00	15.7	0.70	0.13	0.32	0.03			
Lormetazepam	335.2	Neutral	2	4.00	4.0	1.00	0.97	0.83	0.80			
Lovastatin	404.5	Neutral	2	7.20	6.5	0.31	0.22	0.72	0.05	CYP3A4	P-gp	[a]
Mebendazole	295.3	Neutral	2	15.00	14.8	1.00	0.62	0.36	0.22	Esterase		[62]
Medroxalol	372.4	Base	2	16.00								
Meperidine	247.34	Base	2	0.12	0.1	0.90	1.01	0.99	0.90			[a]
Mepivacaine	246.3	Base	2	6.80	6.5							
Metazosin	375.4	Base	2									
Methadone	309.46	Base	2	1.70	1.3	1.00	0.97	0.94	0.92			[a]
Methylphenidate	233.3	Base	2	0.55						Esterase		[b]
Methylprednisolone	374.5	Neutral	2	6.20	5.9	0.82			0.82	CYP3A4	P-gp	[a]+ [26]
Metoclopramide	299.8	Base	2	5.70	4.6	1.00	0.95	0.80	0.76			

(*continued*)

TABLE 13.1 (*Continued*)

Name	M_{wt}	Ionization	ECCS Class	CL_t (mL/min/kg)	CL_h (mL/min/kg)	f_a	f_g	f_h	f	Main Enzymes	Main Uptake Transporters	Main Efflux Transporters	References
Metoprolol	267.4	Base	2	13.00	12.4	0.98	0.84	0.46	0.38	CYP2D6			a
Metronidazole	171.2	Neutral	2	0.85	0.7	1.00	1.02	0.97	0.99				
Mianserin	264.4	Base	2										
Mibefradil	495.6	Base	2	6.60					0.80				
Midazolam	325.8	Neutral	2	6.60	6.2	0.90	0.71	0.69	0.44	CYP3A4			a
Milnacipran	246.3	Base	2	34.50	6.9				0.88				a
Mirtazapine	265.4	Base	2	8.00	7.7	0.80	0.94	0.67	0.50				a
Moclobemide	268.7	Base	2	10.00	10.0	0.98	0.79	0.57	0.44				
Morphine	285.4	Base	2	24.00	23.0	0.85			0.24	UGT			a
Nalmefene	339.4	Base	2	15.00	13.6								
Naloxone	327.4	Base	2	22.00	22.0	0.91			0.02	UGT			[63]
Nebivolol	441.9	Base	2	12.29						CYP2D6			a
Nefazodone	470.0	Base	2	7.50	7.1	1.00	0.28	0.69	0.19	CYP3A4			a
Nefopam	253.3	Base	2	12.00	11.6								
Nevirapine	266.3	Neutral	2	0.30	0.3	1.00	0.94	0.99	0.93				a
Nicardipine	479.5	Base	2	11.00	10.9	0.95	0.21	0.53	0.11	CYP3A4			[64]
Nicotine	162.2	Base	2	18.00	15.0	1.00	0.86	0.35	0.30				
Nifedipine	346.3	Neutral	2	7.30	7.3	1.00	0.73	0.68	0.50	CYP3A4			a
Nimodipine	418.45	Neutral	2	15.00	15.0	0.90	0.38	0.35	0.12	CYP3A4			a
Nisoldipine	388.4	Neutral	2	15.00	15.0	0.90	0.13	0.35	0.04	CYP3A4			[50, 54]
Nitrendipine	360.4	Neutral	2	20.71	20.6	0.88			0.15	CYP3A4			[65]
Nomifensine	238.3	Base	2	22.00	9.5	1.00	0.46	0.59	0.27				
Nortriptyline	263.4	Base	2	10.00	9.8	1.00	0.89	0.57	0.51	CYP2D6			a
Omeprazole	345.4	Neutral	2	8.40	8.3	0.95	0.87	0.64	0.53	CYP2C			a
Ondansetron	293.4	Base	2	5.80	5.5	1.00	0.82	0.76	0.62	CYP3A4			a
Oxazepam	286.7	Neutral	2	1.10	0.2	0.93	1.01	0.99	0.93	UGT			a

Oxprenolol	265.4	Base	2									[a]
Oxycodone	315.4	Base	2	12.40	10.0				0.74	CYP3A4		[a]+ [66]
Oxymorphone hydrochloride	337.8	Base	2							UGT		[a]+ [67]
Palonosetron	296.4	Base	2	2.67	1.6				0.97	CYP2D6		[b]
Pantoprazole	432.0	Neutral	2	2.20	2.2	1.00	0.85	0.90	0.77	CYP2C		[a]
Papaverine	339.4	Neutral	2	11.00	11.0	1.00	0.54	0.52	0.28			
Paroxetine	329.4	Base	2	8.60	8.4					CYP2D6	P-gp	[a]
Pelrinone	241.2	Neutral	2									
Penbutolol	291.428	Base	2									
Pentoxifylline	278.3	Neutral	2	39.00	39.0							
Phencyclidine	243.4	Base	2	5.10	4.3	0.95	0.93	0.81	0.72			
Pilsicainide		Base	2									
Pindolol	248.3	Base	2	7.70	3.8	1.00	1.04	0.83	0.87			[a]
Practolol	266.3	Base	2	2.00	0.3	1.00	0.96	0.99	0.95			
Prasugrel	409.9	Base	2							Esterase		[a]
Primaquine	259.3	Base	2	1.59	1.6				0.88			
Propafenone	341.4	Base	2	17.00	16.9				0.28	CYP2D6		
Propiverine	367.5	Base	2			0.84			0.86	CYP3A4		[a]
Propofol	178.3	Neutral	2	27.00	26.9							
Propoxyphene	367.5	Base	2	7.57					0.50	CYP3A4		[a]
Propranolol	259.4	Base	2	12.00	10.8	0.99	0.50	0.53	0.26	CYP2D6		[a]
Quinidine	324.4	Base	2	4.00	3.4	1.00	0.88	0.85	0.75	CYP3A4		[50, 54]
Quinine	324.4	Base	2	1.90	1.6	0.80	1.02	0.93	0.76	CYP3A4		[50, 54]
Rabeprazole	359.4	Neutral	2	4.00	4.0	0.90	0.70	0.83	0.52	CYP2C		[a]
Reboxetine	313.4	Base	2	0.82	0.8	1.00	0.98	0.97	0.95			
Remoxipride	371.3	Base	2	1.70	1.3	1.00	0.95	0.95	0.90			
Risperidone	410.5	Base	2	5.40	5.2	1.00	0.85	0.77	0.66	CYP2D6	P-gp	[a]+ [44]
Rivaroxaban		Neutral	2	10.70					0.66	CYP3A4	P-gp	[a]
Rivastigmine	250.3	Base	2	0.43	0.4				0.40	Esterase		[a]
Rizatriptan	269.3	Base	2	16.00	13.8					MAO-A		[a]
Ropivacaine	274.4	Base	2	6.30	6.2					CYP3A4		[68]

(continued)

TABLE 13.1 (*Continued*)

Name	M_{wt}	Ionization	ECCS Class	CL_t (mL/min/kg)	CL_h (mL/min/kg)	f_a	f_g	f_h	f	Main Enzymes	Main Uptake Transporters	Main Efflux Transporters	References
Sabeluzole	428.5	Neutral	2										
Scopolamine	303.4	Base	2	16.00	15.0	0.95	0.82	0.35	0.27				
Selegiline	187.3	Base	2	20.00	19.8	1.00	0.00	0.14					[a]
Semaxinib	238.3	Neutral	2	14.00									
Sibutramine	279.86	Base	2							CYP2B6			[a]
Sildenafil	474.6	Base	2	9.10	9.1	0.92	0.74	0.60	0.41	CYP3A4		P-gp	[a]+ [69]
Simvastatin	418.58	Neutral	2	7.60	6.8				0.05	CYP3A4	OATP	P-gp	[a]
Spironolactone	416.58	Neutral	2	93.00	92.5								[a]
Stavudine	224.2	Neutral	2	8.17	5.0	1.00	1.00	0.75	0.82				
Sufentanil	386.6	Base	2	15.00	14.9	0.90	0.00	0.35					
Tapentadol	221	Base	2	21.86					0.32	UGT			[a]
Tebufelone	300.4	Neutral	2										
Temozolomide	194.2	Neutral	2	0.16	0.2				0.98				[a]
Terazosin	387.4	Base	2	1.20	1.1	0.90	0.96	0.95	0.82				
Theophylline	180.2	Neutral	2	0.86	0.8	1.00	0.99	0.97	0.96	CYP1A2			[a]
Thiotepa	189.2	Base	2	0.28									
Ticlopidine	263.79	Base	2	13.20	13.1					CYP2C			[70]
Tilidine	273.4	Base	2	16.00	15.7	1.00	0.19	0.32	0.06				
Timolol	316.4	Base	2	8.50	7.1	0.95	0.99	0.69	0.65	CYP2D6			[a]
Tizanidine	253.7	Base	2	11.00	10.7	1.00	0.39	0.54	0.21	CYP1A2			[71]
Tolterodine	325.5	Base	2	8.40	8.4	0.77	0.53	0.63	0.26				[a]
Tolvaptan	448.95	Base	2	4.70	4.7				0.56	CYP3A4		P-gp	[a]
Tramadol	263.4	Base	2	6.50	5.5	0.90	1.02	0.76	0.70	CYP2D6			[a]
Trazodone	371.9	Base	2	1.40	1.4	1.00	0.82	0.94	0.77	CYP3A4			[a]
Triamterene	253.27	Base	2	63.00	56.7				0.51	CYP1A2			[72, 73]
Triazolam	343.2	Neutral	2	3.00	2.9	0.85	0.74	0.87	0.55	CYP3A4			[a]+ [69]

Trimethoprim	290.3	Base	2	2.10	1.2	0.97	0.83	0.95	0.76				
Urapidil	387.5	Base	2	3.10	2.6	1.00	0.88	0.89	0.78				
Vardenafil	488.6	Base	2	13.00	13.0	0.90	0.37	0.43	0.15	CYP3A4		P-gp	[a]
Venlafaxine	277.4	Base	2	14.00	13.4	0.97	1.03	0.42	0.42				
Verapamil	454.6	Base	2	18.00	16.5	1.00	0.74	0.28	0.21	CYP3A4		P-gp	[a]+ [50, 54]
Viloxazine	237.3	Base	2	2.10									
Voriconazole	349.3	Neutral	2	3.80	3.7				0.96				[a]
Vorinostat	264.3	Neutral	2	4.40					0.43				
Zaleplon	305.3	Neutral	2	16.00	16.0	1.00	1.02	0.30	0.31	Aldehyde oxidase			[a]
Ziprasidone	412.9	Base	2	5.10	5.0	0.90	0.85	0.78	0.60	CYP3A4			[a]
Zolmitriptan	287.4	Base	2	6.70	4.5	0.92	0.53	0.80	0.39	CYP1A2			[a]
Zolpidem	307.4	Base	2	4.30	4.3	1.00	0.88	0.81	0.72	CYP3A4			[a]
Zopiclone	388.8	Base	2	3.30	3.2	1.00	0.93	0.86	0.80	CYP3A4			[a]
Acetazolamide	222.2	Acid	3A	0.65	0.1						OAT	P-gp	[74, 75]
Amifloxacin	334.3	Zwitterion	3A										
Amoxicillin	365.4	Zwitterion	3A	3.30	1.4	0.93	0.88	0.94	0.77		OAT		[76, 77]
Antofloxacin	376.4	Zwitterion	3A	1.17									
Benazeprilat	396.45	Acid	3A										
Betamipron	193.2	Acid	3A	9.20	0.2						OAT		[78]
Bumetanide	364.4	Acid	3A	2.50	1.0	0.90	0.82	0.96	0.70		OAT		[a]+ [79]
Carbenicillin	378.4	Acid	3A	1.90	0.3						OAT		[80]
Cefadroxil	363.4	Zwitterion	3A	2.50	0.2	1.00	0.96	0.99	0.95		OAT	MRP2	[81–83]
Cefetamet	397.4	Acid	3A	1.90	0.4	0.52	0.92	0.98	0.47				
Ceftizoxime	383.4	Acid	3A	2.10		0.72	1.00				OAT		[82]
Cephalexin	347.4	Zwitterion	3A	3.90	0.4	1.00	0.91	0.98	0.90		OAT	MATE1	[80, 82, 84, 85]
Cephalothin	396.4	Acid	3A	1.80	0.9	0.05	1.00				OAT		[82, 83]
Cephradine	349.4	Zwitterion	3A	3.30	0.2	1.00	1.01	0.99	1.00			MATE1	[82, 85]
Cicaprost	374.5	Acid	3A	3.80		1.00	1.00						
Cilastatin	358.5	Zwitterion	3A	2.30	0.7						OAT		[78]
Ciprofloxacin	331.3	Zwitterion	3A	8.30	1.4	0.75	0.98	0.94	0.69		OAT	BCRP	[86, 87]
cis-Cefprozil	389.4	Zwitterion	3A										
Enalaprilat	348.4	Zwitterion	3A	1.60	0.0	0.25	1.00	1.00	0.25				

(continued)

TABLE 13.1 (*Continued*)

Name	M_{wt}	Ionization	ECCS Class	CL_t (mL/min/kg)	CL_h (mL/min/kg)	f_a	f_g	f_h	f	Main Enzymes	Main Uptake Transporters	Main Efflux Transporters	References
Furosemide	330.8	Acid	3A	2.40	0.7	0.61	0.86	0.97	0.51		OAT	BCRP + MRP2	[a]+ [26]
Gabapentin	171.24	Zwitterion	3A	1.70	0.0	0.60	1.00	1.00	0.60				[a]
Gatifloxacin	375.4	Zwitterion	3A	2.80	0.5	0.96	1.02	0.98	0.96		OAT		[86]
Ketorolac	255.7	Acid	3A	0.35	0.2	1.00	1.01	0.99	1.00				[a]
Levofloxacin	361.4	Zwitterion	3A	1.90	0.5	1.00	1.01	0.98	0.99			MAET1+P-gp	[43, 85]
Ofloxacin	361.4	Zwitterion	3A	3.50	1.3	1.00	1.00	0.94	0.98		OAT	BCRP	[86, 87]
Penicillin G	334.4	Acid	3A								OAT		[88, 89]
Pregabalin	159.2	Zwitterion	3A	0.81	0.1	0.90	1.00	1.00	0.90				[a]
Ramiprilat	388.47	Zwitterion	3A										
Sulfadiazine	250.3	Acid	3A	0.55	0.2	0.85	1.01	0.99	0.85				
Ticarcillin	384.4	Acid	3A	1.80	0.4								
Torsemide	348.4	Acid	3A	0.53	0.4	0.96	1.02	0.98	0.96		OAT		[90]
Atrasentan	510.6	Acid	3B								OATP		[91]
Azlocillin	461.5	Acid	3B	2.50	0.9								
Aztreonam	435.4	Acid	3B	1.50	0.5	0.01	1.02	0.98	0.01				
Candesartan	440.47	Acid	3B	0.37	0.2				0.42			P-gp	[a]+ [92]
Cefamandole	462.5	Acid	3B	3.60	0.1						OAT		[83]
Cefazolin	454.5	Acid	3B	0.89	0.1	1.00	0.90	1.00	0.90		OAT		[83, 84]
Cefcanel	478.6	Acid	3B	2.60	0.2	0.42	0.96	0.99	0.40				
Cefepime	481.6	Acid	3B	2.20	0.6	0.96	0.85	0.98	0.80				
Cefixime	453.4	Acid	3B	1.00	0.6	0.50	0.96	0.97	0.47		OAT		[82]
Cefmenoxime	511.6	Acid	3B										
Cefmetazole	471.5	Acid	3B	1.50	0.3						OATP		[82, 84]
Cefodizime	584.7	Acid	3B	0.61	0.1						OATP		[84]
Cefoperazone	645.7	Acid	3B	1.30	0.9						OATP		[84]
Cefotaxime	455.5	Acid	3B	2.70	1.9						OATP		[93, 94]

Cefoxitin	427.5	Acid	3B	3.60	0.5								
Cefpirome	514.6	Acid	3B	1.90									[82]
Ceftriaxone	554.6	Acid	3B	0.26	0.1	0.01	1.00	0.99			OATP		[94]
Cefuroxime	424.4	Acid	3B	2.20	0.1	0.44	1.00	1.00	0.44				[82]
Cephaloridine	416.5	Acid	3B	3.30		0.05	1.00				OAT		[95]
Cephapirin	423.5	Acid	3B	2.70	1.4								[82]
Dexloxiglumide	461.4	Acid	3B	3.70	3.3	0.95	0.59	0.86	0.48			P-gp	[96]
Dicloxacillin	470.3	Acid	3B	2.00	0.8	0.70	1.00	0.97	0.68			P-gp	[82]
Eprosartan	424.5	Zwitterion	3B	1.90	1.2	0.15	0.92	0.95	0.13		OATP		[a]+ [97]
Eptaloprost	402.5	Acid	3B										
Fosinoprilat	435.5	Acid	3B	0.32	0.2								
Lisinopril	405.5	Zwitterion	3B	1.20	0.0	0.28	0.89	1.00	0.25				[a]
Melagatran	429.5	Zwitterion	3B	1.90		0.15						P-gp	[98]
Methotrexate	454.4	Acid	3B	2.10	0.4	0.70	1.02	0.98	0.70		OAT	MRP2 + BCRP	[a]+ [99]
Metrizoate	627.9	Acid	3B	2.70									
Mezlocillin	539.6	Acid	3B	2.10	1.2								
Montelukast	586.2	Acid	3B	0.68	0.7	0.80	0.80	0.97	0.62	CYP2C	OATP		[a]+ [100]
Olmesartan	446.5	Acid	3B								OAT + OATP	MRP2	[a]+ [101, 102]
Oxytetracycline	460.4	Zwitterion	3B	2.00									
Piperacillin	517.6	Acid	3B	4.00	1.2								
Pravastatin	424.5	Acid	3B	14.00	7.7	0.34	0.80	0.67	0.18		OAT + OATP		[a]+ [103, 104]
Quinaprilat	410.5	Zwitterion	3B	0.93	0.0	0.61	0.82	1.00	0.50		OAT		[105]
Rosuvastatin	481.6	Acid	3B	11.00	7.7	0.50	0.60	0.67	0.20		OATP	BCRP	[a]+ [106, 107]
Temocapril	476.6	Acid	3B									BCRP	[28]
Temocaprilat	448.6	Acid	3B								OATP		[108]
Tesaglitazar	408.5	Acid	3B	0.04	0.0	1.00	1.00	1.00	1.00				
Tetracycline	444.4	Zwitterion	3B	1.50	0.6	0.90	0.88	0.97	0.77		OAT		[109]
Valsartan	435.3	Acid	3B	0.49	0.3	0.55	0.72	0.98	0.39		OATP	MRP2 + P-gp	[a]+ [26, 110–112]
Acecainide	277.4	Base	4	4.00	0.8								
Acetylmethadol	353.5	Base	4	5.40									
Acyclovir	225.2	Neutral	4	4.70	1.1	0.23	1.05	0.95	0.23		OAT	MATE1 + MATE2K	[85, 88, 113]
Albuterol®	239.3	Base	4	7.80	3.9					Sulfotransferase	OCT		[114]

(continued)

TABLE 13.1 (*Continued*)

Name	M_{wt}	Ionization	ECCS Class	CL_t (mL/min/kg)	CL_h (mL/min/kg)	f_a	f_g	f_h	f	Main Enzymes	Main Uptake Transporters	Main Efflux Transporters	References
Aliskiren	551.7637	Base	4									P-gp	[a]
Apixaban	459.5035	Base	4	1.20	0.9				0.51			BCRP + P-gp	[115]
Aprepitant	534.4	Neutral	4	1.09	1.1				0.63	CYP3A4		P-gp	[a]
Atenolol	266.3	Base	4	2.50	0.0	0.50	1.00	1.00	0.50				[a]
Azimilide	458.0	Base	4	2.50					1.00				
Bidisomide	408.0	Base	4										
Bisoprolol	325.4	Base	4	3.70	1.4	0.95	0.95	0.94	0.85	CYP3A4		P-gp	[a]
Chloroquine	319.9	Base	4	4.10	1.8	1.00	0.87	0.92	0.80			MAET1+P-gp	[116, 117]
Chlorthalidone	338.7	Neutral	4	1.50	0.5	0.82	0.80	0.98	0.64				
Cimetidine	252.3	Base	4	8.10	0.2	0.68	1.01	0.99	0.68		OAT + OCT	MATE1 + MATE2K + P-gp	[a]+ [26, 85, 114, 118, 119]
Dabigatran	471.52	Base	4	24.97	3.7				0.07				
Digitoxin	764.96	Neutral	4	0.04	0.0							P-gp	[120]
Digoxin	780.96	Neutral	4	1.70	0.0	0.81	0.86	1.00	0.70			P-gp	[121]
Disopyramide	339.5	Base	4	0.90	0.4	0.83	1.02	0.98	0.83				
Dofetilide	441.6	Base	4	5.20	2.1	0.96	1.03	0.91	0.90				
Edoxaban	548.065	Base	4	5.24					0.62			P-gp	[a]+ [122]
Ethambutol	204.3	Base	4	8.60	1.8	0.80			0.77				
Famotidine	337.5	Base	4	6.50	2.1	0.38	1.00	0.89	0.43		OAT + OCT	P-gp	[a]+ [118]
Fenoterol	303.4	Base	4	29.00									
Ganciclovir	255.2	Neutral	4	4.60	0.5	0.09	1.02	0.98	0.09		OAT	MATE1 + MATE2K +	[85, 123]
Hydrochlorothiazide	297	Neutral	4	4.90	0.0	0.67	1.00	1.00	0.71		OAT	P-gp	[a]+ [43]
Hydroflumethiazide	331.3	Neutral	4	9.70	0.7	0.62	1.03	0.97	0.62				

Lamivudine	229.3	Neutral	4	4.95	1.6	1.00	0.94	0.92	0.86		OCT	MATE1 + MATE2K + BCRP	[b]+ [124]
Levoprotiline	293.4	Base	4	12.30	12.1	1.00	0.84	0.47	0.40				
Maraviroc	513.7	Base	4	10.48	9.6				0.23	CYP3A4		P-gp	[a]
Metformin	129.17	Base	4	7.40	0.0	0.52	1.00	1.00	0.52		OCT		[a]
Methylnaltrexone	356.4	Neutral	4	0.44									[a]
Metolazone	365.8	Neutral	4	1.40	0.3	0.26	1.00	0.99	0.26			P-gp	[a]
Nadolol	309.4	Base	4	2.90	1.1	0.36	1.00	0.95	0.34				[a]
Nizatidine	331.5	Base	4	9.80	2.5	1.00	1.05	0.89	0.94			P-gp	[a]+ [125]
Orlistat	495.75	Neutral	4										
Procainamide	235.33	Base	4	10.00	5.5	0.95	1.04	0.76	0.75		OCT	MATE1 + MATE2K	[85, 114]
Ranitidine	314.4	Base	4	9.60	2.2	0.65	1.02	0.90	0.60		OAT + OCT	P-gp	[a]+ [118, 126, 127]
Recainam	263.4	Base	4	4.50	2.2	0.80	1.01	0.90	0.73				
Sematilide	313.4	Base	4	3.70	0.2	0.65	0.95	0.99	0.61				
Sitagliptin	523.3	Neutral	4	6.00	1.3	0.95	0.99	0.95	0.89	CYP3A4	OAT	P-gp	[a]
Sotalol	272.367	Base	4	2.00	0.3	0.95	1.01	0.99	0.95				
Sulpiride	341.4	Base	4	1.83	0.1	0.27	1.00	0.99	0.27				
Terbutaline	225.3	Base	4	3.40	1.2	0.17	1.00	0.95	0.16	Sulfotransferase	OCT		[a]+ [114]
Topotecan	421.4	Base	4	13.00	7.8						OAT	BCRP + P-gp	[a]
Trimetrexate	369.4	Base	4	0.77	0.6								
Ximelagatran	473.5705	Base	4									P-gp	[98]
Zidovudine	267.2	Neutral	4	26.00	25.3	1.00			0.63				

[a]University of Washington Drug Monograph.
[b]Product label.

that have high membrane permeability, ECCS Class 1A, Class 1B, and Class 2 had an average f_g of 92%, 88%, and 77%, respectively (Figure 13.3). However, low-permeability molecules, ECCS Class 3A, Class 3B, and Class 4 had an average f_g of 97%, 89%, and 98%, respectively (Figure 13.3). ECCS Class 2 had the lowest f_g and largest range (105%) relative to other classes. ECCS Class 2 are bases/neutrals drug molecules that have high membrane permeability and metabolism is the driver for their elimination. As summarized in Table 13.1, out of 182 that are ECCS Class 2, 30 drug molecules had research reports indicating that CYP3A4 is the main enzyme driving their metabolism. These results are in concordance with the physicochemical properties of CYP3A4 substrate where they tend to be lipophilic and bases/neutrals with an average of f_g 66%, which is lower from that noted with CYP2C ($f_g = 89\%$) and CYP2D6 ($f_g = 85\%$) compounds (Table 13.1). This is not unexpected considering the significant high expression of CYP3A4 relative to other CYP enzymes in the enterocytes [128]. It is interesting to note that in this data set we have an average f_g of CYP1A2 substrates of 69% (Table 13.1). This data is not in agreement with the poor/lack of expression of CYP1A2 in human small intestine as reported by Groer et al. [128]. Further investigation is warranted to understand the discrepancy between the reported f_g and poor expression of CYP1A2 in the human small intestine.

For low-permeability molecules (ECCS Classes 3A, 3B, and 4), their high intestinal stability is consistent with their high metabolic stability.

13.2.3 f_h *and ECCS*

As a parameter, f_h is calculated using the following equation:

$$f_h = 1 - E_h = 1 - \frac{CL_h}{Q_h} \quad (13.1)$$

where E_h is the hepatic extraction ratio, CL_h the hepatic clearance of drug molecule, and Q_h the hepatic blood flow, which is equal to 20 mL/min/kg in humans. ECCS Classes 3A and 4 have renal elimination as the rate-determining step. Therefore, these molecules would have low hepatic first-pass effect (average $f_h > 90\%$). However, ECCS Classes 1A and 2 have metabolism as the rate-determining step, where their average CL_h is 1.8 and 9.9 mL/min/kg, respectively (Figure 13.4) and have an average f_h value of 91% and 72%, respectively (Figure 13.5). Many nonsteroidal anti-inflammatory drug molecules have high metabolic stability and are acidic in nature [4]. They belong to ECCS Class 1A and these molecules are typically eliminated by CYP2C, UGTs, and esterase enzymes. ECCS Class 2 molecules are bases/neutrals and have high membrane permeability. These molecules are typically metabolized by CYP3A4, CYP2D6, UGTs, esterases, and other enzymes. They have the highest clearance from all classes investigated and this results in their low f_h value relevant to other ECCS classes.

ECCS Class 1B molecules have hepatic uptake mediated by OATP1B1/1B3 as the rate-determining step, with an average CL_h of 4.9 mL/min/kg (Figure 13.4) and have an average f_h value of 91% and 72%, respectively (Figure 13.5). It should be emphasized that these molecules once in the liver undergo extensive metabolism that is mediated by CYP2C, esterases, CYP3A4, and UGT enzymes. Examples on these molecules include statins and sartans [129–131].

13.2.4 f and ECCS

It is clear that characterizing your NMEs based on ECCS classification would enable a proactive handling of potential liabilities of factors that may lead to poor f in humans.

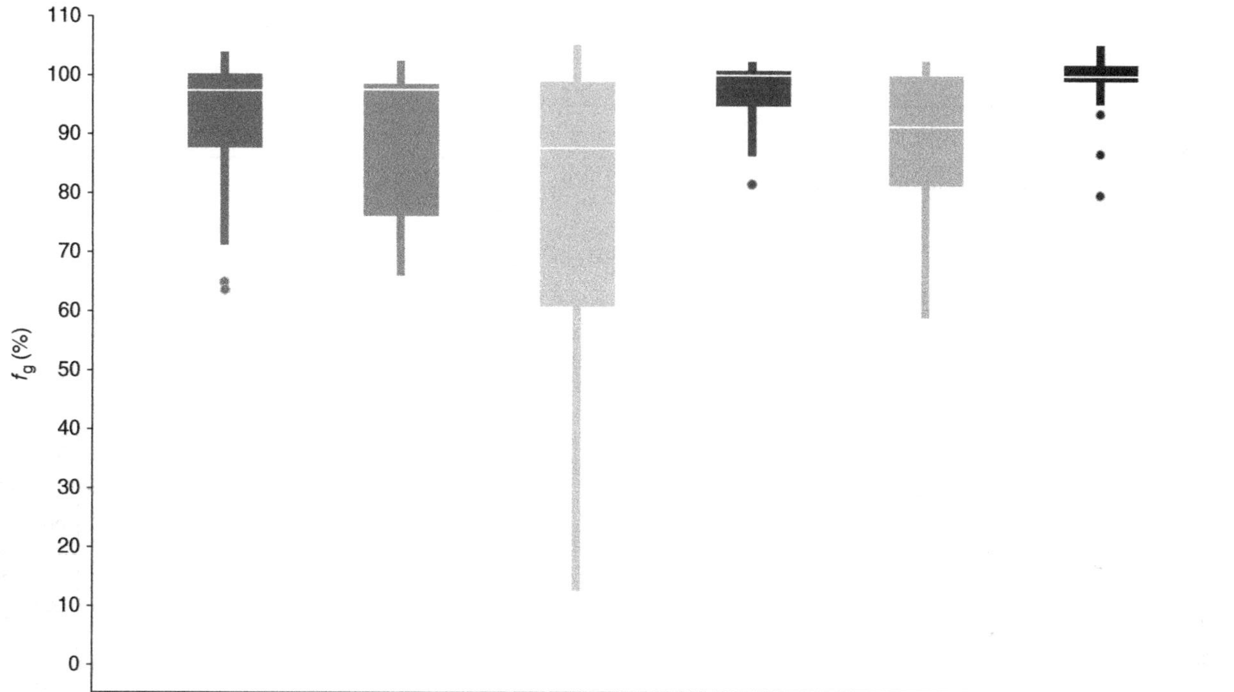

	1A	1B	2	3A	3B	4
Count	18	10	107	20	20	25
Median	97%	97%	87%	100%	91%	100%
Avg	92%	88%	77%	97%	89%	98%
Range	40%	36%	105%	21%	43%	25%

ECCS Class

FIGURE 13.3 The median, average, and range of f_g for compounds classified based on their ECCS.

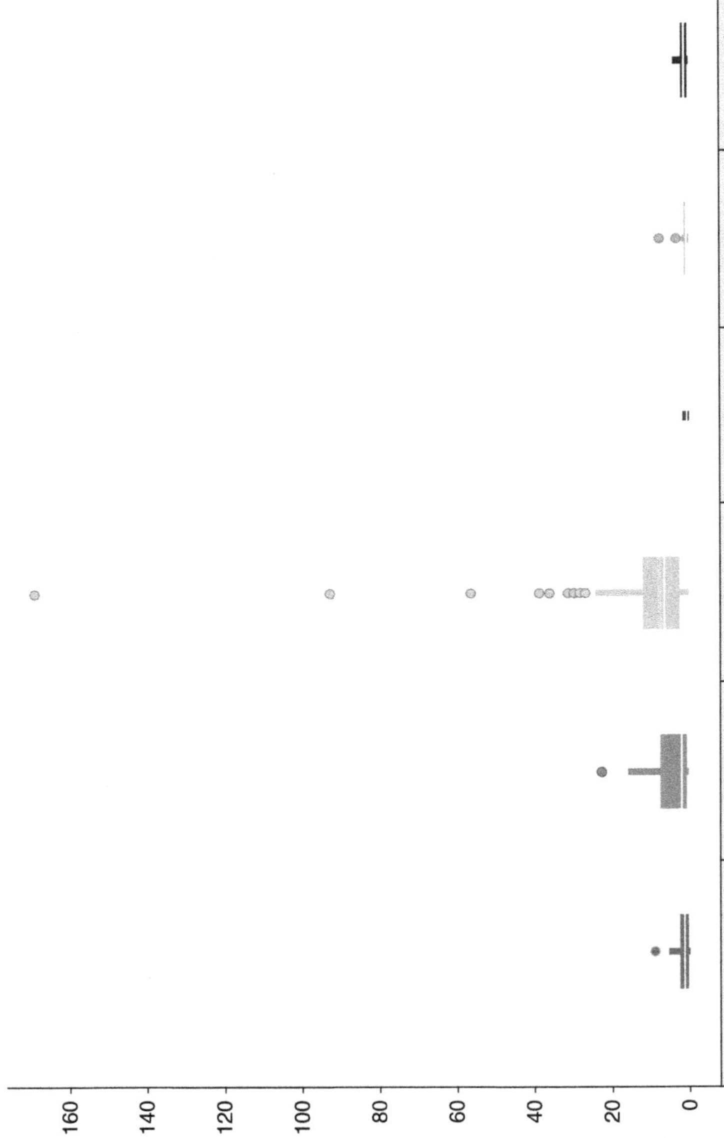

	1A	1B	2	3A	3B	4
Count	28	15	165	23	31	33
Median	1.4	2.1	6.2	0.4	0.5	0.9
Avg	1.9	4.9	9.9	0.5	1.1	1.1

FIGURE 13.4 The median, average, and range of CL_h for compounds classified based on their ECCS.

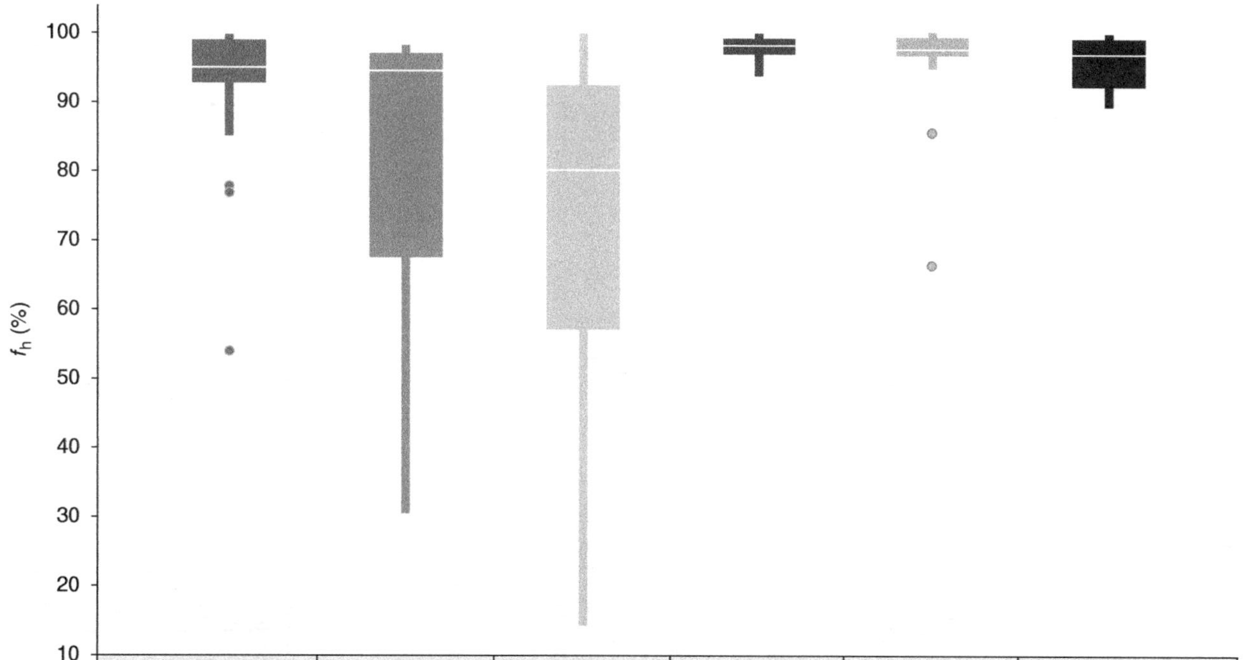

	1A	1B	2	3A	3B	4
	Fh					
Count	18	11	107	17	19	25
Median	95%	95%	80%	98%	98%	97%
Avg	91%	82%	72%	98%	94%	96%
Range	46%	68%	86%	6%	33%	11%

ECCS Class

FIGURE 13.5 The median, average, and range of f_h for compounds classified based on their ECCS.

High-permeability molecules (i.e., ECCS Classes 1A, 1B, and 2) have an average f of 81%, 61%, and 57%, respectively (Figure 13.6). Class 1A had the highest average f due to their overall high permeability associated with high absorption and low CL_h provided that they have no solubility/dissolution-limited absorption. On the other hand, the main driver for the lower f observed with ECCS Class 1B and 2 is mainly attributed to their high CL_h. The poor f can also be exacerbated when these moderate-to-high CL_h molecules are substrates for CYP3A4. This may increase their intestinal first-pass effect and lower the overall f_g. Overall, these high-permeability molecules, if solubility is not a limiting factor, that is, >200 µg/mL in relevant testing matrices, they should have good oral f_a [1, 5].

Low-permeability molecules, that is, ECCS Classes 3A, 3B, and 4, have an average f of 79%, 49%, and 60%, respectively (Figure 13.6). It is evident that other than solubility as a potential limiting factor for the f of various drug molecules, molecules that have low permeability are more prone for poor absorption that is influenced by efflux transporters such as P-gp, BCRP, and MRP2 [6]. However, the same molecules may have good overall f_a and f if they are good substrates for PEPT1 or OATP2B1 intestinal uptake transporters. Therefore, it is pivotal to characterize if your interplay between intestinal uptake and efflux transporters and their impact on the f_a of the NMEs of interest. This should be accompanied with understanding of the projected oral dose in human. For example, OATP2B1 transporters are effective in increasing the overall oral absorption of molecules with low dose (<100 mg), while PEPT1 and monocarboxylate transporter 1 (MCT1) are effective in increasing the overall f_a of drug molecules with high oral dose >100 mg [12].

13.3 f_a MODULATION

As shown by Fick's first law, oral drug absorption (i.e., drug flux (J_{wall})) across intestinal membrane is a product of its effective intestinal permeability (P_{wall}), which is the rate that dissolved drug will cross the enterocytes to reach the portal vein, and C_{int} is the drug concentration in the luminal fluid [2, 3]:

$$J_{wall} = P_{wall} \cdot C_{int} \tag{13.2}$$

An understanding of the permeability and solubility early in drug discovery is pivotal in ensuring that NMEs will have good oral absorption, a prerequisite for oral bioavailability that enables testing the mechanism in humans.

13.3.1 Solubility

To ensure good oral absorption, drug molecules should have optimum solubility to have sufficient drug in solution to be absorbed at the site of administration. To determine if a molecule has poor solubility, our group recommended the assessment of drug solubility in either gastric simulated fluid (pH 1.2) or FaSSIF at pH 6.5, where acidic drug molecule would be assessed in first medium and nonacidic drug molecules would be assessed in the second medium. Drug molecules that have solubility >200 µg/mL are considered high solubility drug molecules and vice versa based on study reported [5]. It should be emphasized that the approach implemented to measure the solubility is by adapting thermodynamic solubility methodology. It is defined as the equilibrium concentration of a compound that is saturated in a given solvent. Many variables that are related to drug physicochemical properties affect its thermodynamic solubility. These include and are not limited to crystal lattice, temperature, and pressure [3]. It should be emphasized that other methodologies can be used to measure drug solubility such as kinetic methodologies where drug solids are predissolved in dimethylsulfoxide, which is then titrated with aqueous medium (i.e., phosphate buffer at pH 6.5–7.4 over time) until precipitation occurs. While high throughput can be used in drug discovery, it cannot be reliably used in predicting

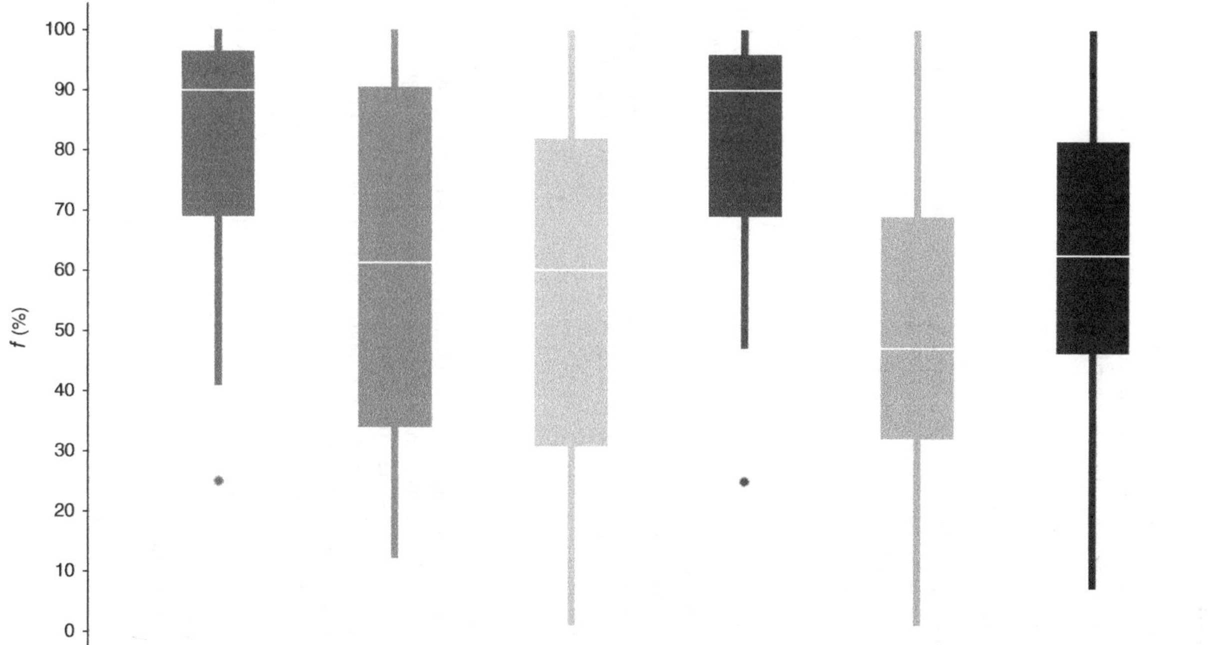

	1A	1B	2	3A	3B	4
Count	22	16	148	17	19	31
Median	90%	61%	60%	90%	47%	63%
Avg	81%	61%	57%	79%	49%	60%
Range	75%	88%	99%	75%	99%	93%

ECCS

FIGURE 13.6 The median, average, and range of f for compounds classified based on their ECCS.

dose and absorption since it does not take into consideration the impact of crystal lattice on the drug solubility.

Chapter 4 has extensive discussion of the variables that can be modulated to improve the solubility of drug molecules. Following is a brief overview.

13.3.1.1 Crystal Lattice and Lipophilicity Impacts on Drug Solubility As illustrated in Equation 13.3, drug solubility (S_W^{Solid}) can be calculated using the compound melting point (MP) and lipophilicity also known as octanol water partition coefficient ($\log K_{ow}$) [3].

$$\log S_W^{solid} = 0.5 - 0.01(MP - 25) - \log K_{ow} \tag{13.3}$$

Drug molecules with highly ordered crystal structure tend to have high melting point and is usually associated with low solubility. Typically, high ordered crystal lattices are a result of either high intermolecular hydrogen bonding and/or short-ranged dispersion forces. These bonds are observed in compounds with high number of hydrogen-bond acceptor (HBA)/hydrogen-bond donor (HBD) groups (i.e., high PSA, planar or inflexible conformation, or a high degree of symmetry) [3]. It is interesting to note that compounds with high melting point (>250 °C) usually have poor thermodynamic solubility. Therefore, altering the crystal lattice by altering the planner structure of molecules would lead to improvement in drug solubility as noted by Goosen et al. and Nikam et al. research reports [132, 133].

Compounds with high lipophilicity ($\log P > 3.5$) usually have poor aqueous solubility [3]. Indeed, reducing lipophilicity by the addition of ionizable and/or polar groups would improve solvation potential. This would increase drug molecules solubility.

Based on Equation 13.3, a decrease of MP by 100 °C or one log unit in log *P* would be associated with 10-fold increase in solubility. While this equation is not appropriate for measuring solubility for ionizable molecules, it still illustrates the significance of optimizing these factors to optimize compound solubility.

13.3.2 Permeability

Drug molecules cross the enterocytes by either passive diffusion or active transport. Passive diffusion occurs by one of the two pathways: the paracellular pathway, where drug molecules diffuse through the aqueous pores between enterocyte tight junctions, or transcellular pathway, in which drug molecules diffuse across the lipid membrane of intestinal enterocytes. The active transport pathways are mediated by active influx and/or efflux transporters. Indeed, drug physicochemical properties and affinity for efflux/influx transporters would determine the significance of each of these routes. For example, drug molecules that use paracellular pathway tend to be hydrophilic ($\log P < 0$) with small molecular weight <250 Da [3]. Compounds offered by this pathway (<0.01% of the total area available for absorption). Drug molecules that are lipophilic with large molecular weight usually use transcellular pathway [3]. This is the main route of absorption for molecules that are absorbed by passive diffusion. The rate of transport across the enterocyte is determined by the physicochemical properties of the molecules, which is captured by the principles of Rule of 5. Developed by Pfizer scientist Christopher Lipinski, this rule outlines that for molecules to be absorbed into the transcellular pathway and be orally bioavailable, they should be unionized, with optimum lipophilicity ($\log P < 5$) and small molecular weight ($M_{wt} < 500$ Da). Furthermore, their polarity measured by the number of hydrogen bond donors (HBD < 10) and acceptors (HBA < 5) should be optimum [134, 135].

As outlined earlier, influx and efflux transporters line the human GIT and play a key role in the absorption of many drug molecules [11, 12]. When we approach these transporters, it is prudent to comprehend their transporter mechanism, regional expression, and substrate specificity as discussed below. Efflux transporters such as P-gp, BCRP, multidrug resistance-associated protein 2 (MRP2) are also expressed at the apical membrane of enterocytes. Efflux transporters play a major role in limiting the absorption of drug molecules especially those with low permeability [2, 3, 11, 12, 14].

13.3.2.1 Permeability and Paracellular Pathway DMPK scientists are usually reluctant about developing paracellular pathway. This is due to multiple reasons: first, paracellular pathway compounds that are absorbed by this pathway are associated with small surface area and typically result in low f_a and high inter- and intraindividual variability. Also, intestinal tight junctions tend to have smaller pore size as we transition from small to large intestines. This would limit the ability of DMPK scientists to target the colon to extend the half-life of lead molecules with short half-life and reduce dosing frequency using traditional sustained release formulations targeting the colon. In addition, species differences were reported for drug molecules absorbed via paracellular pathway. For example, dogs are not recommended to be used as a preclinical model to predict the absorption of drug molecules absorbed via paracellular pathway, since dog tight junctions have larger pore size compared to rat or human. Instead, rat can be an effective preclinical model to predict drug absorption in humans [136–138]. To avoid absorption via paracellular pathway, it is prudent to consider increasing your drug lipophilicity and molecular weight [3].

13.3.2.2 Permeability and Transcellular Pathway Transcellular pathway is the main route of absorption for passively absorbed drug molecules. To improve absorption via this route, DMPK scientists usually consider the following alterations of drug molecules:

1. Reducing hydrogen bond affinity measured as PSA, which is the surface area associated with oxygen, nitrogen, hydroxyl, and amine groups. Compounds with high PSA (>90 Å^2) have stronger hydrophilic shields that are difficult to break to enable compound permeation across the enterocytes [139].
2. Lipophilic compounds have high rate of passive diffusion across the enterocytes through transcellular pathway. Therefore, optimizing compound's hydrophobicity measured as $\log P$ or $\log D$ is a key for effective permeation across the enterocytes [140, 141]. Compounds that are very hydrophilic (i.e., $c \log P$ and $c \log D_{7.4} < -2$) had poor oral absorption, whereas compounds with higher lipophilicity showed improved absorption [139].
3. Most of the drug molecules are either weak acids or bases. The extent of ionization is determined and measured by the principles of Henderson–Hasselbalch equation. This would determine the rate and extent of permeation and absorption across the intestine. Only the unionized fraction will cross the membrane. Therefore, optimizing the drug molecule pK_a is an important option that should be considered when designing bioavailable drug molecules [142].

13.3.2.3 Permeability and Influx Transporters A large number of influx transporters are expressed in the enterocytes and play a major role in the absorption of nutrients and vitamins. In addition, these influx transporters mediate the absorption of some drugs and xenobiotics.

Chapter 7 discussed the involvement of these transporters in intestinal drug absorption. There are two types of influx transporters. The first is high-affinity/low-capacity transporters. Examples include OATP2B1 and system L transporters that are expressed in the small intestine with minimal or no expression in the colon [2, 3, 11, 12, 14]. These transporters would be effective in improving the absorption of drug molecules with low oral dose (<100 mg). The second class are the low-affinity/high-capacity transporters, which are PEPT1 and MCT1 intestinal transporters. Similar to OATP2B1 and System L transporters, PEPT1 is located in the small intestine, with no expression in the colon. Interestingly, MCT1 transporters have high expression in the small intestine and colon. Understanding of the regional expression of transporters at large is important for effective drug development. For example, Xenoport scientists were successful in discovering a prodrug molecule of gabapentin that has affinity for MCT1 transporters. This discovery led to the development of successful sustained release formulations that target the colon using their prodrug gabapentin enacarbil [143–145]. Collectively, unlike low capacity transporters, PEPT1 and MCT1 transporters can transport high oral dose molecules such as penicillin and cephalosporins (oral dose > 1 g).

13.3.2.4 Permeability and Efflux Transporters Efflux intestinal transporters play a key role in limiting the absorption of drug molecules and xenobiotics. They are part of the ATP-binding cassette (ABC) superfamily of transporters and are expressed at the apical surface of the small intestine enterocytes. They include P-gp, BCRP, and MRP2. P-gp is the most studied transporter and is a product of the multidrug resistance (MDR1) gene [3, 12]. As discussed in Chapter 6, P-gp substrates are polar with a median PSA of 88 Å^2, HBD and HBA of 2 and 5, respectively. P-gp substrates also have large molecular weight (median $M_{wt} = 406$ Da) and are lipophilic (median measured $\log P$ and $\log D_{pH\ 7.4}$ of 2.4 and 1.7, respectively). The lipophilic nature enables their interactions with the hydrophobic core of P-gp. As noted by Tachibana et al., the extent of impact of P-gp on its substrate absorption is dependent on their affinity to the transporter (K_m), solubility, and permeability [6]. Therefore, it is prudent to be mindful of these variables to assess the potential impact of P-gp on its substrate absorption. These principles do also apply on compounds that are substrates for BCRP and MRP2. Compounds that are substrates for these transporters are generally lipophilic with large molecular weight and are acidic in nature [2, 3, 11, 12, 14].

13.3.3 *Using* In Vitro *Tools Based Assay to Classify Compound Permeability and Predict f_a*

Drug flux is a product of both drug permeability across the enterocyte and solubility. Our group established a relationship between permeability measured in low efflux-Madin–Darby Canine Kidney Epithelial Cells (LE-MDCK Line) and f_a for a set of molecules ($n = 110$) [5]. The relationship indicated that a value of 5×10^{-6} cm/s can be used as a cutoff value that discerns between high permeability versus low-permeability molecules, where compounds with an LE-MDCK permeability $>5 \times 10^{-6}$ cm/s tend to have high $f_a > 85\%$ and vice versa. It is clear that ECCS Classification of NMEs shed light on the causes of the poor absorption. For example, compound with LE-MDCK permeability $<5 \times 10^{-6}$ cm/s may have low f_a, which may be exacerbated if the molecule is a substrate for efflux transporters and/or have poor solubility and dissolution. On the other hand, compounds that are substrates for influx transporters may have a better profile. The probability will be driven affinity of the drug to either influx or efflux transporters, its solubility, dissolution, and dose.

TIP:

It is prudent for DMPK departments to consider establishing a relationship between permeability using available *in vitro* tool and f_a for a set of molecules with clear ECCS classification. This enables them to define a cutoff value that will predict whether their molecules are high- or low-absorption molecules. This can be a first permeability step in building their ECCS classification using in-house permeability data. The supplementary tables of our paper can be used as a frame of reference to build this important relationship [4].

13.4 f_g

Chapter 5 discussed the parameters that govern drug f_g calculation [146]. These include permeability through the enterocyte membrane and gut intrinsic clearance [147], where NMEs with high permeability and high intestinal metabolic stability (i.e., low intrinsic intestinal clearance) tend to have better intestinal stability. This contributes to higher overall oral bioavailability, provided that they have good absorption. The aforementioned properties are drug-dependent inputs. From system inputs standpoint, close attention should be exercised, when predicting f_g for NMEs. For example, it is well established in the literature that CYP3A4 enzyme has high expression in the intestinal enterocyte, which has overall contribution of ≈82% of the total CYP expression in the small intestine. Therefore, it is not unexpected to observe that compounds that are CYP3A4 substrates are more prone to have lower f_g values relative to other CYP enzymes such as CYP2D6, CYP1A2, CYP2C8, or others [148–150]. It is pivotal early in drug discovery to determine the major metabolizing enzyme that would drive the elimination of NMEs. This would shed light on the potential impact of these enzymes on NMEs f_g. Furthermore, available PBPK models, such as SimCYP, are able to determine f_g based on HLM or hepatocytes inputs and inputs on major enzyme that is driving the metabolism of these molecules in the intestine. Collectively, this would enable an early determination of whether the oral f of the NMEs investigated would be limited by intestinal first-pass effect.

13.5 HEPATIC CLEARANCE

Based on PBPK principles, factors that affect hepatic clearance are determined by system (drug-independent) and drug-dependent parameters. Examples on system-dependent parameters that affect hepatic clearance include hepatic blood flow, protein content of cells, and hepatocyte volume to name a few, whereas drug-dependent parameters include drug plasma protein binding, red blood cells partition, intrinsic clearance, which is the intrinsic ability of the liver to clear the drug. Intrinsic hepatic clearance is driven by the drug affinity to hepatic metabolizing enzymes, basolateral, and canalicular transporters [151–153]. It is important to characterize the transporter(s) metabolizing enzyme(s) interplay to successfully predict hepatic clearance of NMEs [152, 154–156]. The mathematical expression defining this interplay is termed extended clearance concept and shown in Equation 13.4 [151]:

$$\mathrm{CL}_{int,\mathrm{h}} = \frac{(\mathrm{PS}_{\mathrm{influx}} + \mathrm{PS}_{\mathrm{pd}}) \cdot \mathrm{CL}_{\mathrm{int}}}{(\mathrm{PS}_{\mathrm{efflux}} + \mathrm{PS}_{\mathrm{pd}} + \mathrm{CL}_{\mathrm{int}})} \tag{13.4}$$

Herein, total hepatic clearance (CL_{h}) is calculated by assuming well-stirred conditions (Equation 13.5) [153]:

$$\mathrm{CL}_{\mathrm{h}} = Q_{\mathrm{h}} \cdot \frac{f_{\mathrm{b}} \cdot (\mathrm{PS}_{\mathrm{influx}} + \mathrm{PS}_{\mathrm{pd}}) \cdot \mathrm{CL}_{\mathrm{int}}}{Q_{\mathrm{h}} \cdot (\mathrm{PS}_{\mathrm{efflux}} + \mathrm{PS}_{\mathrm{pd}} + \mathrm{CL}_{\mathrm{int}}) + f_{\mathrm{b}} \cdot (\mathrm{PS}_{\mathrm{influx}} + \mathrm{PS}_{\mathrm{pd}}) \cdot \mathrm{CL}_{\mathrm{int}}} = Q_{\mathrm{h}} \cdot E_{\mathrm{h}} \tag{13.5}$$

$\mathrm{PS}_{\mathrm{influx}}$ and $\mathrm{PS}_{\mathrm{efflux}}$ are the active (transporter-mediated) sinusoidal influx and basolateral efflux clearance; $\mathrm{PS}_{\mathrm{pd}}$ represents passive diffusion clearance; $\mathrm{CL}_{\mathrm{int}}$ is the sum of the metabolic and biliary intrinsic clearance ($\mathrm{CL}_{\mathrm{int,met}} + \mathrm{CL}_{\mathrm{int,bile}}$); Q_{h} is the liver blood flow; E_{h} is the hepatic extraction ratio; and $f_{\mathrm{u,b}}$ is the unbound fraction in blood and it is calculated using Equation 13.6.

$$f_{\mathrm{u,b}} = \frac{f_{\mathrm{u,plasma}}}{B/P} \tag{13.6}$$

where $f_{\mathrm{u,plasma}}$ is the unbound fraction in plasma and B/P is the blood to plasma ratio.

It should be emphasized that prediction of the hepatic clearance for drugs that are substrates for combination of metabolizing enzymes and influx or efflux transporters would be biased if only one process is considered [36, 151, 155–158]. However, Equation 13.4 is simplified when "rapid equilibrium" between liver and blood compartments is achieved. Rapid equilibrium is typically realized when the drug molecule is not a substrate for active hepatic uptake transporters (e.g., organic anion transporting polypeptides 1B1/1B3) and the PS_{pd} is significantly higher than CL_{int}. Equation 13.4 is simplified to Equation 13.7 [153]:

$$CL_h = Q_h \cdot \frac{f_b \cdot CL_{int}}{Q_h + f_b \cdot CL_{int}} = Q_h \cdot E_h \tag{13.7}$$

This equation is usually used in drug discovery to predict the hepatic clearance for rapid equilibrium compounds that are only substrates for cytochrome P-450 (CYP) and other metabolizing enzymes [159–161].

For rapid equilibrium compounds, hepatic clearance can be predicted using HLM [162–164]. This is used mainly for drug molecules that are substrates for CYP enzymes. However, to predict the human hepatic clearance for drug molecules that are substrates for UDP glucuronosyltransferases (UGTs), sulfotransferases, aldehyde oxidase, and glutathione transferase (GST), among others, human hepatocytes is the method of choice [162–166].

Hepatic uptake can be the rate-determining step of the systemic clearance of the drug molecule, if PS_{pass} is significantly lower than CL_{int} and the compound is a substrate for uptake transporters (e.g., OATP1B1 and 1B3 transporters) leading to PS_{influx} significantly higher than PS_{pass}. Equation 13.3 can be reduced to Equation 13.8.

$$CL_h = Q_h \cdot \frac{f_b \cdot PS_{influx}}{Q_h + f_b \cdot PS_{influx}} = Q_h \cdot E_h \tag{13.8}$$

Examples on drug molecules that have hepatic uptake as the rate-determining step for their systemic clearance include HMG-CoA reductase inhibitors (statins) and angiotensin II antagonists (sartans) [23, 157, 167–169]. Indeed, a number of studies suggested that active uptake mediated by OATP1B1 and 1B3 transporters expressed at the basolateral membrane of liver hepatocytes could be the rate-determining step in hepatic clearance, not only for compounds that are metabolically stable such as rosuvastatin [170] and pravastatin [171–173], but also for compounds that are extensively metabolized (e.g., atorvastatin [174, 175], glyburide [30, 176], and repaglinide) [37, 177]. This is substantiated by clinically relevant drug–drug interactions (DDIs) leading to changes in systemic exposure of these drugs which are primarily attributed to the inhibition of hepatic uptake mediated by members of the OATP family rather than the inhibition of hepatic metabolism or biliary efflux [156, 174, 178]. Furthermore, polymorphism in *SLCO1B1* (encoding OATP1B1) was reported to lead to major changes in systemic exposure of statins, which, in turn, regulates the relative exposure in peripheral tissue such as muscle and the risk of toxicity such as rhabdomyolysis [131, 179–181]. These examples clearly demonstrate the significant role of active hepatic uptake mediated by OATP transporter family in determining the systemic exposure of these molecules. It should be emphasized that affinity of drug molecules with active hepatic uptake as a rate-determining step to enzymatic metabolism and biliary efflux would determine the hepatic exposure and excretion from the body. Therefore, inhibition or pharmacogenomics variations in these efflux transporters and/or metabolizing enzymes would lead to major changes in the liver exposure and minimal change in the plasma. This is valid provided that no alteration in activity of these transporters and enzymes is realized along the intestine. Following is a discussion of variables that should be considered to modulate hepatic clearance and reduce f_h of NMEs.

13.5.1 Modulation of Metabolism-Mediated Hepatic Clearance

Williams et al. investigated the route of elimination for the top 200 prescribed selling drugs in the United States and reported hepatic metabolism as the dominant mechanism of elimination. This result indicates that it is pivotal to investigate if metabolism is a driver for the elimination of NMEs. Appropriate understanding of system-dependent parameters such as enzyme types, expression, and functional activity, in addition to drug-dependent parameters, such as physicochemical properties that would result into affinity to the metabolizing enzymes is warranted to reduce NMEs hepatic clearance and improve f_h and f at large.

As outlined in Chapter 2, metabolism is divided into phase I and II processes. In phase I, the drug undergoes oxidative metabolism that presents or exposes a polar functional group on the drug molecule. As illustrated in Table 13.1, CYP superfamily is the main driver of the metabolism of many of Class 2 molecules. The main CYP isoforms in human liver are CYP1A2, CYP2B6, CYP2C, CYP2D6, CYP2E1, and CYP3A4 [148, 149]. Lewis et al. investigated drug physicochemical properties that are needed to modulate these enzymes. For example, CYP3A4 substrates typically have wide molecular weight range with large volume of distribution. Compounds are usually acidic or neutral in nature [182]. Compounds that are substrates for CYP1A2 tend to be planar aromatic heterocyclic amines, for example, theophylline and caffeine [183–185]. Substrates for CYP2C enzymes tend to be acidic in nature, that is, ECCS Class 1A/1B, while CYP2D6 substrates are basic in nature [186, 187]. HLM is the main tool to predict human clearance for molecules eliminated by CYP enzymes [4]. In phase II metabolism, attachment of a hydrophilic substituent such as glucuronic acid or amino acid, to a molecule to facilitate the biliary or renal excretion of metabolites is observed. From the enzymes that undergo phase II metabolism, UGTs are responsible for the metabolism of 15% the top 200 drug molecules that are predominantly metabolized. UGT substrates are either acid/amine/zwitterions with no clear SAR has been established on this class of molecules. The tool of choice to measure hepatic clearance mediated by UGTs is human hepatocytes [4].

13.5.2 Modulation of Transporter-Mediated Hepatic Clearance

According to Equation 13.4, various strategies can be adapted to modulate the intrinsic clearance for molecules that have active hepatic uptake as a rate-determining step of molecules elimination. For example, to reduce compound overall $CL_{int,h}$, DMPK scientists may adapt the following strategies:

1. Increase compound passive clearance (PS_{pd}) relative to compound active hepatic uptake mediated by OATPs transporter (PS_{influx}).
2. Reduce PS_{influx} relative to other elimination mechanisms.
3. Interestingly, changes in the compound metabolism and/or biliary elimination should have minimal impact on compound systemic exposure, when the compound has hepatic uptake as the rate-determining step.

It should be emphasized that there is no clear SAR that has been established to modulate each of these parameters. For compounds with active hepatic uptake as the rate-determining step sandwich culture hepatocytes appear to be the golden standard to predict the clearance for this set of molecules [36, 37, 188].

13.6 CONCLUSION

It is evident that ECCS provides a viable road map on characterizing the potential causes of poor bioavailability for NMEs. For example, compounds that are of high permeability, that is, ECCS Class 1 and Class 2 molecules, the factors that would limit their oral bioavailability are predominantly driven by first-pass effect provided that solubility/dissolution are not limiting factors for their oral absorption. The potential impact of intestinal first-pass effect is determined by their affinity to metabolizing enzymes that have high expression in the intestine, such as CYP3A4 and certain UGT isozymes. The molecules absorption would not be limited by efflux transporters provided they have high permeability, solubility >200 μg/mL in relevant mediums as outlined earlier, and/or have oral doses that are higher than 100 mg. Therefore, it is prudent to use the right *in vitro* tools to predict the hepatic metabolic clearance and f_g of these molecules and to assess drug solubility early on in drug discovery. As for low-permeability molecules, that is, ECCS Class 3 and 4, the factors that would limit their oral bioavailability are predominantly absorption. It should be emphasized that for ECCS Class 3B, if hepatic uptake is the major contributor of their overall total CL, their hepatic first pass can also be a factor that should be investigated. Furthermore, due to their low permeability, these molecules absorption is significantly influenced by intestinal uptake and efflux transporters. The tool of choice would be determined by the type of metabolizing enzymes involved in their elimination. Major advancement has been realized in our ability to integrate inputs from various *in vitro* platforms to predict the compound disposition and profile in humans. Still more investment is warranted in this area to improve our ability to better predict human plasma time profile following oral dosing.

13.7 CHAPTER 13 TUTORIALS

1. True or false

Compounds that are of high permeability, that is, ECCS Class 1 and 2, are prone to have poor bioavailability that is a result of solubility-limited absorption.

Compounds that are of ECCS Class 1A have high likelihood to be limited by CYP3A4-mediated intestinal first-pass effect.

Compounds that are of ECCS Class 3 and 4 have a likelihood to have poor absorption that is attributed to poor permeability and higher impact of efflux transporters on their absorption.

ECCS Class 2 compounds are associated with higher likelihood of poor f_g; in part, this is attributed to their higher affinity to CYP3A4 enzyme, which has the highest CYP enzyme expression in the small intestine.

Compounds that are of ECCS Class 1A have the highest median oral bioavailability since they typically are metabolically stable, their metabolism is not usually mediated by CYP3A4 and have high permeability.

2. CYP1B/3B have high propensity for poor oral bioavailability, which is usually attributed to their high active hepatic uptake, that is, hepatic uptake is the rate-determining step of their poor bioavailability. If you are at discovery stage, what is the guidance that you would provide your team to optimize the clearance of your chemical series considering that hepatic uptake is the rate-determining step.

13.8 CHAPTER 13 TUTORIAL ANSWERS KEY

1. True or false

Compounds that are of high permeability, that is, ECCS Classes 1 and 2, are prone to have poor bioavailability that is a result of solubility-limited absorption. **(F)**

Compounds that are of ECCS Class 1A have high likelihood to be limited by CYP3A4-mediated intestinal first-pass effect. **(F)**

Compounds that are of ECCS Classes 3 and 4 have a likelihood to have poor absorption that is attributed to poor permeability and higher impact of efflux transporters on their absorption. **(T)**

ECCS Class 2 compounds are associated with higher likelihood of poor f_g; in part, this is attributed to their higher affinity to CYP3A4 enzyme, which has the highest CYP enzyme expression in the small intestine. **(T)**

Compounds that are of ECCS Class 1A have the highest median oral bioavailability since they typically are metabolically stable, their metabolism is not usually mediated by CYP3A4, and have high permeability. **(T)**

2. CYP1B/3B have high propensity for poor oral bioavailability, which is usually attributed to their high active hepatic uptake, that is, hepatic uptake is the rate-determining step of their poor bioavailability. If you are at discovery stage, what is the guidance that you would provide your team to optimize the clearance of your chemical series considering that hepatic uptake is the rate-determining step.

Answer: As outlined in Section 13.5.2, following are the steps that you need to consider:

1. Increase compound passive clearance (PS_{pd}) relative to compound active hepatic uptake mediated by OATPs transporter (PS_{influx}).

2. Reduce PS_{influx} relative to other elimination mechanisms.

3. Interestingly, changes in the compound metabolism and/or biliary elimination should have minimal impact on compound systemic exposure, when the compound has hepatic uptake as the rate-determining step.

REFERENCES

1. El-Kattan, A. and Varma, M., *Oral absorption, intestinal metabolism and human oral bioavailability*. Top Drug Metab. 2012.
2. Hurst, S., Loi, C.M., Brodfuehrer, J. and El-Kattan, A., *Impact of physiological, physicochemical and biopharmaceutical factors in absorption and metabolism mechanisms on the drug oral bioavailability of rats and humans*. Expert Opin Drug Metab Toxicol, 2007. 3(4): p. 469–489.
3. Thomas, V.H., Bhattachar, S., Hitchingham, L., Zocharski, P., Naath, M., Surendran, N., Stoner, C.L. and El-Kattan, A., *The road map to oral bioavailability: an industrial perspective*. Expert Opin Drug Metab Toxicol, 2006. 2(4): p. 591–608.
4. Varma, M.V., Steyn, S.J., Allerton, C. and El-Kattan, A.F., *Predicting clearance mechanism in drug discovery: extended clearance classification system (ECCS)*. Pharm Res, 2015. 32(12): p. 3785–3802.

5. Varma, M.V., et al., *pH-Dependent solubility and permeability criteria for provisional biopharmaceutics classification (BCS and BDDCS) in early drug discovery.* Mol Pharm, 2012. 9(5): p. 1199–1212.
6. Tachibana, T., Kato, M., Takano, J. and Sugiyama, Y., *Predicting drug–drug interactions involving the inhibition of intestinal CYP3A4 and P-glycoprotein.* Curr Drug Metab, 2010. 11(9): p. 762–777.
7. Tachibana, T., Kato, M. and Sugiyama, Y., *Prediction of nonlinear intestinal absorption of CYP3A4 and P-glycoprotein substrates from their* in vitro *Km values.* Pharm Res, 2012. 29(3): p. 651–668.
8. Niemi, M., et al., *High plasma pravastatin concentrations are associated with single nucleotide polymorphisms and haplotypes of organic anion transporting polypeptide-C (OATP-C, SLCO1B1).* Pharmacogenetics, 2004. 14(7): p. 429–440.
9. Zhang, W., et al., *Role of BCRP 421C>A polymorphism on rosuvastatin pharmacokinetics in healthy Chinese males.* Clin Chim Acta, 2006. 373(1–2): p. 99–103.
10. Igel, S., et al., *Increased absorption of digoxin from the human jejunum due to inhibition of intestinal transporter-mediated efflux.* Clin Pharmacokinet, 2007. 46(9): p. 777–785.
11. Filipski, K.J., Varma, M.V., El-Kattan, A.F., Ambler, C.M., Ruggeri, R.B., Goosen, T.C. and Cameron, K.O., *Intestinal targeting of drugs: rational design approaches and challenges.* Curr Top Med Chem, 2013. 13(7): p. 776–802.
12. Varma, M.V., Ambler, C.M., Ullah, M., Rotter, C.J., Sun, H., Litchfield, J., Fenner, K.S. and El-Kattan, A.F., *Targeting intestinal transporters for optimizing oral drug absorption.* Curr Drug Metab, 2010. 11(9): p. 730–742.
13. Stewart, B.H., Kugler, A.R., Thompson, P.R. and Bockbrader, H.N., *A saturable transport mechanism in the intestinal absorption of gabapentin is the underlying cause of the lack of proportionality between increasing dose and drug levels in plasma.* Pharm Res, 1993. 10(2): p. 276–281.
14. Varma, M.V., Obach, R.S., Rotter, C., Miller, H.R., Chang, G., Steyn, S.J., El-Kattan, A. and Troutman, M.D., *Physicochemical space for optimum oral bioavailability: contribution of human intestinal absorption and first-pass elimination.* J Med Chem, 2010. 53(3): p. 1098–1108.
15. van den Bongard, H.J., Sparidans, R.W., Critchley, D.J., Beijnen, J.H. and Schellens, J.H., *Pharmacokinetic drug–drug interaction of the novel anticancer agent E7070 and acenocoumarol.* Invest New Drugs, 2004. 22(2): p. 151–158.
16. Tang, M., Mukundan, M., Yang, J., Charpentier, N., LeCluyse, E.L., Black, C., Yang, D., Shi, D. and Yan, B., *Antiplatelet agents aspirin and clopidogrel are hydrolyzed by distinct carboxylesterases, and clopidogrel is transesterificated in the presence of ethyl alcohol.* J Pharmacol Exp Ther, 2006. 319(3): p. 1467–1476.
17. Skjodt, N.M. and Davies, N.M., *Clinical pharmacokinetics and pharmacodynamics of bromfenac.* Clin Pharmacokinet, 1999. 36(6): p. 399–408.
18. Lagas, J.S., van der Kruijssen, C.M., van de Wetering, K., Beijnen, J.H. and Schinkel, A.H., *Transport of diclofenac by breast cancer resistance protein (ABCG2) and stimulation of multidrug resistance protein 2 (ABCC2)-mediated drug transport by diclofenac and benzbromarone.* Drug Metab Dispos, 2009. 37(1): p. 129–136.
19. Verbeeck, R.K., *Pharmacokinetic drug interactions with nonsteroidal anti-inflammatory drugs.* Clin Pharmacokinet, 1990. 19(1): p. 44–66.
20. Varma, M.V., Lin, J., Bi, Y.A., Kimoto, E. and Rodrigues, A.D., *Quantitative rationalization of gemfibrozil drug interactions: consideration of transporters–enzyme interplay and the role of circulating metabolite gemfibrozil 1-O-beta-glucuronide.* Drug Metab Dispos, 2015. 43(7): p. 1108–1118.
21. Elliot, D.J., Suharjono, Lewis, B.C., Gillam, E.M., Birkett, D.J., Gross, A.S. and Miners, J.O., *Identification of the human cytochromes P450 catalysing the rate-limiting pathways of gliclazide elimination.* Br J Clin Pharmacol, 2007. 64(4): p. 450–457.
22. Knodell, R.G., Browne, D.G., Gwozdz, G.P., Brian, W.R. and Guengerich, F.P., *Differential inhibition of individual human liver cytochromes P-450 by cimetidine.* Gastroenterology, 1991. 101(6): p. 1680–1691.

23. Kalliokoski, A. and Niemi, M., *Impact of OATP transporters on pharmacokinetics*. Br J Pharmacol, 2009. 158(3): p. 693–705.

24. Hirano, M., Maeda, K., Matsushima, S., Nozaki, Y., Kusuhara, H. and Sugiyama, Y., *Involvement of BCRP (ABCG2) in the biliary excretion of pitavastatin*. Mol Pharmacol, 2005. 68(3): p. 800–807.

25. Fahrmayr, C., Konig, J., Auge, D., Mieth, M., Munch, K., Segrestaa, J., Pfeifer, T., Treiber, A. and Fromm, M., *Phase I and II metabolism and MRP2-mediated export of bosentan in a MDCKII-OATP1B1-CYP3A4-UGT1A1-MRP2 quadruple-transfected cell line*. Br J Pharmacol, 2013. 169(1): p. 21–33.

26. Lin, X., Skolnik, S., Chen, X. and Wang, J., *Attenuation of intestinal absorption by major efflux transporters: quantitative tools and strategies using a Caco-2 model*. Drug Metab Dispos, 2011. 39(2): p. 265–274.

27. Green, B.R. and Bain, L.J., *Mrp2 is involved in the efflux and disposition of fosinopril*. J Appl Toxicol, 2013. 33(6): p. 458–465.

28. Ohura, K., Nozawa, T., Murakami, K. and Imai, T., *Evaluation of transport mechanism of prodrugs and parent drugs formed by intracellular metabolism in Caco-2 cells with modified carboxylesterase activity: temocapril as a model case*. J Pharm Sci, 2011. 100(9): p. 3985–3994.

29. Klatt, S., Fromm, M.F. and Konig, J., *The influence of oral antidiabetic drugs on cellular drug uptake mediated by hepatic OATP family members*. Basic Clin Pharmacol Toxicol, 2013. 112(4): p. 244–250.

30. Varma, M.V., Scialis, R.J., Lin, J., Bi, Y.A., Rotter, C.J., Goosen, T.C. and Yang, X., *Mechanism-based pharmacokinetic modeling to evaluate transporter-enzyme interplay in drug interactions and pharmacogenetics of glyburide*. AAPS J, 2014. 16(4): p. 736–748.

31. Chapy, H., Klieber, S., Brun, P., Gerbal-Chaloin, S., Boulenc, X. and Nicolas, O., *PBPK modeling of irbesartan: incorporation of hepatic uptake*. Biopharm Drug Dispos, 2015. 36(8): p. 491–506.

32. Simons, F.E. and Simons, K.J., *Clinical pharmacology of new histamine H1 receptor antagonists*. Clin Pharmacokinet, 1999. 36(5): p. 329–352.

33. Soldner, A., Benet, L.Z., Mutschler, E. and Christians, U., *Active transport of the angiotensin-II antagonist losartan and its main metabolite EXP 3174 across MDCK-MDR1 and Caco-2 cell monolayers*. Br J Pharmacol, 2000. 129(6): p. 1235–1243.

34. Nishimuta, H., Houston, J.B. and Galetin, A., *Hepatic, intestinal, renal, and plasma hydrolysis of prodrugs in human, cynomolgus monkey, dog, and rat: implications for* in vitro–in vivo *extrapolation of clearance of prodrugs*. Drug Metab Dispos, 2014. 42(9): p. 1522–1531.

35. Hua, W.J., Hua, W.X. and Fang, H.J., *The role of OATP1B1 and BCRP in pharmacokinetics and DDI of novel statins*. Cardiovasc Ther, 2012. 30(5): p. e234–e241.

36. Varma, M.V., Lin, J., Bi, Y.A., Rotter, C.J., Fahmi, O.A., Lam, J.L., El-Kattan, A.F., Goosen, T.C. and Lai, Y., *Quantitative prediction of repaglinide-rifampicin complex drug interactions using dynamic and static mechanistic models: delineating differential CYP3A4 induction and OATP1B1 inhibition potential of rifampicin*. Drug Metab Dispos, 2013. 41(5): p. 966–974.

37. Varma, M.V., Lai, Y., Kimoto, E., Goosen, T.C., El-Kattan, A.F. and Kumar, V., *Mechanistic modeling to predict the transporter- and enzyme-mediated drug–drug interactions of repaglinide*. Pharm Res, 2013. 30(4): p. 1188–1199.

38. Tachibana, M., Tanaka, M., Masubuchi, Y. and Horie, T., *Acyl glucuronidation of fluoroquinolone antibiotics by the UDP-glucuronosyltransferase 1A subfamily in human liver microsomes*. Drug Metab Dispos, 2005. 33(6): p. 803–811.

39. Li, R., Ghosh, A., Maurer, T.S., Kimoto, E. and Barton, H.A., *Physiologically based pharmacokinetic prediction of telmisartan in human*. Drug Metab Dispos, 2014. 42(10): p. 1646–1655.

40. Chang, J.H., Plise, E., Cheong, J., Ho, Q. and Lin, M., *Evaluating the* in vitro *inhibition of UGT1A1, OATP1B1, OATP1B3, MRP2, and BSEP in predicting drug-induced hyperbilirubinemia*. Mol Pharm, 2013. 10(8): p. 3067–3075.

41. Venkatakrishnan, K., von Moltke, L.L., Duan, S.X., Fleishaker, J.C., Shader, R.I. and Greenblatt, D.J., *Kinetic characterization and identification of the enzymes responsible for the hepatic*

biotransformation of adinazolam and N-desmethyladinazolam in man. J Pharm Pharmacol, 1998. 50(3): p. 265–274.

42. Gertz, M., Harrison, A., Houston, J.B. and Galetin, A., *Prediction of human intestinal first-pass metabolism of 25 CYP3A substrates from* in vitro *clearance and permeability data*. Drug Metab Dispos, 2010. 38(7): p. 1147–1158.
43. Poirier, A., Cascais, A.C., Bader, U., Portmann, R., Brun, M.E., Walter, I., Hillebrecht, A., Ullah, M. and Funk, C., *Calibration of* in vitro *multidrug resistance protein 1 substrate and inhibition assays as a basis to support the prediction of clinically relevant interactions* in vivo. Drug Metab Dispos, 2014. 42(9): p. 1411–1422.
44. Zanelli, U., Caradonna, N.P., Hallifax, D., Turlizzi, E. and Houston, J.B., *Comparison of cryopreserved HepaRG cells with cryopreserved human hepatocytes for prediction of clearance for 26 drugs*. Drug Metab Dispos, 2012. 40(1): p. 104–110.
45. Jankovic, S.M., *Pharmacokinetics of selective beta1-adrenergic blocking agents: prescribing implications*. Expert Opin Drug Metab Toxicol, 2014. 10(9): p. 1221–1229.
46. Senda, C., Kishimoto, W., Sakai, K., Nagakura, A. and Igarashi, T., *Identification of human cytochrome P450 isoforms involved in the metabolism of brotizolam*. Xenobiotica, 1997. 27(9): p. 913–922.
47. Wojcikowski, J. and Daniel, W.A., *Perazine at therapeutic drug concentrations inhibits human cytochrome P450 isoenzyme 1A2 (CYP1A2) and caffeine metabolism – an* in vitro *study*. Pharmacol Rep, 2009. 61(5): p. 851–858.
48. Chen, M., Howe, D., Leduc, B., Kerr, S. and Williams, D.A., *Identification and characterization of two chloramphenicol glucuronides from the* in vitro *glucuronidation of chloramphenicol in human liver microsomes*. Xenobiotica, 2007. 37(9): p. 954–971.
49. Niwa, T., Shiraga, T., Mitani, Y., Terakawa, M., Tokuma, Y. and Kagayama, A., *Stereoselective metabolism of cibenzoline, an antiarrhythmic drug, by human and rat liver microsomes: possible involvement of CYP2D and CYP3A*. Drug Metab Dispos, 2000. 28(9): p. 1128–1134.
50. Tseng, E., Walsky, R.L., Luzietti, Jr., R.A., Harris, J.J., Kosa, R.E., Goosen, T.C., Zientek, M.A. and Obach, R.S., *Relative contributions of cytochrome CYP3A4 versus CYP3A5 for CYP3A-cleared drugs assessed* in vitro *using a CYP3A4-selective inactivator (CYP3cide)*. Drug Metab Dispos, 2014. 42(7): p. 1163–1173.
51. Shram, M.J., et al., *Assessment of pharmacokinetic and pharmacodynamic interactions between albumin-fused mutated butyrylcholinesterase and intravenously administered cocaine in recreational cocaine users*. J Clin Psychopharmacol, 2015. 35(4): p. 396–405.
52. Kaivosaari, S., Toivonen, P., Hesse, L.M., Koskinen, M., Court, M.H. and Finel, M., *Nicotine glucuronidation and the human UDP-glucuronosyltransferase UGT2B10*. Mol Pharmacol, 2007. 72(3): p. 761–768.
53. Nakajima, M., Tanaka, E., Kwon, J.T. and Yokoi, T., *Characterization of nicotine and cotinine N-glucuronidations in human liver microsomes*. Drug Metab Dispos, 2002. 30(12): p. 1484–1490.
54. Moaddel, R., Hamid, R., Patel, S., Bullock, P.L. and Wainer, I.W., *Identification of P-glycoprotein substrates using open tubular chromatography on an immobilized P-glycoprotein column: Comparison of chromatographic results with Caco-2 permeability*. Anal Chim Acta, 2006. 578(1): p. 25–30.
55. Marchetti, S., de Vries, N.A., Buckle, T., Bolijn, M.J., van Eijndhoven, M.A., Beijnen, J.H., Mazzanti, R., van Tellingen, O. and Schellens, J.H., *Effect of the ATP-binding cassette drug transporters ABCB1, ABCG2, and ABCC2 on erlotinib hydrochloride (Tarceva) disposition in* in vitro *and* in vivo *pharmacokinetic studies employing Bcrp1-/-/Mdr1a/1b-/- (triple-knockout) and wild-type mice*. Mol Cancer Ther, 2008. 7(8): p. 2280–2287.
56. Li, J., Cusatis, G., Brahmer, J., Sparreboom, A., Robey, R.W., Bates, S.E., Hidalgo, M. and Baker, S.D., *Association of variant ABCG2 and the pharmacokinetics of epidermal growth factor receptor tyrosine kinase inhibitors in cancer patients*. Cancer Biol Ther, 2007. 6(3): p. 432–438.
57. Kassahun, K., et al., *Role of human liver cytochrome P4503A in the metabolism of etoricoxib, a novel cyclooxygenase-2 selective inhibitor*. Drug Metab Dispos, 2001. 29(6): p. 813–820.

58. Munafo, A., Buclin, T., Tuto, D. and Biollaz, J., *The effect of a low dose of quinidine on the disposition of flecainide in healthy volunteers.* Eur J Clin Pharmacol, 1992. 43(4): p. 441–443.

59. Hallifax, D., Rawden, H.C., Hakooz, N. and Houston, J.B., *Prediction of metabolic clearance using cryopreserved human hepatocytes: kinetic characteristics for five benzodiazepines.* Drug Metab Dispos, 2005. 33(12): p. 1852–1858.

60. Agarwal, S., Sane, R., Gallardo, J.L., Ohlfest, J.R. and Elmquist, W.F., *Distribution of gefitinib to the brain is limited by P-glycoprotein (ABCB1) and breast cancer resistance protein (ABCG2)-mediated active efflux.* J Pharmacol Exp Ther, 2010. 334(1): p. 147–155.

61. Yasar, U., Annas, A., Svensson, J.O., Lazorova, L., Artursson, P. and Al-Shurbaji, A., *Ketobemidone is a substrate for cytochrome P4502C9 and 3A4, but not for P-glycoprotein.* Xenobiotica, 2005. 35(8): p. 785–796.

62. Nishimuta, H., Nakagawa, T., Nomura, N. and Yabuki, M., *Significance of reductive metabolism in human intestine and quantitative prediction of intestinal first-pass metabolism by cytosolic reductive enzymes.* Drug Metab Dispos, 2013. 41(5): p. 1104–1111.

63. Soars, M.G., Burchell, B. and Riley, R.J., In vitro *analysis of human drug glucuronidation and prediction of* in vivo *metabolic clearance.* J Pharmacol Exp Ther, 2002. 301(1): p. 382–390.

64. Kadono, K., Koakutsu, A., Naritomi, Y., Terashita, S., Tabata, K. and Teramura, T., *Comparison of intestinal metabolism of CYP3A substrates between rats and humans: application of portal-systemic concentration difference method.* Xenobiotica, 2014. 44(6): p. 511–521.

65. Mitsui, T., Nemoto, T., Miyake, T., Nagao, S., Ogawa, K., Kato, M., Ishigai, M. and Yamada, H., *A useful model capable of predicting the clearance of cytochrome 3A4 (CYP3A4) substrates in humans: validity of CYP3A4 transgenic mice lacking their own Cyp3a enzymes.* Drug Metab Dispos, 2014. 42(9): p. 1540–1547.

66. Hassan, H.E., Myers, A.L., Lee, I.J., Coop, A. and Eddington, N.D., *Oxycodone induces overexpression of P-glycoprotein (ABCB1) and affects paclitaxel's tissue distribution in Sprague Dawley rats.* J Pharm Sci, 2007. 96(9): p. 2494–2506.

67. Coffman, B.L., King, C.D., Rios, G.R. and Tephly, T.R., *The glucuronidation of opioids, other xenobiotics, and androgens by human UGT2B7Y(268) and UGT2B7H(268).* Drug Metab Dispos, 1998. 26(1): p. 73–77.

68. Pere, P.J., Ekstrand, A., Salonen, M., Honkanen, E., Sjovall, J., Henriksson, J. and Rosenberg, P.H., *Pharmacokinetics of ropivacaine in patients with chronic renal failure.* Br J Anaesth, 2011. 106(4): p. 512–521.

69. Choi, M.K. and Song, I.S., *Characterization of efflux transport of the PDE5 inhibitors, vardenafil and sildenafil.* J Pharm Pharmacol, 2012. 64(8): p. 1074–1083.

70. Kinoshita, Y., et al., *Comparison of the effects of omeprazole and rabeprazole on ticlopidine metabolism in vitro.* J Pharmacol Sci, 2011. 117(1): p. 19–26.

71. Karjalainen, M.J., Neuvonen, P.J. and Backman, J.T., *Celecoxib is a CYP1A2 inhibitor* in vitro *but not* in vivo. Eur J Clin Pharmacol, 2008. 64(5): p. 511–519.

72. Goldman, J.L., Leeder, J.S., Van Haandel, L. and Pearce, R.E., *In vitro hepatic oxidative biotransformation of trimethoprim.* Drug Metab Dispos, 2015. 43(9): p. 1372–1380.

73. Fuhr, U., Kober, S., Zaigler, M., Mutschler, E. and Spahn-Langguth, H., *Rate-limiting biotransformation of triamterene is mediated by CYP1A2.* Int J Clin Pharmacol Ther, 2005. 43(7): p. 327–334.

74. Crowe, A. and Teoh, Y.K., *Limited P-glycoprotein mediated efflux for anti-epileptic drugs.* J Drug Target, 2006. 14(5): p. 291–300.

75. Uwai, Y., Saito, H., Hashimoto, Y. and Inui, K.I., *Interaction and transport of thiazide diuretics, loop diuretics, and acetazolamide via rat renal organic anion transporter rOAT1.* J Pharmacol Exp Ther, 2000. 295(1): p. 261–265.

76. Li, M., Anderson, G.D., Phillips, B.R., Kong, W., Shen, D.D. and Wang, J., *Interactions of amoxicillin and cefaclor with human renal organic anion and peptide transporters.* Drug Metab Dispos, 2006. 34(4): p. 547–555.

77. Hill, G., Cihlar, T., Oo, C., Ho, E.S., Prior, K., Wiltshire, H., Barrett, J., Liu, B. and Ward, P., *The anti-influenza drug oseltamivir exhibits low potential to induce pharmacokinetic drug interactions via renal secretion-correlation of* in vivo *and* in vitro *studies*. Drug Metab Dispos, 2002. 30(1): p. 13–19.
78. Takeda, M., Narikawa, S., Hosoyamada, M., Cha, S.H., Sekine, T. and Endou, H., *Characterization of organic anion transport inhibitors using cells stably expressing human organic anion transporters*. Eur J Pharmacol, 2001. 419(2–3): p. 113–120.
79. Hasannejad, H., et al., *Interactions of human organic anion transporters with diuretics*. J Pharmacol Exp Ther, 2004. 308(3): p. 1021–1029.
80. Jariyawat, S., Sekine, T., Takeda, M., Apiwattanakul, N., Kanai, Y., Sophasan, S. and Endou, H., *The interaction and transport of beta-lactam antibiotics with the cloned rat renal organic anion transporter 1*. J Pharmacol Exp Ther, 1999. 290(2): p. 672–677.
81. de Waart, D.R., van de Wetering, K., Kunne, C., Duijst, S., Paulusma, C.C. and Oude Elferink, R.P., *Oral availability of cefadroxil depends on ABCC3 and ABCC4*. Drug Metab Dispos, 2012. 40(3): p. 515–521.
82. Luckner, P. and Brandsch, M., *Interaction of 31 beta-lactam antibiotics with the H+/peptide symporter PEPT2: analysis of affinity constants and comparison with PEPT1*. Eur J Pharm Biopharm, 2005. 59(1): p. 17–24.
83. Khamdang, S., et al., *Interaction of human and rat organic anion transporter 2 with various cephalosporin antibiotics*. Eur J Pharmacol, 2003. 465(1–2): p. 1–7.
84. Nakakariya, M., Shimada, T., Irokawa, M., Maeda, T. and Tamai, I., *Identification and species similarity of OATP transporters responsible for hepatic uptake of beta-lactam antibiotics*. Drug Metab Pharmacokinet, 2008. 23(5): p. 347–355.
85. Tanihara, Y., Masuda, S., Sato, T., Katsura, T., Ogawa, O. and Inui, K., *Substrate specificity of MATE1 and MATE2-K, human multidrug and toxin extrusions/H(+)-organic cation antiporters*. Biochem Pharmacol, 2007. 74(2): p. 359–371.
86. Vanwert, A.L., Srimaroeng, C. and Sweet, D.H., *Organic anion transporter 3 (oat3/slc22a8) interacts with carboxyfluoroquinolones, and deletion increases systemic exposure to ciprofloxacin*. Mol Pharmacol, 2008. 74(1): p. 122–131.
87. Merino, G., Alvarez, A.I., Pulido, M.M., Molina, A.J., Schinkel, A.H. and Prieto, J.G., *Breast cancer resistance protein (BCRP/ABCG2) transports fluoroquinolone antibiotics and affects their oral availability, pharmacokinetics, and milk secretion*. Drug Metab Dispos, 2006. 34(4): p. 690–695.
88. Ye, J., Liu, Q., Wang, C., Meng, Q., Sun, H., Peng, J., Ma, X. and Liu, K., *Benzylpenicillin inhibits the renal excretion of acyclovir by OAT1 and OAT3*. Pharmacol Rep, 2013. 65(2): p. 505–512.
89. Nozaki, Y., Kusuhara, H., Kondo, T., Hasegawa, M., Shiroyanagi, Y., Nakazawa, H., Okano, T. and Sugiyama, Y., *Characterization of the uptake of organic anion transporter (OAT) 1 and OAT3 substrates by human kidney slices*. J Pharmacol Exp Ther, 2007. 321(1): p. 362–369.
90. Vormfelde, S.V., Schirmer, M., Hagos, Y., Toliat, M.R., Engelhardt, S., Meineke, I., Burckhardt, G., Nurnberg, P. and Brockmoller, J., *Torsemide renal clearance and genetic variation in luminal and basolateral organic anion transporters*. Br J Clin Pharmacol, 2006. 62(3): p. 323–335.
91. Katz, D.A., et al., *Organic anion transporting polypeptide 1B1 activity classified by SLCO1B1 genotype influences atrasentan pharmacokinetics*. Clin Pharmacol Ther, 2006. 79(3): p. 186–196.
92. Zhou, L., Chen, X., Gu, Y. and Liang, J., *Transport characteristics of candesartan in human intestinal Caco-2 cell line*. Biopharm Drug Dispos, 2009. 30(5): p. 259–264.
93. Yee, S.W., et al., *Reduced renal clearance of cefotaxime in Asians with a low-frequency polymorphism of OAT3 (SLC22A8)*. J Pharm Sci, 2013. 102(9): p. 3451–3457.
94. Yamaguchi, H., et al., *Screening of antibiotics that interact with organic anion-transporting polypeptides 1B1 and 1B3 using fluorescent probes*. Biol Pharm Bull, 2011. 34(3): p. 389–395.
95. Ueo, H., Motohashi, H., Katsura, T. and Inui, K., *Human organic anion transporter hOAT3 is a potent transporter of cephalosporin antibiotics, in comparison with hOAT1*. Biochem Pharmacol, 2005. 70(7): p. 1104–1113.

96. Tolle-Sander, S., Grill, A., Joshi, H., Kapil, R., Persiani, S. and Polli, J.E., *Characterization of dexloxiglumide* in vitro *biopharmaceutical properties and active transport.* J Pharm Sci, 2003. 92(10): p. 1968–1980.

97. Sun, P., Wang, C., Liu, Q., Meng, Q., Zhang, A., Huo, X., Sun, H. and Liu, K., *OATP and MRP2-mediated hepatic uptake and biliary excretion of eprosartan in rat and human.* Pharmacol Rep, 2014. 66(2): p. 311–319.

98. Eriksson, U.G., Dorani, H., Karlsson, J., Fritsch, H., Hoffmann, K.J., Olsson, L., Sarich, T.C., Wall, U. and Schutzer, K.M., *Influence of erythromycin on the pharmacokinetics of ximelagatran may involve inhibition of P-glycoprotein-mediated excretion.* Drug Metab Dispos, 2006. 34(5): p. 775–782.

99. Chioukh, R., Noel-Hudson, M.S., Ribes, S., Fournier, N., Becquemont, L. and Verstuyft, C., *Proton pump inhibitors inhibit methotrexate transport by renal basolateral organic anion transporter hOAT3.* Drug Metab Dispos, 2014. 42(12): p. 2041–2048.

100. Brannstrom, M., Nordell, P., Bonn, B., Davis, A.M., Palmgren, A.P., Hilgendorf, C., Rubin, K. and Grime, K., *Montelukast disposition: no indication of transporter-mediated uptake in OATP2B1 and OATP1B1 expressing HEK293 cells.* Pharmaceutics, 2015. 7(4): p. 554–564.

101. Yamada, A., et al., *Multiple human isoforms of drug transporters contribute to the hepatic and renal transport of olmesartan, a selective antagonist of the angiotensin II AT1-receptor.* Drug Metab Dispos, 2007. 35(12): p. 2166–2176.

102. Nakagomi-Hagihara, R., Nakai, D., Kawai, K., Yoshigae, Y., Tokui, T., Abe, T. and Ikeda, T., *OATP1B1, OATP1B3, and mrp2 are involved in hepatobiliary transport of olmesartan, a novel angiotensin II blocker.* Drug Metab Dispos, 2006. 34(5): p. 862–869.

103. Choi, M.K., Shin, H.J., Choi, Y.L., Deng, J.W., Shin, J.G. and Song, I.S., *Differential effect of genetic variants of Na(+)-taurocholate co-transporting polypeptide (NTCP) and organic anion-transporting polypeptide 1B1 (OATP1B1) on the uptake of HMG-CoA reductase inhibitors.* Xenobiotica, 2011. 41(1): p. 24–34.

104. Nakagomi-Hagihara, R., Nakai, D., Tokui, T., Abe, T. and Ikeda, T., *Gemfibrozil and its glucuronide inhibit the hepatic uptake of pravastatin mediated by OATP1B1.* Xenobiotica, 2007. 37(5): p. 474–486.

105. Yuan, H., Feng, B., Yu, Y., Chupka, J., Zheng, J.Y., Heath, T.G. and Bond, B.R., *Renal organic anion transporter-mediated drug–drug interaction between gemcabene and quinapril.* J Pharmacol Exp Ther, 2009. 330(1): p. 191–197.

106. Bosgra, S., van de Steeg, E., Vlaming, M.L., Verhoeckx, K.C., Huisman, M.T., Verwei, M. and Wortelboer, H.M., *Predicting carrier-mediated hepatic disposition of rosuvastatin in man by scaling from individual transfected cell-lines* in vitro *using absolute transporter protein quantification and PBPK modeling.* Eur J Pharm Sci, 2014. 65: p. 156–166.

107. Windass, A.S., Lowes, S., Wang, Y. and Brown, C.D., *The contribution of organic anion transporters OAT1 and OAT3 to the renal uptake of rosuvastatin.* J Pharmacol Exp Ther, 2007. 322(3): p. 1221–1227.

108. Maeda, K., et al., *Effects of organic anion transporting polypeptide 1B1 haplotype on pharmacokinetics of pravastatin, valsartan, and temocapril.* Clin Pharmacol Ther, 2006. 79(5): p. 427–439.

109. Babu, E., Takeda, M., Narikawa, S., Kobayashi, Y., Yamamoto, T., Cha, S.H., Sekine, T., Sakthisekaran, D. and Endou, H., *Human organic anion transporters mediate the transport of tetracycline.* Jpn J Pharmacol, 2002. 88(1): p. 69–76.

110. Menochet, K., Kenworthy, K.E., Houston, J.B. and Galetin, A., *Use of mechanistic modeling to assess interindividual variability and interspecies differences in active uptake in human and rat hepatocytes.* Drug Metab Dispos, 2012. 40(9): p. 1744–1756.

111. Poirier, A., Cascais, A.C., Funk, C. and Lave, T., *Prediction of pharmacokinetic profile of valsartan in human based on* in vitro *uptake transport data.* J Pharmacokinet Pharmacodyn, 2009. 36(6): p. 585–611.

112. Yamashiro, W., Maeda, K., Hirouchi, M., Adachi, Y., Hu, Z. and Sugiyama, Y., *Involvement of transporters in the hepatic uptake and biliary excretion of valsartan, a selective antagonist of the angiotensin II AT1-receptor, in humans.* Drug Metab Dispos, 2006. 34(7): p. 1247–1254.

113. Gupta, S.V., Gupta, D., Sun, J., Dahan, A., Tsume, Y., Hilfinger, J., Lee, K.D. and Amidon, G.L., *Enhancing the intestinal membrane permeability of zanamivir: a carrier mediated prodrug approach.* Mol Pharm, 2011. 8(6): p. 2358–2367.

114. Hendrickx, R., Johansson, J.G., Lohmann, C., Jenvert, R.M., Blomgren, A., Borjesson, L. and Gustavsson, L., *Identification of novel substrates and structure–activity relationship of cellular uptake mediated by human organic cation transporters 1 and 2.* J Med Chem, 2013. 56(18): p. 7232–7242.

115. Zhang, D., et al., *Characterization of efflux transporters involved in distribution and disposition of apixaban.* Drug Metab Dispos, 2013. 41(4): p. 827–835.

116. Muller, F., Konig, J., Glaeser, H., Schmidt, I., Zolk, O., Fromm, M.F. and Maas, R., *Molecular mechanism of renal tubular secretion of the antimalarial drug chloroquine.* Antimicrob Agents Chemother, 2011. 55(7): p. 3091–3098.

117. Polli, J.W., Wring, S.A., Humphreys, J.E., Huang, L., Morgan, J.B., Webster, L.O. and Serabjit-Singh, C.S., *Rational use of* in vitro *P-glycoprotein assays in drug discovery.* J Pharmacol Exp Ther, 2001. 299(2): p. 620–628.

118. Tahara, H., Kusuhara, H., Endou, H., Koepsell, H., Imaoka, T., Fuse, E. and Sugiyama, Y., *A species difference in the transport activities of H2 receptor antagonists by rat and human renal organic anion and cation transporters.* J Pharmacol Exp Ther, 2005. 315(1): p. 337–345.

119. Motohashi, H., Uwai, Y., Hiramoto, K., Okuda, M. and Inui, K., *Different transport properties between famotidine and cimetidine by human renal organic ion transporters (SLC22A).* Eur J Pharmacol, 2004. 503(1–3): p. 25–30.

120. Hughes, J. and Crowe, A., *Inhibition of P-glycoprotein-mediated efflux of digoxin and its metabolites by macrolide antibiotics.* J Pharmacol Sci, 2010. 113(4): p. 315–324.

121. Ebner, T., Ishiguro, N. and Taub, M.E., *The use of transporter probe drug cocktails for the assessment of transporter-based drug–drug interactions in a clinical setting–proposal of a four component transporter cocktail.* J Pharm Sci, 2015. 104(9): p. 3220–3228.

122. Mikkaichi, T., Yoshigae, Y., Masumoto, H., Imaoka, T., Rozehnal, V., Fischer, T., Okudaira, N. and Izumi, T., *Edoxaban transport via P-glycoprotein is a key factor for the drug's disposition.* Drug Metab Dispos, 2014. 42(4): p. 520–528.

123. Takeda, M., Khamdang, S., Narikawa, S., Kimura, H., Kobayashi, Y., Yamamoto, T., Cha, S.H., Sekine, T. and Endou, H., *Human organic anion transporters and human organic cation transporters mediate renal antiviral transport.* J Pharmacol Exp Ther, 2002. 300(3): p. 918–924.

124. Muller, F., Konig, J., Hoier, E., Mandery, K. and Fromm, M.F., *Role of organic cation transporter OCT2 and multidrug and toxin extrusion proteins MATE1 and MATE2-K for transport and drug interactions of the antiviral lamivudine.* Biochem Pharmacol, 2013. 86(6): p. 808–815.

125. Dahan, A., Sabit, H. and Amidon, G.L., *The H2 receptor antagonist nizatidine is a P-glycoprotein substrate: characterization of its intestinal epithelial cell efflux transport.* AAPS J, 2009. 11(2): p. 205–213.

126. Ashiru-Oredope, D.A., Patel, N., Forbes, B., Patel, R. and Basit, A.W., *The effect of polyoxyethylene polymers on the transport of ranitidine in Caco-2 cell monolayers.* Int J Pharm, 2011. 409(1–2): p. 164–168.

127. Bourdet, D.L., Pritchard, J.B. and Thakker, D.R., *Differential substrate and inhibitory activities of ranitidine and famotidine toward human organic cation transporter 1 (hOCT1; SLC22A1), hOCT2 (SLC22A2), and hOCT3 (SLC22A3).* J Pharmacol Exp Ther, 2005. 315(3): p. 1288–1297.

128. Groer, C., Busch, D., Patrzyk, M., Beyer, K., Busemann, A., Heidecke, C.D., Drozdzik, M., Siegmund, W. and Oswald, S., *Absolute protein quantification of clinically relevant cytochrome P450 enzymes and UDP-glucuronosyltransferases by mass spectrometry-based targeted proteomics.* J Pharm Biomed Anal, 2014. 100: p. 393–401.

129. Shitara, Y., Maeda, K., Ikejiri, K., Yoshida, K., Horie, T. and Sugiyama, Y., *Clinical significance of organic anion transporting polypeptides (OATPs) in drug disposition: their roles in hepatic clearance and intestinal absorption*. Biopharm Drug Dispos, 2013. 34(1): p. 45–78.

130. Watanabe, T., Kusuhara, H., Maeda, K., Kanamaru, H., Saito, Y., Hu, Z. and Sugiyama, Y., *Investigation of the rate-determining process in the hepatic elimination of HMG-CoA reductase inhibitors in rats and humans*. Drug Metab Dispos, 2010. 38(2): p. 215–222.

131. Ieiri, I., Higuchi, S. and Sugiyama, Y., *Genetic polymorphisms of uptake (OATP1B1, 1B3) and efflux (MRP2, BCRP) transporters: implications for inter-individual differences in the pharmacokinetics and pharmacodynamics of statins and other clinically relevant drugs*. Expert Opin Drug Metab Toxicol, 2009. 5(7): p. 703–729.

132. Goosen, C., Laing, T.J., du, P.J., Goosen, T.C. and Flynn, G.L., *Physicochemical characterization and solubility analysis of thalidomide and its N-alkyl analogs*. Pharm Res, 2002. 19(1): p. 13–19.

133. Nikam, S.S., et al., *Design and synthesis of novel quinoxaline-2,3-dione AMPA/GlyN receptor antagonists: amino acid derivatives*. J Med Chem, 1999. 42(12): p. 2266–2271.

134. Lipinski, C.A., Solubility in Water and DMSO: Issues and Potential Solutions, in *Pharmaceutical Profiling in Drug Discovery for Lead Selection*, Borchardt, R.T., Kerns, E.H., Lipinski, C.A., Thakker, D.R., and Wang, B.W., Editors. 2004, AAPS Press: Washington DC.

135. Lipinski, C.A., *Drug-like properties and the causes of poor solubility and poor permeability*. J Pharmacol Toxicol Methods, 2000. 44(1): p. 235–249.

136. He, Y.L., Murby, S., Warhurst, G., Gifford, L., Walker, D., Ayrton, J., Eastmond, R. and Rowland, M., *Species differences in size discrimination in the paracellular pathway reflected by oral bioavailability of poly(ethylene glycol) and D-peptides*. J Pharm Sci, 1998. 87(5): p. 626–633.

137. Stevenson, C.M., Radulovic, L.L., Bockbrader, H.N. and Fleisher, D., *Contrasting nutrient effects on the plasma levels of an amino acid-like antiepileptic agent from jejunal administration in dogs*. J Pharm Sci, 1997. 86(8): p. 953–957.

138. Teuscher, N.S., Shen, H., Shu, C., Xiang, J., Keep, R.F. and Smith, D.E., *Carnosine uptake in rat choroid plexus primary cell cultures and choroid plexus whole tissue from PEPT2 null mice*. J Neurochem, 2004. 89(2): p. 375–382.

139. Hou, T.J., Zhang, W., Xia, K., Qiao, X.B. and Xu, X.J., *ADME evaluation in drug discovery. 5. Correlation of Caco-2 permeation with simple molecular properties*. J Chem Inf Comput Sci, 2004. 44(5): p. 1585–1600.

140. Caron, G., Ermondi, G., Scherrer, R.A., John, B.T. and David, J.T., Lipophilicity, Polarity, and Hydrophobicity, in *Comprehensive Medicinal Chemistry II*. 2007, Elsevier: Oxford. p. 425–452.

141. Chandler, D., *Hydrophobicity: Two faces of water*. Nature, 2002. 417(6888): p. 491.

142. van Niel, M.B., et al., *Fluorination of 3-(3-(piperidin-1-yl)propyl)indoles and 3-(3-(piperazin-1-yl)propyl)indoles gives selective human 5-HT1D receptor ligands with improved pharmacokinetic profiles*. J Med Chem, 1999. 42(12): p. 2087–2104.

143. Sivam, S. and Yee, B.J., *Role of gabapentin enacarbil XR in restless legs syndrome*. Ther Clin Risk Manag, 2012. 8: p. 201–208.

144. Bolger, M.B., Lukacova, V. and Woltosz, W.S., *Simulations of the nonlinear dose dependence for substrates of influx and efflux transporters in the human intestine*. AAPS J, 2009. 11(2): p. 353–363.

145. Cundy, K.C., Sastry, S., Luo, W., Zou, J., Moors, T.L. and Canafax, D.M., *Clinical pharmacokinetics of XP13512, a novel transported prodrug of gabapentin*. J Clin Pharmacol, 2008. 48(12): p. 1378–1388.

146. Yang, J., Jamei, M., Yeo, K.R., Tucker, G.T. and Rostami-Hodjegan, A., *Prediction of intestinal first-pass drug metabolism*. Curr Drug Metab, 2007. 8(7): p. 676–684.

147. Galetin, A., Gertz, M. and Houston, J.B., *Potential role of intestinal first-pass metabolism in the prediction of drug–drug interactions*. Expert Opin Drug Metab Toxicol, 2008. 4(7): p. 909–922.

148. Wang, H. and Tompkins, L.M., *CYP2B6: new insights into a historically overlooked cytochrome P450 isozyme*. Curr Drug Metab, 2008. 9(7): p. 598–610.

149. Paine, M.F., Hart, H.L., Ludington, S.S., Haining, R.L., Rettie, A.E. and Zeldin, D.C., *The human intestinal cytochrome P450 "pie"*. Drug Metab Dispos, 2006. 34(5): p. 880–886.

150. Hisaka, A., Nakamura, M., Tsukihashi, A., Koh, S. and Suzuki, H., *Assessment of intestinal availability (FG) of substrate drugs of cytochrome p450s by analyzing changes in pharmacokinetic properties caused by drug–drug interactions*. Drug Metab Dispos, 2014. 42(10): p. 1640–1645.

151. Liu, L. and Pang, K.S., *The roles of transporters and enzymes in hepatic drug processing*. Drug Metab Dispos, 2005. 33(1): p. 1–9.

152. Pang, K.S., Maeng, H.J. and Fan, J., *Interplay of transporters and enzymes in drug and metabolite processing*. Mol Pharm, 2009. 6(6): p. 1734–1755.

153. Pang, K.S. and Rowland, M., *Hepatic clearance of drugs. I. Theoretical considerations of a "well-stirred" model and a "parallel tube" model. Influence of hepatic blood flow, plasma and blood cell binding, and the hepatocellular enzymatic activity on hepatic drug clearance*. J Pharmacokinet Biopharm, 1977. 5(6): p. 625–653.

154. Yoshida, K., Maeda, K. and Sugiyama, Y., *Hepatic and intestinal drug transporters: prediction of pharmacokinetic effects caused by drug–drug interactions and genetic polymorphisms*. Annu Rev Pharmacol Toxicol, 2013. 53: p. 581–612.

155. Barton, H.A., Lai, Y., Goosen, T.C., Jones, H.M., El-Kattan, A.F., Gosset, J.R., Lin, J. and Varma, M.V., *Model-based approaches to predict drug–drug interactions associated with hepatic uptake transporters: preclinical, clinical and beyond*. Expert Opin Drug Metab Toxicol, 2013. 9(4): p. 459–472.

156. Shitara, Y., Horie, T. and Sugiyama, Y., *Transporters as a determinant of drug clearance and tissue distribution*. Eur J Pharm Sci, 2006. 27(5): p. 425–446.

157. Shitara, Y. and Sugiyama, Y., *Pharmacokinetic and pharmacodynamic alterations of 3-hydroxy-3-methylglutaryl coenzyme A (HMG-CoA) reductase inhibitors: drug–drug interactions and interindividual differences in transporter and metabolic enzyme functions*. Pharmacol Ther, 2006. 112(1): p. 71–105.

158. Camenisch, G. and Umehara, K., *Predicting human hepatic clearance from* in vitro *drug metabolism and transport data: a scientific and pharmaceutical perspective for assessing drug–drug interactions*. Biopharm Drug Dispos, 2012. 33(4): p. 179–194.

159. De Buck, S.S., Sinha, V.K., Fenu, L.A., Nijsen, M.J., Mackie, C.E. and Gilissen, R.A., *Prediction of human pharmacokinetics using physiologically based modeling: a retrospective analysis of 26 clinically tested drugs*. Drug Metab Dispos, 2007. 35(10): p. 1766–1780.

160. Jones, H.M., et al., *Simulation of human intravenous and oral pharmacokinetics of 21 diverse compounds using physiologically based pharmacokinetic modelling*. Clin Pharmacokinet, 2011. 50(5): p. 331–347.

161. Obach, R.S., Baxter, J.G., Liston, T.E., Silber, B.M., Jones, B.C., Macintyre, F., Rance, D.J. and Wastall, P., *The prediction of human pharmacokinetic parameters from preclinical and* in vitro *metabolism data*. J Pharmacol Exp Ther, 1997. 283(1): p. 46–58.

162. Hosea, N.A. and Jones, H.M., *Predicting pharmacokinetic profiles using in silico derived parameters*. Mol Pharm, 2013. 10(4): p. 1207–1215.

163. Hosea, N.A., *Drug design tools –* in silico, in vitro *and* in vivo *ADME/PK prediction and interpretation: is PK in monkey an essential part of a good human PK prediction?* Curr Top Med Chem, 2011. 11(4): p. 351–357.

164. Hosea, N.A., et al., *Prediction of human pharmacokinetics from preclinical information: comparative accuracy of quantitative prediction approaches*. J Clin Pharmacol, 2009. 49(5): p. 513–533.

165. Williams, J.A., Hyland, R., Jones, B.C., Smith, D.A., Hurst, S., Goosen, T.C., Peterkin, V., Koup, J.R. and Ball, S.E., *Drug–drug interactions for UDP-glucuronosyltransferase substrates: a pharmacokinetic explanation for typically observed low exposure (AUCi/AUC) ratios*. Drug Metab Dispos, 2004. 32(11): p. 1201–1208.

166. Di, L., Feng, B., Goosen, T.C., Lai, Y., Steyn, S.J., Varma, M.V. and Obach, R.S., *A perspective on the prediction of drug pharmacokinetics and disposition in drug research and development*. Drug Metab Dispos, 2013. 41(12): p. 1975–1993.

167. Fenner, K.S., Jones, H.M., Ullah, M., Kempshall, S., Dickins, M., Lai, Y., Morgan, P. and Barton, H.A., *The evolution of the OATP hepatic uptake transport protein family in DMPK sciences: from obscure liver transporters to key determinants of hepatobiliary clearance*. Xenobiotica, 2012. 42(1): p. 28–45.

168. International Transporter, C., et al., *Membrane transporters in drug development*. Nat Rev Drug Discov, 2010. 9(3): p. 215–236.

169. Shimizu, K., et al., *Whole-body distribution and radiation dosimetry of [11C]telmisartan as a biomarker for hepatic organic anion transporting polypeptide (OATP) 1B3*. Nucl Med Biol, 2012. 39(6): p. 847–853.

170. Bergman, E., Forsell, P., Tevell, A., Persson, E.M., Hedeland, M., Bondesson, U., Knutson, L. and Lennernas, H., *Biliary secretion of rosuvastatin and bile acids in humans during the absorption phase*. Eur J Pharm Sci, 2006. 29(3–4): p. 205–214.

171. Yamazaki, M., Kobayashi, K. and Sugiyama, Y., *Primary active transport of pravastatin across the liver canalicular membrane in normal and mutant Eisai hyperbilirubinemic rats*. Biopharm Drug Dispos, 1996. 17(7): p. 607–621.

172. Varma, M.V., Lai, Y., Feng, B., Litchfield, J., Goosen, T.C. and Bergman, A., *Physiologically based modeling of pravastatin transporter-mediated hepatobiliary disposition and drug–drug interactions*. Pharm Res, 2012. 29(10): p. 2860–2873.

173. Watanabe, T., Kusuhara, H., Maeda, K., Shitara, Y. and Sugiyama, Y., *Physiologically based pharmacokinetic modeling to predict transporter-mediated clearance and distribution of pravastatin in humans*. J Pharmacol Exp Ther, 2009. 328(2): p. 652–662.

174. Maeda, K., Ikeda, Y., Fujita, T., Yoshida, K., Azuma, Y., Haruyama, Y., Yamane, N., Kumagai, Y. and Sugiyama, Y., *Identification of the rate-determining process in the hepatic clearance of atorvastatin in a clinical cassette microdosing study*. Clin Pharmacol Ther, 2011. 90(4): p. 575–581.

175. Lau, Y., Huang, Y., Frassetto, L. and Benet, L., *Effect of OATP1B transporter inhibition on the pharmacokinetics of atorvastatin in healthy volunteers*. Clin Pharmacol Ther (St Louis, MO, USA), 2006. 81(2): p. 194–204.

176. Zheng, H., Huang, Y., Frassetto, L. and Benet, L., *Elucidating rifampin's inducing and inhibiting effects on glyburide pharmacokinetics and blood glucose in healthy volunteers: unmasking the differential effects of enzyme induction and transporter inhibition for a drug and its primary metabolite*. Clin Pharmacol Ther (St Louis, MO, USA), 2008. 85(1): p. 78–85.

177. Kalliokoski, A., Backman, J.T., Kurkinen, K.J., Neuvonen, P.J. and Niemi, M., *Effects of gemfibrozil and atorvastatin on the pharmacokinetics of repaglinide in relation to SLCO1B1 polymorphism*. Clin Pharmacol Ther, 2008. 84(4): p. 488–496.

178. Prueksaritanont, T., et al., *Pitavastatin is a more sensitive and selective OATP1B clinical probe than rosuvastatin*. Br J Clin Pharmacol, 2014. 78(3): p. 587–598.

179. Group, S.C., et al., *SLCO1B1 variants and statin-induced myopathy – a genomewide study*. N Engl J Med, 2008. 359(8): p. 789–799.

180. Niemi, M., Neuvonen, P.J., Hofmann, U., Backman, J.T., Schwab, M., Lutjohann, D., von Bergmann, K., Eichelbaum, M. and Kivisto, K.T., *Acute effects of pravastatin on cholesterol synthesis are associated with SLCO1B1 (encoding OATP1B1) haplotype *17*. Pharmacogenet Genomics, 2005. 15(5): p. 303–309.

181. Nishizato, Y., et al., *Polymorphisms of OATP-C (SLC21A6) and OAT3 (SLC22A8) genes: consequences for pravastatin pharmacokinetics*. Clin Pharmacol Ther, 2003. 73(6): p. 554–565.

182. Lewis, D.F., *On the recognition of mammalian microsomal cytochrome P450 substrates and their characteristics: towards the prediction of human p450 substrate specificity and metabolism*. Biochem Pharmacol, 2000. 60(3): p. 293–306.

183. Ke, A.B., Nallani, S.C., Zhao, P., Rostami-Hodjegan, A., Isoherranen, N. and Unadkat, J.D., *A physiologically based pharmacokinetic model to predict disposition of CYP2D6 and CYP1A2 metabolized drugs in pregnant women*. Drug Metab Dispos, 2013. 41(4): p. 801–813.

184. Mise, M., Hashizume, T. and Komuro, S., *Characterization of substrate specificity of dog CYP1A2 using CYP1A2-deficient and wild-type dog liver microsomes*. Drug Metab Dispos, 2008. 36(9): p. 1903–1908.

185. Brosen, K., *Drug interactions and the cytochrome P450 system. The role of cytochrome P450 1A2*. Clin Pharmacokinet, 1995. 29(Suppl 1): p. 20–25.

186. Bosch, T.M., Meijerman, I., Beijnen, J.H. and Schellens, J.H., *Genetic polymorphisms of drug-metabolising enzymes and drug transporters in the chemotherapeutic treatment of cancer*. Clin Pharmacokinet, 2006. 45(3): p. 253–285.

187. Jones, B.C., Hyland, R., Ackland, M., Tyman, C.A. and Smith, D.A., *Interaction of terfenadine and its primary metabolites with cytochrome P450 2D6*. Drug Metab Dispos, 1998. 26(9): p. 875–882.

188. Varma, M.V., Lai, Y., Feng, B., Litchfield, J., Goosen, T.C. and Bergman, A., *Physiologically based modeling of pravastatin transporter-mediated hepatobiliary disposition and drug–drug interactions*. Pharm Res, 2012. 29(10): p. 2860–2873.

INDEX

Oral Bioavailability Assessment: Basics and Strategies for Drug Discovery and Development, First Edition.
Ayman F. El-Kattan.